# The Scientific Foundations of Neurology

# The Scientific Foundations of Neurology

*edited by*

A. Norman Guthkelch, MD
*Former Professor, Department of Neurological Surgery*
*University of Pittsburgh*
*Pittsburgh, Pennsylvania*
*Former Research Professor*
*University of Arizona*
*Tucson, Arizona*

Karl E. Misulis, MD, PhD
*Associate Clinical Professor of Neurology*
*Vanderbilt University School of Medicine*
*Nashville, Tennessee*
*Neurologist*
*Semmes-Murphey Clinic*
*Jackson, Tennessee*

**Blackwell
Science**

**Blackwell Science**

EDITORIAL OFFICES:
238 Main Street, Cambridge, Massachusetts 02142, USA
Osney Mead, Oxford OX2 0EL, England
25 John Street, London WC1N 2BL, England
23 Ainslie Place, Edinburgh EH3 6AJ, Scotland
54 University Street, Carlton, Victoria 3053, Australia

OTHER EDITORIAL OFFICES:
Arnette Blackwell SA, Boulevard St. Germaine, 75007 Paris, France
Blackwell Wissenschafts-Verlag GmbH Kurfürstendamm 57, 10707 Berlin, Germany
Zehetnergasse 6, A-1140 Vienna, Austria

DISTRIBUTORS:
USA
Blackwell Science, Inc.
238 Main Street
Cambridge, Massachusetts 02142
(Telephone orders: 800-215-1000 or 617-876-7000; Fax orders: 617-492-5263)

*Canada*
Copp Clark, Ltd.
2775 Matheson Blvd. East
Mississauga, Ontario
Canada, L4W 4P7
(Telephone orders: 800-263-4374 or 905-238-6074)

*Australia*
Blackwell Science Pty., Ltd.
54 University Street
Carlton, Victoria 3053
(Telephone orders: 03-9347-0300; Fax orders: 03-9349-3016)

*Outside North America and Australia*
Blackwell Science, Ltd.
c/o Marston Book Services, Ltd.
P.O. Box 269
Abingdon Oxon OX14 4YN
England
(Telephone orders: 44-01235-465500; Fax orders: 44-01235-465555)

Acquisition: Chris Davis
Development: Kathleen Broderick
Production: Colophon
Manufacturing: Lisa Flanagan
Typeset by Modern Graphics
Printed and bound by BookCrafters

© 1996 by Blackwell Science, Inc.
Printed in the United States of America.
96  97  98  99  5  4  3  2  1

*Library of Congress Cataloging-in-Publication Data*

Scientific foundations of neurology / edited by A. Norman Guthkelch, Karl E. Misulis.
        p.    cm.
    Includes bibliographical references and index.
    ISBN 0-86542-408-X
    1. Neurosciences.  2. Neuropathology.  3. Neurology.
I. Guthkelch, A. Norman.  II. Misulis, Karl E.
    [DNLM:  1. Nervous System Diseases.  2. Nervous System—physiology.
3.  Nervous System—physiopathology.  WL 140 S416 1996]
RC341.S35    1996
616.8—dc20
DNLM/DLC
for Library of Congress                                      96-14099
                                                                  CIP

*To Our Families*

# Contents

# Preface

Some senior clinical neuroscientists will recognize that our chosen title, *The Scientific Foundations of Neurology*, was used by Critchley, Jennett, and O'Leary in a successful marriage of basic science with clinical practice, which appeared in 1973. We thank Dr. Hal Rekate for suggesting that the time was ripe for a revival of the concept. Clinical specialities are passing into a state of veritable revolution in the manner in which they address the pathophysiology of disease processes, and specifically in the progression toward understanding both normal function and disease at a molecular level. This movement seems certain to have effects on practice that will be no less radical and far reaching than nineteenth century medicine's transition from thinking in terms of gross appearances and humors to the acceptance of cellular pathology and the germ theory. Although vast amounts of data are accumulating daily, there nonetheless seems to be a greater degree of unity in the basic concepts of neuroscience than was previously possible. Less than a generation ago, the responses of the nervous system to trauma, stroke, neoplasia, and infection appeared to be separate biologic phenomena, with relatively little to connect them. Now, whether we are considering congenital malformations, infections, tumors, or even degenerative conditions, we can think of them all in similar terms—as manifestations of disturbed enzyme systems and genetic programming. Surgeons need not fear that all of this biochemical progress will relegate them to a minor role in the future. On the contrary, advancing technology, not least the development of frameless stereotactic methods, will surely increase the accuracy and scope of intracranial and intraspinal operations by a whole order of magnitude, while minimizing the disturbance of normal tissue that used to accompany them.

The book is designed to be read through, rather than as a work of reference. Therefore, to keep its length within appropriate limits and maximize the space allotted to the text itself, we have restricted all bibliographies to the minimum necessary to give guidance for further study, foregoing exhaustive lists of references that in this electronic age are readily available elsewhere. Rather, for those who are considering or in training for a career in any branch of clinical neuroscience, we have aimed to provide a glimpse of some of the scientific ideas on which the discipline depends. For physicians already established in a neurologic specialty, the presentation of their own particular field may seem all too short, but we hope they will find recompense and enjoyment in the opportunity to survey the wider picture. There also is a wide range of approaches to the subjects covered, varying from general reviews to closely argued expositions of particular points of view. We believe this diversity will prove stimulating.

As we go to press, we have learned with sadness of the death of one of our contributors, Bernard Williams, in a motor vehicle accident. For many years he had devoted much of his formidable intellectual energy to understanding and treating syringomyelia; in this book, he has set out his final views with the vigor and candor that were characteristic of all of his writing.

A.N.G.
K.E.M.

# Contributors

**Mark J. Alberts, MD**
Associate Professor of Medicine
Division of Neurology
Director, Stroke Acute Care Center
Duke University Medical Center
Durham, North Carolina

**James R. Burke, MD, PhD**
Assistant Professor of Medicine
Division of Neurology
Duke University Medical Center
Durham, North Carolina

**J. Robert Cassady, MD**
Chairman
Department of Radiation Oncology
Lahey Clinic
Burlington, Massachusetts

**Phillip F. Chance, MD**
Division of Neurology Research
The Children's Hospital of Philadelphia
Departments of Pediatrics and Neurology
University of Pennsylvania School of Medicine
Philadelphia, Pennsylvania

**Paul R. Cooper, MD**
Professor of Neurosurgery
New York University School of Medicine
Attending Neurosurgeon
New York University Bellevue Medical Center
New York, New York

**Robert M. Crowell, MD**
Professor of Neurosurgery
University of Massachusetts Medical School
Worcester, Massachusetts
Neurosurgeon
Berkshire Medical Center
Pittsfield, Massachusetts

**Julio Cruz, MD, PhD**
Associate Professor of Neurosurgery
Medical College of Pennsylvania and
Hahnemann University
Philadelphia, Pennsylvania

**John M. Dawson, PhD**
Assistant Professor
Department of Orthopaedics and Rehabilitation
Vanderbilt University Medical Center
Nashville, Tennessee

**Russell G. Durkovic, PhD**
Professor of Physiology
State University of New York Health Science
Center at Syracuse
Syracuse, New York

**Kenneth H. Fischbeck, MD**
Department of Neurology
University of Pennsylvania School of Medicine
Philadelphia, Pennsylvania

**Damirez Fossett, MD**
Department of Neurological Surgery
The George Washington University Medical
Center
Washington, DC

**Thomas A. Gennarelli, MD**
Department of Neurosurgery
Medical College of Pennsylvania and
Hahnemann University
Philadelphia, Pennsylvania

**Saadi Ghatan, MD**
Resident, Department of Neurological Surgery
University of Washington Medical Center
Seattle, Washington

**A. Norman Guthkelch, MCh (Oxon), FRCS (Eng.)**
Former Professor, Department of Neurological Surgery
University of Pittsburgh
Pittsburgh, Pennsylvania
Former Research Professor
University of Arizona
Tucson, Arizona

**Allan J. Hamilton, MD**
Chief, Section of Neurosurgery
Department of Surgery
Executive Director, Minimally Invasive-Stereotactic Program
University of Arizona Health Sciences Center
Tucson, Arizona

**Olle J. Hoffstad, MA**
Department of Neurosurgery
Medical College of Pennsylvania and
Hahnemann University
Philadelphia, Pennsylvania

**Peter J. Jannetta, MD**
Department of Neurological Surgery
Presbyterian Hospital
Pittsburgh, Pennsylvania

**J. Philip Kistler, MD**
Chief, Stroke Service
Massachusetts General Hospital
Associate Professor of Neurology
Harvard Medical School
Boston, Massachusetts

**Fred C. Krebs, PhD**
Research Associate
Department of Microbiology and Immunology
The Pennsylvania State University College of Medicine
The Milton S. Hershey Medical Center
Hershey, Pennsylvania

**Olle Lindvall, MD, PhD**
Professor of Neurology
Laboratory for Experimental Brain Research
Restorative Neurology Unit
University Hospital
Lund, Sweden

**Patricia A. Lodge, PhD**
Multiple Sclerosis Research Laboratory
Vanderbilt Stallworth Rehabilitation Hospital
Nashville, Tennessee

**Bruce A. Lulu, PhD**
Department of Radiation Oncology
University of Arizona Health Sciences Center
Tucson, Arizona

**James R. Lupski, MD, PhD**
Department of Molecular and Human Genetics
Department of Pediatrics
Baylor College of Medicine
Houston, Texas

**Robert J. Maciunas, MD, FACS**
Associate Professor of Neurological Surgery and
Director, Vanderbilt Brain Tumor Center
Vanderbilt University Medical Center
Chief of Neurosurgical Service
Veterans Administration
Hospital
Nashville, Tennessee

**Kim H. Manwaring, MD**
Director, Pediatric Neurosurgery
Phoenix Children's Hospital
Phoenix, Arizona

**Marc R. Mayberg, MD**
Associate Professor of Neurological Surgery
Chief of Clinical Services
University of Washington Medical Center
Seattle, Washington

**Karl E. Misulis, MD, PhD**
Associate Clinical Professor of Neurology
Vanderbilt University
Nashville, Tennessee
Neurologist
Semmes-Murphey Clinic
Jackson, Tennessee

**Blaine S. Nashold, Jr., MD**
Professor Emeritus
Department of Surgery
Section of Neurosurgery
Duke University Medical Center
Durham, North Carolina

**James R. B. Nashold, MD**
Senior Resident in Neurosurgery
Department of Surgery
Duke University Medical Center
Durham, North Carolina

**Christopher S. Ogilvy, MD**
Assistant Professor of Surgery
Harvard Medical School
Assistant in Neurosurgery
Massachusetts General Hospital
Boston, Massachusetts

**John B. Penney, Jr., MD**
Professor of Neurology
Harvard University School of Medicine
Neurologist
Massachusetts General Hospital
Boston, Massachusetts

**Bryan Philbrook, MD**
Instructor in Neurology
International Center for Epilepsy
University of Miami School of Medicine
Veterans Administration Medical Center
Miami, Florida

**David E. Pleasure, MD**
Division of Neurology
The Children's Hospital of Philadelphia
Departments of Pediatrics and Neurology
University of Pennsylvania School of Medicine
Philadelphia, Pennsylvania

**Jeanette Pleasure, MD**
Division of Neurology
The Children's Hospital of Philadelphia
Department of Pediatrics
University of Pennsylvania School of Medicine
Philadelphia, Pennsylvania

**R. Eugene Ramsay, MD**
Professor of Neurology and Psychiatry
International Center for Epilepsy
University of Miami School of Medicine
Veterans Administration Medical Center
Miami, Florida

**Oscar M. Reinmuth, MD**
Clinical Professor of Neurology
University of Arizona Health Sciences Center
Tucson, Arizona

**Harold L. Rekate, MD**
Chairman, Section of Neurosurgery
Director, Pediatric Neurosurgical Research
Laboratory
Barrow Neurological Institute
St. Joseph's Hospital and Medical Center
Phoenix, Arizona

**Benjamin B. Roa, PhD**
Department of Molecular and Human Genetics
Baylor College of Medicine
Houston, Texas

**Laligam Sekhar, MD, FACS**
Department of Neurological Surgery
The George Washington University Medical
Center
Washington, DC

**Joan Rankin Shapiro, PhD**
Director, Neuro-Oncology Research
Barrow Neurological Institute
St. Joseph's Hospital and Medical Center
Phoenix, Arizona

**William R. Shapiro, MD**
Chairman, Division of Neurology
Barrow Neurological Institute
St. Joseph's Hospital and Medical Center
Phoenix, Arizona

**Bo K. Siesjö, MD, PhD**
Professor and Chairman
Laboratory for Experimental Brain Research
Restorative Neurology Unit
University Hospital
Lund, Sweden

**Dan M. Spengler, MD**
Professor and Chairman
Department of Orthopaedics and Rehabilitation
Vanderbilt University Medical Center
Nashville, Tennessee

**Robert F. Spetzler, MD, FACS**
Director, Barrow Neurological Institute
J. N. Harber Chairman of Neurological Surgery
Professor, Section of Neurosurgery
St. Joseph's Hospital and Medical Center
Phoenix, Arizona

**Subramaniam Sriram, MD**
Multiple Sclerosis Research Laboratory
Vanderbilt Stallworth Rehabilitation
Hospital
Nashville, Tennessee

**Baldassarre Stea, MD, PhD**
Associate Professor of Radiation Oncology
Director of Stereotactic Radiotherapy
University of Arizona Health Sciences Center
Tucson, Arizona

**B. Gregory Thompson, MD**
Director of Cerebrovascular Surgery
Department of Neurosurgery
University of Utah Medical Center
Salt Lake City, Utah

**Jeffrey M. Vance, PhD, MD**
Associate Professor of Medicine
Division of Neurology
Duke University Medical Center
Durham, North Carolina

**Ronald E. Warnick, MD**
Assistant Professor
Director of Surgical Neuro-Oncology
Department of Neurosurgery
University of Cincinnati Medical Center
The Mayfield Neurological Institute
Cincinnati, Ohio

**Brian Wigdahl, PhD**
Professor of Microbiology and Immunology
The Pennsylvania State University College of
Medicine
The Milton S. Hershey Medical Center
Hershey, Pennsylvania

**Bernard Williams, MD, ChM, FRCS
(Deceased)**
Senior Neurosurgeon
Midland Centre for Neurosurgery and
Neurology
West Midlands, United Kingdom

**Nicholas T. Zervas, MD**
Professor and Chair
Department of Neurosurgery
Massachusetts General Hospital
Boston, Massachusetts

# I APPLICATIONS OF MOLECULAR BIOLOGY

# 1 Degeneration and Regeneration in the Brain

Bo K. Siesjö

Olle Lindvall

Premature cell death in the brain occurs in acute and chronic neurodegenerative disorders. Among the former, ischemia and trauma are the leading causes of morbidity and mortality, with stroke having the largest medical and socioeconomic impact on Western societies. Less frequent, albeit of considerable medical importance, are cardiac arrest, cardiopulmonary bypass surgery, and hypoglycemic coma. Chronic neurodegenerative diseases encompass Parkinson's disease, Huntington's disease, Alzheimer's disease, and other forms of dementia, as well as motor neuron diseases such as amyotrophic lateral sclerosis.

The causes of neuronal degeneration are being intensely explored by workers using a wide variety of approaches, ranging from neuroimaging techniques and cerebral blood flow (CBF) measurements to molecular genetics and molecular biology. The most comprehensive information on mechanisms of cell dysfunction and cell death exists for acute degenerative disorders, particularly ischemia. A fascinating development is the realization that the same cellular and molecular mechanisms that lead to "acute" cell death also operate in chronic neurodegenerative disease. Common mediators encompass excitatory amino acids (EAAs), calcium, and free radicals.

This chapter concentrates on cellular and molecular mechanisms of brain damage in acute neurodegenerative disease, as well as on measures that enhance the trophic support of cells at risk. It ends by summarizing current knowledge of mechanisms that are believed to be involved in chronic neurodegenerative diseases.

## Degeneration

Three major factors serve as triggers of metabolic cascades that cause cell death: mitochondrial dysfunction, enhanced production of free radicals, and changes in transcription and translation. These factors are by no means mutually exclusive; for example, energy failure and mitochondrial dysfunction can trigger production of free radicals, and these in turn may aggravate the mitochondrial dysfunction and cause DNA damage.

3

TRIGGERING FACTORS

## Loss of Mitochondrial Function

Cell damage due to ischemia/hypoxia or hypoglycemic coma is preceded, and probably triggered, by failure of mitochondrial adenosine triphosphate (ATP) production. Figure 1.1 schematically outlines the characteristics of mitochondrial failure in disease, and the consequences of such failure. In *ischemia/hypoxia*, shortage of oxygen primarily blocks electron flow at the cytochrome a-a₃ step, thereby decreasing oxidation of pyruvate. This secondarily reduces oxidative phosphorylation, that is, ATP production. Furthermore, pyruvate is reduced to lactate, which accumulates with a stoichiometrical amount of $H^+$. A similar series of events (i.e., failure of ATP production and lactic acidosis) is encountered in conditions that cause dysfunction of the pyruvate dehydrogenase (PDH) complex (e.g., thiamine deficiency). The lactate dehydrogenase (LDH) equilibrium is shifted to yield an increased $NADH/NAD^+$ ratio. This reflects a generalized reduction of cellular oxidation/reduction systems.

In *hypoglycemia*, insufficient glucose delivery reduces production of pyruvate and, thereby, oxidative phosphorylation. Because oxygen delivery is maintained, however, endogenous substrates are oxidized, including lactate and pyruvate, as well as citric acid cycle intermediates and associated amino

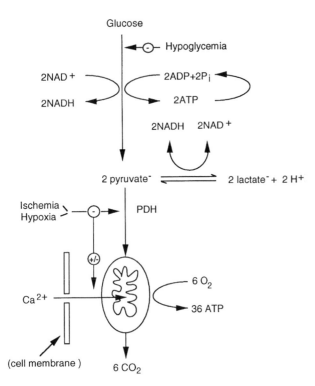

**Figure 1.1**
Schematic diagram illustrating ATP production from glucose metabolism via aerobic and anaerobic pathways. Under normal aerobic conditions, glucose is broken down to pyruvate, which is oxidized to $CO_2$ (and $H_2O$) in the mitochondria, the initial reaction being catalyzed by the PDH complex. Under anaerobic conditions, pyruvate is reduced to lactate, which accumulates with a stoichiometrical amount of $H^+$. In hypoglycemia, glucose delivery and pyruvate production are reduced, but lactic acidosis does not develop. Both ischemia/hypoxia and hypoglycemia allow influx into the cell of calcium, which may be sequestered in the mitochondria. (Modified with permission from Siesjö BK. Pathophysiology and treatment of focal cerebral ischemia. I. Pathophysiology. J Neurosurg 1992;77:169–184.)

acids. The consumption of anions of organic acids occurs with a stoichiometrical amount of $H^+$, which tends to shift intracellular pH ($pH_i$) in the alkaline direction; however, because hydrolysis of nucleoside triphosphates (such as ATP) and phospholipids tends to acidify the cell, $pH_i$ remains essentially unchanged. Thus, ischemia/hypoxia and hypoglycemia share in common energy failure, but only ischemia/hypoxia is accompanied by acidosis. Another difference is that, in hypoglycemia, cellular redox systems are oxidized rather than reduced. This means that ischemia and recirculation are accompanied by an anaerobic/aerobic transition, while hypoglycemic coma is not. The difference in oxidation/reduction state is one factor that can explain why free radicals seem to play a smaller role in hypoglycemia than in ischemia.

Common to ischemia/hypoxia and hypoglycemic coma is a derangement of cell calcium homeostasis. Conditions that lead to depolarization of cell membranes are accompanied by translocation of calcium from extra- to intracellular fluids, with an ensuing rise in the free cytosolic cell calcium concentration ($Ca^{2+}_i$). If loss of energy failure is less than complete, and a membrane potential is retained across the inner mitochondrial membranes, the mitochondria can limit the rise in $Ca^{2+}_i$ by sequestering calcium; however, this occurs at the expense of their capacity to produce ATP. Conversely, if the mitochondrial membrane potential is lost, as in complete ischemia, the mitochondria can no longer sequester calcium, and $Ca^{2+}_i$ can rise to excessive values.

During *seizures*, perturbation of mitochondrial metabolism is of a different type, since it primarily involves enhanced consumption of ATP. Seizure disorders, particularly if leading to status epilepticus, can cause neuronal loss affecting selectively vulnerable brain regions. However, this damage usually occurs first after periods of status epilepticus of 1 to 2 hours or longer, and it then affects relatively few cells. Because overall cerebral energy state is well maintained, even after 2 hours of status epilepticus, such damage may reflect collapse of energy metabolism of a small number of neurons, and be caused by excessive calcium entry.

### Enhanced Production of Free Radicals

The free radical hypothesis is outlined in Figure 1.2. All aerobic cells produce reactive oxygen species (ROS) by univalent reduction of oxygen. Most electrons accepted by respiratory carriers in the mitochondria end up reducing $O_2$ to $H_2O$ at the cytochrome a-$a_3$ step. A few percent of them leak out of the chain, however, reducing $O_2$ in ROS univalent reactions to $O_2^-$ and $H_2O_2$. Reactive oxygen species are also produced in enzymatic reactions, such as those catalyzed by cyclo-oxygenase, lipoxygenase, and xanthine oxidase, or are formed by auto-oxidation of reduced molecular species.

There are no enzymatic reactions catalyzing a three-electron reduction of $O_2$ to yield ·OH, and the uncatalyzed reaction between $O_2^-$ and $H_2O_2$ to yield ·OH is extremely slow. It has been recognized for decades, however, that

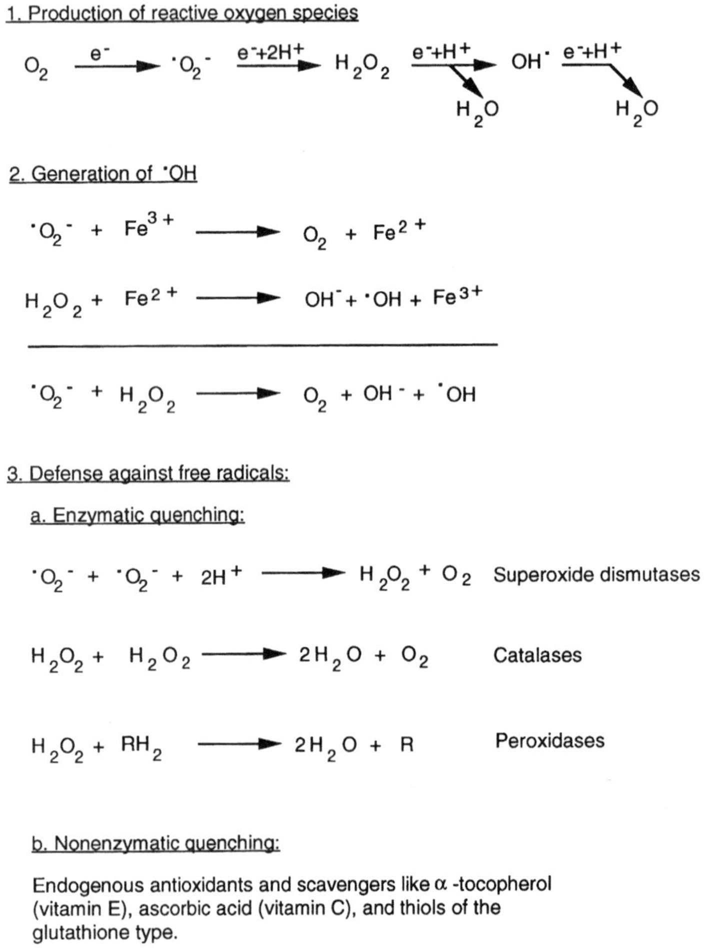

1. Production of reactive oxygen species

$$O_2 \xrightarrow{e^-} \cdot O_2^- \xrightarrow{e^- + 2H^+} H_2O_2 \xrightarrow{e^- + H^+} OH^\cdot \xrightarrow{e^- + H^+}$$
$$H_2O \qquad\qquad H_2O$$

2. Generation of $\cdot OH$

$$\cdot O_2^- + Fe^{3+} \longrightarrow O_2 + Fe^{2+}$$

$$H_2O_2 + Fe^{2+} \longrightarrow OH^- + \cdot OH + Fe^{3+}$$

$$\cdot O_2^- + H_2O_2 \longrightarrow O_2 + OH^- + \cdot OH$$

3. Defense against free radicals:

a. Enzymatic quenching:

$$\cdot O_2^- + \cdot O_2^- + 2H^+ \longrightarrow H_2O_2 + O_2 \quad \text{Superoxide dismutases}$$

$$H_2O_2 + H_2O_2 \longrightarrow 2H_2O + O_2 \qquad \text{Catalases}$$

$$H_2O_2 + RH_2 \longrightarrow 2H_2O + R \qquad \text{Peroxidases}$$

b. Nonenzymatic quenching:

Endogenous antioxidants and scavengers like $\alpha$-tocopherol (vitamin E), ascorbic acid (vitamin C), and thiols of the glutathione type.

**Figure 1.2**
(1) Schematic diagram illustrating production of reactive oxygen species (ROS) by univalent reduction of $O_2$. Uni-and divalent reduction of $O_2$ yields $\cdot O_2^-$ and $H_2O_2$, but there are no reactions directly yielding the extremely toxic OH species. (2) OH can be formed in the iron-catalyzed Haber-Weiss reaction. (3) Cellular defense systems against free radicals. (Modified from Fridovich I. The biology of oxygen radicals. The superoxide radical is an agent of oxygen toxicity; superoxide dismutase provides an important defense. Science 1978;201:875–80; and from Siesjö BK. Cell damage in the brain: a speculative synthesis. J Cereb Blood Flow Metab 1981;1:155–185.)

iron can catalyze the formation of $\cdot$OH according to the reactions shown in Figure 2.2. Such catalysis requires that $Fe^{3+}$ is delocalized from its bindings to proteins such as transferrin and ferritin.

The production of $H_2O_2$, $O_2^-$, and particularly $\cdot$OH represents a threat to the viability of cells, because free radicals can cause lipid peroxidation, oxidize proteins, and damage DNA. Normally, the antioxidative defense systems of cells, which are both enzymatic and nonenzymatic, suffice to scavenge the ROS produced. Certain conditions, however, enhance the production to the extent that the antioxidative systems are overwhelmed; alternatively, the concentrations of scavengers or the activity of enzymes such as superoxide dismutase (SOD) may be reduced.

Recent research has identified nitric oxide (NO) as a molecule with a Janus face, normally fulfilling important physiologic functions but sometimes, when homeostasis is lost, contributing to cell damage. The adverse effects of NO are illustrated by the following reactions:

$$\text{Arginine} + O_2 + \text{NADPH} \rightarrow \text{Citrulline} + NO^{\cdot} + \text{NADP} \qquad (1)$$

$$NO^{\cdot} + O_2^{-} \rightarrow ONOO^{-} + H^{+} \rightarrow ONOOH \qquad (2)$$

$$ONOOH \rightarrow {}^{\cdot}OH + NO_2 \qquad (3)$$

Two putatively adverse events can result from this reaction cascade. First, peroxynitrate ($ONOO^{-}$) can alter protein function by nitrosylation. Second, breakdown of the protonated form of peroxynitrate yields $^{\cdot}OH$.

### Changes in Gene Expression Leading to "Programmed" Cell Death

Insults such as ischemia with reperfusion are known to cause changes in the intracellular signal transduction pathway which encompass alterations in transcription and translation. It has been known for decades that cell death in vertebrate and invertebrate systems conforms to one of two major types: necrosis and apoptosis. By tradition, cell death due to mitochondrial failure has been considered to be of the necrotic type, common features being inhibition of protein synthesis and early plasma membrane failure. More recent results, however, suggest that cell death, particularly when it is of a delayed type, shows features characteristic of apoptotic cell death. A possible sequence is that signals arising as a result of the insults induce changes in mRNA expression and translation of a type that orchestrates cell death, for example, by causing expression of an endonuclease that degrades DNA at internucleosomal bridges ("programmed cell death").

   Recent results, however, suggest that, in many cellular systems, the events leading to apoptotic cell death start with a permeable transition of the inner mitochondrial membrane, sometimes described in terms of the opening of a megapore for $H^{+}$ and other ions (including $Ca^{2+}$), continues with the generation of ROS, and ends with damage to nuclear DNA and plasma membranes. This sequence of events can be blocked by immunosuppressants such as cyclosporine A.

### Transient Global Ischemia: The Triggering Factors Exemplified

Studies in rats have shown that transient cerebral ischemia is followed by a phase of increased flow (reactive hyperemia) followed by a phase of subnormal flow. This secondary hypoperfusion is paralleled by a corresponding reduction in metabolic rate. These changes are accompanied by a reduction in the activity of the PDH complex and by an attenuated metabolic and circulatory response to sensory stimulation. The changes, however, do not imply a reduced metabolic capacity of the postischemic tissues, but rather reflect a perturbed coupling between stimulus, metabolism, and blood flow.

Dense, sudden ischemia leads to rapid depletion of cellular high-energy phosphate stores, and to gradual accumulation of anaerobically formed lactate (Fig. 1.3). When the ATP stores are reduced to less than 50% of control, ion homeostasis is lost, as reflected in massive efflux of $K^+$ from cells, and their uptake of $Ca^{2+}$, $Na^+$, and $Cl^-$. The rapid dissipation of ions is preceded by a gradual rise in extracellular $K^+$ concentration. A likely explanation for this rise is activation of $K^+$ channels, modulated by $Ca^{2+}$ or ATP, while the subsequent "shock" opening of cation conductances could reflect the sudden release of EAAs from presynaptic endings, and their activation of postsynaptic, ionotropic

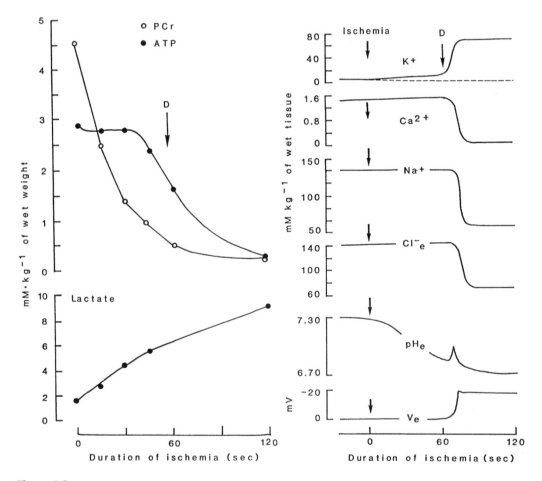

**Figure 1.3**

Graphs illustrating the influence of complete ischemia on the tissue concentrations of phosphocreatine (PCr), ATP, and lactate, as well as on cellular ion homeostasis, as the latter is reflected in extracellular concentrations of $K^+$, $Ca^{2+}$, $Na^+$, and $Cl^-$, as well as in extracellular pH ($pH_e$) and tissue direct-current potential ($V_e$). D, time of massive depolarization. (Modified with permission from Siesjö BK. Basic mechanisms of traumatic brain damage. Ann Emerg Med 1993;22:959–969.)

glutamate receptors. Increased extracellular glutamate is probably mediated by reversal of the $Na^+$-dependent mechanism, which is normally responsible for the cellular reuptake of EAAs, notably glutamate.

Changes in intracellular ions include a decrease in $K^+$ and increases in $Na^+$, $Cl^-$, and $Ca^{2+}$ concentrations. Uptake of $Na^+$ and $Cl^-$ occurs along with an osmotically obligated influx of water, leading to cell swelling.

The fall in ATP concentration and the rise in $Ca^{2+}$ act synergistically to degrade macromolecules and, thereby, the structure of plasma and intracellular membranes. This is because the fall in ATP concentration arrests resynthesis of polysaccharides, phospholipids, proteins, and nucleic acids, following their spontaneous or enzyme-catalyzed degradation, and because the rise in $Ca^{2+}_i$ enhances this degradation by activating enzymes such as phosphorylase, lipases, proteases, and endonucleases.

Figure 1.4 highlights some of the key biochemical events during and following a 15-minute period of forebrain ischemia. Apart from causing depolarization, release of EAAs, and loss of ion homeostasis, the fall in adenylate energy charge leads to accumulation of lactate and a fall in pHi, while activation of phospholipases is reflected in a rise in free fatty acid concentrations (FFAs) and in diacylglycerides (DAGs.) Protein synthesis is also arrested. With return of circulation, the resupply of oxygen at the time when FFAs have not yet returned to control levels triggers a spurt of reactions in which arachidonic acid (AA) is metabolized to biologically active metabolites, which are putative mediators of microvascular damage.

Despite the suppression of overall protein synthesis, new mRNA transcripts are expressed and novel proteins are synthesized, such as heat shock and stress proteins. Among the mRNA transcripts expressed are those encoding for neurotrophins and for SOD.

Thus, primary mitochondrial failure leads to depletion of high-energy phosphate compounds and to loss of ion homeostasis. This results in degradation of membrane structure and accumulation of molecules with potentially adverse biologic activities. There are changes in transcription and translation, some of which may serve the purpose of helping the cell survive the insult; others, however, may be detrimental.

## CALCIUM HOMEOSTASIS AND CALCIUM-TRIGGERED REACTIONS

Calcium has been considered a mediator of cell death in a variety of tissues in several different disease conditions. Loss of calcium homeostasis may play a major role in causing neuronal necrosis in both acute and chronic neurodegenerative disease.

### Calcium as a Second Messenger

Calcium plays a ubiquitous role in the transmission of information from the external environment to the cell interior. The messenger calcium is derived

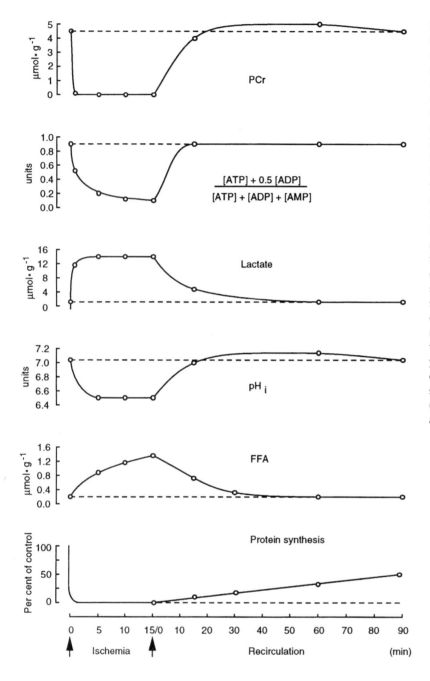

**Figure 1.4**
Some key metabolic events elicited by complete forebrain ischemia of 15 minutes' duration. During ischemia, energy failure occurs within the first 2 to 3 minutes. This triggers production of lactate and leads to a fall in pH$_i$. Furthermore, phospholipid hydrolysis causes free FFAs to accumulate and protein synthesis to be arrested. Recirculation rapidly restores cellular energy state. However, FFA concentrations are maintained for some time after return of oxygen supply, while overall protein synthesis only slowly recovers (resistant areas), if at all (vulnerable areas).

from extracellular sources and internal pools. External calcium enters the cells by voltage-sensitive calcium channels (VSCCs) (i.e., those opened in response to depolarization) and by agonist-operated calcium channels (AOCCs), the most important of which are gated by glutamate and related EAAs. The rise in $Ca^{2+}_i$, occurring in response to influx or release of calcium, is attenuated by intracellular binding and sequestration and terminated by extrusion of calcium from the cell. A delicate balance exists between influx/release and binding/sequestration, in which $Ca^{2+}_i$ is kept within physiologic limits, while allowing calcium to fulfill its second-messenger role.

## Calcium-Mediated Cell Damage

By employing calcium as a second messenger, cells carry the seeds of their own destruction. When calcium accumulates to pathophysiologic levels, it can trigger a host of potentially destructive reactions. One of these encompasses functional and structural damage to mitochondria. Normally, a "pump-leak" relationship for calcium exists at the level of mitochondrial membranes, where the "pump" is an uptake driven by the potential across the inner mitochondrial membrane (inside negative), and the "leak" is provided by a $Ca^{2+}/2Na^+$ exchanger. At normal $Ca^{2+}_i$ values, calcium cycling across the inner membrane probably serves to regulate intramitochondrial dehydrogenases, but if $Ca^{2+}_i$ increases above the set point for net calcium influx into the mitochondria, these can accumulate massive amounts of $Ca^{2+}$ with ensuing functional and structural damage (Fig. 1.5). The potentially adverse effects of excessive enzyme activation can be described as follows.

*Lipases*    Activation of phospholipase $A_2$ ($PLA_2$) gives rise to lysocompounds, including lyso-PAF (platelet activating factor), and AA (Fig 1.6). The corresponding activation of phospholipase C (PLC) produces inositol triphosphase ($IP_3$) and DAGs, and yields another source of AA. Platelet activating factor potentiates an inflammatory response, leading to activation of platelets as well as to activation and adherence of leukocytes to endothelial cells.

The breakdown of AA by cyclo-oxygenase and lipoxygenase, triggered by the resupply of oxygen in the setting of AA accumulation, may be a major source of free radicals during ischemia-reperfusion.

*Kinases and Phosphatases*    Activation or deactivation of kinases (and phosphatases) modulates important membrane response elements, such as ion channel and receptor proteins, as well as intracellular elements regulating transcription of mRNAs and initiation of protein synthesis. Clearly, excessive activation (or down-regulation) of kinases (or phosphatases) can alter membrane function over long periods, modulate gene expression, and inhibit protein synthesis.

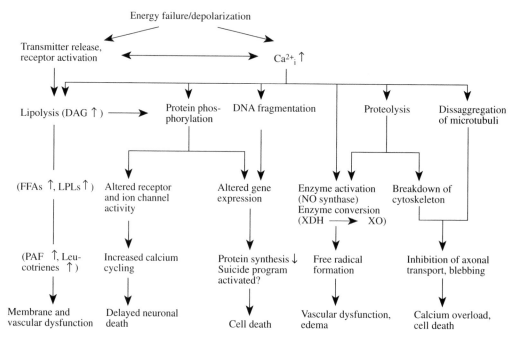

**Figure 1.5**
Diagram illustrating the effect of depolarization, transmitter release, receptor activation, and rise in $Ca^{2+}i$ on reactions that potentially lead to cell dysfunction and cell death. This diagram suggests that the major adverse effects of a massive rise in $Ca^{2+}i$ involve enhanced lipolysis, changes in protein phosphorylation, activation of endonucleases, activation of enzymatic cascades catalyzing the formation of free radicals, activation of proteolytic enzymes, and disaggregation of microtubules. The secondary effects of these events, many of which are speculative, are shown. (Modified with permission from Siesjö B. *The role of calcium in cell death.* In: Price D, Aguayo A, Thoenen H, eds. *Neurodegenerative disorders: mechanisms and prospects for therapy.* Chichester: John Wiley & Sons, 1991:35–59.)

*Free Radical Production*   Activation of $PLA_2$, and other enzymes like that of NO synthetase, is triggered by a rise in calcium concentration. Furthermore, it has been suggested that xanthine dehydrogenase (XDH) can be converted to xanthine oxidase (XO) by limited, $Ca2+$-activated proteolysis, XO then triggering production of $O_2^-$ and $H_2O_2$ by catalyzing breakdown of xanthine and hypoxanthine. Thus, a clear coupling exists between influx/release of $Ca^{2+}$ and production of ROS.

*Endonuclease Activation*   Endonucleases break down DNA into smaller fragments. Extensive DNA breakdown is believed to occur by $Ca^{2+}$- and $Mg^{2+}$-dependent endonuclease activity, and probably reflects a general breakdown of cell structure. DNA degradation reflects overt or impending cell death.

*Breakdown of Cytoskeleton*   This occurs by two processes, both catalyzed by calcium. One is the cleavage of cytoskeletal proteins by neutral proteases

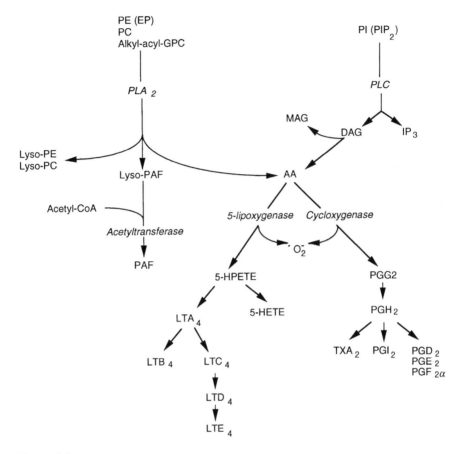

**Figure 1.6**
Schematic diagram illustrating the principal pathways leading to the production of lipid mediators as a result of activation of phospholipases of the $A^2$ and C types. PE, phosphatidyl ethanolamine; PC, phosphatidyl choline; MAG, monoacylglyceride; LT = leukotriene; PG = prostaglandin. (Modified with permission from Siesjö BK. Pathophysiology and treatment of focal cerebral ischemia. I. Pathophysiology. J Neurosurg 1992;77:169–184.)

(calpains), and the second is the dissolution of the microtubular structure, also a calcium-dependent event. Breakdown of the cytoskeleton leads to two potentially detrimental consequences: Axoplasmic transport is arrested, and the anchorage of the cytoskeleton to the plasma membrane is severed, causing a rise in membrane permeability to calcium.

## Pathways of Calcium Entry: The Excitotoxic Hypothesis

Figure 1.7 provides a schematic overview of VSCCs and AOCCs. The former differ with respect to threshold characteristics, duration of channel opening, and modulation by calcium blockers. The major types of VSCCs have been named T (transient), L (long-lasting), N (neuronal), and P (which stands for

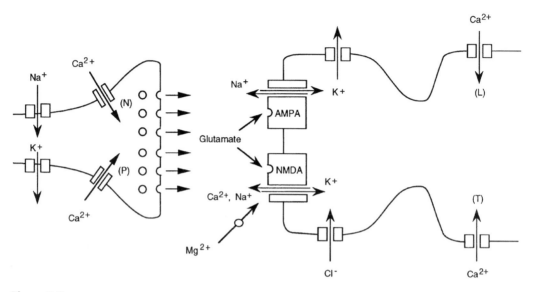

**Figure 1.7**
Schematic diagram illustrating pathways of Ca²⁺ entry by voltage-sensitive and agonist-operated channels at pre- and postsynaptic sites. For explanations, see text. (Modified with permission from Siesjö BK. Basic mechanisms of traumatic brain damage. *Ann Emerg Med* 1993;22:6.)

Purkinje cells in which the channel was first described). Figure 1.7 makes the assumption that the N and P types of VSCCs abound at presynaptic endings and that they modulate the release of transmitters. The L and T types of VSCCs may be mainly localized to postsynaptic membranes.

Much interest has been focused on AOCCs, particularly those gated by glutamate receptors. Interest is focused mainly on two "ionotropic" receptors, selectively activated by α-amino-hydroxy-5-methyl-4-isoxazole propionic acid (AMPA) and N-methyl-D-aspartate (NMDA), and on a metabotropic, quisqualate-preferring receptor, selectively activated by (+)-amino-1,3-cyclopentane-trans-dicarboxylic acid (trans-ACPD). The first two ionotrophic receptors gate ion channels, while the third (metabotropic) is coupled with PLC. Glutamate activation of ionotropic and metabotropic receptors leads to further reactions, because a coupling exists between $IP_3$-triggered release of calcium and influx of calcium via membrane channels. Activation of PLC triggers not only release of calcium from the endoplasmic reticulum (ER) and other intracellular stores, but also calcium influx via membrane channels, which are in close contiguity with the ER.

The AMPA receptor gates a channel that is unselectively permeable to monovalent cations (see Fig. 1.7). Because of these properties, receptor activation leads to Na⁺ influx and, thereby, to membrane depolarization. The NMDA receptor gates a channel that is permeable to $Ca^{2+}$. When open, this channel is blocked by $Mg^{2+}$ in physiologic concentrations. Because this block is voltage-dependent, however, the block is relieved when the membrane depolarizes. This means that AMPA receptor activation, with an ensuing depolarization, allows calcium influx via the NMDA-gated channel. Furthermore, depolarization also allows calcium influx via different types of VSCCs.

## The Excitotoxic Hypothesis Re-evaluated

Recent results obtained in vivo and in vitro suggest that the excitotoxic hypothesis, as applied to ischemic conditions, should be re-evaluated. Neither competitive nor noncompetitive NMDA antagonists ameliorate damage due to transient global ischemia, or have a weak effect only. It also remains to be explained why an AMPA receptor antagonist or SNX-111, an N-type calcium channel blocker, is efficacious when given many hours after the ischemic transient.

## ACIDOSIS-RELATED DAMAGE

Pre-ischemic hyperglycemia exaggerates brain damage due to transient global or forebrain ischemia, typically leading to rapidly developing brain lesions, gross edema, and epileptic seizures. Similarly, damage is accentuated when hyperglycemic rats are exposed to relatively brief periods of focal ischemia.

## The Role of Acidosis

Because pre-ischemic hyperglycemia enhances extra- and intracellular lactic acidosis during ischemia and in the immediate recovery period, it has been tempting to consider acidosis as the aggravating factor. But acidosis also retards calcium influx by blocking translocation of calcium via the NMDA-receptor gated channels. Why, then, is the acidosis not protective? In vivo, the detrimental effect of acidosis probably overwhelms any protective effect of reducing calcium influx.

## The Molecular Mechanisms Involved

The molecular mechanisms whereby acidosis exerts its effects are not known, but acidosis enhances iron-catalyzed production of free radicals. Two possible reactions may be responsible. The first of these reactions is the protonation of $O_2^-$ which yields a species that is more lipid-soluble and a stronger pro-oxidant than the relatively innocuous $O_2^-$. The second reaction is the dislocation of $Fe^{3+}$ from its $HCO_3^-$ (or $CO_3^{2-}$)-dependent binding to transferrin-like proteins, promoting iron-catalyzed free radical production.

## ACUTE NEURODEGENERATIVE DISEASE: PATHOPHYSIOLOGY AND MECHANISMS OF DAMAGE

### Ischemia

The past 5 to 10 years have produced rapid advances in our knowledge of mechanisms responsible for ischemic brain damage. This area, however, is burdened with controversies and seemingly disparate findings. To a large extent, though, these discrepancies can be explained by differences in pathophysiology between global/forebrain and focal ischemia.

*Global* or *forebrain ischemia* is usually severe, and it is always transient if the animal survives. In the clinical setting, ischemia of this type is observed when cardiac arrest is followed by successful resuscitation. Models of global ischemia are sometimes used in research, but most investigators use forebrain ischemia, because circulatory and respiratory systems are not compromised in the recovery period. Ischemia followed by reperfusion typically gives rise to neuronal damage of a delayed nature, affecting vulnerable regions.

In *focal ischemia* caused by occlusion of a major artery, the ischemia is typically more sustained and CBF is usually less drastically reduced than in global ischemia. There is a relatively densely ischemic focus, supplied by the occluded artery, and perifocal "penumbra" areas that are less densely ischemic because they receive collateral blood supply. It is commonly held that the focus is so poorly perfused that the affected tissues are doomed unless reperfusion can be instituted quickly. In contrast, cells in the penumbra can survive for hours before they are recruited in the infarction process. Cells in the penumbra are potentially salvageable.

In one sense, this description is oversimplified. Primary damage to structures directly affected by the ischemia can lead to secondary, transynaptic necrosis of neurons, which are deprived of their trophic support or their inhibitory tone. Transynaptic cell damage can be produced not only in focal ischemia, but also in forebrain ischemia.

The effects of forebrain ischemia depend on the specific characteristics of the ischemic episode. Three types are considered:

*Ischemia of brief duration, with adequate reflow.* There is selective neuronal damage, with sparing of glial cells and microvessels. The characteristic metabolic accompaniments include a long-lasting reduction in metabolic rate, and in protein synthesis, with expression of mRNAs for immediate early genes (IEGs) and neurotrophins, and synthesis of heat shock and stress proteins. Neither initial mitochondrial failure nor perfusion defects are responsible for the neuronal necrosis. Damage has usually been considered to be due to $Ca^{2+}$ overload, persistent depression of protein synthesis, and/or a change in gene expression with induction of a cell suicide program. It is not unlikely, however, that a gradual rise in $Ca^{2+}$ triggers a permeability transition of the mitochondrial membrane potential, with production of ROS; that is, changes that are presumed to give rise to apoptotic cell death.

*Global* or *forebrain ischemia with less optimal characteristics.* This model better represents the clinical situation of cerebral ischemia than the previous model; ischemia is of longer duration, or perfusion pressure is not promptly restored to adequate levels. The result is frequent involvement of glial cells and microvessels in the ischemic lesions, and neuronal necrosis conforming to a vascular pattern. In the laboratory, such pannecrotic lesions are often studied by induction of pre-ischemic hyperglycemia or of hyperthermia in animals that are subjected to standard ischemic insults. Hyperglycemia aggravates the lactic acidosis that occurs during ischemia. As is shown in Figure 1.8, however, the reduction in $pH_e/pH_i$ following 10 minutes of forebrain ischemia is resolved within 60 minutes. Because tissue lesions, seizures, and secondary edema develop later, aggravation of brain damage obviously represents the maturation of acidosis-mediated damage incurred during ischemia or in the immediate recirculation period.

*Focal ischemia.* Conventional wisdom is that cells in a densely ischemic focus can be salvaged only if reperfusion is instituted within a short period (e.g., 20–40 min in rats and 30–60 min in monkeys), but that the less densely ischemic penumbra remains viable for longer periods. Thus, reduction of total infarct volume can be achieved even if reperfusion is accomplished after the specified times.

Because it has not been possible to salvage the densely ischemic core in conditions of permanent middle cerebral artery (MCA) occlusion, attempts to unravel mechanisms of focal ischemic damage have been concentrated on events in penumbral tissues. Current discussions are centered on two major pathophysiologic events: (1) the occurrence of irregularly occurring depolarization waves and (2) a gradual compromise of the capillary circulation. Areas outside the focal ischemic core are known to show irregular *depolarization events.* These may take the form of transient and recurrent anoxic depolarizations or

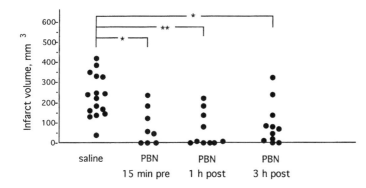

**Figure 1.8**

Infarct volumes after 2 hours of MCA occlusion, as evaluated after 48 hours of recovery. The spin trapping agent α-phenyl-N-tert-butyl nitrone (PBN) was given either before ischemia or after 1, 3, or 6 hours of recirculation. (Modified with permission from Smith M-L, Hanwehr RV, Siesjö BK. Changes in extra- and intracellular pH in the brain during and following ischemis in hyperglycemic and in moderately hypoglycemic rats. J Cereb Blood Flow Metab 1986;6:574–583.)

of conventional spreading depression waves. The depolarization is accompanied by cellular release of $K^+$ and uptake of $Ca^{2+}$.

In normal tissues, spreading depression waves do not give rise to tissue damage, even if repeated over many hours. However, because CBF is reduced in the penumbra zone, and because the capacity of the tissue to increase its ATP production is curtailed, the strain imparted on cellular energy metabolism by recurrent depolarizations or by the associated calcium transients may cause a gradual extension of the infarct into the penumbra zone.

Focal ischemia may be associated with *progressive microvascular failure*. Once the MCA has been occluded, blood flow rates in the focal and penumbral areas remain relatively stable, at least over the next few hours. There is evidence, however, for plugging of a proportion of the capillaries, whether the ischemia is permanent or transient.

It has recently become clear that two distinctly different therapeutic windows exist in focal ischemia. Experiments with glutamate antagonists and other membrane-active anti-ischemic drugs reveal an unexpectedly narrow therapeutic window. Other results, however, suggest that the real window of opportunity is in fact considerably wider. Phenyl-tert-butyl nitrone (PBN) ameliorates damage in permanent and transient focal ischemia in rodents. Infarct size was reduced by about 50% when PBN was given after occlusion or after transient ischemia with reperfusion (Fig. 1.9). Although it is not known how PBN acts, one must assume that scavenging of free radicals is involved. The target of attack may be the microvessels or the mitochondria.

It is possible to formulate a speculative synthesis of available data that would explain the wide therapeutic window of PBN. The hypothesis is that the initial period of ischemia triggers the production of chemical mediators, which leads to the expression of adhesion molecules on endothelial cells and polymorphonuclear neutrophilic leukocytes (PMNs). The mediators trigger

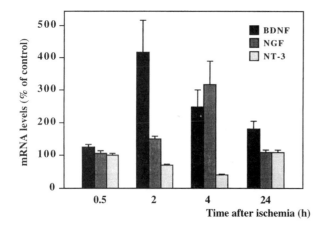

**Figure 1.9**
Levels of BDNF, NGF, and NT-3 mRNAs in dentate granule cell layer at different time points following 2 minutes of transient forebrain ischemia in rats. Means ± SEM. Quantitative image analysis of in situ hybridization autoradiograms. Values are expressed as percentage of controls subjected to sham procedure. (Modified with permission from Lindvall O, Ernfors P, Bengzon J, et al. Differential regulation of mRNAs for nerve growth factor, brain-derived neurotrophic factor and neurotrophin-3 in the adult rat brain following cerebral ischemia and hypoglycemic coma. Proc Natl Acad Sci USA 1992;89:648–652.)

synthesis/release of inflammatory cytokines such as interleukin-1, interferon, or tumor necrosis factor. The latter could then initiate production of adhesion molecules in the sluggishly perfused penumbra zone.

In this scenario, drugs that block depolarization and $Ca^{2+}$ influx would be inefficacious once the mediators have triggered the expression of adhesion molecules, while the secondary microvascular damage would be delayed by the time required for transcription and translation. This proposed series of events would explain the beneficial effect of antibodies to adhesion molecules. This hypothesis postulates that PBN acts on the reactions in which activation of the adhesion molecules leads to microvascular "plugging." However, an alternative explanation of the experiments with PBN is that the secondary damage reflects damage to the mitochondria.

## Hypoglycemia

Profound hypoglycemia leads to many of the events that are observed in transient ischemia. For example, hypoglycemia is accompanied by deterioration of cellular energy state and by loss of ion homeostasis. Furthermore, transient hypoglycemic insults are followed by a post-ischemic decrease in CBF and glucose consumption. The pathophysiology of hypoglycemic brain damage, however, is somewhat different. First, the blood flow is maintained or increased; hence, oxygen-dependent mechanisms may operate already during the insult. Second, hypoglycemia does not lead to the anaerobic/aerobic transition that is observed in ischemia. Third, shortage of glucose prevents accumulation of lactate plus $H^+$, hence acidosis does not develop. Hypoglycemic brain damage affects only neurons, sparing glial cells and vascular tissue.

Although acidosis does not develop, ion homeostasis is lost, the changes in $K^+_e$ and $Ca^{2+}_e$ being similar to those observed in ischemia. Activation of ion conductances seems to precede measurable hydrolysis of ATP. A likely scenario is that depolarization at a circumscript tissue locus triggers a wave of depolarization, and that the sudden activation of ion transport causes ATP concentrations to fall. Conceivably, this leads to a state in which the rate of production of ATP is insufficient to allow repumping of ions and repolarization of membranes. Thus, although some ATP is produced, it is wasted in the futile cycling of ions across the "leaky" membranes.

Predictably, the ATP hydrolysis and the rise in $Ca^{2+}_i$, which occur during hypoglycemia, will activate phospholipases. Unique to hypoglycemic coma is extensive breakdown of phospholipids. This may reflect an attempt by the tissue to derive energy from alternative (endogenous) substrates, but the fate of the degradation products is not known. Recovery of cerebral energy state is prompt and extensive, even after long periods of hypoglycemic coma. Lingering alterations encompass a post-insult decrease in metabolic rate and blood flow, and a reduction in the rate of protein synthesis.

In summary, hypoglycemic coma represents a condition in which extensive damage to neurons occurs in the absence of infarction. Because coma is accompanied by relatively extensive energy failure, it seems likely that the neuronal necrosis is, at least in part, an excitotoxic lesion in which calcium is the mediator. Microvascular damage does not develop.

### Status Epilepticus

Although brain damage is observed in many patients with frequent or sustained seizures, there is controversy about whether the damage is due to the seizures, per se, or to concomitant hypoxia, hypotension, or perhaps temperature alterations. Animal experiments have produced conflicting results. In rats, sustained seizures showed patchy neuronal damage in the neocortex, hippocampus, and thalamus, along with pannecrotic lesions in the substantia nigra and globus pallidus. The etiology of these latter lesions is not known, but mitochondrial failure may play a role. The patchy cortical lesions probably reflect a moderate metabolic disturbance, with small changes in electrolyte composition.

### Trauma

Trauma implies deformation of tissue, creating sheer stress forces of a type that causes functional or structural damage to cell membranes, axons, and microvessels. Damage to axons can either be in the form of primary membrane breaks (physical tearing) or involve dissolution of the cytoskeleton with secondary transection of the axons and formation of so-called retraction balls. Evidence exists that such breakdown of the cytoskeleton is triggered by calcium influx, which causes proteolysis and disassembly of microtubules.

Primary vascular damage is reflected in two alterations: (1) Pial vessels are dilated and show reduced responsiveness to hypocapnia, and (2) blood-brain barrier (BBB) permeability to proteins and low-molecular-weight compound is increased. Evidence exists that vascular damage is mediated by free radicals released during the breakdown of AA. Trauma is known to lead to massive cellular release of EAAs, which accumulate extracellularly, and to release of $K^+$ and efflux of $Ca^{2+}$.

It seems likely, therefore, that the brain damage due to trauma is caused by two pathophysiologic events. One is release of EAAs and the activation of glutamate receptors, giving rise to so-called excitotoxic brain damage. The other is the damage to microvessels, which may be responsible for secondary tissue damage of the pannecrotic type. We assume that the initial insult gives rise to membrane changes that, by themselves, activate membrane conductances and/or membrane receptors, some of which may be connected to phospholipases of the $A_2$ and C types.

## CHRONIC DEGENERATIVE DISEASE

Brain damage in chronic neurodegenerative disease may be caused by mechanisms akin to those operating in acute disorders. These mechanisms encompass defects in mitochondrial ATP production, excitotoxic neuronal damage due to enhanced release of or reduced uptake of EAAs, reactions triggered by a pathologic rise in $Ca^{2+}_i$, and cell damage caused by an enhanced production of or a reduced capacity to quench ROS.

One of the earliest hypotheses in this field incriminates calcium in the pathogenesis of neuronal damage in aging and Alzheimer's disease. This postulate was later coupled to the excitotoxic hypothesis of cell death in that dysfunction of glutamate metabolism was made responsible for the calcium-related cell death. Some evidence exists that the underlying cause is a defect in mitochondrial energy metabolism. A somewhat speculative relationship exists between production of free radicals and brain damage in chronic neurodegenerative disease.

## Repair

Until three decades ago, the prevailing belief was that the adult mammalian brain is fixed and immutable. With the advent of new and more sensitive morphologic techniques, it has been clearly demonstrated that collateral sprouting and reorganization of damaged axons is possible in the adult mammalian brain after nerve injury. Also, degeneration of damaged central neurons can be prevented by an extra supply of neurotrophic molecules normally produced by the brain.

A large number of neurotrophic substances have been identified that support the survival of different neuronal populations in the brain, and several factors are known to influence central nervous system (CNS) regeneration, including neurotrophic and/or neurotropic molecules, adhesive molecules, extracellular matrix, basement membranes, and cell surfaces. Axons fail to regenerate because of inhibitory influences in the adult mammalian CNS environment and can regrow if such inhibitors of neurite growth are neutralized.

## NEUROTROPHIC FACTORS IN ACUTE NEURODEGENERATIVE DISEASE

Neurotrophic factors are of undisputed importance in the developing brain, but their role in the mature brain is poorly understood. A large number of neurotrophic factors increase their expression in response to brain injury. For example, cerebral ischemia leads to increased synthesis of nerve growth factor (NGF), brain-derived neurotrophic factor (BDNF), platelet-derived growth factor (PDGF), basic fibroblast growth factor (bFGF), and transforming growth factor β1 (TGFβ1). Two major working hypotheses have been proposed: First, the increased production of neurotrophic factors may constitute an intrinsic neuroprotective mechanism. Second, these factors may regulate regenerative responses such as sprouting and synaptic reorganization. The most thoroughly studied neurotrophic factors belong to the neurotrophin family of structurally related proteins, which comprises NGF, BDNF, neurotrophin-3 (NT-3), and neurotrophin-4/5 (NT-4/5).

The neurotrophins are expressed by neurons in the adult brain, with highest levels in the hippocampus. The neurotrophins show high-affinity interactions with the tyrosine kinase receptors, TrkA, TrkB, and TrkC. Nerve growth factor mediates its effects via TrkA; BDNF via TrkB; NT-3 interacts mainly with TrkC; and NT-4/5 activates mainly the TrkB receptor. Nerve growth factor is necessary for the survival and maintenance of the basal forebrain cholinergic system.

## Expression of Neurotrophins and Trk Receptors

In rats, global forebrain ischemia and hypoglycemic coma lead to markedly increased BDNF and NGF mRNA levels in dentate granule cells (Fig. 1.9). Neurotrophin-3 mRNA is reduced in these cells and in medial CA1 and CA2 pyramidal layers. Expression returns to control levels at 24 hours. Following 20 minutes of ischemia, there are similar gene changes and, in addition, increased BDNF mRNA expression in CA3. Maximum increases of NGF and BDNF mRNAs seem to occur later than after brief insults. Whether the altered gene expression following brief ischemic and hypoglycemic insults leads to the presumed changes in the levels of the corresponding proteins remains to be elucidated.

Focal ischemia of 2 hours' duration induces increased BDNF mRNA expression in the ipsilateral cerebral cortex outside the lesion and bilaterally in the hippocampus. In the cerebral cortex, the BDNF gene is probably activated through a spreading depression–like process, dependent on glutamatergic mechanisms.

Traumatic injury also leads to changes of neurotrophin gene expression. Trauma to the hippocampus produces transient elevation of NGF, BDNF, TrkB, and TrkC mRNAs in dentate granule cells. Smaller increases of BDNF, TrkB, and TrkC mRNA expression are observed in the piriform cortex.

Seizure activity has been found to transiently increase BDNF and NGF mRNA expression in cortical and hippocampal neurons (Fig. 1.10). When seizure activity originates in the hippocampus, NGF and BDNF mRNA levels increase most rapidly in dentate granule cells. In contrast, NT-3 mRNA expression in these cells is markedly reduced. The changes of gene expression are triggered by brief periods of seizure activity. Recurrent seizures lead to increased neurotrophin mRNA levels also in CA1 and CA3 pyramidal layers, the amygdala, the piriform cortex, and the neocortex. The temporal and regional pattern

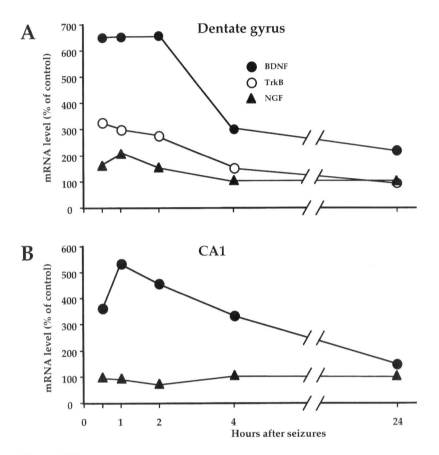

**Figure 1.10**
Time course for the expression of BDNF, TrkB, and NGF mRNAs in the (A) dentate gyrus granule cell layer and (B) CA1 pyramidal layer after 40 recurrent seizures produced during approximately 3.5 hours by rapid hippocampal kindling stimulations. Note the similar time course for (A) the changes of BDNF and TrkB mRNAs in dentate granule cells and (B) the increase of BDNF but not of NGF mRNA in CA1 pyramidal neurons. Quantitative image analysis of *in situ* hybridization autoradiograms. Values are expressed as percentage of levels in controls stimulated at low frequency. (Modified from Ernfors P, Bengzon J, Kokaia Z, et al. Increased levels of messenger RNAs for neurotrophic factors in the brain during kindling epileptogenesis. Neuron 1991;7:165–176; and from Merlio J-P, Ernfors P, Kokaia Z, et al. Increased production of the trkB protein tyrosine kinase receptor after brain insults. Neuron 1993;10:151–164.)

of induction of BDNF mRNA in these regions is different from that of NGF mRNA (see Fig. 1.10). NGF and BDNF mRNA expression largely returns to basal level within 24 to 48 hours, whereas NT-3 mRNA remains below control levels for several days. Seizure activity also induces increased mRNA levels for the high-affinity receptors TrkB and TrkC in dentate granule cells (see Fig. 1.10). These changes of gene expression after seizures seem to lead to the presumed alterations of the levels of neurotrophin and Trk proteins.

## Regulation of Neurotrophin and Trk Gene Expression

Studies on hippocampal and cortical neurons in vitro have shown that elevated BDNF and NGF mRNA levels can be induced by potassium depolarization, kainic acid, glutamate, and calcium influx (Fig. 1.11). Epileptic seizures, cerebral ischemia, and other brain insults are associated with glutamate release, depolarization, and calcium influx, which most likely trigger the changes of neurotrophin and Trk gene expression (see Fig. 1.11). Gene expression is likely influenced by other factors as well: Increased activity of GABAergic and cholinergic neurons is thought to reduce and enhance, respectively, neurotrophin mRNA expression under physiologic conditions.

The BDNF gene has a complex structure, with four short 5' exons and one 3' exon encoding the mature BDNF protein. A separate promoter is present upstream of each 5' exon, and alternative usage of these promoters results in transcription of eight different BDNF mRNAs. Various transcripts could give rise to different amounts of BDNF protein, and the differential use of multiple BDNF promoters may, therefore, provide additional flexibility and possibilities for fine tuning in the regulation of BDNF synthesis. Following brain insults, the pattern of increases of the various exon mRNAs is region-specific and insult-specific: Seizures, global ischemia, and hypoglycemia produce different patterns of promoter activation.

## Functional Effects of Neurotrophins

It is not yet known in detail which neurons respond to the changes of neurotrophin levels induced by brain insults. There is good evidence, however, that neurotrophins act not only as classical target-derived neurotrophic factors, but also locally via autocrine or paracrine mechanisms. Virtually all BDNF mRNA–expressing neurons in the hippocampus, the amygdala, the piriform cortex, and the neocortex also express TrkB mRNA. Seizure activity increases the expression of BDNF and TrkB mRNAs within the same cell. In the dentate gyrus, the subsequent increase of both BDNF protein and the functional TrkB receptor should lead to strongly enhanced BDNF signaling.

If the neurotrophins act locally on the same or neighboring neurons, what are the functional consequences evoked by the insult-induced changes? One major working hypothesis is that the increased levels of neurotrophins after seizures and other insults prevent neuronal death (see Fig. 1.11). Protective effects after administration of neurotrophins have been observed both in vitro and in vivo after various insults (Fig. 1.12).

These data support the idea that administration of neurotrophins might counteract neuronal necrosis in acute neurodegenerative disease. Studies suggest that endogenous production of neurotropins has a protective effect as well. The protective action of NGF and BDNF seems to be mediated via stabilization of calcium homeostasis (see Fig. 1.11). The stabilization of calcium homeostasis might be mediated through up-regulation of calcium-binding pro-

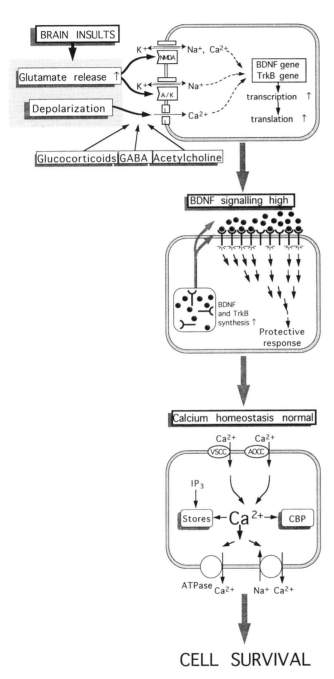

**Figure 1.11**
Hypothetical scheme for the protective action of neurotrophins as illustrated for BDNF. Brain insults rapidly induce increased transcription and translation of the BDNF and, in dentate granule cells, the TrkB gene, through glutamate release acting on NMDA or non-NMDA receptors and through depolarization, leading to $Ca^{2+}$ influx via L-type VSCCs (L). The magnitude of the change in gene expression may be modified by GABAergic and cholinergic neural activity and glucocorticoids. Elevated levels of BDNF secreted by the neuron and acting on the same or a neighboring neuron with an increased number of TrkB receptors lead to markedly enhanced BDNF signaling; calcium homeostasis is preserved and the neuron survives. This figure also illustrates the major mechanisms for $Ca^{2+}$ homeostasis, which could be the targets of neurotrophin action: The level of cytosolic calcium is dependent on influx via VSCCs and agonist-operated (NMDA receptor-gated) calcium channels (AOCCs) and on $Ca^{2+}$ release from intracellular stores, triggered by inositol triphosphate ($IP_3$). Efflux of calcium occurs by ATP-dependent $Ca^{2+}$ pumps in plasma membrane and intracellular stores, and through a $Na^+/Ca^{2+}$ exchanger in the plasma membrane. Calcium-binding proteins (CBP) also buffer $Ca^{2+}$ entering or released within the cell. BDNF and NT-3 have been shown to increase the number of neurons expressing the calcium-binding protein, calbindin, in hippocampal cultures. (Modified with permission from Lindvall O, Kokaia Z, Bengzon J, et al. Neurotrophins and brain insults. TINS 1994;17:490–496.)

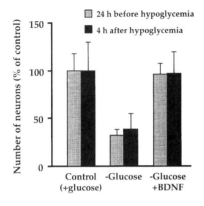

**Figure 1.12**
Neuronal survival in cell cultures of dentate gyrus subjected to glucose deprivation with BDNF added either 24 hours before or 4 hours after the onset of hypoglycemia. BDNF prevents the hypoglycemia-induced neuronal death. Means ± SEM. Values are expressed as percentage of control cultures containing glucose. (Modified with permission from Kokaia Z, Othberg A, Kokaia M, Lindvall O. BDNF makes cultured dentate granule cells more resistant to hypoglycemic damage. Neuro-Report 1994;5:1241–1244.)

teins (see Fig. 1.11). Another possibility is that neurotrophins induce systems that detoxify free radicals.

Endogenous production of neurotrophins not only may have an effect in preventing neuronal death, but also may promote sprouting, axonal regeneration, and synaptic reorganization. If this were the case, neurotrophins could be used to promote regeneration after CNS injury. Unfortunately, regenerative responses leading to the formation of aberrant connections may have negative consequences. Neurotrophins actually may promote connections that enhance seizure susceptibility after insults.

## Perspectives

The dramatic and rapid changes of the synthesis of several neurotrophic factors after brain insults imply important functions for repair and regeneration. Available data provide evidence that neurotrophic factors can prevent neuronal death and cause nerve cells to form new fiber connections. Clearly, further investigation is needed to delineate the effect of various insults on neurotrophins and the functional effects of the neurotrophins.

## CELL TRANSPLANTATION

Grafting of fetal neural tissue into the mammalian CNS has emerged as a widely used experimental tool with which to study a diversity of neurobiologic problems (e.g., mechanisms of neural development, plasticity, and regeneration). Much basic research has been devoted to the morphologic and functional analysis of neural grafts in animal models of chronic neurodegenerative disorders such as Parkinson's disease, Huntington's disease, and dementia. This has led to the first clinical trials with intrastriatal implantation of fetal dopamine-rich mesencephalic tissue in patients with Parkinson's disease. The results from these studies indicate that the basic principles of cell replacement, established in animal experiments, are valid also in the diseased human brain. Considerably less work has been performed with neural grafts to restore brain function in acute neurodegenerative disease.

It has been demonstrated that neural grafts can survive in brain lesions produced by focal or global ischemia, establish afferent and efferent connections with the recipient's brain, and also reverse some functional deficits in the host. Functional effects of neural grafts in ischemic brain lesions may be exerted in several different ways:

1. *Trophic action:* The graft stimulates recovery mechanisms such as sprouting from intrinsic neurons.
2. *Biologic minipump:* The graft establishes no or very few synaptic contacts with host neurons but releases transmitters into the surrounding paren-chyma, as would a paracrine gland.
3. *Synaptic release:* Grafted neurons re-innervate the host brain, providing host neurons with afferent synaptic contacts and a tonic unregulated (or autoregulated) supply of transmitter sufficient to restore activation, inhibition, or disinhibition of host circuitry.
4. *Integration into the host brain:* The grafted neurons establish extensive afferent and efferent synaptic connections with host neurons.

### Grafting in Focal Ischemia

Fetal cortical tissue has been implanted in the region of the cortical infarct in normal and spontaneously hypertensive rats subjected to middle cerebral artery occlusion. Both solid and cell suspension grafts survive well in the ischemic cavity and can fill the area of infarction (Fig. 1.13A). The morphology of fetal cortical suspension grafts placed in the infarcted cortex differs from the normal neocortical structure; the laminar organization is not preserved (Fig. 1.13A).

Cortical grafts are richly innervated by the host brain. The density of cholinergic (Fig. 1.13B), noradrenergic (Fig. 1.13C), and serotonergic fibers of host origin increases in the graft over time. Furthermore, retrograde tracing of host projections to the graft shows labeled cells in the nucleus basalis, ventral pallidum, thalamus, dorsal raphe, and locus coeruleus, as well as in the ipsilateral and contralateral neocortex. In contrast, there is only a sparse axonal outgrowth from cortical cell suspension grafts into the host brain. In the appropriate circumstances, transplanted cortical neurons can be functionally integrated with host neural circuitry after a focal ischemic insult (Fig. 1.14). Given the paucity of efferent connections from the graft to the host, it seems most likely that if symptomatic recovery occurs after transplantation, it is secondary to a trophic influence by the graft on the host and not determined by a reconstruction of neuronal circuitry.

Attempts have been made to implant fetal striatal tissue into infarcted striatal regions following MCA occlusion. Striatal grafts survive well in the

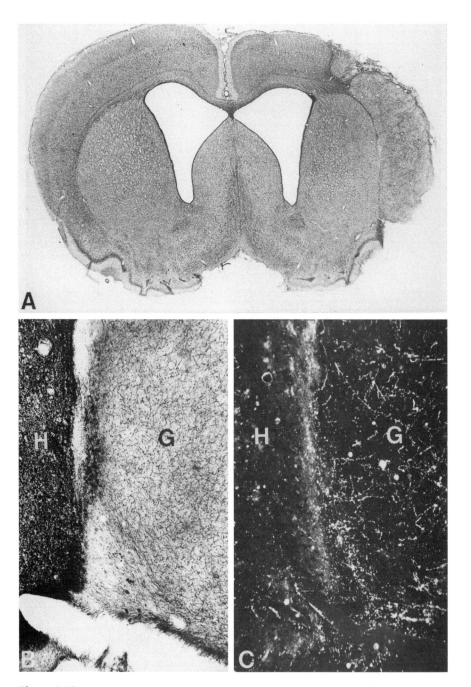

**Figure 1.13**

(A) Photomicrograph of a coronal section through the brain of a rat first subjected to MCA occlusion and then, about 1 week later, grafted with fetal cortical tissue in the right hemisphere. At 10 weeks after implantation, two separate lobules of graft tissue of rather homogenous appearance fill the infarct cavity. There is no normal cortical lamination. (B, C) Photomicrographs showing, in a coronal section, (B) cholinergic and (C) noradrenergic fibers of host origin in a cortical graft (G); the host caudate-putamen (H) is seen to the left. (Modified with permission from Grabowski M, Brundin P, Johansson BB. Fetal neocortical grafts implanted in adult hypertensive rats with cortical infarcts following a middle cerebral artery occlusion: ingrowth of afferent fibers from the host brain. Exp Neurol 1992;116:105–121.)

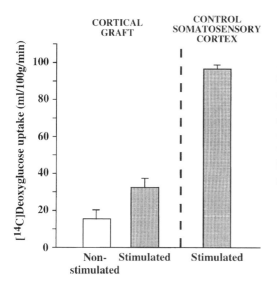

**Figure 1.14**

[14C]-Deoxyglucose utilization in cortical grafts (placed in the infarct cavity after MCA occlusion) of nonstimulated and vibrissae-stimulated groups, and in vibrissae-stimulated somatosensory cortex of a normal control group. There is a significant 110% increase of glucose utilization in the grafts after stimulation, providing evidence for functional integration of the graft in host neuronal circuitry. Means ± SEM. (Modified with permission from Grabowski M, Brundin P, Johansson BB. Functional integration of cortical grafts placed in brain infarcts of rats. Ann Neurol 1993;34:362–368.)

ischemically damaged striatum and improve some functional deficits. The grafts contain both presumed cholinergic and GABAergic neurons.

## Grafting in Global Ischemia

Grafts of fetal hippocampal tissue survive well in the hippocampus of rats subjected to global ischemia. The grafts contain cells with morphology, electrical firing patterns, and neurotransmitter receptor binding similar to those of normal adult CA1 pyramidal neurons and interneurons. The fetal hippocampal grafts establish both afferent and efferent connections with the host brain. Many of the connections are normal, but there are abnormal connections with the host as well. Studies in rodents indicate that hippocampal grafts can afford some measure of functional recovery on learning and memory tasks. The degree of functional recovery also may depend on the cells of origin; cells of CA1 origin do better than cells of basal forebrain origin for CA1 ischemia.

## Perspectives

It has been clearly demonstrated in rodents that cortical and striatal grafts can survive in the area of a focal ischemic infarction and become innervated by the host brain. From the clinical point of view, however, this research is still at a very early stage. Of major importance will be to clarify why the efferent projections from the graft to the host are so few. Strategies must be developed to overcome growth inhibitory mechanisms preventing axonal elongation along the efferent pathways of normal cortical projection neurons. It also seems possible that exogenous administration of various neurotrophic factors to the grafts could stimulate both survival and fiber outgrowth.

## Acknowledgments

The work of the authors was supported by the Swedish MRC, the U.S. Public Health Service via the NIH (NINDS), the Royal Swedish Academy of Sciences, the Segerfalk Foundation, and the Medical Faculty of the University of Lund. We thank Dr. Martin Grabowski for valuable help with Figure 1.13 and Marie Lundin and Katarina Manson for secretarial help.

## References/Further Reading

Auer RN, Siesjö BK. Biological differences between ischemia, hypoglycemia and epilepsy. Ann Neurol 1988;24:699–707.

Beal MF. Mitochondrial dysfunction and oxidative damage in neurodegenerative diseases. Heidelberg: Springer-Verlag, 1995.

Choi DW, Rothman SM. The role of glutamate neurotoxicity in hypoxic/ischemic neuronal death. Annu Rev Neurosci 1990;13:171–182.

Choi DW. Glutamate neurotoxicity and diseases of the nervous system. Neuron 1988;1:623–634.

del Zoppo GJ. Microvascular changes during cerebral ischemia and reperfusion. Cerebrovasc Brain Metab Rev 1994;6:47–96.

Deng H, Hentati A, Tainer J, et al. Amyotrophic lateral sclerosis and structural defects in CU, Zn superoxide dismutase. Science 1993;261:1047–1051.

Dunnett SB, Björklund A, eds. Functional neural transplantation. New York: Raven, 1994:1–587.

Faden AI, Demediuk P, Panter SS, Vink R. The role of excitatory amino acids and NMDA receptors in traumatic brain injury. Science 1989;244:798–800.

Feuerstein GZ, Liu T, Barone FC. Cytokines, inflamation, and brain injury: role of tumor necrosis factor-α. Cerebrovasc Brain Metab Rev 1994;6:341–360.

Floyd RA, Carney JM. Free radical damage to protein and DNA: mechanisms involved and relevant observations on brain undergoing oxidative stress. Ann Neurol 1992;32:S22–S27.

Halliwell B. Reactive oxygen species and the central nervous system. J Neurochem 1992;59:1609–1623.

Hossmann K-A. Glutamate-mediated injury in focal cerebral ischemia: the excitotoxin hypothesis revised. Brain Pathol 1994;4:23–36.

Isackson PJ. Trophic factor response to neuronal stimuli or injury. Current Op Neurobiol 1995;5:350–357.

Khatchaturian Z, et al., eds. Calcium hypothesis of aging and dementia. New York: Annals of the New York Academy of Sciences, 1995.

Lindsay RM, Wiegand SJ, Altar CA, DiStefano PS. Neurotrophic factors: from molecule to man. TINS 1994;17:182–190.

Lindvall O, Kokaia Z, Bengzon J. Neurotrophins and brain insults. TINS 1994;17:490–496.

Mattson M, Scheff S. Endogenous neuroprotection factors and traumatic brain injury: mechanisms of action and implications for therapy. J Neurotrauma 1994;11:3–33.

Olanow C. A radical hypothesis for neurodegeneration. TINS 1993;16:439–444.

Siesjö B, Wieloch T, eds. Cellular and molecular mechanisms of brain damage in *Advances in Neurology*. New York: Raven, 1996 (in press).

Siesjö B. The role of calcium in cell death. In: Price D, Aguayo A, Thoenen H, eds. Neurodegenerative disorders: mechanisms and prospects for therapy. Chichester: John Wiley & Sons, 1991:35–59.

Siesjö BK, Katsura K, Mellergård P, et al. Acidosis-related brain damage. Prog Brain Res 1993;96:23–48.

Siesjö BK. Basic mechanisms of traumatic brain damage. Ann Emerg Med 1993;22:959–969.

Siesjö BK. Pathophysiology and treatment of focal cerebral ischemia. I. Pathophysiology. J Neurosurg 1992;77:169–184.

Siesjö BK. Pathophysiology and treatment of focal cerebral ischemia. II. Mechanisms of damage and treatment. J Neurosurg 1992;77:337–354.

Thoenen H. Neurotrophins and neuronal plasticity. Science 1995;270:593–598.

Wieloch T, Bergstedt K, Hu BR. Protein phosphorylation and the regulation of mRNA translation following cerebral ischemia. Prog Brain Res 1993;96:179–191.

Zoratti M, Szabó I. The mitochondrial permeability transition. Biochim Biophys Acta 1995;1241:139–176.

# Neuroimmunology

Patricia A. Lodge
Subramaniam Sriram

Although the CNS is considered to be in an immunologically privileged site, close interactions and trafficking of immune cells occur continuously. The surveillance provided by the lymphocytes against pathogenic organisms to the brain is essentially no different than those in other organs. However, the unique features of the blood-brain barrier and the differential regulation of immune molecules make the elements of immune function unique. This chapter focuses on the structure of the immune components and how these functions are altered in pathologic states.

## Players of the Immune System

Components of the immune system can be divided into two categories: those that are antigen-specific and those that are nonspecific. Antigen-specific components respond only to individual antigens and form the basis of memory responses. Memory responses confer the organism with an increase in the potency, speed, and specificity of the response upon re-challenge with the same antigen. Nonspecific components of the immune system, such as polymorphonuclear cells and macrophages, have rapid deployment kinetics that clear the invading organism, but lack memory and specificity to the targeted pathogen; they clear infectious agents that have been targeted by the specific immune components and also respond to general features shared by all pathogens, such as bacterial lipopolysaccharide, by phagocytosis and cytokine production. In this way, the two mechanisms work in concert to protect the body against infection.

### ANTIGEN-SPECIFIC COMPONENTS

#### T Lymphocytes

T lymphocytes (so called because their development is dependent on the thymus) form the linchpin of the immune response, because all antigen-specific responses are uniquely initiated by the binding of antigenic peptides to a specific T cell receptor (TCR) in the context of self major histocompatibility molecules (discussion follows). Congenitally athymic individuals are deficient in T cells, and removal of the thymus in the early postnatal period also renders an individual T cell deficient. T-lymphocyte precursors arise from hematopoietic stem cells in the bone marrow and migrate to the thymus. On their arrival in the thymus, T cell precursors begin to rearrange their TCR genes. The TCR is a noncovalently linked heterodimer that belongs to the immunoglobulin superfamily. There are two forms of TCR, one composed of alpha-beta dimers and the other composed of gamma and delta chains. Expression of one form of receptor excludes the expression of the other so that each T cell expresses either alpha-beta or gamma-delta, but not both. The alpha-beta TCR is far more common in peripheral lymphoid organs, while gamma-delta T cells appear to localize in certain organs, such as skin, intestinal epithelium, and lungs, and are thought to play a sentinel role in the immune response. Many different receptors are needed to recognize the wide variety of antigens that the organism may encounter, and both T and B lymphocytes generate receptors by gene rearrangement (see section, Receptor Diversity). Once the TCR is expressed, T cells are positively and negatively selected on the basis of this receptor.

T cells are further divided into two groups on the basis of the expression of two mutually exclusive co-receptors, CD4 and CD8. These two subsets of T cells also serve different functions; they are called helper (CD4) and cytotoxic (CD8) T cells. However, the division of functions based on CD4 and CD8 expression is not absolute.

## Major Histocompatibility Antigens and Antigen Presentation

The major histocompatibility complex (MHC) is a genetic locus which contains many important immune-response genes; they are grouped into three classes. The class I and class II genes encode the cell surface molecules, called *human leukocyte antigens* (HLAs). The class III region contains genes for complement components. There are three class I molecules—HLA-A, B, and C—encoded on each chromosome. Class I MHC antigens are heterodimers composed of an invariant light chain and a polymorphic heavy chain. The two polypeptides are noncovalently associated. The light chain, beta-2-microglobulin, is encoded by a gene outside of the MHC. All nucleated cells, with the exception of neurons and oligodendrocytes, express MHC class I antigens.

Class II molecules are also heterodimers, but both alpha and beta chains are transmembrane proteins encoded in the MHC. There are three class II antigens in humans, DR, DQ, and DP. Class I and class II molecules share a common structure characterized by two alpha helical loops separated by a beta pleated sheet, forming a floor. Antigens and peptides that are processed by antigen-presenting cells are delivered to the surface bound in the groove between the two alpha helical loops. Class I molecules bind peptides derived from proteins such as viral proteins synthesized within the cell (Fig. 2.1). Class II molecules present peptides of extracellular proteins, as illustrated in Figure 2.2. Class II

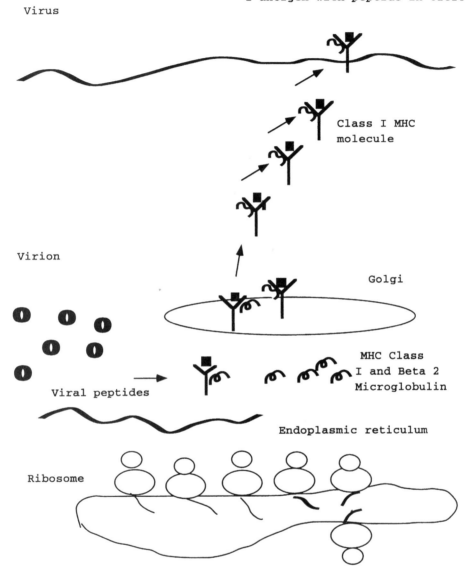

**Figure 2.1**

Class I antigen processing and presentation. Class I heavy chain and beta 2 microglobulin are synthesized into the lumen of the endoplasmic reticulum. Peptide transporter molecules translocate peptides from the cytoplasm to the lumen of the endoplasmic reticulum where they are bound by MHC class I and transported to the surface. These peptides in the cytoplasm can be derived from normal cell proteins or from intracellular parasites such as viruses.

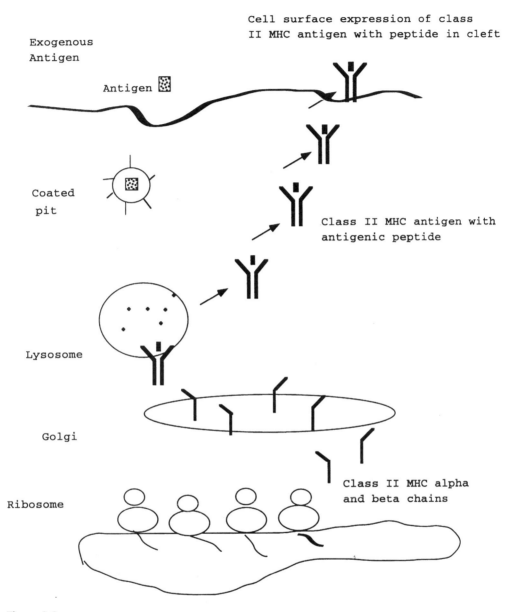

**Figure 2.2**

Class II antigen processing and presentation. The alpha and beta chains of class II MHC are synthesized into the lumen of the endoplasmic reticulum and thence transported to the Golgi. In a post-Golgi compartment, the vesicle containing class II MHC fuses with a lysosome containing digested proteins from exogenous proteins. These peptides bind in the groove of class II MHC and are transported to the surface.

molecules are expressed mainly by macrophages and B cells. However, other cells, such as astrocytes and endothelial cells, can also be induced to express MHC class II antigens on stimulation by cytokines (e.g., gamma interferon), which enhance the expression of immune molecules. It is not difficult, therefore, to imagine the development of an immune response in the brain, which is normally bereft of immune cells.

T lymphocytes possess either CD4 or CD8 molecules, which act as co-receptors. CD4 binds to a determinant on class II, so CD4-positive T cells recognize peptides bound by class II. CD8 binds to the alpha-3 domain of class I, so CD8-positive T cells are specific for antigen bound by class I molecules (Fig. 2.3). This division of specificity results in predictable differences in function. CD4 cells are targeted to extracellular antigens and provide help to B cells; they are the cells responsible for delayed-type hypersensitivity responses. In contrast, CD8 cells recognize peptides derived from intracellular pathogens (e.g., viruses, lyse-infected cells).

Most autoimmune diseases are associated with particular MHC alleles. For example, HLA-DR2 is seen in 60% of multiple sclerosis (MS) patients of Northern European Caucasian ancestry but is found in only 20% of the unaffected population. A similar association is observed between myasthenia gravis and HLA-DR3; this may be due to the ability of a given MHC allele to bind and present self-antigens. Therefore, individuals expressing the disease-associated allele would be more likely to be sensitized to self-antigens and develop autoimmune disease.

## B Lymphocytes

B lymphocytes secrete immunoglobulin and also can serve as antigen-presenting cells, because they express MHC class II molecules. They develop in the bone marrow and share certain developmental features with T lymphocytes. The B cell receptor (BCR) is a membrane-bound form of immunoglobulin and is generated by recombination of gene segments in a manner analogous to the TCR. Like T lymphocytes, self-reactive B lymphocytes are deleted or rendered unresponsive during development. Unlike the TCR, which recognizes short peptides, BCR/Ig recognizes intact antigens based on their three-dimensional structure. Mature B cells that have not encountered antigen express immunoglobulin M (IgM) on their cell surfaces. Following an encounter with antigen, activated B cells differentiate further into plasma cells, which secrete immunoglobulin G (IgG). B cells can make antibodies on contact with antigen only in the presence of requisite "help" from T cells. This help may be derived by direct binding of T cells to MHC/peptide complexes on B cells, also referred to as *cognate recognition,* or by humoral factors such as cytokines secreted by T cells, also called *noncognate help.*

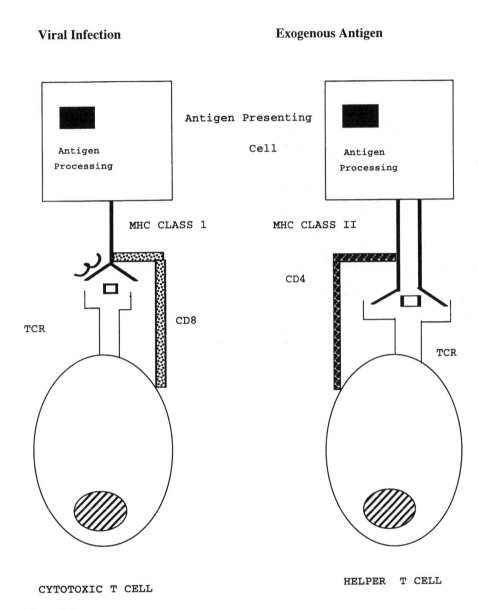

**Viral Infection**          **Exogenous Antigen**

Antigen Presenting

Cell

Antigen Processing

Antigen Processing

MHC CLASS 1          MHC CLASS II

CD4

TCR

CD8

TCR

CYTOTOXIC T CELL          HELPER  T CELL

**Figure 2.3**
T cell receptor-ligand interactions. Co-receptors CD4 and CD8 direct T cells to recognize antigens presented by class II and class I MHC molecules, respectively. CD8 binds to class I MHC, directing CD8 + T cells to recognize (viral) peptides bound to class I. CD4 binds to class II MHC; therefore, CD4 + lymphocytes recognize exogenous peptides associated with class II MHC.

## Receptor Diversity

For the T and B lymphocytes to recognize the enormous number of potential antigens in a specific fashion, a wide array of receptors with different specificity is required. If separate genes specifically encoded each receptor, a huge burden would be placed on the genome. Instead, receptor diversity is generated by genetic rearrangement. Both TCRs and BCRs can be divided into different regions: variable (V), joining (J), and constant (C). An additional diversity (D) region is present in immunoglobulin heavy chains and TCR beta chains. There are many gene segments that can encode these regions. During T and B cell development, one V gene segment will be juxtaposed with one of the available J segments, and the intervening gene segments are spliced out by recombinase. A second rearrangement brings V-J to a D segment when present. In addition to the many combinations possible through gene rearrangement, additional diversity is generated at the joining regions. The sites recognized by the recombinase may be either degraded or nucleotides added before the two segments are ligated. This is referred to as *N-region diversity*, the consequence of which is that even though two cells may express the same V, J, and C genes, the N region is likely to differ between them. B cells have two additional mechanisms for generating further diversity. Following an encounter with antigen, somatic mutation occurs within immunoglobulin genes at a relatively high frequency. The antibodies generated following somatic mutation may have a higher affinity for antigen than the "original" antibody. During isotype switching, further rearrangements lead to recombination of the same variable-region gene with new constant-region genes.

## NONSPECIFIC COMPONENTS OF THE IMMUNE SYSTEM

### Monocytes and Macrophages

Bone marrow–derived myeloid progenitor cells give rise to monocytes. These constitute about 4% of the peripheral blood leukocytes and are morphologically identified by more abundant cytoplasm and a kidney-shaped nucleus. Following migration into tissues, these cells, referred to as *macrophages*, play important roles in both nonspecific and specific immune responses. They do not possess antigen-specific receptors as do T and B lymphocytes, but their activities are often directed by these cell types. Monocytes/macrophages express receptors for the Fc region of Ig so that pathogens coated with antibodies are thereby tagged for phagocytosis. Cytokines secreted by T cells directly activate macrophages to enhance their uptake of extracellular antigens. In turn, macrophages affect the specific immune response by presenting antigen to T cells.

### Granulocytes

Granulocytes also arise from the myeloid lineage and share some functions with monocytes. They are chiefly phagocytic cells, but they also secrete important inflammatory mediators. Granulocytes are subdivided into three types—neutrophils, eosinophils, and basophils. Neutrophils account for 60% to 70% of peripheral blood leukocytes and are usually the first cell type to accumulate at an inflammatory site; they express receptors for the Fc region of IgG and the C3a component of complements, which direct phagocytosis of microorganisms. Eosinophils are prominent in parasitic infections, but the significance of this association is not clear. Basophils, which are termed *mast cells* when in tissue, express Fc receptors specific for IgE, an isotype of immunoglobulin commonly generated in response to allergens. Binding of antigen to IgE on the mast cell surface triggers the release histamines, resulting in the symptoms of allergy.

## Natural Killer Cells

Natural killer cells make up about 2.5% of peripheral blood lymphocytes and are called *large granular lymphocytes* because of their large intracytoplasmic azurophilic granules and high cytoplasm-nuclear ratio. Natural killer cells are of the lymphocytic lineage but lack the cell surface markers characteristic of B cells and T cells. Unlike cytotoxic CD8+ T cells, natural killer cells lack immunologic memory and have the ability to kill a wide variety of tumor- and virus-infected cells without any evidence of MHC restriction or prior activation. The biologic function of the natural killer cell is uncertain. In view of its in vitro function of lysing tumor cells, it may play a role in tumor immunity.

## Complement

The complement system is composed of many proteins with a wide range of functions. It can be activated by bound or aggregated immunoglobulin and by some bacterial cell surfaces. A cascade of events involving proteolytic cleavage and binding of multiple components leads to the formation of a membrane attack complex. This structure forms a doughnut-shaped lesion in the target cell membrane, which results in lysis of the cell. Several of the products of the complement cascade act as chemoattractants and ligands for granulocytes and monocytes. In particular, C3a and C5a bind to receptors on basophils/mast cells, triggering release of histamine and other mediators of anaphylaxis.

## Cytokines

Cytokines can be thought of as the paracrine, autocrine, and endocrine hormones of the immune system. The availability of recombinant cytokines has enhanced immeasurably our understanding of these factors. The list of effects of cytokines on both immune and nonimmune cells continues to grow, and the complex interactions between cytokines are only beginning to be appreciated. Table 2.1 lists some of the important cytokines and their most prominent effects. Cytokines are broadly divided into the following categories, which are not mutually exclusive:

**Table 2.1**  Sources and Functions of Cytokines

| Cytokine | Major Functions |
| --- | --- |
| T Cell–Derived Cytokines | |
| IL-2 | T and B cell growth stimulation |
| IL-3 | Stimulates growth of bone marrow stem cells |
| IL-4 | B cell growth and differentiation |
| IL-5 | B cell growth and differentiation |
| Interferon-$\gamma$ | Induces/up-regulates expression of MHC class I and II adhesion molecules |
| Macrophage-Derived Cytokines | |
| IL-1 | Co-stimulates T and B cell activation; enhances adhesion molecule expression |
| TNF-$\alpha$ | Cytotoxic; induces adhesion molecule expression; pyrogenic |
| TGF-$\beta$ | Inhibits T cell proliferation and migration |
| Multiple Sources | |
| IL-12 | Co-stimulates T cell activation |
| Interferon-$\alpha$ | Antiviral, antiproliferative |
| Interferon-$\beta$ | Antiviral, antiproliferative |
| IL-6 | Stimulates production of acute-phase proteins |
| IL-8 | Chemoattractant cytokine |

1. Growth factors: interleukins (IL) 1, 2, 3, and 4, and colony-stimulating factors
2. Activation factors, such as interferons (alpha, beta, and gamma): these are also antiviral.
3. Regulatory/cytotoxic factors, including IL-10, IL-12, tumor growth factor-$\beta$ (TGF-$\beta$), lymphotoxins, and tumor necrosis factor (TNF)
4. Chemotactic inflammatory factors, also called *chemokines*, such as IL8, MIP-1a, and MIP-1b

## ORCHESTRATION OF IMMUNE RESPONSES

The organization of the immune response is somewhat dependent on the agent or injury that initiates it. Bacteria, viruses, tumors, and parasites all elicit different types of immune responses, which are adapted to effectively meet the challenge of those particular pathogens. Both antigen-specific and nonspecific components of the immune system contribute to the overall response, but the relative contribution of each component varies.

The entry of immune cells into tissue is one of the earliest events in the immune response. The regulation of this process has been studied intensely in the past 10 years, but the application of this information to the CNS has just begun. Table 2.2 lists adhesion molecules and their counterreceptors. Three families of molecules participate in the process of adhesion and migration. The selectins—P-selectin and E-selectin on the surface of endothelium and L-selectin on the leukocyte cell membrane—have a lectin-like domain that binds carbohydrate residues. Integrins form a large family of proteins composed of two chains, alpha and beta. Lymphocyte function antigen (LFA-1) and very late antigen 1 (VLA-1) are two examples of this family. Many other integrins may be important in extravasation but are not listed here because their role is still ill defined.

Most members of the integrin family bind to members of the immunoglobulin superfamily, which includes intercellular adhesion molecules 1 and 2 (ICAM-1, ICAM-2) and the vascular cell adhesion molecule (VCAM). The sequence of events leading to extravasation has been described for neutrophils, and the same or a similar process is thought to regulate lymphocyte adhesion and migration. The initial interaction between lymphocytes and endothelium is mediated by the selectins, which tether the lymphocytes to the endothelium so that they roll along it. Integrin-type receptors then bind to immunoglobulin family ligands, causing the lymphocyte to stop rolling and form a tighter association with endothelial cells, at which stage, influenced by chemokines, it exits the vessel and enters the tissue. Interleukin-8 is an example of a chemokine, and its action has been described as follows: IL-8 is produced by endothelial cells and is secreted both into the vessel lumen and the basement membrane; it causes leukocytes to down-regulate expression of selectins, presumably aiding the cell in detaching from the endothelium and migrating through the vessel wall. The association of IL-8 with connective tissue elements is thought to direct leukocyte migration into tissue. In addition, integrins such

**Table 2.2**  Adhesion Molecules: Receptors and Counterreceptors

Selectins
  Leukocyte receptor
    L-selectin
    (Carbohydrate moeity?)
    Sialylated Lewis X
  Endothelial counterreceptor
    (Carbohydrate moeity?)
    P-selectin
    E-selectin
Integrins-Immunoglobulin Superfamily
  Leukocyte integrins
    CD11/CD18 (LFA1, Mac1)
    VLA-4
  Endothelial Ig superfamily members
    ICAM-1, ICAM-2
    VCAM

as VLA-4 also have the capacity to bind connective tissue proteins, enhancing the migration of leukocytes through tissue.

Migration is regulated in part by the pattern of expression of the various adhesion molecules. This varies between normal and activated endothelium and also between resting and activated lymphocytes. Normal endothelium does not express P-selectin, E-selectin, or VCAM, but does express some level of ICAM-1. On activation by cytokines or other soluble factors (e.g., histamine and thrombin), these molecules are up-regulated. Transport of P-selectin to the cell surface is induced rapidly by endothelial cell activators, and expression is transient. IL1 and TNF-α induce E-selectin expression over a period of hours, so P-selectin may be involved in the earliest extravasation of granulocytes, while E-selectin directs migration somewhat later. Expression of VCAM by endothelium occurs still later and also is controlled by cytokines, principally IL-1, IL-4, and TNF. The counterreceptors on leukocytes also are regulated by cytokines and the activation status of the cell. L-selectin expression is limited to resting lymphocytes, where it is thought to mediate entry to lymph nodes. Following activation, lymphocytes shed L-selectin from their surfaces. In contrast, integrin expression is increased on activated lymphocytes. Activated lymphocytes produce cytokines which increase levels of integrin ligands, ICAM-1, ICAM-2, and VCAM, thereby optimizing the ability of activated cells to reach sites of infection or injury.

A foreign object is first phagocytosed by dendritic cells and tissue macrophages. Following ingestion, pathogens are degraded by proteases within the cell. The resulting peptide fragments are bound by class II MHC and transported to the cell surface. The phagocytes then migrate to draining lymphoid organs wherein the antigens are presented to T cells. Here, CD4+ T cells recognize peptides bound by class II MHC and respond by secreting cytokines and proliferating. Interaction of B cells with antigen, plus the cytokines produced by T helper cells, promote growth and differentiation of B cells into plasma cells. The immunoglobulins produced by B cells contribute to killing pathogens by directing complement-mediated lysis and enhancing phagocytosis by granulocytes and macrophages. During viral infections, cytolytic T cells (CTLs) of the CD8+ phenotype are important to viral clearance. As discussed earlier, CD8+ CTLs recognize proteins synthesized within cells. Generation of CTL specific for viral proteins synthesized by infected cells limits the spread of infection by killing infected cells before release of viral progeny. Following resolution of the infection, some T and B cells specific for the pathogen are retained. The presence of increased numbers of T and B lymphocytes allows for more rapid response if the pathogen is re-encountered.

## Immune Responses in the Central Nervous System

The CNS has been described as an immunologically privileged site because it seemed to be inaccessible to the immune system. The blood-brain barrier

(BBB), formed by endothelial cells joined by tight junctions, effectively excluded immune cells and immunoglobulin from the CNS. In addition, the lack of lymphatic drainage of the CNS limits effective circulation of cells through nervous tissue. Expression of MHC molecules in the brain is extremely low, as compared with other tissues, which restricts the potential for antigen presentation. These features have been invoked to explain the reduced response to grafts placed within the CNS. It is clear, however, that immune responses do take place within the CNS, and the special features of these have been clarified by study of animal models of disease, in particular, the research into experimental allergic encephalomyelitis (EAE).

In normal brain tissue, MHC class I molecules are expressed only on endothelium, and class II antigens are absent. During inflammation (e.g., MS or viral encephalitis), these antigens are induced on several cell types. Major histocompatibility complex expression is most prominent on endothelium and microglia, has some expression on astrocytes, but apparently none on oligodendrocytes or neurons. Therefore, T cell activation and function must be confined to interactions with endothelium, microglia, and astrocytes. These cell types have been examined for their ability to act as antigen-presenting cells, and all may contribute to antigen presentation in varying degrees. The antigen-presenting capabilities of endothelial cells and astrocytes are variable and may depend on the species or the method of testing. Microglial cells serve as the antigen-presenting cells of the brain. In addition, they respond to cytokines released by activated T cells and release proinflammatory cytokines (IL-1, TNF, IL-6, and IL-12) that further augment the immune response to the antigen. Some of the CNS damage incurred as a result of inflammation may be due to the secretion of cytokines by macrophages and T cells. In particular, TNF-$\alpha$ has been shown to lyse oligodendrocytes in vitro. In addition, levels of TNF are elevated in patients with MS, suggesting that TNF may play a role in CNS demyelination.

As stated, the expression of adhesion molecules in the brain has only recently been examined. Microvessels isolated from lesion areas in the brains of MS patients expressed ICAM-1, VCAM, and E-selectin, but these molecules were absent from normal brain tissue. In EAE, which is the animal model of MS, expression of the integrin VLA-4 is an important determinant of the ability of T cells to cause disease. Treatment with antibodies to either VLA-4 or its counterreceptor, VCAM, delayed onset of paralysis. Anti-adhesion molecule antibodies also have been used in animal models of ischemia-reperfusion injury and bacterial meningitis. In these situations, much of the tissue injury is caused by neutrophils via oxygen-free radicals. Antibodies to integrins block adhesion of neutrophils to endothelium and reduce tissue damage.

The disadvantage of immune responses within the CNS is the potential for irreparable damage to tissue. This is especially true of the nonspecific effector mechanisms mentioned earlier. Neutrophils, macrophages, and microglia generate highly toxic oxygen-free radicals. Tumor necrosis factor-alpha has been shown to lyse oligodendroglia. Breach of the BBB allows fluid and immunoglobulin to enter. If autoreactive antibodies enter the CNS, they may direct attack tissue by complement and phagocytic cells. It is logical, therefore, that entry of lymphocytes into the CNS is limited. Experiments in the EAE model have revealed that migration into the CNS is limited to activated cells. If these activated T cells encounter antigen in the CNS, they respond by producing cytokines, which induce expression of the adhesion molecules described earlier and allow entry of both resting and activated cells.

## Immune-Mediated Diseases of the Nervous System and Putative Mechanisms

Autoimmune diseases can be divided into those that are mediated by antibodies to self-proteins and those that are thought to be the result of damage directed by T lymphocytes. The undeniable evidence of a causal relationship between an autoantigen and autoimmune disease rests on the fulfillment of Koch's postulates. In humans, this is proven mainly in autoantibody-mediated autoimmune diseases such as myasthenia gravis. Here, the antigen is known, the antibody is seen at the end organ, removal of the autoantibody improves the disease, and transfer of immune antisera into naive animals results in the clinical signs of myasthenia gravis. A causal relationship, however, between an autoantigen and autoimmune disease has not yet been shown for what are believed to be T cell–mediated diseases, such as rheumatoid arthritis, autoimmune uveitis, and MS. This is, in part, because the determinants recognized by B and T cells are different; that is, T cells recognize peptides of 10 to 20 amino acids, while B cells recognize a whole antigenic molecule. Thus, for large molecules, the peptides recognized by T cells are likely to be numerous, as is perhaps the case in myasthenia gravis.

Antigen-specific therapy cannot be accomplished successfully without knowing all of the antigenic epitopes, which presents a monumental problem. In contrast to autoantibody-mediated autoimmune diseases, evidence of a causality between antigen and disease has been even scantier. This is, in part, because the presence of autoreactivity does not necessarily guarantee disease. Although presence of antibody titers to autoantigens in autoantibody-mediated disease is usually indicative of disease, T cell reactivity to autoantigens does not carry the same implications. Hence, the only conclusive evidence of causality between an antigen and T cell–mediated autoimmune disease will be the reversal of the disease process following the removal of the putative autoreactive T cell repertoire. Although this has been feasible in autoimmune diseases such as EAE, it has been very difficult in most human T cell–mediated autoimmune diseases.

The mechanism by which an autoimmune disease is initiated is not known. One hypothesis, referred to as *molecular mimicry*, holds that viral or bacterial antigens have antigenic epitopes that resemble self-antigens. In mounting an immune response to the pathogen, an autoimmune response is inadvertently initiated, leading to autoimmune disease. This may be the case with rheumatic chorea, in that autoantibodies to streptococcal cell-wall protein cross-react with antigens on the caudate nucleus. A second possibility is the development of an immune response to superantigens. Superantigens are usually of bacterial or viral origin and bind as intact molecules to MHC. They have the property of stimulating all T cells that express a given TCR variable gene family, regardless of their exact specificity, by direct Vβ superantigen interaction. Superantigen stimulation leads to expansion of T cells that express a particular Vβ gene. If T cells in this particular expanded Vβ gene population include T cells that have autoreactivity to self-antigens, it is likely that an autoimmune disease may be initiated. Finally, autoimmune diseases may be associated with neoplasms, the immune response to tumor antigens cross-reacting with self-antigens and leading to autoantibody formation. All of the paraneoplastic neurologic syndromes are thought to occur in this manner.

## AUTOANTIBODY-MEDIATED DISEASES

In the nervous system, autoantibody has been proven conclusively to be pathogenic only in myasthenia gravis, in the paraneoplastic syndromes, and in chorea resulting from streptococcal disease.

### T Cell–Mediated Diseases

Only indirect evidence points to the association between autoantigens and autoimmune disease when the diseases are thought to be T cell–mediated. Multiple sclerosis has been thought to represent a T cell–mediated disease, in view of the similarities between MS and EAE. There are numerous reports of increased frequency of T cell reactivity to neural antigens, in particular myelin basic protein (MBP), in patients with MS; of abnormalities in T cell function in vitro; and of the presence of unique TCRs in their brains. A similar autoimmune pathology also is considered to occur in chronic inflammatory demyelinating polyneuropathy, polymyositis, dermatomyositis, and autoimmune vasculitis, but in these diseases, the likely candidate antigens are currently unknown.

## Experimental Models of Autoimmune Disease of the Nervous System

A number of immune-mediated autoimmune disorders of the nervous system are available for study in laboratory animals. In addition to allowing the analysis of the immunoregulatory network, they have been important models for designing immunotherapy. Although these model systems show many similarities to their analogous human diseases, they are induced by known procedures, while the inciting event in human diseases is unknown. Hence, extrapolation to human disease should be undertaken with caution.

### EXPERIMENTAL ALLERGIC ENCEPHALOMYELITIS

Experimental allergic encephalomyelitis is used widely as a model of MS. Inflammatory demyelination causes paralysis, which can exhibit a relapsing-remitting course. Lesions within the spinal cord consist mainly of CD4+ T cells and macrophages. The disease can be induced in many species by immunization with spinal cord homogenate or purified CNS proteins. Of these, MBP and proteolipoprotein (PLP) are the most extensively studied. Experimental allergic encephalomyelitis also can be initiated by transfusion of T cells specific for MBP or PLP, demonstrating that these cells are the pathogenic agents. The similarity of EAE to MS has raised a question regarding whether MS also might represent an autoallergic process to MBP.

### EXPERIMENTAL AUTOIMMUNE MYASTHENIA GRAVIS

Experimental autoimmune myasthenia gravis (EAMG) is the archetypal model of autoimmune disease, resembling human myasthenia gravis. It can be induced in animals by repeated injection of acetylcholine receptor (AChR) isolated from the electric organ of eels. This disease is mediated by antibodies directed against the AChRs. Transfer of hyperimmune sera or monoclonal antibodies directed to the receptor has been sufficient to induce it. As in the human disease, muscular weakness results from destruction of AChRs by antigenic modulation and complement-mediated lysis, and improves after treatment with acetylcholinesterase inhibitors or with immunosuppression. The main differences from the human counterpart are that the murine disease is monophasic and no thymic abnormalities are seen.

### EXPERIMENTAL AUTOIMMUNE NEURITIS

Among various peripheral nerve myelin proteins, the P2 fraction is neuritogenic in experimental animals. Injection of P2 in adjuvant results in the development of an acute monophasic polyneuropathy. This, like EAE, is T cell–mediated, and transfer has been performed successfully with T cell lines and clones, but its relationship to acute inflammatory polyneuropathy is uncertain.

## Immunotherapy

The immunosuppressive therapies currently available affect beneficial immune responses as well as the disease-associated activity and therefore carry inherent risks. The immunosuppressive agents conventionally used in treating neurologic disease include glucocorticoids, cyclophosphamide, and azathioprine. Plasma pheresis is useful in the management of myasthenia gravis; it removes the autoantibodies responsible for tissue damage but does not eliminate the plasma cells producing the antibody or control the immune response to AChR, and is therefore a temporary, palliative therapy.

Intravenous immunoglobulin (IVIG) has gained attention recently as an immunomodulating agent. Because IVIG consists of a pool of immunoglobulins from multiple donors, the mechanism by which it affects immune function is difficult to determine. It is thought that among the pool of antibodies are some that serve an immunoregulatory function by blocking activity of autoreactive cells and antibodies, but these regulatory antibodies have not been isolated and identified. Nevertheless, IVIG has been useful in the treatment of myasthenia gravis, acute and chronic inflammatory polyneuropathy, and dermatomyositis-polymyositis.

Beta interferon is the first drug approved for use in the treatment of relapsing-remitting MS. This cytokine is known to have antiproliferative effects, which may account for its ability to slow progression of disease.

Potential immunotherapeutic agents now in clinical trials make use of the advances in understanding of the regulation of the immune system. An example of this is the use of TGF-β for the treatment of MS. Tumor growth factor-beta is an immunoregulatory cytokine that can suppress T cell activation, proliferation, and migration. Other approaches target the antigen presentation process. Peptides have been designed that compete with autoantigenic peptides for binding to MHC molecules. Co-polymer 1, a random polymer of four basic amino acids, is an example of this kind of strategy. T cell receptors also are targets for immunotherapy. Peptides corresponding to the antigen-binding site of autoantigen-specific T cells have been synthesized and have been used to immunize MS patients. The rationale is that the TCR peptide will provoke an immune response to the autoreactive TCR and down-regulate activity of the autoreactive T cells. Monoclonal antibodies also are being tested for their ability to affect autoimmune disease. These include antibodies to class II MHC, TCRs, and CD4. These antibodies either bind to antigen and block function or can eliminate the cells expressing the antigen. With further advances in our understanding of autoimmune disease and the regulation of the immune system, more specific, less toxic therapies will be designed.

## References/Further Reading

Dinarello CA, Mier JW. Current concepts: lymphokines. N Engl J Med 1987;317:940–946.

Fabry Z, Raine CS, Hart MN. Nervous tissue as an immune compartment: the dialect of the immune response in the CNS. Immunol Today 15;218–224.

Hedrick SM. T lymphocyte receptors. In Paul WE, ed. Fundamental immunology. New York: Raven, 1989:291–313.

Kincade PW, Gimble JM. B lymphocytes. In Paul WE, ed. Fundamental immunology. New York: Raven, 1989.

Perry VH. Macrophages and the nervous system. Austin, TX: R.G. Landes, 1994.

Schonbeck S, Chrestel S, Hohlfeld R. Myasthenia gravis: prototype of the antireceptor autoimmune diseases. Int Rev Neurobiol 1990;32:175–200.

Springer T. Adhesion receptors of the immune system. Nature 1990;346:425–432.

Steinman LS. The development of rational strategies for selective immunotherapy against autoimmune disease. Adv Immunol 1991;49:357–379.

Trichineri G. Biology of natural killer cells. Adv Immunol 1989;47:187–206.

Washington R, Burton J, Todd RF, et al. Expression of immunologically relevant endothelial cell activation antigens on isolated central nervous system microvessels from patients with multiple sclerosis. Ann Neurol 1994;35:89–97.

Zamvil SS, Steinman LS. The T lymphocyte in experimental allergic encephalomyelitis. Annu Rev Immunol 1990;8:579–621.

# Neurogenetics

Phillip F. Chance      David E. Pleasure

Benjamin B. Roa      James R. Lupski

Jeanette Pleasure      Kenneth H. Fischbeck

## Inherited Neuropathy: Charcot-Marie-Tooth Disease and Related Neuropathies

### CHARCOT-MARIE-TOOTH NEUROPATHY: BACKGROUND AND CLINICAL SUBTYPES

Charcot-Marie-Tooth (CMT) neuropathy, also known as hereditary motor and sensory neuropathy (HMSN), is a heterogeneous group of inherited diseases of the peripheral nerve. Charcot-Marie-Tooth neuropathy is a common disorder affecting both children and adults, and may cause significant neuromuscular impairment. An estimated 1 in 2500 persons has a form of CMT, making it one of the largest diagnostic categories of neurogenetic disease.

Motor and sensory nerve function is affected in CMT. The clinical findings include distal muscle weakness and atrophy, impaired sensation, and diminished or absent deep tendon reflexes. CMT1 (HMSNI) denotes individuals with a hypertrophic demyelinating neuropathy ("onion bulbs") and reduced nerve conduction velocities, whereas CMT2 (HMSNII) refers to individuals with an axonal neuropathy and normal or near-normal nerve conduction velocities. The nosologic classification within CMT and related disorders has advanced with the application of molecular genetic methods for the purpose of categorizing the various subtypes of CMT. Table 3.1 summarizes a current classification scheme based on recent genetic studies.

### LINKAGE STUDIES IN CHARCOT-MARIE-TOOTH NEUROPATHY

In the majority of CMT1 and CMT2 pedigrees, the mode of inheritance is autosomal dominant; there are, however, pedigrees exhibiting X-linked inheritance and even rarer ones with autosomal recessive inheritance. Initial genetic linkage studies identified pedigrees in which the CMT1 gene was linked to the Duffy blood group locus on chromosome 1. However, subsequently it was

**Table 3.1**   Charcot-Marie-Tooth Neuropathy (Hereditary Motor and Sensory Neuropathy) and Related Disorders

|  | Locus | Gene | Mechanism |
|---|---|---|---|
| Charcot-Marie-Tooth type 1 (HMSNI) | | | |
| CMT1A | 17p11.2–12 | PMP22 | Duplication/point mutation |
| CMT1B | 1q22–23 | $P_0$ | Point mutation |
| CMT1C | Unknown | Unknown | Unknown |
| CMTX | Xq13.1 | Cx32 | Point mutation |
| CMT4 | 8q | Unknown | Unknown |
| Charcot-Marie-Tooth type 2 (HMSNII) | | | |
| CMT2A | 1p36 | Unknown | Unknown |
| CMT2B | Unknown | Unknown | Unknown |
| CMT2C | Unknown | Unknown | Unknown |
| Déjérine-Sottas disease (HMSNIII) | | | |
| DSDA | 17p11.2–12 | PMP22 | Point mutation |
| DSDB | 1q22–23 | $P_0$ | Point mutation |
| Hereditary neuropathy with pressure palsies | | | |
| HNPPA | 17p11.2–12 | PMP22 | Deletion/point mutation |
| HNPPB | Unknown | Unknown | Unknown |

found that the majority of CMT1 pedigrees demonstrate linkage to chromosome 17p11.2–12 and are designated as having CMT1A. Pedigrees that link to Duffy are designated as CMT1B.

Autosomal dominant CMT1 pedigrees that do not map to the region of the Duffy locus on chromosome 1q (CMT1B) or the proximal chromosome 17p (CMT1A) are designated as CMT1C. The CMT1C locus (or perhaps even loci) remains unassigned; as myelin genes are identified in the future, however, they will be strong candidates.

## CHROMOSOME 17p11.2–12 DUPLICATION ASSOCIATED WITH CMT1A

Attempts to further refine the localization of CMT1A through positional cloning led to the identification of a tandem DNA duplication in chromosome 17p11.2–12 associated with CMT1A. The physical size of the duplication is approximately 1.5 megabases (Mb), which represents a genetic distance of approximately 6 centimorgans (cM). The duplication is identified frequently in CMT1A patients of many ethnic groups. Surprisingly, the duplication can arise as a de novo event. Nine of ten patients with sporadic CMT1 had evidence of de novo genesis of the 17p11.2–12 duplication, suggesting that the de novo duplication may account for some cases of CMT1 that were thought to occur on the basis of autosomal recessive inheritance. The de novo duplication appears to result from errors in spermatogenesis. The duplication also is associated with the generation of a novel 500-kb SacII fragment, detected by marker VAW409, and is found in the vast majority of patients with inherited and de novo CMT1A. There have been two reports suggesting that rare patients may have duplications in 17p11.2–12 with different breakpoints.

## GENETIC MECHANISMS IN CMT1A

The finding of the duplication in 17p11.2–12 raised questions regarding the molecular mechanism underlying CMT1A. The two principal models proposed were gene disruption, whereby a critical gene is altered by a duplication breakpoint, and gene dosage, in which the phenotype results from having an extra copy of the gene (trisomic overexpression). Identification of a critical CMT1A gene affected by duplication remained of great import. Observations in humans and other organisms have helped to explain how the duplication results in the CMT1A phenotype. A small number of patients have been identified who have total or partial trisomy 17p, including bands 17p11.2–12, the region of the DNA duplication in CMT1A patients. The consistent phenotypic features of these patients are mental retardation and multiple somatic anomalies, including micrognathia, hypoplastic low-set ears, and foot deformities. Features consistent with CMT1A were detected in three patients with either partial or complete trisomy 17p, supporting the hypothesis that the duplication in CMT1A may have phenotypic consequences through a gene dosage effect.

Identification of a critical gene for CMT1A followed observations in a murine model, the *trembler* mutation (*Tr*), a dominant disorder resulting in a hypomyelinating neuropathy. The *Tr* locus maps to mouse chromosome 11, which has conserved synteny with human chromosome 17p, in the region of the CMT1A locus. The *Tr* locus was found to result from a point mutation in the peripheral myelin protein-22 (*PMP22*) gene. Observations in *Tr* suggested that the human *PMP22* gene might map to chromosome 17p11.2–12 in the region of the CMT1A gene, and might be the CMT1A gene. To test this hypothesis and further characterize the 1.5-Mb DNA duplication most commonly associated with CMT1A, several laboratories mapped *PMP22* to the CMT1A gene region on chromosome 17p11.2–12. Furthermore, the *PMP22* gene is also duplicated in the two CMT1A patients mentioned earlier who have alternative DNA duplications that appear to be less that 1.5 Mb in size.

CMT1A patients with point mutations in *PMP22* also have been identified. A CMT1 pedigree linked to 17p11.2–12 markers, in which the duplication was not present, was detected, suggesting that forms of CMT1A may exist in which other mechanisms are present. This hypothesis was recently confirmed when a missense mutation within the *PMP22* gene was found in this same nonduplicated pedigree. Interestingly, the point mutation in this family is identical to that found in the *Tr^J* mouse, a variant of *Tr*. Furthermore, in another family, the CMT1A phenotype was found to result from a de novo point mutation in the *PMP22* gene. In addition to dominant point mutations, an apparent recessive *PMP22* mutation also was identified. A compound heterozygote CMT1A patient was found to carry this *PMP22* mutation, paired with deletion on the 1.5-Mb region (including *PMP22*) on the homologous chromosome 17.

The *PMP22* gene encodes a 160–amino acid membrane-associated protein with a predicted molecular weight of 18 kDa that is increased to 22 kDa by glycosylation. The *PMP22* protein is localized to the compact portions of peripheral nerve myelin, contains four putative transmembrane domains, and is highly conserved in evolution. Duplication of the *PMP22* gene apparently leading to trisomic overexpression leads to the demyelinating phenotype through a gene dosage mechanism. Alternatively, point mutations resulting in nonconservative amino acid substitutions that presumably alter the native *PMP22* structure and function also lead to CMT1A. To reconcile these two mechanisms, it was postulated that increased gene dosage and expression of the normal *PMP22* gene, or altered function of the mutated protein, may disrupt the stoichiometry of *PMP22* on the peripheral nerve membrane that is critical to normal function. A structural model of the *PMP22* protein with known sites of mutation is shown in Figure 3.1. Specific mutations found in the *PMP22* gene in patients with neuropathies are listed in Table 3.2.

The mechanism underlying generation of the DNA duplication deserves careful investigation. The detection of frequent de novo cases of duplication CMT1A suggests that a common, recurring mechanism may be responsible for the genesis of the duplication. Unequal crossing-over is a likely mechanism for generating the DNA duplication in CMT1A. As the duplication in CMT1A spans an estimated 1.5 Mb, two homologous regions, widely separated within chromosome 17p11.2–12, are required for unequal crossing-over. Recently, a low-copy-number repeat sequence (CMT1A-REP) that flanks the proximal and distal duplication breakpoint regions on normal (nonduplicated) chromosomes was identified. The CMT1A-REP sequence is present in three copies on the CMT1A-duplicated chromosome. The CMT1A-REP sequence, which is an intrinsic property of a normal chromosome 17, appears to mediate misalignment of homologous chromosomal segments during meiosis with subsequent crossing-over to produce the duplicated CMT1A chromosome.

It has been found that 70% to 80% of patients with a clinical diagnosis of CMT1 carry the 17p11.2–12 duplication, implying that an assay for the duplication provides a powerful marker for screening suspected patients and at-risk family members. Clinical studies have detected uniform slowing of nerve conduction velocities in patients with the 17p11.2–12 duplication. As DNA testing for CMT1A becomes more widely available, it may become an accepted part of the evaluation of any patient with suspected hereditary neuropathy.

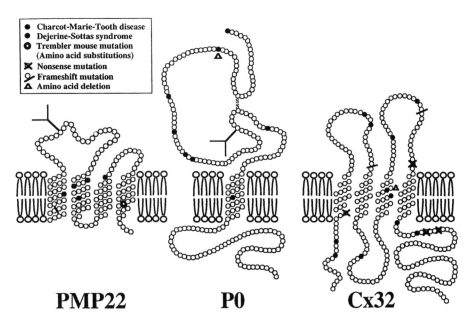

**Figure 3.1**

Structural models of the *PMP22*, $P_0$, and Cx32 membrane-associated myelin proteins and mutations associated with inherited peripheral neuropathies. Individual amino acid residues in each protein are indicated by circles, and the glycosylation sites are shown for *PMP22* and $P_0$. Mutations affecting *PMP22* are associated with CMT1A (filled circles), Déjérine-Sottas syndrome (shaded circles), and HNPP (slashes), indicating a frameshift mutation. The trembler mouse mutation in *PMP22* is represented by a bold circle. Mutations affecting Po are associated with CMT1B (filled circles) and Déjérine-Sottas syndrome (shaded circles), and mutations affecting Cx32 lead to CMTX. Filled circles represent amino acid substitutions, nonsense mutations are indicated by X, deletions are indicated by the delta symbol beside the specific amino acid, and frameshift mutations are indicated by a slash preceding the first amino acid affected in the reading frame.

## MUTATIONS IN THE MYELIN PROTEIN ZERO GENE IN CMT1B

The human myelin protein zero gene ($P_0$) was isolated and mapped to chromosome 1q22–q23 in the region of the CMT1B locus. $P_0$ became an especially attractive candidate for CMT1B, as it was known to be the major structural component of peripheral nervous system myelin ($\sim 50\%$ by weight) and about 7% of Schwann cell message. Analysis of $P_0$ as a candidate gene for CMT1B did detect different point mutations in six pedigrees with this disorder. The point mutations in these families fully cosegregated with the CMT1B phenotype, suggesting that abnormalities in the $P_0$ gene are responsible for CMT1B. Three additional pedigrees that carry mutations in $P_0$ associated with CMT1B have been described. A structural model of the $P_0$ protein with known sites

**Table 3.2** Mutations Associated with Inherited Peripheral Neuropathies

| Disease | Mutation |
| --- | --- |
| *PMP22* | |
| CMT1A | Leu(16)Pro |
| | Ser(79)Cys |
| | Thr(118)Met |
| DSS | Met(69)Lys |
| | Ser(72)Leu |
| HNPP | Ser(7)FS |
| *MPZ* | |
| CMT1B | Asp(90)Glu |
| | Lys(96)Glu |
| | Ser(63)Del |
| | Arg(98)His |
| | Tyr(82)His |
| | Ile(30)Met |
| DSS | Gly(167)Arg |
| | Ser(63)Cys |
| *Cx32* | |
| CMTX | Gly(12)Ser |
| | Val(139)Met |
| | Arg(142)Trp |
| | Leu(156)Arg |
| | Pro(172)Ser |
| | Asn(175)FS |
| | Glu(186)Lys |
| | Arg(15)Gln |
| | Cys(60)Phe |
| | Val(63)Ile |
| | His(73)FS |
| | Leu(143)Del |
| | Glu(208)Lys |
| | Arg(215)Trp |
| | Arg(220)stop |
| | Arg(22)stop |
| | Glu(102)Gly |
| | Glu(186)stop |
| | Cys(217)stop |

of mutation is shown in Figure 3.1. Specific mutations found in the $P_0$ gene in patients with neuropathies are listed in Table 3.2.

## DÉJÉRINE-SOTTAS DISEASE: EXPANSION OF THE CMT1 PHENOTYPIC SPECTRUM

Déjérine-Sottas disease (DSD), also called hereditary motor and sensory neuropathy type III (HMSNIII), is a severe, infantile- and childhood-onset, hypertrophic demyelinating polyneuropathy. The clinical features of DSD overlap with those of severe CMT1. Many patients with DSD appear to represent sporadic cases, and the disease is usually thought to result from an autosomal recessive gene. Recent molecular genetic studies have revealed that DSD may be associated with point mutations in either the $P_0$ or the *PMP22* gene. Interestingly, in patients with the DSD phenotype studied to date, all mutations have been present in the heterozygous state, suggesting that DSD actually may be caused by dominantly acting genetic defects. Molecular studies have not disclosed evidence for recessive inheritance in DSD.

## HEREDITARY NEUROPATHY WITH LIABILITY TO PRESSURE PALSIES: GENETIC RELATIONSHIP TO CMT1A

Hereditary neuropathy with liability to pressure palsies (HNPP; also called *tomaculous neuropathy, recurrent pressure-sensitive neuropathy*, and *"bulb digger's palsy"*) is an autosomal dominant disorder that produces an episodic, recurring demyelinating neuropathy. Peroneal palsies, carpal tunnel syndrome, and other entrapment neuropathies are manifestations of HNPP. Motor and sensory nerve conduction velocities may be reduced in clinically affected patients, as well as in asymptomatic gene carriers. Pathologic changes observed in the peripheral nerves of HNPP patients include segmental demyelination and, in some pedigrees, tomaculous or sausage-like formations.

The HNPP locus has been assigned to chromosome 17p11.2–12 and is associated with a 1.5-Mb deletion. All DNA markers known to map to the region in 17p11.2–12 associated with the CMT1A duplication, including the *PMP22* gene, are deleted in HNPP. The deletion breakpoints in HNPP have been found to map to the same intervals that include the CMT1A duplication breakpoints. De novo deletion of paternal or maternal origin has been detected as a basis for sporadic HNPP. HNPP likely results from deletion of the *PMP22* gene and underexpression of this locus. Further support for the hypothesis that HNPP results from reduced expression at the *PMP22* locus was provided by the identification of a nondeleted HNPP kindred in which a 2-base pair deletion and early termination codon within exon 1 of *PMP22* was present.

It has been proposed that the deleted chromosome in HNPP and the duplicated chromosome in CMT1A are the reciprocal products of unequal crossing-over. The apparent homogeneity for size of the duplication/deletion in unrelated patients and detection of de novo duplication/deletion events suggest that a common mechanism may account for the generation of the duplicated CMT1A chromosome and the deleted HNPP chromosome. As mentioned earlier, the CMT1A-REP repeat represents a complex low-copy repeat sequence that flanks the CMT1A monomer unit and is present in three copies on the CMT1A duplicated chromosome. Recently, the CMT1A-REP repeat was found to be present in only one copy on HNPP deletion chromosomes, strengthening the model that the duplicated chromosome in CMT1A and the deleted chromosome in HNPP are reciprocal products of unequal crossover. The possibility of genetic heterogeneity in HNPP was raised by the recent identification of an HNPP pedigree that did not demonstrate linkage to the region of 17p11.2–12.

## CHARCOT-MARIE-TOOTH NEUROPATHY TYPE 2

Charcot-Marie-Tooth neuropathy type 2 (CMT2) is a less common disorder than CMT1. Generally CMT2 has a later age of onset and less involvement of the small muscles of the hands, and does not have palpably enlarged nerves. Extensive demyelination with "onion bulb" formation is not present in CMT2. Motor nerve conduction velocities are normal, or only slightly prolonged, in affected persons. CMT2 is genetically distinct from all mapped forms of CMT1. A CMT2 locus was assigned by linkage studies to the short arm of chromosome 1 (1p36) and designated as CMT2A. Additional families fulfilling the diagnostic criteria for CMT2 did not have evidence of linkage to this region on chromosome 1, suggesting genetic heterogeneity within CMT2. Further genetic heterogeneity within CMT2 is likely, as kindreds with the features of axonal neuropathy, diaphragm weakness, and vocal cord paralysis have been described and are designated as having CMT2C.

## X-LINKED CHARCOT-MARIE-TOOTH NEUROPATHY

The clinical features of X-linked Charcot-Marie-Tooth neuropathy (CMTX) include demyelinating neuropathy, absence of male-to-male transmission, and a generally earlier onset and faster rate of progression of illness in males. The initial regional assignment for CMTX was made to the proximal long arm of the X chromosome (Xq), by demonstration of linkage to marker DXYS1. Recent refinements in the localization of CMTX to the region of Xq13–q21 have led to mapping of the CMTX gene to a 5-cM interval, flanked proximally by the phosphoglycerate kinase pseudogene (PGKP1) (Xq11.2) and distally by marker DXS72 (Xq21.1). The connexin 32 (Cx32) gene, which encodes a major component of gap junctions, mapped to the CMTX candidate region and was found to be expressed in the peripheral nerve. An analysis of Cx32 in unrelated CMTX pedigrees showed multiple-point mutations associated with the CMTX phenotype. At least 24 different Cx32 mutations have now been found in 27 CMTX families. A structural model of the Cx32 protein with known sites of mutation is shown in Figure 3.1. Specific mutations found in the Cx32 gene in patients with neuropathies are listed in Table 3.2.

Cx32 has a pattern of expression in the peripheral nerve similar to that of other myelin protein genes; immunohistochemical studies, however, show a different localization. Unlike $PMP22$ and $P_0$, which are present in compact myelin, Cx32 is located at uncompacted folds of Schwann cell cytoplasm around the nodes of Ranvier and at Schmidt-Lanterman incisures. This localization suggests a role for gap junctions composed of Cx32 in providing a pathway for the transfer of ions and nutrients around and across the myelin sheath.

### AUTOSOMAL RECESSIVE NEUROPATHY

Rare autosomal recessive families with motor and sensory neuropathy have been reported, particularly in Tunisian families with parental consanguinity. Both demyelinating and axonal types have been described, and given the tentative designation of CMT4. One form of autosomal recessive demyelinating neuropathy has been mapped to chromosome 8q (CMT4A). Other families with CMT4 do not show linkage to chromosome 8q; the other chromosomal loci have not been determined.

## DNA Expansion Syndromes

### KENNEDY'S DISEASE

Spinobulbar muscular atrophy (SBMA), or Kennedy's disease, is an X-linked form of adult-onset bulbar muscular atrophy. This disorder has a variable onset, usually after the age of 30, with bulbar and extremity muscle weakness, atrophy, and fasciculations. The usual presenting complaints are muscle cramps and weakness in the hip and shoulder muscles. There is a lack of upper motor neuron findings, which helps to distinguish patients with this disease from those who have amyotrophic lateral sclerosis. Patients with SBMA may have subclinical sensory findings. Affected males frequently, but not always, have gynecomastia and other signs of androgen insensitivity, and the disease has a progressive, often fatal course.

In the mid-1980s, SBMA was mapped to the proximal long arm of the X chromosome, implicating the androgen receptor as a candidate gene, because it was known to map to this region. In 1991, the mutation responsible for SBMA was identified as an expanded trinucleotide repeat located within the first exon in the androgen receptor gene. A CAG repeat, which encodes a polyglutamine tract in the receptor protein, was previously known to vary in length in the normal population; in patients with SBMA, however, the repeat length was found to be roughly doubled beyond that even seen in the normal population. The expanded CAG repeat is unstable and shifts in length as it is passed from one generation to the next. In patients with SBMA, there is an inverse correlation between repeat length and age of onset: The longer the repeat length, the earlier the onset of symptoms.

The androgen receptor is a member of the steroid and thyroid hormone receptor family. These are not membrane receptors, but intracellular receptors that bind the hormone and then bind to DNA, turning off and on various target genes. The CAG repeat is in a part of the gene separate from the hormone-binding and DNA-binding domains. Patients with mutations in those domains, and patients who lose function of the entire androgen receptor, have a different syndrome—testicular feminization syndrome—without the weakness and motor neuron degeneration of SBMA. Because of this, it is hypothesized that the CAG repeat expansion does not cause loss of androgen receptor function, but a toxic gain of function. That is, it alters the androgen receptor protein in such a way that it becomes toxic to motor neurons.

## FRAGILE X SYNDROME

Fragile X syndrome is the most common hereditary form of mental retardation, accounting for about 15% of the mentally retarded population. Fragile X patients tend to have large ears, a prominent jaw, and large testicles, but the physical findings are usually not very striking. The mental retardation is more common and more severe in males than in females. About 1 in 1500 males are affected, compared with 1 in 3000 females; as many as 1 in 200 females may be fragile X carriers. The disease derives its name from a fragile site on the X chromosome, which can be induced by culturing cells in folate-deficient medium. There is an unusual inheritance pattern differing from other X-linked disorders, in that penetrance in males is less than 100%, but about 30% of females show signs of the disease.

The fragile X syndrome was mapped to the site of chromosome fragility, near the end of the long arm of the X chromosome, and when the defective gene (FMR-1) was isolated, it was found to contain an expanded trinucleotide repeat. The expanded trinucleotide in the fragile X syndrome is a CGG repeat and may become much longer in affected individuals (up to thousands of CGGs), as compared with the CAG expansions in the androgen receptor gene associated with SBMA. Asymptomatic carriers in fragile X families have more modest repeat expansions, 50 to 200 CGGs, which are unstable and may become fully expanded when passed on to affected offspring. The CGG repeat is localized to a noncoding part of the FMR-1 gene, which is an RNA-binding protein active in the brain. It is now known that the effect of the repeat expansion is a lost gene transcription and a failure to produce the gene product.

## MYOTONIC DYSTROPHY

Myotonic dystrophy is the most common form of muscular dystrophy. Typically affected patients have progressive weakness of the face, neck, and extremities. Occasional patients, almost always the offspring of affected women, have a severe, congenital-onset form of the disease: They are floppy, with feeding difficulties at birth, and are left with marked weakness and mental retardation.

Other patients are only minimally affected late in life. Myotonic dystrophy is an autosomal dominant disease with highly variable expression. Additional clinical features include myotonia, cataracts, cardiac arrhythmia, and insulin resistance in addition to weakness. There is a phenomenon of increasing severity as the disease is passed down through the generations, known as anticipation.

The myotonic dystrophy gene was mapped originally to chromosome 19 by linkage to the Lutheran and secretor blood types. The disease locus was eventually narrowed to a segment of the long arm of the chromosome near the apolipoprotein complex and the gene for creatine kinase. When the gene (myotonin protein kinase) was isolated, it was found to have a trinucleotide repeat, a CTG repeat, that is expanded in patients with the disease. As with the CGG repeat seen in fragile X, the myotonic dystrophy repeat also does not fall within the coding region of the gene. It is not clear whether the expansion causes a loss or a gain of function in the gene product, a member of the protein kinase gene family. As with fragile X, the repeats may become dramatically expanded, up to thousands of CTGs in severely affected patients. More moderately affected individuals have fewer marked repeat expansions, usually hundreds of CTGs. The repeat length generally correlates with disease severity, with a better correlation in muscle than in blood.

## HUNTINGTON'S DISEASE

Huntington's disease is an autosomal dominant disorder with a high penetrance and low mutation rate. It is well known to cause chorea and intellectual decline with onset most commonly in adulthood and gradual progression to death. Pathologically, there is a loss of neurons, which is most prominent in the striatum. Until the gene was identified, there were no definite biochemical abnormalities to suggest a cause for the disease. Because of the true dominant inheritance, a "toxic gain of function" mechanism was proposed.

In 1984, the gene for Huntington's disease was mapped to chromosome 4 and, through an extensive mapping effort, the gene locus was narrowed to a segment near the tip of the short arm. Recently, the disease gene was found to contain a trinucleotide repeat that is expanded in patients with the disease. As in SBMA, the expanded repeat is an expansion of CAGs that encodes a polyglutamine tract. The range of CAG repeats in Huntington's patients is essentially the same as that seen in the androgen receptor repeat in SBMA.

## SPINOCEREBELLAR ATAXIA AND DENTATORUBRAL-PALLIDOLUYSIAN ATROPHY

In 1993, a fifth trinucleotide expansion disease was reported. This disease is the chromosome 6–linked form of spinocerebellar ataxia, and again the repeat is a polyglutamine-encoding run of CAGs. As with SBMA and Huntington's disease, a toxic gain of function mechanism has been proposed. Recently, another disease has been found to fit into this pattern—a form of dentatorubral degeneration with a CAG expansion in a gene on chromosome 12. This disorder, dentatorubral-pallidoluysian atrophy (DRPLA), has features overlapping Huntington's disease and spinocerebellar ataxia.

Therefore, at least six neurologic diseases are known to have expanded trinucleotide repeats. These diseases fall in two categories: 1) fragile X and myotonic dystrophy, where the repeats are in noncoding regions and become quite large before the patients become symptomatic through loss (or gain) of gene function; and 2) SBMA, Huntington's disease, spinocerebellar ataxia, and DRPLA, wherein the expanded repeats are CAGs encoding polyglutamine tracts. In the latter category, the expansions are more modest, and the likely mechanism is a toxic gain of function by the altered protein.

## POLYGLUTAMINE EXPANSION DISEASES

The four diseases with polyglutamine expansion have similar mutations and are listed in Table 3.3. In each case, the repeat is near the 5' end of the gene and the amino terminal end of the protein. The androgen receptor is the best-known gene product. The protein products of the other genes—huntingtin, ataxin, and the DRPLA gene product—do not have known functions.

In retrospect, these diseases also are remarkably similar in their clinic and pathologic manifestations. Each causes adult onset and gradual progression of neuronal loss in a different portion of the CNS. These diseases are part of a series of dominantly inherited neurodegenerative disorders that affect different populations of neurons along the neuraxis. With these four sharing the same mechanism, it is likely that there will be others.

The common features of these disorders are that they are all chronic, progressive neurodegenerative diseases with similar ages of onset and rates of progression. All are caused by unstable expansions of CA polyglutamine tracts in widely expressed genes. In each case, increasing tract length correlates with

**Table 3.3**  Hereditary Diseases with Expanded Trinucleotide Repeats

| Disease | Gene | Repeat | Normal | Repeat Length Disease |
|---------|------|--------|--------|-----------------------|
| Kennedy's disease | Androgen receptor | CAG | 11–33 | 40–66 |
| Fragile X syndrome | FMR-1 | CGG | 6–54 | 100–1000 + |
| Myotonic dystrophy | Myotonin-protein kinase | CTG | 5–40 | 200–1000 + |
| Huntington's disease | Huntingtin | CAG | 10–34 | 37–121 |
| Spinocerebellar ataxia | Ataxin-1 | CAG | 19–36 | 42–81 |
| Dentatorubral-pallidoluysian atrophy | (CTG-B37) | CAG | 7–34 | 49–83 |

earlier age of onset. Finally, for each of these diseases, a toxic gain-of-function mechanism has been proposed.

The goal is to define this mechanism and to determine how expanded glutamine tracts lead to neuronal degeneration. Determining this mechanism is the best hope for effective treatment of these progressively disabling and usually fatal neurodegenerative diseases.

# Neurofibromatosis Types 1 and 2

Neurofibromatosis type 1 (NF1, von Recklinghausen neurofibromatosis, peripheral neurofibromatosis) is the most common dominantly inherited disorder affecting the peripheral nervous system (PNS), with a prevalence of 1 in 3500. Clinical features include Schwann cell tumors and focal skin and iris hyperpigmentation. Neurofibromatosis type 2 (NF2, central neurofibromatosis) is a rare, dominantly inherited disorder characterized by bilateral acoustic nerve Schwann cell tumors. NF1 and NF2 are caused by mutations of tumor suppressor genes on chromosomes 17 and 22, respectively.

## CLINICAL AND PATHOLOGIC FEATURES OF NEUROFIBROMATOSIS TYPE 1

Clinical diagnosis of NF1 requires the presence of two or more of the following: (1) six or more cafe-au-lait macules, each over 5 mm in greatest diameter (if prepubertal) or over 15 mm (if postpubertal); (2) two or more dermal neurofibromas, or at least one plexiform neurofibroma; (3) axillary or inguinal freckling; (4) optic glioma; (5) iris melanomatous hamartomas (Lisch nodules); (6) cranial sphenoid bone dysplasia or long-bone cortical thinning with or without pseudarthrosis; and (7) a first-degree relative with NF1.

Cafe-au-lait spots and axillary freckles in NF1 are caused by increased synthesis of melanin by skin melanocytes. Lisch nodules are hamartomas enriched in iris melanocytes. Dermal neurofibromas, which arise in continuity with small subcutaneous nerves, contain Schwann cells interspersed with unmyelinated axons, fibroblasts, small blood vessels, mast cells, and collagen fibrils. Neurofibroma Schwann cells resemble Schwann cells in traumatic neuromas; they are bi- to multipolar and stain immunohistologically with antibodies against the low-affinity nerve growth factor receptor, $2',3'$-cyclic nucleotide-$3'$-phosphohydrolase, S-100 protein, or cell adhesion molecules expressing HNK-1 carbohydrate epitopes.

Plexiform neurofibromas arise in deeper nerves, and resemble dermal neurofibromas in cellular composition, but may also contain regions in which Schwann cells have formed onion bulb–like arrays around en passant axons. Sarcomatous transformation of plexiform neurofibroma may occur; this should be suspected when there is a rapid increase in size, and can be confirmed histologically by the demonstration of aggregates of pleomorphic Schwann cells with frequent mitotic figures.

Patients with NF1 are at increased risk of developing spinal root and peripheral nerve schwannomas. Fibroblasts are far less common in schwannomas than in neurofibromas, and axons are rare or absent. The Schwann cells in schwannomas are arranged in two patterns: bundles of bipolar, spindle-shaped cells, sometimes in palisades (Antoni type A tissue); and loosely packed whorls (Antoni type B tissue).

Pilocytic astrocytomas of the optic nerves and third ventricular region are common in NF1. Even more frequent are dysplastic lesions of the CNS, including subependymal glial nodules, regions of proliferative gliosis, and neuronal heterotopias. Patients with NF1 also have an increased incidence of pheochromocytoma, carcinoids, and other neural crest-derived neuroendocrine tumors.

## MOLECULAR GENETICS OF NEUROFIBROMATOSIS TYPE 1

With a high estimated rate of mutation accounting for 30% to 50% of cases, it could be speculated that the NF1 gene would be unusually large or might have one or more so-called hotspots for mutation. It appears that the former case is more likely for NF1. The initial gene assignment by linkage analysis for NF1 was made to the region of chromosome 17q11.2 through the use of linked anonymous DNA markers. The attempt to identify the NF1 gene through positional cloning strategies was expedited greatly by the recognition of two NF1 patients with associated cytogenetically detectable lesions involving chromosome 17q11.2. One patient had a t(1;17), and the other patient had a t(17;22). In both situations, the translocation breakpoint fell within band 17q11.2, the region of the NF1 locus as defined by studies with tightly linked flanking markers. Subsequently, it was shown that the translocation breakpoints actually could be localized to the same 600-kb NruI DNA fragment. Further cloning and mapping of the translocation breakpoint region showed that the two breakpoints in the NF1 patients fell within a 60-kb interval and detected small deletions in this region in other patients, leading to the identification of the NF1 gene. Sequence analysis using transcripts from the breakpoint region detected single-base change mutations, predicted to inactivate the NF1 gene in patients, providing additional evidence that the NF1 gene was in hand. Independently, two CDNA clones were isolated from the translocation breakpoint region and demonstrated sequence identity to that reported associated with point mutations in NF1 patients.

Interestingly, during the process of investigating the translocation breakpoint region for NF1 candidate genes, three additional genes (EV12A, EV12B, and OMgp) mapping to the translocation breakpoint region were detected. These three genes are entirely contained within the NF1 gene and are transcribed in the opposite direction for which transcription occurs in the NF1 gene. EV12A and EV12B are human homologs of mouse proto-oncogenes and are thought to play a possible role in the genesis of murine leukemias. OMgp (oligodendrocyte-myelin glycoprotein) was briefly an exciting candidate for NF1, because it is an abundantly expressed myelin protein. However, no OMgp-specific mutations were found to be associated with NF1. The significance, if any, of the location and orientation of these three genes within the NF1 gene is not known.

The NF1 gene is unusually large, spanning approximately 350 kb and encoding a 13-kb messenger RNA. The predicted NF1 protein is composed of 2818 amino acids and has a molecular mass of 327 kd. It is composed of 49 exons, with two alternately spliced forms. The NF1 gene is ubiquitously expressed, and its protein product is known as neurofibromin. Neurofibromin carries sequence homology with the GTPase-activating protein (GAP) superfamily of genes, including mammalian GAP, yeast IRA1, IRA2, and sar1 and *Drosophila* Gap1.

Because of the large size of the NF1 gene and the apparent degree of heterogeneity in type of mutations, no abnormalities have been detected in the overwhelming majority (>90%) of patients. Furthermore, the high spontaneous mutation rate for NF1 may increase the difficulties of molecular genetic diagnosis in many cases. For the majority of presumed de novo mutations in NF1, most are of paternal origin, implicating problems during male germ cell meiosis.

Aside from the two previously mentioned chromosomal translocations involving band 17q11.2, various deletions in the NF1 gene have been found. A 320-bp Alu repetitive element insertion resulting in a splice-site disruption and frameshift mutation has been described. Point mutations have included Lys to Glu or Lys to Gln substitutions at position 1423, and an Arg to stop codon at positions 1610 or 365.

With direct analysis for specific NF1 mutations available for only a small portion of patients, DNA-based diagnosis relying on polymorphic intragenic and closely linked markers becomes a mainstay for families seeking either presymptomatic or prenatal diagnosis of NF1. The currently available DNA markers for diagnosis, however, are only informative in approximately 75% of pedigrees, and linkage disequilibrium, which serves to limit the usefulness of a clustered set of markers, has been found for the NF1 region.

The often extreme variability in expression of NF1 within a family contributes to the difficulty of genetic counseling for this disorder. Environmental factors may contribute to this variability. However, analysis of the correlation between numbers of cafe-au-lait spots or neurofibromas and the degree of relatedness within a family suggests that as yet unidentified genes influence penetrance of NF1 mutations.

## CELL BIOLOGY OF NEUROFIBROMATOSIS TYPE 1

NF1 is characterized by multifocal Schwann cell hyperplasia. Neurofibromas are frequent and are often excised for cosmetic reasons. Despite this abundance of tissue for study, and the previously summarized advances in the molecular genetics of NF1, the cell biology of NF1 is poorly understood.

Because of the admixture of cell types within neurofibromas, it is unclear whether the Schwann cells in these benign tumors are monoclonal or polyclonal. In either case, a local change in Schwann cells or their environment must be present to explain the occurrence of foci of Schwann cell hyperplasia in some regions, but not others, of the PNS. The nature of this local factor is not known. It has been established, however, that neurofibrosarcoma arising from a plexiform neurofibroma or schwannoma is monoclonal, and that the tumor cells lack a normal copy of the NF1 gene because of a "second hit."

Although there are obviously too many Schwann cells in neurofibromas and schwannomas of patients with NF1, and these Schwann cells have a mutation affecting one copy of the NF1 gene, tissue culture studies have failed to demonstrate any abnormal behaviors of NF1 Schwann cells. There is one report that Schwann cells isolated from NF1 neurofibromas and transplanted to a chick egg chorioallantoic membrane are better able to induce angiogenesis and to penetrate basement membranes than are normal Schwann cells, but this observation has not been confirmed. While it might seem self-evident that an abnormality in NF1 Schwann cells causes neurofibromas and schwannomas to develop, it is too early to discard an alternative scenario in which there is a failure by NF1 neurons to drive Schwann cells to their terminally differentiated states.

Neurofibromin catalyzes the conversion of guanosine triphosphate (GTP) bound to the signal transduction protein (ras) to guanosine diphosphate (GDP). This causes inhibition of the interaction between ras and the next protein in this signaling cascade (raf) and prevents ras from influencing cell proliferation and differentiation. It is of interest that the gene mutated in another phakomatosis, tuberous sclerosis, also may have GTPase activity. Neurofibrosarcomas, which have markedly depressed levels of neurofibromin, demonstrate increased ratios of activated to inactivated ras. Neurofibromin deficiency and increased levels of activated ras also have been observed in some human neuroblastoma and melanoma cell lines. Proliferation of these cultured tumor cells is inhibited, and their differentiation enhanced, by transfection with complementary DNA encoding normal neurofibromin.

Most embryonic tissues express neurofibromin, but later in development, expression becomes progressively restricted to the CNS, PNS, and adrenal medulla. The mechanism of this tissue-specific regulation of neurofibromin protein expression is not known.

Neurofibromin GAP enzyme activity in the various tissues is also regulated at the level of RNA processing. Two forms of neurofibromin RNA are formed by alternative splicing. Neurofibromin type 1, which predominates during the first 20 weeks of human gestation, is a more potent inactivator of p21-*ras* than is neurofibromin type 2, which is the preferred transcript later in development. The level of expression of neurofibromin type 2 by Schwann cells is subject to modulation by the environment; agents that markedly raise intracellular 3′,5′-cyclic adenosine-3′-monophosphate (cAMP) levels and induce the synthesis of myelin lipids and proteins also induce expression of neurofibromin type 2. An additional way in which neurofibromin GAP activity is modulated is by association with microtubules. Tubulin partially inhibits neurofibromin GAP. Targeting to microtubules is conferred by regions of the neurofibromin molecule in or adjacent to the GAP domain.

The GAP domain accounts for only about 10% of the neurofibromin sequence. Most of the mutations of the NF1 gene thus far recognized do not directly affect this region of the protein. It is reasonable to speculate, therefore, that some of the tissue abnormalities in NF1 result from defects in non–GAP-related functions of neurofibromin. This speculation is supported by two recent observations: (1) that overexpression of neurofibromin in 3T3 fibroblasts inhibits growth of the cells without altering the level of activity of ras in the cells; and (2) that transformation of cells with *v-ras*, an oncogenic member of the *ras* family resistant to inactivation by neurofibromin (GAP), is suppressed by transfection with neurofibromin.

## NEUROFIBROMATOSIS TYPE 2

Neurofibromatosis type 2 (NF2) is a dominantly inherited disorder, with an incidence of 1 in 40,000. Symptoms and signs of vestibular schwannoma (tinnitus, deafness) usually appear in early adulthood. Bilateral vestibular schwannomas eventually appear in 95% of patients. In some instances, the first tumor to be recognized is a spinal root schwannoma, meningioma, or glioma. The individual tumors do not have histologic features that permit them to be distinguished from sporadic tumors of the same type. Posterior lenticular opacities may be noted during childhood in NF2, but the multiple cutaneous lesions and dermal neurofibromas characteristic of NF1 are lacking.

Linkage studies permitted assignment of the NF2 gene to chromosome 22, and then to 22q12. Somatic mutations affecting this region of chromosome 22 also have been documented in sporadic acoustic neuromas and ependymomas. Identification of the NF2 gene was facilitated by study of NF2 families with non-overlapping germ line deletions in this chromosomal segment.

The NF2 gene encodes a widely expressed protein, merlin, with a sequence homologous to that of moesin, ezrin, radixin, talin, and protein 4.1. Members of this family of proteins are believed to link cytoskeletal elements with the plasma membrane in erythrocytes and other cell types. Protein 4.1, for example, links the erythrocyte actin-spectrin network with the plasma membrane anion channel and glycophorin, and a protein-4.1 mutation is one of the causes of hereditary elliptocytosis.

Meningiomas from patients with NF2 demonstrate loss of heterozygosity on chromosome 22. This observation suggests that these tumors arise when both copies of the NF2 gene are mutated or lost, and that the NF2 gene, like the NF1 gene, encodes a protein with tumor-suppressor activity.

## References/Further Reading

Ballester R, Marchuk D, Boguski M, et al. The NF1 locus encodes a protein functionally related to mammalian GAP and yeast IRA proteins. Cell 1990;63:851–859.

Barker DF, Wright E, Nguyen K, et al. Gene for NF1 von Recklinghausen neurofibromatosis is in the pericentromeric region of chromosome 17. Science 1987;236:1100–1102.

Ben Othmane K, Middleton LT, Loprest LJ, et al. Localization of a gene for autosomal dominant Charcot-Marie-Tooth type to chromosome 1p and evidence for genetic heterogeneity. Genomics 1993;17:370–375.

Bergoffen J, Trofatter J, Pericak-Vance MA, et al. Linkage localization of X-linked Charcot-Marie-Tooth disease. Am J Hum Genet 1993;52:312–318.

Brook JD, McCurrach ME, Harley HG, et al. Molecular basis of myotonic dystrophy: expansion of a trinucleotide (CTG) repeat at the 3' end of a transcript encoding a protein kinase family member. Cell 1992;68:799–808.

Chance PF, Matsunami N, Lensch MW, et al. Analysis of the DNA duplication 17p11.2 in Charcot-Marie-Tooth neuropathy type 1 pedigrees: additional evidence for a third autosomal CMT1 locus. Neurology 1992;42:2037–2041.

Chance PF, Bird TD, Matsunami N, et al. Trisomy 17p associated with Charcot-Marie-Tooth neuropathy I phenotype: evidence for gene dosage as a mechanism in CMT1A. Neurology 1992;42:2295–2299.

Chance PF, Alderson MK, Leppig KA, et al. DNA deletion associated with hereditary neuropathy with liability to pressure palsies. Cell 1993;72:143–151.

Chance PF, Abbas N, Lensch MW, et al. Two autosomal dominant neuropathies result from reciprocal duplication/deletion of a region of chromosome 17. Hum Mol Genet 1994;3:223–228.

Dyck PJ, Chance PF, Lebo RV, Carney JA. Hereditary motor and sensory neuropathies. In: Dyck PJ, Thomas PJ, Griffin JW, et al., eds. Peripheral neuropathy. 3rd ed. Philadelphia: WB Saunders, 1993:1094–1136.

Fu YH, Pizzuti RG, Fenwick J, et al. An unstable triplet repeat in a gene related to myotonic muscular dystrophy. Science 1992;255:1256–1258.

Gutmann DH, Collins FS. The neurofibromatosis type 1 gene and its protein product, neurofibromin. Neuron 1993;10:335–343.

Huntington's Disease Collaborative Research Group. A novel gene containing a trinucleotide repeat that is expanded and unstable on Huntington's disease chromosomes. Cell 1993;72:971–983.

Kolde R, Ikeuchi T, Onodera O, et al. Unstable expansion of CAG repeat in hereditary dentatorubral-pallidoluysian atrophy (DRPLA). Nature Genet 1994;6:9–13.

La Spada A, Wilson EM, Lubahn DB, et al. Androgen receptor gene mutations in X-linked spinal and bulbar muscular atrophy. Nature 1991;352:77–79.

La Spada AR, Roling D, Harding AE, et al. Meiotic stability and genotype-phenotype correlation of the expanded trinucleotide repeat in X-linked spinal and bulbar muscular atrophy. Nature Genet 1992;2:301–304.

Lupski JR, Chance PF, Garcia CA. Inherited peripheral neuropathies: molecular genetics and clinical applications. JAMA 1993;270:2326–2330.

Nagafuchi S, Yanagisawa H, Sato K, et al. Dentatorubral and pallidoluysian atrophy expansion of an unstable CAG trinucleotide on chromosome 12p. Nature Genet 1994;6:14–18.

Orr HT, Chung M, Banfi S, et al. Unstable expansion of CAG repeat in hereditary dentatorubral-pallidoluysian atrophy (DRPLA). Nature Genet 1994;6:9–13.

Ponder BAJ. Neurofibromatosis: from gene to phenotype. Cancer Biol 1992;3:115–120.

Siomi H, Siomi MC, Nussbaum RL, Dreyfuss G. The protein product of the fragile X gene, FMP has characteristics of an RNA binding protein. Cell 1993;74:291–298.

Upadhyaya M, Fryer A, MacMillan J, et al. Prenatal diagnosis and presymptomatic detection on neurofibromatosis type 1. J Med Genet 1992;29:180–183.

Verkerk AJMH, Pieretti M, Sutcliffe JS, et al. Identification of a gene (FMR-1) containing a CGG repeat coincident with a breakpoint cluster region exhibiting length variation in fragile X syndrome. Cell 1991;65:905–914.

Windebank AJ. Inherited recurrent focal neuropathies. In: Dyck PJ, Thomas PJ, Griffin JW, et al., eds. Peripheral neuropathy. 3rd ed. Philadelphia: WB Saunders, 1993:1137–1148.

Wise CA, Garcia CA, Davis SN, et al. Molecular analysis of unrelated Charcot-Marie-Tooth disease patients suggests a high frequency of the CMT1A duplication. Am J Hum Genet 1993;53:853–863.

# Retrovirus Infection of the Central Nervous System

Fred C. Krebs

Brian Wigdahl

A retrovirus, as the name implies, is a virus that reverses the natural flow of genetic information (DNA to RNA to protein) during its life cycle. The genetic material of a retrovirus is RNA; the virus utilizes a viral enzyme (reverse transcriptase) to direct the synthesis of a complementary DNA copy from the RNA genome. This DNA intermediate can then integrate into the genome of the host cell. Once the viral DNA has been integrated into the host chromosome, it remains as part of the cell's genetic complement for the life of the cell. These aspects of retroviruses make it difficult, if not impossible, to successfully eliminate them from an infected host.

Retroviruses of animals and humans belong to the *Retroviridae* virus family, two subfamilies of which are known to be pathogenic: Oncoviruses transform cells in vitro and cause tumors in vivo while "slow" viruses (lentiviruses) fuse and destroy cells in vitro and cause infections in vivo. Both are characterized by long latency periods, which result in immunodeficiency and/or neurologic diseases (Fig. 4.1).

The first human retrovirus to be identified was the etiologic agent of adult T-cell leukemia (ATL) and is designated human T-cell lymphotropic virus type I, or HTLV-I. The HTLVs can be further classified according to virion structure into types A through D. HTLV-I and HTLV type II (HTLV-II), the prototypic type C human retroviruses, have been associated with leukemia and neurologic disorders.

About the time of the discovery of HTLV-I and HTLV-II, cases of *Pneumocystis carinii* pneumonia and Kaposi's sarcoma began to be reported in homosexual men, along with a variety of other unusual opportunistic infections. It soon became apparent that a common clinical finding in these patients was a selective depletion of the CD4-positive subset of T lymphocytes, resulting in a defective cellular immune response. As a consequence, the syndrome was named the acquired immunodeficiency syndrome (AIDS). In 1984, a new retrovirus, the human immunodeficiency virus (HIV), was identified as the etiologic agent of AIDS and over the next several years

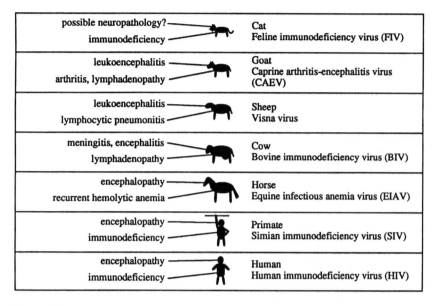

**Figure 4.1**
Human and nonhuman lentiviruses linked to neurologic disease. Examples of systemic and neurologic pathologies are given for each virus.

was referred to by several names, including lymphadenopathy-associated virus, human T-cell lymphotropic virus type III (HTLV-III), and AIDS-associated virus.

Not long after the appearance of the first cases of what would later be called AIDS, reports began to appear that described CNS dysfunctions and neurologic disorders in approximately 40% of adult AIDS patients. Initial speculation concerning the underlying cause of this neurologic dysfunction focused on the numerous opportunistic agents that were observed to infect the CNS during the course of AIDS (Fig. 4.2). As the epidemic evolved, however, it became apparent that only 30% of the neurologic disorders could be explained as the result of opportunistic infections. With the identification of HIV-1 in 1984 as the etiologic agent of AIDS, a connection was postulated between the presence of HIV-1 in the CNS and the symptoms observed.

Approximately 50% of terminal AIDS patients exhibit neurologic dysfunction, and 80% to 90% are found to have neurologic damage at autopsy. One of the most common clinical syndromes of HIV-1-associated neurologic disease is a progressive neurologic syndrome termed *AIDS dementia complex* (ADC). ADC may be the sole indication of HIV-1 infection, and can occur in the absence of any immunologic dysfunction.

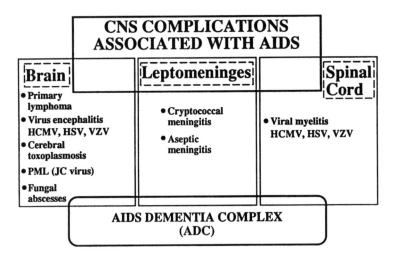

**CNS COMPLICATIONS
ASSOCIATED WITH AIDS**

| Brain | Leptomeninges | Spinal Cord |
|---|---|---|
| • Primary lymphoma | | |
| • Virus encephalitis HCMV, HSV, VZV | • Cryptococcal meningitis | • Viral myelitis HCMV, HSV, VZV |
| • Cerebral toxoplasmosis | • Aseptic meningitis | |
| • PML (JC virus) | | |
| • Fungal abscesses | | |

**AIDS DEMENTIA COMPLEX
(ADC)**

**Figure 4.2**
Complications associated with HIV-1 infection of the CNS. A large number of individuals suffer from primary cancers and opportunistic infections of the CNS caused by a large body of agents, including several previously characterized viruses. Although these conditions may arise independently of the onset of ADC, they also may complicate the progression and clinical presentation of HIV-1–associated dementia. The resulting neurologic abnormalities may arise as a consequence of the presence of both HIV-1 and the agents depicted. HCMV, Human cytomegalovirus; HSV, Herpes simplex virus; VZV, varicella zoster virus; PML, progressive multifocal leukoencephalopathy; JC, Jacob-Creutzfeldt virus (a human polyomavirus).

## Retroviral Life Cycle

### RETROVIRAL STRUCTURE

HIV-1 and HTLV-I are similar to other members of the retrovirus family with respect to virion structure and mode of replication. The mature virions (Fig. 4.3) are enveloped and range from 110–140 nm in diameter, with dense,

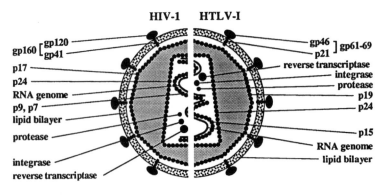

**Figure 4.3**
Structure of the HIV-1 and HTLV-I virions. Although depicted as identical, the structures of these two viruses differ in some respects. The most striking difference is the appearance of the core. In electron micrographs of the mature virions, the core of HIV-1 appears as a dense, cylindrical cone, while the HTLV-I core has a more spherical appearance.

centrally located nucleoid structures. The cores of the viruses are ribonucleo-protein complexes composed of two copies of genomic RNA, nucleic acid–bind-ing proteins, magnesium-dependent reverse transcriptase, the transfer RNA (tRNA) molecules required for the initiation of reverse transcription, and internal core proteins forming the capsids. Additional components of the mature virion cores include the retroviral-encoded protease and integrase en-zymes, which are responsible for cleavage of structural proteins and for integra-tion of the viruses into the host genome, respectively. The virions are enveloped by host-derived lipid bilayers studded with short virus-encoded glycoprotein spikes composed of two protein components.

## OVERVIEW OF RETROVIRAL REPLICATION

HIV-1 and HTLV-1 replication cycles can be divided into two phases (Fig. 4.4). The first phase encompasses entry of the virus into the cell, reverse transcription of the retroviral RNA genome into DNA, transport of the proviral genome into the nucleus, and integration of the provirus into the host chromo-somal DNA. These events depend on proteins located within the retroviral virion. The retrovirus attaches and enters susceptible cells by a receptor-mediated mechanism. HIV-1 uses the CD4 cell surface molecule as its primary receptor. The primary target of HTLV-I infection in the genesis of ATL, and possibly tropical spastic paraparesis (TSP), is also the CD4-positive lymphocyte, although the CD4 cell surface antigen is not the receptor for HTLV-1.

Subsequent to viral entry, reverse transcription results in the transcription of viral DNA utilizing the single-stranded viral RNA as a template. The double-stranded proviral DNA intermediate is then transported to the nucleus where integration into the host chromosomal DNA takes place via activity of a viral endonuclease. The integrated provirus may remain latent (i.e., not producing

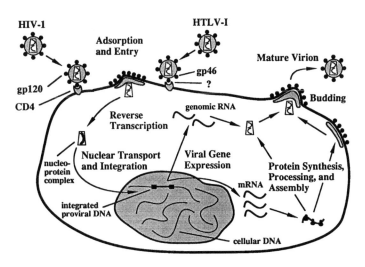

**Figure 4.4**
Retroviral life cycle. The replicative cycles of both HTLV-I and HIV-1 are depicted. The HTLV-I surface glyco-protein gp46 interacts with a cell-surface molecule encoded at least in part on the distal arm of chromo-some 17. HIV-1 enters host cells pre-dominantly through the interaction of gp120 and the cell-surface mole-cule CD4.

viral RNA or proteins) or it may serve as a template for the production of viral mRNAs, leading to the synthesis and processing of viral genomes, transcription of retroviral mRNAs, translation of viral proteins, virion assembly, and release of progeny virus. These processes are accomplished by utilizing elements of the host cell transcriptional machinery in conjunction with virus-specific products. Following the integration event, both HIV-1 and HTLV-1 exert their influence by producing viral products that subsequently disrupt normal cellular events (Fig. 4.5).

## RETROVIRAL REGULATORY PROTEINS

Retroviruses as a family can be grouped into two categories, noncomplex (or basic) and complex, based on the viral genes necessary for replication. Basic retroviruses require the expression of three virus-specific genes: (1) *gag*, which encodes the structural proteins; (2) *pol*, which encodes the virus-associated enzymes; and (3) *env*, which encodes the viral envelope proteins. Basic retroviruses rely solely on host cellular transcription factors for efficient transcription of the viral genome. These host cell–virus interactions are sufficient to support

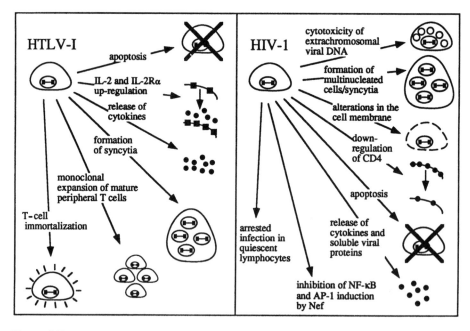

**Figure 4.5**

Effects of HTLV-I and HIV-1 infection on host cells. The effects depicted are those observed during infection of CD4-positive T lymphocytes. HIV-1 infection of its target cells results primarily in depletion of those cell populations. Although HTLV-I infection can result in target-cell depletion, it can also cause polyclonal and monoclonal expansion of the infected cell types, as well as target-cell immortalization.

high proviral DNA transcription levels. In contrast, the complex retroviruses, such as HTLV-I and HIV-1, require the expression of additional viral regulatory proteins for efficient replication (Fig. 4.6), including virus-specific *trans*-activators of retroviral transcription; only the viral regulatory proteins for HIV-1 are discussed here.

Tat, which is the *trans*-activating protein for HIV-1, is critical for efficient viral transcription. Moderate levels of Tat production engage a positive-feedback loop in which *trans*-activation by this protein enhances expression of the viral genes (including the Tat coding region) by interacting with specific *cis*-acting viral regulatory sequences. Also critical for successful viral replication is the regulatory protein Rev, which directs the replicative cycle toward progeny virus assembly and release. Nef, an HIV-1 regulatory protein, is not essential for viral replication, but has an apparent role in regulating its course.

Tat is an 86 amino acid protein encoded by multiply spliced viral mRNA. It comprises three domains: a region with seven cysteine residues, which may be involved in protein-protein interactions; a C-terminal lysine- and arginine-rich domain responsible for nuclear localization and RNA binding, and a central region, which has been proposed to be the activation domain. Tat binds to and acts through the *trans*-activation response (TAR) element, an RNA stem-loop structure encoded by 59 nucleotides in the R region of the long-terminal repeat (LTR) (Fig. 4.7). Because the TAR is encoded by sequences just downstream of the start site of transcription, the TAR is found on the 5' end

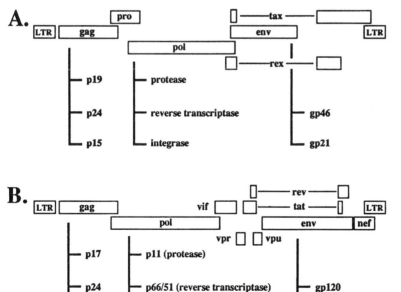

**Figure 4.6**
Viral genomes. Schematic illustrations of (A) the HTLV-I and (B) HIV-1 genomes. The products of the *gag, pol,* and *env* genes are indicated below the coding regions. The coding regions of the HTLV-I (Tax and Rex) and HIV-1 (Tat and Rev) regulatory proteins also are depicted. The HIV-1 genome also encodes several accessory proteins: Vpu, Vpr, Vif, and Nef.

**A.**

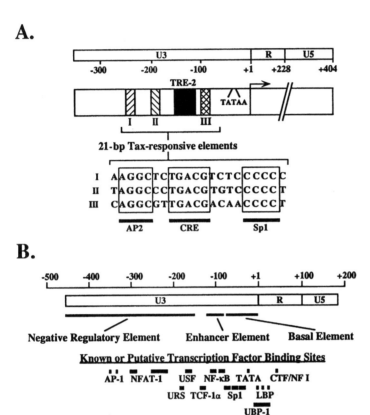

**Figure 4.7**
HTLV-I and HIV-1 LTRs. (A) The
HTLV-1 LTR is illustrated with the
TFIID binding site (TATAA), three
21-bp Tax-responsive elements
(collectively referred to as TRE-1),
and TRE-2. The sequences of the
21-bp repeat elements are shown
with respect to consensus sequences that are putatively recognized by cellular *trans*-acting
factors. Homologies to transcription factor binding consensus sequences are indicated below the
21-bp repeats. The sequence for
Sp1 binding is not an exact consensus site, but is nevertheless GC-
rich. (B) The HIV-1 LTR is depicted
to scale with the locations of
known or putative *cis*-acting binding sites for cellular transcription
factors.

of all HIV-1 transcripts. The function of the TAR is both orientation- and location-dependent. The double-stranded stem serves a support function, while the specificity of Tat binding is in the 6-nucleotide terminal loop and 3-nucleotide bulge. There is also evidence to support a role for one or more cellular factors that bind in a sequence-specific manner to the TAR and assist Tat in its *trans*-activation function. The presence of Tat causes an approximately 100-fold increase in HIV-1 LTR activity. Tat functions to increase the steady-state level of HIV-1 mRNA transcripts due to an increase in transcription rate rather than in mRNA stability. Investigations into the mechanism of Tat *trans*-activation have shown that Tat functions predominantly to promote transcript elongation by increasing the rate at which transcription complex is processed. Evidence also exists that indicates that transcription increases may be the result of an increase in initiation. In addition, some observations have suggested that Tat may have a posttranscriptional effect that results in increased protein production from TAR-containing mRNA.

Rev is a 116 amino acid protein that includes an arginine-rich region that may be involved in RNA recognition and binding, and nuclear localization. This core domain is flanked by two regions involved in Rev multimerization.

An additional region, C-terminal to the core domain, is postulated to interact with nuclear RNA transport or splicing proteins, because Rev molecules with mutations in this region bind the Rev response element (RRE) and each other, but are defective in function. Rev binds to the RRE, a large RNA element (234 nucleotides) with a complex secondary structure encoded within the *env* gene. Unlike Tat, which binds one molecule of Tat to one TAR, Rev binds to the RRE as an oligomerized aggregate of Rev molecules. Rev functions to temporally regulate HIV-1 gene expression. In the absence of Rev, multiply spliced transcripts of approximately 2 kb in size, which encode the regulatory proteins Tat, Rev, and Nef, are produced in the early phase of replication. The subsequent accumulation of Rev (as a consequence of HIV-1 LTR-dependent Tat *trans*-activation and increased transcription rates) directs a switch to the production of the late class of unspliced and singly spliced (approximately 9 and 4 kb, respectively) transcripts, which encode the structural proteins Gag, Pol, and Env, and the accessory proteins Vif, Vpu, and Vpr. At the molecular level, the presence of Rev has no effect on nuclear expression of HIV-1 mRNA. During the early phase of infection, only the multiply spliced transcripts reach the cytoplasm, despite the expression of all classes of messages. As Rev accumulates, it interacts with the RREs on nascent HIV-1 late transcripts to facilitate their export from the nucleus before they are completely spliced.

Nef is a 27-kD myristylated phosphoprotein encoded by an open reading frame that overlaps both the *env* gene and the 3' LTR. It is expressed early in replication and is localized to cytoplasmic membrane structures. While Nef is not required for viral expression and replication in vitro, its conservation in the primate lentiviruses suggests a role in the replicative cycle, but negative and positive effects, and none at all, have all been reported. This may be the consequence of several variables, including the variability of the *nef* gene in vivo. A less controversial function of Nef is the down-regulation of the cell surface molecule CD4, possibly by endocytosis and degradation. A suggestion that Nef exhibits properties characteristic of G proteins, a family of proteins involved in signal transduction, has not been substantiated. Nevertheless, Nef likely functions through interactions with membrane, cytoskeletal, or cytosolic proteins, thus modulating transcription factors important to the LTR function.

## RETROVIRAL LONG-TERMINAL REPEAT AND REGULATION ASPECTS OF VIRAL TRANSCRIPTION

The HIV-1 genome is flanked at either end by LTR nucleic acid sequences composed of three regions (U3, R, and U5), which comprise integral components of the viral regulatory system (see Fig. 4.7). The LTR sequences contain information essential to the regulation of reverse transcription, proviral DNA integration, viral transcription, and other virus-specific replication events. The U3 region contains sequences necessary for transcriptional regulation of viral gene expression, as well as the binding site for the cellular transcriptional machinery.

Overall expression of the HIV-1 genome also is controlled by the LTR (see Fig. 4.7B). Transcription of the HIV-1 genome is regulated in *cis* by promoter sequences localized within the HIV-1 LTR, and in *trans* by Tat. Additionally, host cell factors regulate LTR activity and therefore viral expression, because the LTR directs viral transcription through the utilization of host transcriptional machinery. Early functional studies subdivided the LTR into the basal element, from $-1$ through approximately $-79$; the enhancer element, spanning nucleotides $-80$ through $-109$; and the negative regulatory element, or NRE, defined broadly as sequences upstream of nucleotide $-157$. Factors that interact with the LTR through the basal or enhancer elements generally maintain or induce transcription. Transcription is initiated in the vicinity of the TATA box (nucleotides $-22$ to $-28$) by the binding of TFIID and subsequent host cellular transcription factors. Additional factors, including LBP-1, CTF/NF-1, and UBP-1, may assist in the maintenance of transcription promoted by the basal element by binding to sites located on both sides of the U3-R border. Furthermore, at least one of three consecutive Sp1 binding sites, located at the 5′ end of the basal element, is required for maintenance of RNA transcription. The binding of Sp1 to these sites activates RNA synthesis five- to eightfold. Within the enhancer element, the predominant factor that affects HIV-1 transcription is NF-B, which can induce HIV-1 transcription up to 50-fold in activated T lymphocytes. Mutational analyses of the NF-B binding site in the HIV-1 LTR eliminated inducibility of LTR-mediated gene expression. With respect to the more distal regulatory sequences within the HIV-1 LTR, the NRE also interacts with various host cell transcription factors and down-regulates LTR activity. In transient expression analyses, the NRE has been shown to be involved in down-regulation of LTR-driven expression by two- to threefold in lymphocytes and HeLa cells. Similarly, in experimentation utilizing infectious molecular clones, deletion of the NRE resulted in increased viral replication. However, the overall contribution of the NRE to LTR activity and the involvement of the NRE in cell type–specific transcriptional control are still under investigation.

## HTLV-I and Neurologic Dysfunction

### EPIDEMIOLOGY OF HTLV-I INFECTION

Adult T-cell leukemia is endemic to several areas of the world, including southern Japan, the Caribbean basin, Central and South America, and Africa. More than 90% of the affected individuals have been demonstrated to be seropositive for HTLV-I. In general, pediatric infection with HTLV-I results in ATL, while adult infection results in TSP. This phenomenon is thought to be a consequence of the mechanism of transformation of T lymphocytes, which results in ATL. Accumulation of required events occurring over a long period has been hypothesized to be required to reach the transformed state. The mechanisms involved in the genesis of TSP are unclear; it is known, however, that the period between infection by HTLV-I and the development of disease symptomatology is shorter for TSP than for ATL. Several determinants for lymphoproliferative versus neurodegenerative disease have been proposed, including host factors, environmental influences, genetic predisposition, and/or viral factors.

Because HTLV-I is primarily a cell-associated virus, with infection occurring via the passage of infected cells rather than via cell-free virus, several routes of transmission exist, all involving the passage of infected cells, including (1) vertical transmission, either prenatally via transplacental routes, natally during the birthing process, or postnatally via breast feeding; (2) sexual contact from male to female or male to male, rarely from female to male; (3) transfusion of contaminated cell-containing blood products; and (4) contaminated needles or syringes associated with intravenous drug abuse.

## HTLV-I INFECTION AND ADULT T-CELL LEUKEMIA

Adult T-cell leukemia has a latency period of approximately 20 to 30 years, but once manifested, is rapidly fatal. This disorder consists of a monoclonal expansion of mature peripheral T lymphocytes that contain copies of retroviral sequences. When exposed to immature human hematopoietic cells, infected cells fuse with and transform T lymphocytes by passage of infectious virus. By direct cell-to-cell fusion, the HTLV-I–transformed cells also may transfer infectivity to a wide variety of other cell types. A large majority of the HTLV-I–infected population is asymptomatic, but these individuals are still capable of transmitting HTLV-I via passage of infected cells because the proviral genome is integrated into host cellular DNA.

Acute or subacute ATL is characterized clinically by generalized lymphadenopathy, hypercalcemia due to high osteoclast activity and cytokine release from malignant cells, hepatosplenomegaly, bone marrow involvement, lytic bone lesions, and skin involvement due to infiltrating leukocytes. The prevalence of nervous system complications in ATL patients is approximately 60%.

## HTLV-I INFECTION AND TROPICAL SPASTIC PARAPARESIS

More than 1500 cases of TSP have been documented since 1985. Tropical spastic paraparesis exhibits a shorter latency period than ATL and presents as a slowly progressive disease. Although variable degrees of HTLV-I–infected lymphocytic infiltrates have been observed in the CNS of TSP patients, the cellular target for viral infection of neuroglial elements remains unclear. Tropical spastic paraparesis involves slowly progressive demyelination of long motor neuron tracts in the spinal cord and results in paraparesis, impotence, muscle weakness, and difficulty with gait, ultimately culminating in virtual immobilization. The disease is characterized by a gradual onset of lower-extremity weakness, stiffness, and disturbance of gait, which progresses to bladder dysfunction and often parasthesia, ataxia, and back pain. Pain is continuous over several months, with only mild improvement as the disease progresses. The course of this disease is chronic, occurs without remission or exacerbation, and averages 8 years. On physical examination, TSP patients present with spasticity and proximal muscle weakness in the lower extremities, along with hyperactive knee and ankle deep tendon reflexes. Histologically, there is occasional atypical lymphocytosis in the peripheral blood and CSF. Serologically, patients are positive for antibodies to HTLV-I in serum and CSF. The risk of developing neurologic complications for HTLV-I–infected individuals has been determined to be 1 in 2000 to 3000 carriers, exhibiting an age-dependent increase.

HTLV-I has been isolated from the CSF of patients with TSP, and HTLV-I DNA sequences have been detected in peripheral blood mononuclear cells (PBMCs) and in several regions of the brain and spinal cord from a group of patients with TSP who had HTLV-I–specific antibodies. HTLV-I displays a polyclonal integration pattern in PBMCs from TSP patients, in contrast to the monoclonal integration of HTLV-I in acute or subacute ATL patients. In addition, studies have demonstrated that TSP patients have ten- to 100-fold higher levels of proviral DNA on a per HTLV-I–infected cell basis than do asymptomatic HTLV-I carriers, and the percentage of PBMCs polyclonally infected by HTLV-I is approximately threefold higher in TSP patients than in asymptomatic carriers. While evidence has been presented indicating that HTLV-I infects the nervous system, studies determining the neuropathogenic role of HTLV-I infection in patients suffering from TSP have been hindered due to the inability to detect viral DNA, RNA, or antigens in the CNS tissue of TSP patients.

A number of cell types have been infected by HTLV-I in vitro, including CD4-positive and CD8-positive T lymphocytes, B cells, fibroblasts, monocytes and macrophages, and a number of nervous system tissue cell types. The cellular target of HTLV-I in the hematopoietic system in ATL is known to be the CD4-positive T lymphocyte, but the target of HTLV-I infection in CNS tissues remains undefined. Although assault on CNS tissue may result indirectly via immunologic mechanisms or directly from HTLV-I invasion of resident CNS cell types, the possibility exists that a combination of both indirect and direct assaults on CNS tissues results in the neuropathology observed in HTLV-I–associated myelopathy.

Damage to resident cells of the CNS may arise as a result of immune-mediated processes triggered by the presence of HTLV-I–infected cells. Demyelination mediated by cytolytic T cells (CTLs) could occur by direct assault on HTLV-I–infected glial cells or by the release of cytokines from CTLs activated by HTLV-I infection. Alternatively, an autoimmune mechanism could be responsible, autoreactive T lymphocytes becoming activated by infection with HTLV-I or by infected immunoregulatory T lymphocytes, resulting in a T-lymphocyte–mediated destruction of myelin. Furthermore, neural tissue damage could result indirectly from cytokines released by microglia cells activated by lymphokines released from HTLV-I–infected T lymphocytes. The neuropathology associated with HTLV-I infection may result from direct infection of neuroglial elements by HTLV-I, resulting in cellular dysfunction. Alternatively, infection of neuroglial cells by HTLV-I could render them susceptible to immune surveillance cells such as HTLV-I–specific CTLs. Constituent cells of the CNS also may harbor HTLV-I in patients suffering from TSP. The infection of resident CNS cells by HTLV-I may directly result in cellular dysfunction, cause the production and/or release of toxic substances, or make them targets of immune system assault.

## HIV-1 and Neurologic Dysfunction

### CLINICAL ASPECTS

The clinical symptoms attributed to ADC encompass a broad constellation of cognitive, behavioral, and motor abnormalities. The most distinctive clinical aspect of ADC is a progressive loss of cognitive functions. Affected patients initially demonstrate a slowing in both mental capacity and motor control. Early behavioral changes include apathy, social withdrawal, and personality changes. Later, nearly all aspects of cognition, motor function, and behavior are affected. Severe cases of ADC may be characterized by near or absolute mutism, incontinence, and severe dementia. Price has formulated a scaled system for defining the severity and progression of ADC based on an assessment of the patient's cognitive and motor functions (Fig. 4.8). HIV-1–infected patients in stage 0 are usually in the clinical latency period of the systemic infection. Patients with ADC stages 0.5 and 1 are generally nonprogressive for several years and have CD4-positive counts in the range of 200 to 500 cells/$\mu$L. AIDS dementia complex stage 2 (or greater) generally coincides with CD4-positive counts below 200 cells/$\mu$L, immunosuppression, increased systemic viral load, and systemic and nervous system opportunistic infections.

There are exceptions to these general observations, however. Some patients, plagued with numerous systemic opportunistic infections, have normal CNS function. Other HIV-infected patients develop nervous system abnormalities in the absence of any immunologic dysfunction or other symptoms related to the systemic presence of HIV. In addition, the severity of ADC does not always correlate either with the CD4-positive cell count or with the viral

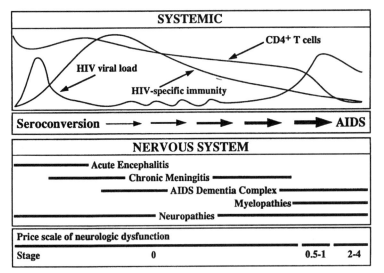

**Figure 4.8**
Significant events during the course of HIV-1 infection. Major systemic events are presented in parallel with neurologic observations. Price's clinical stages for ADC also are depicted. The timelines for the events are suggestive of temporal relationships and do not necessarily represent defined onsets and durations.

burden in the brain. Given the delay between the entry of the virus into the CNS and the appearance of progressive neurologic symptoms, the progression of systemic immunologic dysfunction and the onset of ADC are not necessarily temporally linked. The apparent lag between the initial seeding of the CNS and the onset of observable neurologic dysfunction may be the result of controls imposed by an early host immune response.

At least 75% of infants and children infected with HIV-1 develop a static or progressive encephalopathy caused by HIV-1 CNS infection. Clinical symptoms, which generally appear within the first 2 years of life, include impaired brain growth, progressive motor function deterioration, and the loss of developmental milestones and cognitive abilities.

## NEUROPATHOLOGY OF HIV-1 CENTRAL NERVOUS SYSTEM INFECTION

Of those patients who die of AIDS, roughly 20% to 30% exhibit postmortem CNS disease caused by HIV-1 infection. Pathologic indications of ADC are heterogeneous, but can be categorized under three major observations: white matter pallor and gliosis (most common), multinucleated-cell encephalitis, and vacuolar myelopathy. In cases where white matter pallor and gliosis are observed, the presence of virus may not always be detected, even with the added sensitivity of the polymerase chain reaction, suggesting a low level of viral burden in the brain. Multinucleated-cell encephalitis is also observed in HIV-1–infected patients with CNS disease. Multinucleated giant cells (syncytia), fused as a consequence of HIV infection, comprise macrophages and microglial cells and are sometimes clustered around blood vessels. When multinucleated-cell encephalitis is observed, the presence of virus and virus-specific antigens and nucleic acids can be readily detected. Vacuolar myelopathy, the third pathologic observation, may occur throughout the spinal cord, but often affects the cervical and thoracic sections more severely. A causal relationship between vacuolar myelopathy and HIV-1 replication in the affected tissues has yet to be established. Atrophy of the HIV-1–infected brain has been demonstrated by computed tomography, magnetic resonance imaging, and postmortem examination in patients with ADC, and increases with the severity of the condition.

The neuropathology of pediatric HIV-1 infection of the CNS differs somewhat from observations made in HIV-1–infected adults. Children may develop either static or progressive encephalopathy, but pediatric HIV-1 infection of the CNS results in an increased frequency of the appearance of inflammatory infiltrates and multinucleated giant cells, and the relatively rare appearance of vacuolar myelopathy. The fact that, in general, HIV-I–infected children suffer from fewer opportunistic infections than do adults supports the view that the presence of HIV-1 is the direct cause of the neurologic abnormalities in all age groups.

## HIV-1 TROPISM AND ENTRY INTO THE CENTRAL NERVOUS SYSTEM

HIV-1 may gain access to the brain relatively soon after the initial infection; it has been observed in the CSF as early as seroconversion and during the period of clinical latency. It is not clear, however, whether continued HIV-1 infiltration is required for the development of ADC. HIV-1 must circumvent the blood-brain barrier (BBB) to initiate neuropathogenic change (Fig. 4.9), possibly by crossing the BBB as cell-free virus, but there is very little evidence for this suggestion. A more generally accepted means for entry is referred to

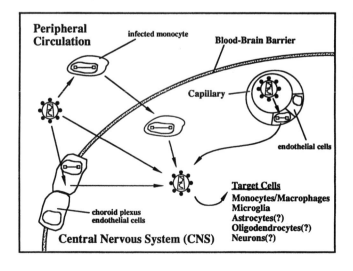

**Figure 4.9**

Proposed mechanisms for HIV-1 entry into the CNS. HIV-1 may gain entry into the brain as free virus through a compromised BBB; within infected monocytes and/or lymphocytes, which traffic across the BBB; or through infected choroid plexus or capillary endothelial cells.

as the Trojan horse mechanism, in which HIV-1 infects the CNS in the form of infected cells, crossing the BBB as viral particles or integrated provirus within monocytes that have been infected in the periphery. There is a precedent for immune-cell circulation in the CNS in observations made in animal models.

HIV-1 also may enter the brain by infection of endothelial and choroid plexus cells. HIV-1 replication in brain capillary endothelial cells may lead to infection of adjacent neuroglial cells by direct cell-to-cell contact or the spread of cell-free viral progeny. HIV-1 also may enter through replication in cells of the choroid plexus. This port of entry explains why HIV-1 observed within the brain is found particularly within deep white matter, and diencephalic and mesencephalic structures, but not the cortex.

Once the virus enters the CNS, it mainly infects macrophages and microglia. In addition, some evidence also suggests infection of vascular endothelial cells. There is little evidence of detectable HIV-1 expression in cells of neuroglial origin, even late in the disease course. In vitro investigations, conducted in our laboratory as well as in others, have shown that human neuroglial cells in both primary cultures and continuous cell lines are susceptible to HIV-1 infection. Infection of these cells results in low levels of HIV replication, characterized by limited expression of viral antigens and nucleic acids, no changes to cell mortality or morphology, and very low or undetectable levels of progeny virus production. Although glial cells are devoid of the HIV-1 cellular receptor molecule CD4, alternative receptors mediate HIV-1 entry into a variety of cells of neuroglial origin. For example, in vitro HIV-1 entry into neuronal cell lines is mediated by galactosyl ceramide (galC), a cell-surface glycolipid.

Descriptions of strains of HIV-1 that infect cells within the CNS usually involve the term *neurotropic*. Poliovirus, a member of the picornovirus family, is an example of a human neurotropic virus that causes CNS dysfunction as a direct consequence of viral infection of the CNS and the subsequent cytolytic demise of spinal motor neurons. In contrast, HIV-1 causes ADC through infection of cells of bone marrow origin and not cells originating from the neuroectoderm. Neurotropic strains of virus isolated from the CNS are, in fact, macrophage-tropic. The importance of HIV-1 macrophage tropism is underscored by studies with simian immunodeficiency virus that suggest that macrophage tropism is necessary but not sufficient for CNS infection.

Viral functions potentially involved in neuroglial tropism may map to other parts of the viral genome. Recent studies suggest a role for the HIV-1 LTR in determining cellular tropism. Their experiments used transgenic mice that contained a reporter gene expressed by LTRs derived from a lymphotropic virus or from a virus isolated from the CNS. Only the CNS-derived LTR produced reporter expression in the brain, especially in neurons. Other investigators are currently examining the effect of HIV-1 LTR sequence variations on the activity of the LTR in cells susceptible to HIV-1 infection within the immune and nervous systems. During the course of an in vivo infection, nucleotide changes are made in the *cis*-acting sequences of the LTR, as well as in the rest of the genome, as a result of the error-prone activity of reverse transcriptase.

## Mechanisms of HIV-1–Mediated Central Nervous System Damage

As previously noted, the primary cell type infected within the CNS is CD4-positive cells of macrophage and microglial origin. Although some evidence exists to support HIV-1 infection of neuroglial cells, the apparent level of infection seems insufficient to account for more than a small part of the clinical and pathologic effects of HIV-1 CNS infection.

The one-cell model of HIV-I infection of the CNS is similar to the consequences of neuronal poliovirus infection, in which neurons undergo cytolytic damage as a result of infection, damage to cells of both bone marrow and neuroectodermal origin being caused directly by cellular perturbation and functional disruptions during HIV-1 replication. Because little evidence is available to support productive in vivo infection of neurons and the subsequent disruption of neuronal function, this model is used primarily as the foundation of the multiple-cell models, in which macrophages and microglia are affected by HIV-1 infection, and subsequently indirectly affect other cells of the CNS.

Models using two or more cell types incorporate mechanisms that may account for observations made in the brains of ADC patients. In the two-cell model, an infected initiator cell releases soluble mediators that adversely affect the function of an uninfected target cell. These extracellular molecules might include virus-coded mediators, such as gp120 and Tat, or cell-encoded molecules, such as neopterin, quinolinic acid, or beta-2-microglobulin $\beta_2M$, which are released as a result of the perturbation of the initiator cell's metabolism during HIV-1 infection. The three-cell model includes an intermediary called an amplifier cell, the role of which is to respond to the signaling products of the initiator cell by producing cell-encoded mediators (i.e., cytokines), which, in turn, disrupt the function of the target cell. The advantage of this model is that the damage potential of a limited number of HIV-1–infected cells can be amplified to account for the disproportionate relationship between the limited number of infected cells observed in the CNS and the appearance of widespread pathologic effects. This model also incorporates elements of the two-cell model, in that the initiator cell can bypass the amplifier cells and affect the target cell directly. In this way, HIV-1–infected initiator cells can directly affect uninfected cells in close proximity.

The four-cell model expands on the three-cell model to account for the extended time course of ADC (Fig. 4.10). A modulator cell is added which functions primarily to regulate viral replication in the initiator cell. Acting as an arm of the immune system, the ability of the modulator cell to keep viral

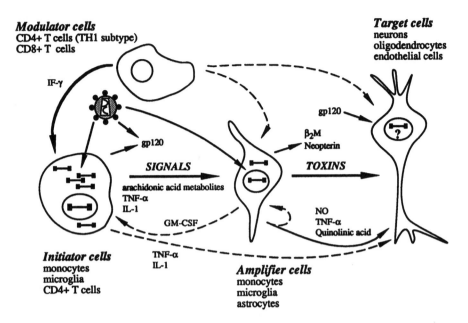

**Modulator cells**
CD4+ T cells (TH1 subtype)
CD8+ T cells

**Target cells**
neurons
oligodendrocytes
endothelial cells

IF-γ

gp120

gp120

β₂M
Neopterin

**SIGNALS**

arachidonic acid metabolites
TNF-α
IL-1

**TOXINS**

NO
TNF-α
Quinolinic acid

GM-CSF

**Initiator cells**
monocytes
microglia
CD4+ T cells

TNF-α
IL-1

**Amplifier cells**
monocytes
microglia
astrocytes

**Figure 4.10**

A model for HIV-1–induced neuroglial damage. HIV-1 infection of the initiator cells causes a cascade of cellular effects mediated by soluble signaling molecules and temporally regulated by the action of modulator cells. Solid lines indicate primary signaling pathways; broken lines depict secondary pathways and feedback loops. IL-1, interleukin-1; NO, nitric oxide; TNF-α, tumor necrosis factor-α. (Modified by permission from Price, RW. Understanding the AIDS dementia complex (ADC): the challenge of HIV and its effects on the central nervous system. In: Price RW, Perry SW, eds. *HIV, AIDS, and the brain.* New York: Raven, 1994:1–45.)

replication in check diminishes over time as the immune system is suppressed. The complete model also includes feedback loops in which any cell type in the model can affect any other cell type, as well as itself.

The candidate cell types for each of the cells in the model are as follows:

- The modulator cells may be CD4-positive T lymphocytes in their helper cell role; CD8-positive T lymphocytes that direct cytotoxicity toward other cell types; and B cells, which may produce HIV-1–specific antibodies. Generally, the function of all of these cell types centers on virus clearance or neutralization. Alternatively, these cells could affect viral expression through the release of cytokines and the subsequent alteration of the initiator cell's intracellular environment and the up- or down-regulation of HIV-1 LTR through changes in transcription factor availability and function.
- The initiator cells are primarily the CD4-positive T lymphocytes, especially early in the infection when the CNS is first seeded by activated T lymphocytes. During the progression of ADC, infected

macrophages and microglia may take on the role of maintaining viral replication and a macrophage-tropic viral quasispecies. Effector molecules include gp120, Tat, and Nef, released from chronically infected cells or released extracellularly after cell lysis. These gene products could serve as signaling molecules or as antigens that elicit an immune response.

- Candidate amplifier cells include cells of macrophage/monocyte/microglial lineage, and astrocytes. Both are likely amplifier cells because they can respond to and produce cytokines, and their numbers are increased during the course of the disease. Indeed, astrogliosis is the earliest pathologic finding during the development of ADC.
- The target cells are primarily neurons and oligodendrocytes, but vascular endothelial cells and astrocytes also may be involved. Cells may be killed or functionally altered as a result of the presence of soluble mediators produced in the cascade of cellular stimulation and response.

The central instruments of damage in these models are the soluble mediators released in response to HIV-1 infection or molecules released from other cells as a result of infection. Macrophages secrete soluble products that are toxic to neuroglial cells.

Recent studies have begun to examine the production and presence of cytokines in the brain during HIV-1 infection and have shown increased expression of cytokines as well as increases in several indicators of cytokine activity. Increases in $\beta_2M$, neopterin, and quinolinic acid, all considered markers of immune activation, have been observed in the CSF of HIV-1–infected patients with ADC.

Soluble viral products also may mediate cellular perturbation and damage. Gp120 may function as an intracellular signaling molecule, as well as a cytotoxic agent. The mechanism of damage is mediated indirectly by microglial and macroglial cells within the primary cultures. Tat and Nef also may act as neurotoxins, which cause brain dysfunction through neuronal death.

## Summary and Perspectives

The complex interplay between retroviruses and the cells that they infect within the nervous system warrants further investigation. The multifaceted relationship between the host cells and HTLV-I and HIV-1 results in not only the replication of the virus and propagation of the infection, but also the perturbation of the architecture and functions of the cells resident to the CNS. Many aspects of the direct effects of retroviral CNS infection, such as the roles played by gp120 neurotoxicity and the transforming ability of HTLV-I, remain to be fully characterized. Additionally, the indirect effects, such as those caused by HIV-1–induced cytokine cascades and the subsequent cellular perturbations, also require further study to reveal their roles in causing CNS damage. Furthermore, it will be necessary to integrate the results of these investigations with the clinical and pathologic observations, as well as with the findings related to other determinants in the course of virus-associated neurologic disease, such as opportunistic infections and the host immune response.

The complex nature of retroviral infection of the CNS also complicates the task of developing strategies for combating the infection and its effects. The integration of the viral genome into the host cell DNA makes eradication of the virus an almost impossible task; as yet, no existing therapies will excise the proviral DNA without killing the host cell. Treatments envisioned in the near future will focus on the influence of viral replication on cellular metabolism. Potential therapeutic targets include the regulatory proteins of both viruses, as well as other aspects of the retroviral replicative life cycle. Future treatment strategies also will focus on the host immune response, in an attempt to negate the damaging effects of immune system activation and virus-induced cytokine release in the CNS.

The goal of investigations into retroviral infections of the CNS is to understand the processes that affect the progression of neurologic disease to the degree required for the development of effective treatments for those afflicted. The research completed to date suggests that this is a worthwhile goal, and that further efforts in this field will reward us with tenable solutions to the problems faced by individuals with nervous systems infected with retroviral agents like HTLV-I and HIV-1.

## References/Further Reading

Anders KH, Guerra WF, Tomiyasu U, et al. The neuropathology of AIDS. UCLA experience and review. Am J Pathol 1986;124:537–558.

Blattner WA. Epidemiology of HTLV-I and associated diseases. In: Blattner WA, ed. Human retrovirology: HTLV. New York: Raven, 1990:251–265.

Budka H, Costanzi G, Cristina S, et al. Brain pathology induced by infection with the human immunodeficiency virus (HIV). A histological, immunocytochemical and electron microscopical study of 100 autopsy cases. Acta Neuropathol (Berl) 1987;75:185–198.

Dewhurst S, Sakai K, Zhang XH, et al. Establishment of human glial cell lines chronically infected with the human immunodeficiency virus. Virology 1988;162:151–159.

Epstein LG, Sharer LR, Oleske JM, et al. Neurologic manifestations of human immunodeficiency virus infection in children. Pediatrics 1986;78:678–687.

Everall I, Luthert P, Lantos P. A review of neuronal damage in human immunodeficiency virus infection: its assessment, possible mechanism and relationship to dementia. J Neuropathol Exp Neurol 1993;52:561–566.

Harouse JM, Kunsch C, Hartle HT, et al. CD4-independent infection of human neural cells by human immunodeficiency virus type 1. J Virol 1989;63:2527–2533.

Kunsch C, Hartle HT, Wigdahl B. Infection of human fetal dorsal root ganglion cells with human immunodeficiency virus type 1 involves an entry mechanism independent of the CD4 T4A epitope. J Virol 1989;63:5054–5061.

Kunsch C, Wigdahl B. Analysis of nonproductive human immunodeficiency virus type 1 infection of human fetal dorsal root ganglia glial cells. Intervirology 1990;31:147–158.

McArthur JC, Becker PS, Parisi JE, et al. Neuropathological changes in early HIV-1 dementia. Ann Neurol 1989;26:681–684.

Murphy EL, Figueroa JP, Gibbs WN, et al. Sexual transmission of human T-lymphotropic virus type I (HTLV-I). Ann Intern Med 1989;111:555–560.

Navia BA, Petito CK, Gold JW, et al. Cerebral toxoplasmosis complicating the acquired immune deficiency syndrome: clinical and neuropathological findings in 27 patients. Ann Neurol 1986;19:224–238.

Price RW, Brew B, Sidtis J, et al. The brain in AIDS: central nervous system HIV-1 infection and AIDS dementia complex. Science 1988;239:586–592.

Spencer DC, Price RW. Human immunodeficiency virus and the central nervous system. Annu Rev Microbiol 1992;46:655–693.

Tillmann M, Krebs FC, Wessner R, et al. Neuroglial-specific factors and the regulation of retrovirus transcription. Adv Neuroimmunol 1994;4:305–318.

Vaishnav YN, Wong-Staal F. The biochemistry of AIDS. Annu Rev Biochem 1991;60:577–630.

Vernant J-C, Maurs L, Gessain A, et al. Endemic tropical spastic paraparesis associated with human T-lymphotropic virus type I: a clinical and seroepidemiological study of 25 cases. Ann Neurol 1987;21:123–130.

Wigdahl B, Guyton RA, Sarin PS. Human immunodeficiency virus infection of the developing human nervous system. Virology 1987;159:440–445.

# II NORMAL AND ABNORMAL NEUROPHYSIOLOGY

# 5 Hippocampal Long-Term Potentiation

## A Model of the Cellular Processes Underlying Learning and Memory

Russell G. Durkovic

One of the most important goals in neurobiology has been to develop an understanding of the brain mechanisms responsible for learning and memory processes. Although even early investigators proposed that memory involved modifications in the structure of neurons, it is only within the past two to three decades that preliminary evidence for such an idea has been developed. Synaptic alterations that are thought to form the bases for neural plasticity appear to be brought about by a complex cascade of events that are by no means totally understood. Nevertheless, a general picture of the cellular processes underlying activity-dependent synaptic plasticity, of which learning and memory are important examples, can be represented (Fig. 5.1). For vertebrate systems, details of the various steps presented in Figure 5.1 are best understood for a phenomenon in the hippocampus known as long-term potentiation (LTP), which is the primary topic of this chapter. Long-term potentiation is currently the subject of intensive study because its properties are those that would be expected of a neural process involved in learning and memory.

Long-term potentiation is a long-lasting increase in the magnitude of the postsynaptic response to a single shock to the presynaptic fibers following repetitive activation (a tetanus) of these same fibers. Its discovery followed observations of prolonged synaptic enhancement of the perforant fiber pathway to hippocampal dentate granule cells following repetitive stimulation of their afferent fibers. Subsequent research revealed the presence of LTP phenomena in many regions of the CNS, including neocortex, amygdala, brainstem structures, and several sensory relay nuclei.

Thus, just as there is evidence for various brain locations for different kinds of learning and memory, so LTP-type phenomena appear to be widely distributed throughout the nervous system. Furthermore, substantial empirical evidence exists for a role for hippocampal LTP in certain forms of learning

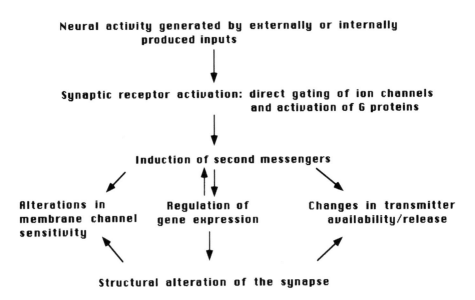

**Figure 5.1**
The generalized cascade of cellular events thought to underlie mechanisms of activity-dependent neural plasticity.

and memory. Examples include recordings of the development of LTP-like phenomena in animals during the learning of behavioral responses, decrements in the magnitude of LTP related to amount of forgetting, and the fact that drugs and genetic manipulations affect learning and LTP or hippocampal synaptic transmission in parallel ways. In addition, like behavioral memory, LTP has temporal stages, and the later stages of both processes require modification of gene expression. Finally, hippocampal LTP exhibits associativity: When a weak input is paired with a separate but convergent strong input, potentiation of the weak input results. This can be viewed as a cellular level representation of classical (Pavlovian) conditioning.

Long-term potentiation–like phenomena also may be related mechanistically to other activity-dependent plasticity phenomena. These include kindling (an animal model of epilepsy), the development of ocular dominance columns in visual cortex, and some central chronic pain syndromes. Thus, neural processes similar to those responsible for LTP may have wide-ranging significance in the CNS. Compared with what is known about hippocampal LTP, however, understanding of the cellular mechanisms underlying these other processes is currently minimal.

A number of excellent reviews of LTP and its potential relationship to learning and memory have been published in the past few years. Nevertheless, the explosive nature of progress on this topic has led to the attempt to provide an updated survey and a synthesis of prevailing ideas about the cellular mechanisms underlying LTP.

## Hippocampal Long-Term Potentiation: How It Is Produced, Measured, and Defined

Bliss and Lømo (1973), using anesthetized rabbits, and Bliss and Gardner-Medwin (1973), using chronically implanted, unanesthetized rabbits, were the first to describe the details of LTP in the hippocampus. They placed bipolar-stimulating electrodes in the perforant fiber pathway (pp, the axonal projection from the nearby entorhinal cortex to the granule cells of the dentate gyrus of the hippocampus) and recorded compound potentials from the dentate induced by single shocks applied to the pp at 2- to 3-second intervals. In general, recordings of potentials elicited by individual shocks were stable for an hour or more. Following one or more tetani (10–20 Hz for 10–15 sec, or 100 Hz for 3–4 sec), however, the extracellularly recorded compound potentials (excitatory postsynaptic potentials [EPSPs] and spikes) were potentiated for periods varying from 30 minutes to 10 hours in acute preparations to at least 16 weeks in chronic preparations. Bliss and Lømo called potentiation lasting 30 minutes "long lasting potentiation."

While additional studies of LTP using in vivo preparations have subsequently been conducted in various laboratories, the use of the hippocampal slice preparation has been a real boon to understanding the cellular mechanisms underlying hippocampal LTP. Not long after the initial in vivo demonstrations of LTP, similar results were obtained from the slice preparation (Fig. 5.2), first

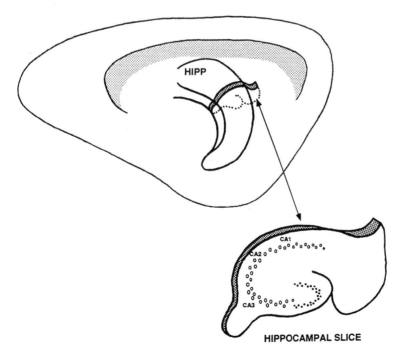

**Figure 5.2**
Orientation of the hippocampus (HIPP) and the hippocampal slice within the rodent brain. The hippocampus proper is also called *cornu Ammonis* (CA) or *Ammon's horn* and can be divided into three principal subdivisions, CA1 through CA3.

from area CA1, and then from the dentate gyrus and area CA3. Figure 5.3 shows the organization of this trisynaptic circuitry in the hippocampal slice. Note that this circuitry represents only a portion of the complex afferent and efferent neural connectivity within the hippocampus and between the hippocampus and other brain regions. Extracellular field potential recordings from the cell body layer of area CA1 before and, at various times, after a tetanus applied to the Schaffer collateral/commissural pathway are displayed in Figures 5.4 and 5.5.

While all three commonly recorded variables (EPSP slope, EPSP height, and spike height) generally exhibit long-term increases following the tetanus, the most commonly presented data are those of the changes in EPSP slope. This is because the EPSP height is sometimes difficult to measure due to influences from the compound spike, and the spike height can reflect effects in addition to those produced by the synaptic changes under investigation.

## Cellular Mechanisms of Long-Term Potentiation

### INDUCTION

The trigger for induction of LTP is an increase in free intracellular calcium concentration in those postsynaptic cells that have been activated by the tetanic stimuli. The first evidence of a calcium requirement came from experi-

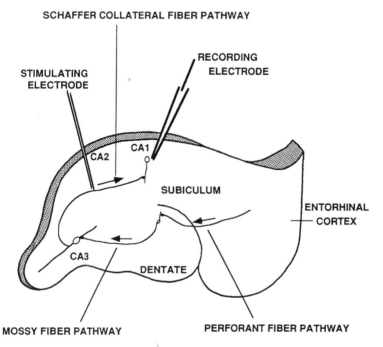

**Figure 5.3**
Components of the three major excitatory pathways in the hippocampal slice: 1) perforant fibers to dentate granule cells, 2) mossy fibers to CA3 pyramidal cells, and 3) Schaffer collateral fibers to CA1 pyramidal cells. Stimulating and recording electrodes are positioned for obtaining LTP in area CA1 (see Fig. 5.4). Note that there are many other circuits within the hippocampal slice that are not displayed in this illustration. The dentate gyrus, CA, and the subiculum compose the hippocampal formation.

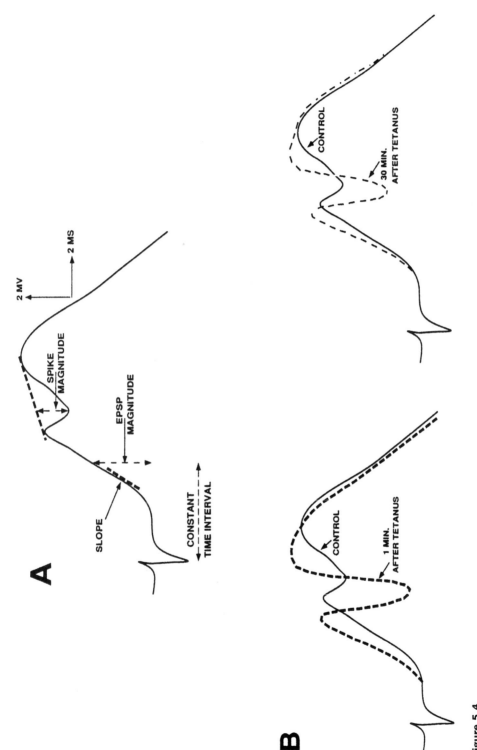

**Figure 5.4**

(A) Typical field potential recordings from area CA1 produced by constant-sized single shocks to Schaffer collateral/commissural fibers (see Fig. 5.3). Note the various factors that are commonly measured (compound spike and compound EPSP magnitudes, EPSP slope). (B) Recordings of potentiated field potentials (dashed lines) 1 minute and 30 minutes following a tetanus.

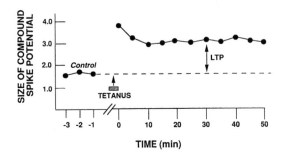

**Figure 5.5**
Data plotted from an experiment such as that shown in Figure 5.4B. Compound action potential magnitude measurements (millivolts), recorded from area CA1 in response to single shocks applied to Schaffer collateral/commissural fibers before and after a tetanus, were applied to the same fibers.

ments employing a $Ca^{2+}$-free solution in which tetani failed to induce LTP in area CA1. On the other hand, simply increasing the extracellular $Ca^{2+}$ concentration of the solution bathing hippocampal slices produced an LTP in area CA1 and the dentate gyrus, and in area CA3. Intracellular injection of $Ca^{2+}$ chelators into CA1 pyramidal cells and into CA3 cells receiving mossy fiber inputs have been reported to prevent LTP induction. Furthermore, intracellular release of $Ca^{2+}$ by photoactivation of a caged calcium complex results in LTP.

How do the commonly used stimulation techniques (tetani) that produce LTP result in increases in intracellular $Ca^{2+}$ concentration? The presynaptic activity generated as a consequence of the tetani results in release of excitatory synaptic transmitter (most likely glutamate) and the activation of at least three subtypes of amino acid receptors on the postsynaptic cells: amino-3-hydroxy-5-methyl-4-isoxazole propionic acid (AMPA), an ionotropic glutamate receptor (iGluR) agonist; N-methyl-D-aspartate (NMDA), a compound that specifically activates the NMDA iGluR; and a metabotropic glutamate receptor (mGluR).*

AMPA and NMDA receptors are known as iGluRs because, when they are activated by glutamate, the alterations of their protein conformations lead to an increase in the conductance of their intrinsic ion channels, which are permeable to $Na^+$ and $K^+$ ions. In the case of NMDA receptors, the channels are also permeable to $Ca^{2+}$ ions. In contrast, mGluRs do not contain ion channels, but are coupled to membrane-bound cellular effectors via GTP-binding proteins that influence multiple second-messenger systems. All three receptor subtypes are thought to be directly or indirectly involved in influencing the rise in intracellular free $Ca^{2+}$ following glutamate activation.

The AMPA receptor (so named because of its selective ligand AMPA, and previously known as the quisqualate/kainate receptor) is responsible for fast EPSP generation at many excitatory synapses in the mammalian brain, including those of the hippocampus. This membrane depolarization is involved

*The mossy fiber termination on CA3 cells appears to be an exception in that there are very few NMDA receptors at this synaptic site.

indirectly in the increase in intracellular $Ca^{2+}$ because of two voltage-dependent influences: (1) the voltage-dependent $Ca^{2+}$ channels and (2) the relief of the voltage-dependent $Mg^{2+}$ block of the $Ca^{2+}$-permeant NMDA receptor. Because intracellular free $Ca^{2+}$ levels are normally quite low, increases in $Ca^{2+}$ permeability of the neuronal membrane as a consequence of depolarization produces a $Ca^{2+}$ influx from the extracellular matrix into the neuron. Finally, there is a form of the AMPA receptor channel that is itself $Ca^{2+}$ permeable, but its role is currently unknown.

The NMDA receptor is particularly sensitive to glutamate, but the NMDA ionophore is blocked in a voltage-dependent manner by $Mg^{2+}$. Thus, without membrane depolarization, the increase in $Ca^{2+}$ permeability of NMDA channels in response to synaptic activation is largely blocked. Also, the depolarization brought about by single inputs does not provide a great enough depolarization for a long enough time to open many NMDA channels. However, the greater and longer-lasting depolarization induced by tetanic stimulation removes the $Mg^{2+}$ block and results in $Ca^{2+}$ influx through the NMDA receptors, with additional $Ca^{2+}$ influx occurring via voltage-gated $Ca^{2+}$ channels.

Metabotropic glutamate receptor activation also has important influences on intracellular free $Ca^{2+}$ levels. The most widely accepted mechanism for the mGluR activation effect involves stimulation of phosphoinositide (PI) hydrolysis; phospholipase C (PLC), a G protein–activated lipase, cleaves phosphatidylinositol-biphosphate ($PIP_2$) to generate inositol triphosphate ($IP_3$) and diacylglycerol (DAG)*. Inositol triphosphate is a small water-soluble molecule that causes release of $Ca^{2+}$ from intracellular calcium-sequestering compartments, possibly endoplasmic-reticulum–associated structures within dendritic spines. Support for a role for this mechanism in LTP production comes from studies showing that activation of mGluRs can induce LTP via a release of $[Ca^{2+}]_i$, and that an inhibitor of amino acid–stimulated PI hydrolysis in the hippocampus, AP3, blocks hippocampal LTP. Furthermore, DAG, by activating protein kinase C (PKC), enhances $Ca^{2+}$ currents through NMDA receptor channels. In addition to these receptor-mediated $[Ca^{2+}]_i$ increases, other $Ca^{2+}$ regulating mechanisms, such as $Ca^{2+}$-induced $Ca^{2+}$ release processes, have been suggested as role players in regulating $[Ca^{2+}]_i$.

The importance of activation of the three receptor subtypes for the normal induction of LTP is exemplified by the effects of specific pharmacologic blocking agents: 6,7-dinitroquinoxaline-2,3-dione (DNQX) and 6-cyano-7-nitroquinoxaline-2,3-dione (CNQX) block AMPA receptors; 2-aminophosphonovalerate (APV) blocks NMDA receptors; and (RS)-α-methyl-4-carboxyphenylglycine (MCPG) blocks certain mGluRs. Application of any one of these receptor antagonists, with few exceptions, blocks induction of LTP.

---

*Additional evidence suggests that other mGluR activated G-protein systems may be required to obtain LTP.

The effect of blocking AMPA receptors is primarily an indirect one based on the role that activated AMPA receptors have in producing membrane depolarization, and thus relieving the blockage of NMDA receptors by $Mg^{2+}$ under normal conditions. However, hippocampal slices maintained in DNQX and low $Mg^{2+}$ (reducing the $Mg^{2+}$ block of NMDA receptors) exhibited LTP following tetani once the DNQX was washed out of the slice. Occasionally, with intense stimulation in slices with DNQX and normal $Mg^{2+}$ levels, LTP can be produced if NMDA currents are large enough (Muller D, personal communication, 1994). Another exception to receptor blocking effects is that mossy fiber/CA3 LTP is not blocked by NMDA antagonists, an expected finding based on the relative absence of NMDA receptors at these synaptic sites.* Furthermore, when multiple trains of 200-Hz stimuli (i.e., stronger than usual tetani) were employed in the presence of APV, a voltage-dependent $Ca^{2+}$ channel–induced LTP was obtained in CA1.

In summary, induction of LTP in both CA1 and the dentate gyrus is generally thought to require the combined glutamate activation of AMPA, NMDA, and mGlu receptors to obtain the necessary increases in intracellular free $Ca^{2+}$ levels. Mechanisms underlying mossy fiber/CA3 LTP are unresolved but clearly do not involve activation of NMDA receptors. The question here is whether mossy fiber/CA3 LTP depends on a postsynaptic $Ca^{2+}$ influx for induction, or whether it is an entirely presynaptic phenomenon. The apparently contradictory experimental results are difficult to reconcile but may be a consequence of the underlying complexity of the neuronal circuitry.

## EXPRESSION

Despite intensive study, the factors responsible for the increased response of postsynaptic neurons following tetanic stimuli of their presynaptic inputs are still poorly understood. The major controversy concerns the locus of the synaptic alterations underlying LTP; evidence for exclusively postsynaptic alterations, exclusively presynaptic alterations, or a combination of these two processes is abundant. This is frustrating, but probably not surprising, given that the research derives from multiple laboratories that apply a variety of techniques and conditions to a very complex process.

Evidence for a presynaptic locus comes from measurements of increased levels of glutamate release following LTP induction. Both LTP and the sustained increase in glutamate release have been found to be blocked by APV. Evidence suggesting a purely postsynaptic locus for LTP based on differential enhancement of non-NMDA receptor currents versus NMDA receptor currents was soon countered by experiments indicating that both NMDA and non-NMDA components were persistently increased during LTP, suggesting that LTP involves sustained increases in release of glutamate from presynaptic fibers. Furthermore, increased release of preloaded, labeled glutamate was observed following tetani, yet there was no evidence of changes in high-affinity uptake or in total $Na^+$-independent binding of $[H^3]Glu$. Other experiments, however, suggested that this was only short-lived. A recent study linking changes in paired-pulse facilitation with LTP in area CA1 provides strong evidence for including increases in presynaptic transmitter release as a mechanism for LTP expression. Results of tests of postsynaptic sensitivity to ionophoretic ejection of AMPA or quisqualate in rat hippocampal slices indicated a slowly developing postsynaptic change following LTP induction, suggesting that mechanisms involved in the rapid onset of LTP include some that are presynaptic.

Much of the evidence derived from studies of quantal analysis also supports the idea of a concomitant increase in transmitter release with LTP. However, data in support of increased postsynaptic responses using this methodology have also been obtained, and it must be realized that the assumptions underlying quantal analysis may not be valid when applied to the CNS.

Finally, recordings of spontaneous miniature EPSPs have shown an increase in their size and frequency during LTP—the first suggesting a postsynaptic locus of change, and the second, a presynaptic one. Overall, one must at least consider the possibility that the mechanisms underlying LTP maintenance involve both increased transmitter release and increased sensitivity of postsynaptic structures.

## RETROGRADE MESSENGERS

Given that the trigger for the induction of LTP involves increases in free calcium levels in postsynaptic cells, it seemed reasonable to propose that the presynaptic alterations thought to contribute to LTP maintenance are brought about by a retrograde messenger released from the postsynaptic cell. Initially, evidence for two different diffusible substances, arachidonic acid (AA) and nitric oxide (NO), was provided, each fulfilling many of the criteria for retrograde messengers. Exogenous application of AA, however, delayed the onset of potentiation in dentate neurons and had no consistent effect on CA1 neurons. Furthermore, APV (an NMDA blocker) blocked enhancement by AA when exogenous application was paired with a weak tetanus. These results favor a postsynaptic locus of effect of AA and cast doubt on the role of AA as a retrograde messenger, particularly in area CA1.

While incomplete, data supporting this retrograde messenger role for NO and another diffusible gas, carbon monoxide (CO), are quite strong. Either gas, applied simultaneously with weak tetanic stimulation, produces pathway-specific LTP, even in the presence of APV. Neither the tetanus alone nor either NO or CO alone produced a long-term effect. Blockers of the enzyme NO synthase (NOS), which generates NO from the amino acid l-arginine, prevented LTP induction, as did blocking of heme oxygenase (HO), the enzyme for CO generation. Hemoglobin, which binds NO and CO, also blocked LTP when added to the extracellular medium.

Differences between NO and CO effects and LTP have been found that complicate the picture, and not all studies have found a consistent block of LTP by NOS inhibitors, although this may be related in part to the intensity of tetanus employed. However, one of the greatest objections to the hypothesis that NO is a retrograde messenger has been resolved; until recently, NOS could not be detected in appropriate postsynaptic cells. Recent studies have shown both the brain-specific NOS isoform and, more important for LTP, the endothelial NOS isoform in CA1 pyramidal cells and dentate granule cells.

One of the molecular actions of retrograde messengers may be on soluble guanylate cyclase (GC). Arachidonic acid, NO, and CO all activate soluble GC, and an inhibitor of GC has been shown to block induction, but not maintenance, of LTP. Because GC catalyzes the production of GMP from GTP, investigators have also tested cGMP analogues and found that pairing them with weak tetanic stimuli produced LTP. The potentiation was not prevented by NMDA receptor blockage. This and other evidence suggested a presynaptic locus of effect and mediation by cGMP-dependent protein kinase (PKG). Nitric oxide also may influence levels of basal adenosine diphosphate (ADP) ribosylation activity. Further complicating the picture is evidence implicating NO as an NMDA receptor inhibitor.

Recently, an additional substance, platelet-activating factor (PAF), has been determined to be another strong candidate for a retrograde messenger. The fact that specific blockers of NO, CO, or PAF can each inhibit LTP induction implies that combined actions of multiple retrograde messengers may be required to produce presynaptic LTP effects in certain cases.

## Ca$^{2+}$-DEPENDENT PROTEIN KINASES AND LONG-TERM POTENTIATION

Because Ca$^{2+}$ entry into the postsynaptic cell is the critical step in beginning events responsible for LTP, what are the biochemical processes influenced by Ca$^{2+}$ entry? Most studies have focused on the roles of several Ca$^{2+}$-dependent protein phosphorylating enzymes (protein kinases) in the cascade of biochemical processes leading to LTP, in particular PKC, Ca$^{2+}$-calmodulin–dependent protein kinase II (CaMKII), and protein tyrosine kinases (PTKs). Protein tyrosine kinase involvement is suggested both by LTP-like synaptic enhancement by PKC activators and by its potentiation by intracellular injection of the catalytic subunit of PKC. A decrement to baseline of the potentiated response was produced by intracellularly injected PKC blockers if they were injected before, but not 5 minutes after, LTP induction. Extracellular application of PKC inhibitors up to 30 minutes after LTP induction, and presumably acting presynaptically, also resulted in a decrement to baseline of the synaptic response. These results suggest that a narrow time window exists for PKC influences on LTP postsynaptically, but a longer time window prevails presynaptically.

Postsynaptically, PKC is associated with enhancement of NMDA currents, an effect of reduced $Mg^{2+}$ block of the NMDA receptors. Presynaptically, growth-associated protein 43 (GAP-43), a PKC substrate, may play a role in LTP. Phosphorylation of GAP-43 has been correlated with neurotransmitter release in hippocampal slices and with LTP in the dentate gyrus and CA1. These observations suggest that presynaptic PKC activation and its phosphorylation of GAP-43 may play a role in the early maintenance of LTP.

In addition to PKC, CaMKII has been implicated in the regulation of transmitter release due to its phosphorylation of synapsin I, a protein associated with presynaptic vesicles. CaMKII, because of its heavy concentration in the postsynaptic density, also may play a role in postsynaptic transmission processes by enhancing AMPA currents. Like PKC, blockers of CaMKII activity reduce the duration of LTP, and knockout of the gene encoding CaMKII (the isoform heavily enriched in the postsynaptic densities) hampers formation of LTP in hippocampal slices.

Evidence is also accumulating that suggests involvement of PTKs in LTP. For example, PTK inhibitors block LTP induction, but not established LTP, when injected intracellularly or when added to the bath. Furthermore, in fyn PTK mutant mice, LTP is blunted. Postsynaptically, a PTK has been shown to increase currents through NMDA receptors.

Another kinase, cAMP-dependent protein kinase (PKA), or protein kinase A, is involved in a later stage of LTP. These inhibitors have been shown to block a stage of LTP (L-LTP, beginning about 1 to 3 hr after tetani) that requires protein synthesis. Furthermore, cAMP analogs produced a potentiation that interfered with subsequent attempts to induce LTP by tetanic stimulation, and inhibitors of protein synthesis blocked the potentiating effects of these cAMP analogs. How PKA might function in LTP has not been totally resolved, but it may play an ion channel–modulating role by phosphorylating AMPA channels, thereby increasing channel opening frequency and open time. Furthermore, PKA has been implicated in the induction of gene activity by activation of the transcription factor cAMP response-element binding protein (CREB).

## ALTERED GENE EXPRESSION AND PROTEIN SYNTHESIS

Enduring changes in neuronal function and structure require alterations in gene expression and protein synthesis. Initial studies examining LTP-related alterations in gene expression focused on induction of immediate early genes (proto-oncogenes) such as *fos*, *jun*, and *Zif268*, but their roles, if any, in the production of LTP are as yet uncertain. It may be more than coincidence, however, that the induction of these immediate early genes involves a cellular protein (CREB) that induces gene transcription and requires (1) PKA activation or (2) $Ca^{2+}$ influx and CaMK activation to do so. Furthermore, the serum response element on the c-*fos* promotor is particularly sensitive to $Ca^{2+}$ influx via NMDA receptors. Recently, increases in neurotrophin mRNAs (brain-derived neurotrophic factor [BDNF] and neurotrophin-3 [NT-3]) and the immediate early gene tissue-plasminogen activator (tPA) have been found during LTP; both BDNF and NT-3 may be involved in the regulation of neural connections. Because tPA is released from growth cones and is correlated with morphologic differentiation, it has also been hypothesized to play a role in the structural changes that accompany LTP.

Another factor potentially related to structural alterations in the hippocampus is microtubule-associated protein 2 (MAP-2) mRNA, which increases following dentate injection of NMDA, sin-1 molsidomine (a substance that releases NO), or cGMP. Evidence of alterations in the synthesis of other proteins following LTP is also beginning to be developed. The requirement for synthesis of new proteins in the later stage of LTP (L-LTP) has been demonstrated by using blockers of both transcription and translation. A summary diagram representing the molecular events currently thought to be involved in LTP, as described in this chapter, is presented as Figure 5.6.

## STRUCTURAL MODIFICATIONS

The results of electron microscopy (EM) also suggest that both modifications in existing synapses and formation of new synapses are associated with LTP. In tetanized synapses, increases in dendritic spine area and spine width and surface area of the postsynaptic density have been observed. Furthermore, dendritic spines having spinules (protuberances of the postsynaptic membrane into the presynaptic element) were increased in frequency. Results of examination of tetani-induced alterations in numbers and distributions of synaptic vesicles have been variable; increases, decreases, and no change in vesicle numbers have all been reported. Redistribution of vesicles toward the active zone of the synapse also has been reported as a consequence of LTP.*

One of the more interesting studies suggesting structural remodeling and potential formation of new synapses related to LTP was carried out on the dentate gyrus. Following tetani of the medial perforant pathway given once each day for 4 days, EM data showed no statistically significant changes in total number of synapses compared with electrode-implanted or low-frequency–stimulated controls. However, the number of a specific type of synapse characterized by seg-

---

*It should be noted that while transmitter release by vesicular exocytosis remains the most popular model of chemical transmission (so much so as often to be represented as a fact rather than a model), a number of experimental observations seem incompatible with this hypothesis. Considerations of the totality of data have led several investigators to propose other synaptic transmission mechanisms.

mented postsynaptic densities was doubled. A similar result was obtained in a later study on aged rats. Without additional data in which different times after tetani are employed, it is difficult to interpret these results, but they may represent evidence for the formation of new synapses (Fig. 5.7).

Earlier work on CA1 synapses had indicated that LTP was associated with increases in certain types of synapses (synapses on dendritic shafts as well as sessile spine synapses). Structural changes at different intervals after tetani (10–15 min, 2 hr, and 8 hr) were examined in this latter study. Changes in shape of dendritic spines proved to be transient, declining at 2 hours and disappearing after 8 hours. On the other hand, the increased density of dendritic shafts and sessile synapses was observed at all time intervals examined (see Fig. 5.7). Furthermore, the results were correlated with LTP in that an overstimulation control group (40 or 100 Hz for 10 min) that exhibited no LTP also exhibited no increase in numbers of synapses. This study shows that the processes of synaptogenesis can begin so rapidly that altered gene expression is probably not required. Rather, the structural alterations are likely to be brought about by assembly of preexisting macromolecules.

A reasonable guess, given the correlation of structural change and LTP, is that the initial synaptogenesis evolves from interactions of activated protein kinases and other enzymes with synaptic structural proteins. Later, synthesis of new proteins may be required to solidify these early structural modifications. Nevertheless, to more conclusively establish a link between structural change and LTP, future morphologic studies correlated with different magnitudes of LTP and long-term depression are necessary, and these could be coupled with investigations using pharmacologic agents to augment or block the changes in synaptic efficacy.

## PRIMED-BURST POTENTIATION

One of the problems in trying to relate the aforementioned LTP mechanisms to learning and memory is that the stimulus parameters (frequency × duration) commonly used to evoke the potentiation are outside the range observed for neuronal activity in the hippocampus of intact animals. Thus, it was significant when a long-lasting potentiation effect was induced in area CA1 by a stimulus pattern composed of a single pulse followed 170 milliseconds (i.e., one theta-wave period) later by a brief burst of four pulses at 100 Hz. These patterns (Fig. 5.8) are similar to two endogenous patterns of hippocampal activity in mammals: complex spike activity and hippocampal theta rhythms. This potentiation phenomenon, called primed-burst potentiation (PBP), appears to be governed by the same mechanisms as LTP, as evidenced by studies showing LTP-PBP occlusion and PBP sensitivity to NMDA receptor blockage.

Investigations of the interval between the pulse and burst showed that only 140- and 170-millisecond intervals were capable of inducing PBP (10–70

**Figure 5.6**

A broad summary diagram indicating cellular processes that may be involved in the production of LTP in the CA1 area of the hippocampus. Nerve terminals (top) release glutamate, which activates AMPA, NMDA, and mGlu receptors on the dendrites of the postsynaptic cells (bottom). Activation of these receptors triggers an increase in intracellular free Ca$^{2+}$ levels postsynaptically, causing a cascade of molecular events: activation of various protein kinases that enhance receptor sensitivity; release of retrograde transmitters (NO, CO, PAF?) that travel to the presynaptic terminals and result in increased transmitter release; retrograde transport of transcription factors to the nuclei of the neurons activating gene expression, protein synthesis, and stabilization of and/or new structural changes.

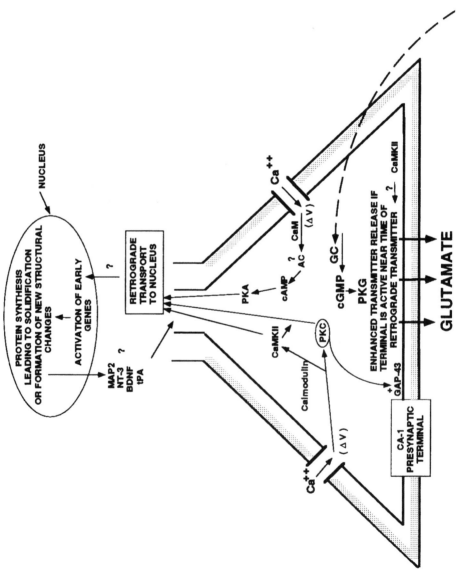

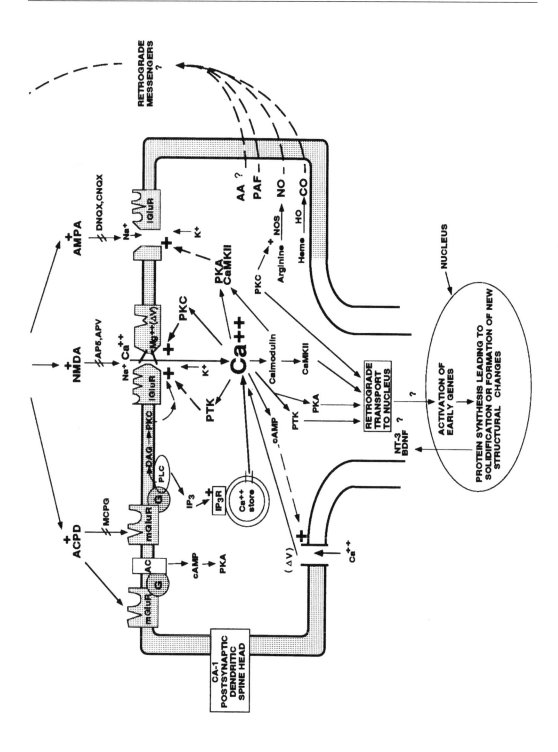

**A**

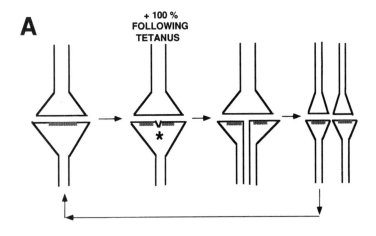

**+ 100 %
FOLLOWING
TETANUS**

**B**

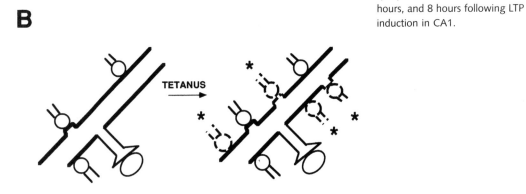

TETANUS

**Figure 5.7**
Representations of EM results showing structural synaptic changes associated with LTP. (A) A model of synapse turnover. Note that tetanic stimulation (4 days, once a day) was associated with an approximate doubling of synapses with segmented post-synaptic densities (asterisk) in the dentate gyrus. These results may be related to the observations of increases in dendritic spines that develop spinules following LTP. (B) Representations of increases (asterisks) in dendritic shaft synapses and sessile (stubby; no discernable spine head or neck) spine synapses 10 to 15 minutes, 2 hours, and 8 hours following LTP induction in CA1.

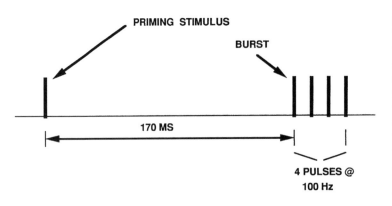

PRIMING STIMULUS

BURST

170 MS

4 PULSES @
100 Hz

**Figure 5.8**
The stimulus pattern applied to Schaffer collateral/commissural fibers that produces an LTP-like effect in area CA1, called primed-burst potentiation. This stimulus pattern is similar to endogenous patterns of hippocampal unit activity recorded during learning and memory tasks. Neither the burst by itself nor the pulse by itself results in potentiation.

msec and 500–5000 msec were without effect), suggesting a close correlation with physiologically observed activity patterns. This is thought to be a result of the influence of a GABA receptor-mediated decrease in postsynaptic inhibition of pyramidal cells at the time of burst presentation. Primed-burst potentiation appears to be efficient in evoking potentiation, in that the number of stimulus pulses required to evoke a given potentiation is 4 to 20 times less than that required to produce the same amount of LTP.

If patterns of cell activity like those producing PBP were recorded from the hippocampus during learning and memory tasks, this would be additional empirical evidence for a role for hippocampal LTP as a mechanism underlying learning and memory. In fact, in just such a study of unit activity recorded during olfactory and spatial learning tasks, firing patterns of many hippocampal cells exhibited the features of patterned stimulation found to be optimal for PBP induction.

## Summary

This review updates the substantial progress that has been made in understanding the cellular mechanisms of one form of activity-dependent plasticity of the nervous system: hippocampal LTP. The general features of the phenomenon and its cellular mechanisms constitute the most well understood example of the inner workings of a portion of the distributed neural networks of the brain that function in the processes of vertebrate learning and memory. Many significant questions on this topic, however, remain unanswered. Some of these pertain to the underlying processes of hippocampal LTP. Others are concerned with how the cellular processes in the hippocampus compare with LTP-type mechanisms elsewhere in the nervous system, and how the totality of these distributed networks function to produce the behavioral abilities that we call learning and memory. It is clear that the knowledge gained due to research on hippocampal LTP has provided an important foundation for future studies dealing with these issues. For the clinician, the relevance of an understanding of the cellular mechanisms underlying learning and memory lies in the potential to develop pharmacologic treatments that augment these processes. For example, knowledge gained from an understanding of the cellular mechanisms of hippocampal LTP has led to experiments suggesting that an AMPA receptor agonist can increase certain types of memory in animals. Such results could lead to future pharmacologic manipulations designed to enhance human memory and learning capabilities.

## Acknowledgments

This work was supported in part by a grant from the National Science Foundation (IBN 9220206). The author wishes to thank Ms. Terri Brown for typing, and Drs. Clancy Leahy and Tim Rigney for their help in reviewing this chapter.

## References/Further Reading

Aramori I, Nakanishi S. Signal transduction and pharmacological characteristics of a metabotropic glutamate receptor, mGluR1, in transfected CHO cells. Neuron 1992;8:757–765.

Bliss TVP, Gardner-Medwin, AR. Long-lasting potentiation of synaptic transmission in the dentate area of the unanaesthetized rabbit following stimulation of the perforant path. J Physiol (Lond) 1973;232:357–374.

Bliss TVP, Lømo T. Long-lasting potentiation of synaptic transmission in the dentate area of the anesthaetized rabbit following stimulation of the perforant path. J Physiol (Lond) 1973;232:331–356.

Bliss TVP, Lynch MA. Long-term potentiation of synaptic transmission in the hippocampus: properties and mechanisms. In: Landfield PW, Deadwyler SA, eds. Long-term potentiation: from biophysics to behavior. New York: AR Liss, 1988:3–72.

Davis HP, Squire LR. Protein synthesis and memory: a review. Psychol Bull 1984;96:518–559.

Fazeli MS. Synaptic plasticity: on the trail of the retrograde messenger. Trends Neurosci 1992; 15:115–117.

Goelet P, Castellucci VF, Schacher S, Kandel ER. The long and the short of long-term memory—a molecular framework. Nature 1986;322:419–422.

Kandel ER. A cell—biological approach to learning. Bethesda, MD: Society for Neuroscience, 1978.

Kriebel ME, Vautiin J, Holsapple J. Transmitter release: prepackaging and random mechanism or dynamic and deterministic process. Brain Res Rev 1990;15:167–178.

Kupfermann I. Learning and memory. In: Kandel ER, Schwartz JH, Jessell TM, eds. Principles of neuroscience. 3rd ed. New York: Elsevier, 1991:997–1008.

Levy VVB, Steward O. Synapses as associative memory elements in the hippocampal formation. Brain Res 1979;175:233–245.

Lømo T. Frequency potentiation of excitatory synaptic activity in the dentate area of the hippocampal formation. Acta Physiol Scand 1966;68:128.

Malenka RC, Kauer JA, Perkel DJ, et al. An essential role for postsynaptic calmodulin and protein kinase activity in long-term potentiation. Nature 1989;340:554–557.

Staubli U, Rogers G, Lynch G. Facilitation of glutamate receptors enhances memory. Proc Natl Acad Sci USA 1994;91:777–781.

Traub RD, Miles R. Neuronal networks of the hippocampus. New York: Cambridge University Press, 1991.

Vautrin J. Vesicular or quantal and subquantal transmitter release. News Physiol Sci 1994;9: 59–64.

Wallace C, Hawrylak N, Greenough WT. Studies of synaptic structural modifications after long-term potentiation and kindling: context for a molecular morphology. In: Baudry M, Davis JL, eds. Long-term potentiation: a debate of current issues. Cambridge, MA: MIT Press, 1991:189–232.

# 6 Pain and Its Surgical Relief

Blaine S. Nashold, Jr.

James R. B. Nashold

*When I was young I used to say "All pain can be relieved, if it is really required, by cutting the proper sensory nerves or nerve tracts in the proper place at the proper time," but it is not always as simple as that.*

Wilder Penfield

## Historical Aspects of the Study of Pain

The first modern theory of the nature of pain was the *theory of specificity*. Following the discovery of the various sensory receptors found in tissues, pain was considered a specific sensation transmitted by a specific pain receptor. Later, rather than being related to a specific receptor, pain was thought to be a central *summation phenomenon* due to excessive stimulation of a group of receptors. This could explain pain of central origin. Next, the *pattern theory* suggested that all sensations are dependent on spatial and temporal patterns, including central reverberating circuits, which would explain the occurrence of phantom pain after an amputation. A fourth theory suggested that the *perception* of pain and *reactions* to pain were linked.

Recent theories of pain and its treatment have been dominated by the idea that the input from any specific type of sensory nerve fibers may interact with that of other types of fibers in the spinal cord, and that this interaction influences the final sensory experience. Noordenbos was the first to suggest that the input of the larger-diameter sensory fibers inhibits that of the smaller-diameter (pain) fibers. Later, Melzack and Wall proposed that the inhibitory activity of the large myelinated nerve fibers of the posterior nerve roots upon the action of the smaller pain fibers occurred in specific neurons of the dorsal horn; these acted like a gate to inhibit the central transmission of painful impulses. Wall and Sweet stimulated their own infraorbital nerves and, as expected, produced an analgesia over the skin. The physiologic details of the gate control theory are still disputed, but its prediction that direct electrical stimulation of the nerve trunk in patients with painful nerve lesions will sometimes give relief has been amply fulfilled.

Other evidence that events occurring in the dorsal horns of the spinal cord are important in the production of pain soon accumulated. Loeser and Ward recorded hyperactivity of the neurons at the level of injury from the spinal cord of a paraplegic patient. These hyper-irritable neurons may be responsible for the generation of pain in paraplegia, and denervation hypersensitivity of central neurons may be the physiologic mechanism that causes deafferentation pain.

## THE NEUROCHEMISTRY OF PAIN

From studies of neurotransmitter and receptor localizations in the nociceptive primary afferents and the dorsal horn, it is evident that there are both ascending neural systems conveying pain messages and descending systems that control and modulate these. There also is an increasing realization of the large number of chemical agents that function as neurotransmitters that convey or modulate nociception. Forty percent of the small dorsal root ganglion cells (small A and C nerve fibers) contain substance P, which is released in response to painful stimulation. Stimulation of afferent fibers also results in an increase in substance P in certain dorsal horn neurons, and substance P is inhibited by enkephalins in the dorsal horn. Sectioning the afferent fibers results in a decrease of substance P in the same neurons. In experimental dorsal root avulsion experiments in animals, there is an immediate decrease of both substance P and enkephalins. Over time (weeks), the levels of both neurotransmitters recover but never to normal limits; there is always an excess of substance P over enkephalin. The presence of this excess may be responsible for the hyperactivity of the secondary nociceptive neurons in Rexed layer 5 of the posterior horn, and this is believed to be responsible for the pain experienced by patients with brachial plexus avulsion and traumatic paraplegia.

Similarly, somatostatin, cholecystokinin (CCK), and vasoactive intestinal polypeptide are all located in the small terminal afferents of the dorsal horn. Noxious thermal stimuli may release somatostatin, which in turn can be inhibited by morphine, while the nociceptive spinothalamic tract cells are inhibited by CCK. Glutamate, which is a possible neurotransmitter, also may be released by afferent neurons associated with light touch and pressure. Metenkephalin, a naturally occurring opiate, has been located in dorsal horn neurons, and other opiate receptors are found on the central terminals of the primary afferents. The smaller interneurons, which may be sensitized by substance P, contain dynorphin. Presynaptic and postsynaptic inhibition may be due to gamma-aminobutyric acid and glycine. In the brain stem, descending modulation involves catecholamines, norepinephrine, dopamine, and serotonin (5-HT). The descending influences act presynaptically or through enkephalin-containing interneurons (serotonin) of the dorsal horn. Further elucidation of these complex chemical systems and their relationship to central pain mechanisms remains a challenge for the future.

## Acute Versus Chronic Pain

Acute and chronic pain are the result of different pathologic conditions and cannot be explained by the same physiologic mechanisms. Acute pain is self-limiting and can be treated by establishing a diagnosis and then applying the appropriate medical or surgical treatment. On the other hand, chronic pain involves psychological, financial, and personal issues. Half of all patients with chronic pain are clinically depressed and many abuse drugs. The main concern of a patient complaining of phantom pain following traumatic amputation may be the loss of his or her job or the question of compensation for disability. Similarly, although everyone may understand what the diagnosis of cancer means to their life, the questions that are most often asked are, "Doctor, will I have pain?" and "How can I avoid it?"

### MANAGEMENT OF THE PATIENT WITH CHRONIC PAIN

The International Association for the Study of Pain, founded by John J. Bonica, is a worldwide organization of pain specialists. It publishes the journal *Pain*, and arranges symposia in the basic science, physiology, and medical treatment of pain. Bonica was also instrumental in the development of multidisciplinary pain clinics. The broader the approach, the more effective are these clinics.

### DEAFFERENTATION PAIN

Many patients who suffer from injury or disease that results in a physical or physiologic disconnection of the sensory input of some portion of their body from the CNS continue to complain of pain in the anesthetic region. This phantom sensation so common in amputees has been recognized by physicians since ancient times, yet we still have not discovered an adequate physiologic explanation. Similar painful sensations are commonly experienced by paraplegics, who, following their injury, may experience pain in their paralyzed limbs, and by patients suffering a traumatic brachial plexus avulsion with complete disconnection of the sensory as well as the motor roots, who experience severe pain in their useless arms.

Several explanations have been proposed to explain this phenomenon. It has been regarded as a a psychologic disturbance due to a change in the body image or as being due to the development of reverberating circuits in those parts of the CNS that represent the missing body part. Phantom pain has been designated as "central" in origin, because it seems clear that the loss of peripheral input alters the central physiologic processes. The painful sensations are not obliterated by the removal of the sensory cerebral cortex, but can be significantly reduced by lesions that involve the spinothalamic and reticular pathways, either at a spinal level or higher in the midbrain.

Currently, this pain is termed *the deafferentation pain syndrome*. Indeed, interruption of the sensory pathways at any level from the skin receptors to the cerebral cortex is capable of inducing pain. Wall believes that dysesthetic (deafferentation) pain is due to "not only a loss of pain input but actual degeneration so the spinal cord cells [are] free to act in a pathologic way." Biochemical and anatomical rearrangements may well occur in the dorsal horn or at higher levels of the CNS and lead to hyperactivity of those central neurons that are responsible for the transmission of pain but have been deprived of their peripheral input.

Deafferentation pain is usually unresponsive to the use of narcotics and other forms of medical treatment, but it can be treated surgically. Aware that the gate control theory of pain proposed that the large sensory fibers exhibited an inhibitory influence on the smaller pain fibers, Sindou studied the anatomic arrangement of these in the posterior roots as they enter the spinal cord, and concluded that the smaller pain fibers were grouped together ventrolaterally, while the larger proprioceptive fibers were situated dorsolaterally. He therefore sectioned the ventrolateral dorsal root fibers and made a small cut into the adjacent spinal cord. Patients suffering from spinal cord disease, with a combination of pain and spasticity, were relieved of their symptoms.

## Nerve Plexus Avulsion Injury

The incidence of pain following avulsion of the brachial plexus is high. Estimates vary between 20% and 90% of patients, many of them young males with a normal life expectancy. This type of deafferentation pain presents a unique anatomic problem. The trauma avulses the dorsal roots from the spinal cord, leaving the deafferented secondary neurons isolated in the deeper Rexed layers of the dorsal horn (Rexed layers I, II, and V). In experimental studies of avulsion injuries in animals, these secondary neurons become hyperactive within a few hours after injury, but at the same time, the behavior of the animal suggests the onset of pain.

Prior to the 1970s, cordotomy and midbrain tractotomy controlled the pain in a few patients. The therapeutic lesions in the spinal cord involved the spinothalamic tract, while in the midbrain, a combination lesion of the spinothalamic tract and the adjacent ascending reticular pathway was made. Some years ago, however, and based on the histologic evidence referred to earlier, the senior author decided to lesion the isolated Rexed layers of the dorsal horn using electrocoagulation. Later a special dorsal root entry zone (DREZ) electrode, having a thermocouple in its tip, was designed to produce radiofrequency (RF) thermal lesions. (The DREZ is that portion of the dorsal horn that contains the secondary neurons from Rexed zones I–V.) The lesion is made at a temperature of 75°C, maintained for a period of 15 seconds, and produces an area of coagulation that destroys the Rexed layers I through V, and hence the secondary neurons of the afferent pain pathways.

In the first group of patients with pain from brachial plexus avulsion injury, 60% were relieved, so we now recommend the DREZ operation if the pain does not respond to medical treatment within 6 months. Large series of DREZ operations for avulsion report 60% to 90% good relief of pain for periods of over 5 years. ("Good pain relief" here means little or no pain, no medication, and the resumption of a normal life within the confines of the patient's disability.) If the pain can be relieved within the first year, an attempt should be made to restore the function of the biceps muscle by transplanting an intercostal nerve into the motor nerve of the biceps. This may restore some flexion at the elbow. If, however, the avulsion injury is years old, amputation should be done, but only *after* the pain has been relieved.

Avulsion injuries involving the lumbosacral plexus are less common than brachial plexus avulsion and often go unrecognized; they may be treated by exposing the conus medullaris and performing the DREZ operation on the involved segments of the spinal cord. Usually, fewer nerve roots are avulsed in this group of patients than in those with brachial plexus avulsion, but our first patient with a lower limb avulsion had all the dorsal roots from L1 and S5 missing when the conus medullaris was surgically explored. Post-amputation phantom pain, but not painful amputation stumps, also can be relieved by the DREZ operation.

## Pain in Paraplegia

Paraplegic patients complain of a variety of pains, ranging from a diffuse burning in both legs and the perineum to pain of segmental distribution at or just above the level of the sensory loss. This is referred to as "end zone" pain; patients complain that touching the skin in the end zone region activates their pain and may cause electrical shooting pains that radiate into their paralyzed legs. This type of pain is thought to originate from the segment of the spinal cord just rostral to the point of injury to the spinal cord. Its pathophysiology is not understood completely, but it appears that injuries with local compression at the site of an acute vertebral fracture or dislocation cause less involvement of the adjacent segments of the spinal cord than those with the widespread disruption that is found in many cases of missile injury, where damage often extends several segments above the injury site. Pain in paraplegia occurs more often when the trauma involves the lower lumbar and sacral segments of the spinal cord than when the more rostral parts of the cord are injured.

The onset of the pain in the paraplegic is usually immediate, as is the case with other deafferentation pain syndromes. There is, however, a second group of paraplegics in whom the onset of pain is delayed, usually for years after the injury; 65% of them harbor a spinal syrinx.

If deafferentation pain persists 6 months after the injury in paraplegic patients, surgical treatment should be considered. Following the success of the DREZ operation in relieving the avulsion pain, it was decided to produce DREZ lesions in the dorsal horns of the segments of the spinal cord just above the injury site. The lesions are made (bilaterally if the pain is bilateral) for at least three to four spinal segments rostral to the injury. This procedure relieves end zone pain. It also raises the sensory level, but because the change most often involves the lower thoracic and abdominal dermatomes, no functional disability occurs. In a study of 120 paraplegics who had the DREZ operation, Friedman and Bullitt reported a relief rate of 78% (94/120), while Sampson et al. recently conducted a 3-year follow-up of 39 patients with painful lesions of the conus medullaris and upper cauda equina and found that 29 (74%) were relieved after the DREZ operation.

## POSTHERPETIC PAIN

Herpes zoster (shingles) is a common cause of intractable pain, especially in the elderly. Pain is an early symptom, often appearing before the cutaneous eruption. It is usually described as a burning and painful tingling in the involved region of the skin. When the skin lesions heal, leaving scattered areas of scarring, these may or may not continue to be painful. The reason why one patient's pain subsides while another's continues and becomes chronic is unknown.

Medical treatment should be started early, but if the pain persists for longer than 6 months, it is unlikely to resolve spontaneously. Surgical treatment has its limits and should be considered only when nonsurgical therapies have failed. Dorsal rhizotomy, performed when the herpes has involved one or more thoracic roots, has had limited success because only two or three dorsal roots were cut. Better results occur with the intraspinal sectioning of four to five roots.

Friedman and Bullitt performed the DREZ operation on a group of 32 patients with postherpetic pain, of whom 29 (90%) were initially relieved. Within 6 months, only 8 (25%) were pain-free, but 19 (60%) experienced milder pain. The burning and electrical shock–like pains were relieved, but some patients complained of deeper, aching pain. Twenty-two (69%) of Friedman and Bullitt's patients developed some residual ipsilateral weakness of the leg, of whom eight required assistance with ambulation. The patient must be advised of this risk prior to surgery.

We believe that those patients suffering from postherpetic burning and electrical shock–like pains who were relieved by the DREZ operation were suffering from deafferentation pain, while in those patients who were not relieved, the deep, aching pain may have been due to postherpetic changes in the peripheral nerves themselves.

## CRANIOFACIAL PAIN AND THE TRIGEMINAL NUCLEUS CAUDALIS

The sensory supply to the head and neck is complex due to the overlap between the cranial nerves (V,VII,IX,X) and the upper cervical nerves (C2,C3,C4). The trigeminal nerve supplies the anterior portions of the face, as well as the inside of the mouth, including the teeth, gums, and anterior two-thirds of the tongue and soft palate. Within the oral cavity, however, there is overlap with the lower cranial nerves, one or more of which may be involved with lesions in this region of the cranium.

Trigeminal neuralgia, or painful tic, is a common disorder that often responds to medication, but when this fails to give relief, the simplest surgical option is neurectomy of the branch of the trigeminal nerve supplying the painful region; the pain is usually relieved, but may return if the nerve regenerates. When larger regions of the face are involved, a percutaneous RF neurolysis of the trigeminal nerve at the level of the Gasserian ganglion is the most effective. Alternatively, the ganglion may be injected with glycerol, with the advantage that facial sensation is preserved. Intracranial vascular decompression of the trigeminal nerve as it exits from the pons also is often successful.

Anesthesia dolorosa (trigeminal dysesthesia) occurs in 2% to 4% of the patients who have undergone a standard trigeminal operation for tic pain (neurolysis or decompression). These patients complain of an area of skin on the face that is painful despite the loss of feeling. The pain is said to have a burning quality, unlike the original tic pain, and may involve one or more divisions of the trigeminal nerve. It conforms occasionally to an onion-skin pattern. We believe that this is a deafferentation pain syndrome because similar pain may be associated with central lesions (vascular infarcts, angiomas, tumors, and multiple sclerosis) of the trigeminal pathways, whether in the midbrain, pons, or medulla.

Trigeminal dysesthesia is severe and difficult to treat. Fifty years ago, Sjøquist, and much more recently Kerr, reported that the most caudal nucleus of the trigeminal nuclei received the majority of the pain afferents from the trigeminal system. Kerr therefore suggested that a lesion of this nucleus, made in a fashion similar to the DREZ lesions in the dorsal horn, might relieve facial pain, noting that the anatomy of the nucleus caudalis and the dorsal horn were similar and that the nucleus caudalis extended into the dorsal horn of the C-IV or C-V roots.

The nucleus caudalis is easily accessible at the cervicomedullary junction through a small suboccipital craniectomy and a hemilaminectomy of C-I–C-II. We have therefore made DREZ caudalis lesions for patients with anesthesia dolorosa and postherpetic facial pain since the 1980s. Altogether, 100 patients have been operated on, with the best results in persons with trigeminal postherpetic neuralgia. Good results also occur in patients with facial pain due to brain stem lesions of the trigeminal pathways (infarcts, angiomas, MS plaques), while fair-to-poor results occurred in patients with anesthesia dolorosa and facial pain resulting from traumatic lesions of the trigeminal nerve. The main complication is postoperative ataxia of the ipsilateral arm, which is seen initially in 30% of patients, but quickly diminishes and is not disabling.

## THE TREATMENT OF CANCER PAIN

Pain is an early and frequent symptom of cancer; it may be due not only to the cancer itself, but also to the effects of the various treatment modalities that have been used. For example, following ablative surgery, patients may suffer from radicular nerve pain (e.g., incisional pain, postthoracotomy pain, and localized pelvic pain), while women who undergo radiation treatments for cancer of the breast may develop intractable pain in the arm due to the late effects of the radiation on the brachial plexus. The decision to use surgical treatment should be considered at that point in treatment when strong narcotics appear to be required, particularly if it is considered that the patient may live more than a year.

### Nerve Blocks in the Control of Cancer-Related Pain

The use of temporary and/or permanent nerve blocks is often helpful in the control of cancer-related pain. Small cutaneous sensory nerves can be injected with alcohol, with good relief of pain, but the injection of alcohol or other destructive chemicals into large somatic nerves rarely results in complete analgesia, and may be followed by a secondary dysesthesia that distresses the patient more than the original pain. If the patient is suffering from pain of peripheral origin, blocking the involved nerves at the level of the vertebral foramen may predict the effectiveness of a rhizotomy and allows the patient some idea of the extent of the area of numbness that will follow a permanent nerve sectioning.

Anesthetic blocks of the major ganglia that supply the abdominal viscera are effective in the control of abdominal pain. Pain from pancreatic cancer is effectively relieved by a nerve block of the celiac ganglion with local anesthetic. The period of relief can be extended for up to a year by injection of alcohol, whereafter the injection can be repeated. Pelvic and perineal pain due to extensive pelvic cancer can be relieved by blocks of the superior hypogastric plexus and/or the ganglion impar or by the use of intrathecal alcohol or phenol. Intrathecal therapy may cause bladder or bowel dysfunction, but in the final stages of extensive pelvic and rectal cancer, those functions are often already compromised and the patient may be willing to sacrifice sphincter control for pain relief.

## Surgical Treatment of Cancer Pain

Pain resulting from cancer may be relieved by spinal cordotomy. This was originally performed through a laminectomy, but the lesions are now made by using a special electrode, carried inside a needle introduced under radiographic control, that delivers an RF current at its exposed tip. Stimulation is performed first, and the subsequent responses of the patient are used to determine the location of the needle tip before a lesion is made. The quality of pain relief after cordotomy can be predicted by the extent of analgesia produced by the spinal lesion. This operation can be done unilaterally or bilaterally, according to the distribution of the pain, but the risk of complications after bilateral cordotomy is not trivial. The mortality rate after bilateral cordotomy ranges from 2% to 11%, respiratory failure being the major cause of death after cervical cordotomy. Mild motor weakness occurs in 20% of patients (the incidence is 10% after unilateral cordotomy), and loss of bladder control is frequent. Initially, relief of pain occurs in 80% to 90% of patients, but diminishes to around 60% after 2 years. Pain from nerve and visceral involvement responds well, but bone pain is less well controlled.

To avoid the complications of bilateral cordotomy, a midline myelotomy can be performed, the spinal cord being sectioned longitudinally over several segments so as to interrupt the pain fibers crossing the central portion of the spinal cord. This reduces the incidence of motor deficits and results in bilateral pain relief that lasts from 2 to 5 years in 60% to 70% of patients. In patients with cancer who suffer from both pain and spasticity, the pain can be relieved by Sindou and Fobe's selective posterior rhizotomy and "drezotomy."

Cancer of the head and neck is difficult to treat because it often involves multiple cranial nerves, but two surgical operations are helpful in reducing pain. If the patient's condition contraindicates general anesthesia, a stereotactic mesencephalotomy can be done under local anesthesia through a small burr hole. The therapeutic lesions are made in the upper midbrain at the level of the superior colliculus and involve the medial trigeminothalamic tract and the adjacent reticular formation at the edge of the central gray matter surrounding the aqueduct of Sylvius. Bilateral relief of pain may result; even when this is incomplete, it is often possible to significantly reduce the intake of morphine, because the lesion in the central gray matter mimics the effect of a cingulotomy (discussion follows). Because the lesions are located near the superior colliculus, there may be loss of upward gaze, but this does not usually cause the patient a problem.

## Surgical Relief of Suffering

Stereotactic bilateral cingulotomy reduces not only the suffering of cancer patients, but also their need for narcotics. The physiologic mechanism responsible for this relief is poorly understood. The cingulum has connections with both cerebral hemispheres and is a functional part of the limbic system, which is thought to be associated with emotional states of mind, sending connections to the medial temporal lobe structures and the upper midbrain reticular formation. The reactive emotion to painful experiences is absent in postcingulotomy patients, but there are no changes in the patients' other emotions. It is of interest that their pain threshold remains normal, although they do not overtly complain of pain. In a recent report of the effects of cingulotomy, using stereotactic MRI guidance, 7 (58%) of 12 cancer patients experienced significant relief.

The introduction of opioids into either the ventricles or the subarachnoid space has been used to a limited extent for the relief of cancer pain, but many physicians who treat cancer patients believe standard medical treatment is more useful.

## Hormonal Manipulation for Pain Relief

Two forms of cancer are responsive to hormonal manipulation: cancer of the prostate in men and cancer of the breast in women. In 1953, Luft and Olivecrona performed the first hypophysectomies to relieve the severe metastatic bone pain associated with prostatic and breast cancers. Since then, techniques to destroy the pituitary gland have ranged from transnasal surgical removal, through stereotactic implantation of radioactive materials and RF lesions of the gland, to the intrasellar injection of absolute alcohol. Overall, 60% of the patients are relieved of their pain for periods of 6 months to a year. Some, who were bedridden with extensive bone involvement, have resumed limited physical activity. The physiologic mechanism responsible for relief of this pain is unknown. Originally, it was thought to be due to removal of the pituitary hormones, but current evidence implicates the hypothalamus and its relationship to the endorphins.

### CHRONIC PAIN DUE TO PERIPHERAL NERVE INVOLVEMENT

For many years, dorsal rhizotomy was the standard operation for pain originating in peripheral nerves. This procedure initially relieves the pain, but this relief diminishes, being replaced in some patients by a secondary dysesthetic pain of probable central origin. Nevertheless, dorsal rhizotomy continues to have a role in the treatment of the localized pain of cancer.

According to White and Sweet, secondary dysesthetic pain occurs because surgeons commonly section only one or two roots. By contrast, they recommend sectioning at least five dorsal roots, two above and two below the painful dermatome, and report that in their series of patients, 64% were relieved for as long as 10 years.

As proposed by White and Sweet, dorsal rhizotomy may be used to relieve pain due to postherpetic involvement of the chest and abdomen, incisional pain after thoracotomy or hernia repair, and radicular pain with specific dermatomal involvement. Pain due to lesions of peripheral nerves does not respond as readily to DREZ lesions as does chronic deafferentation pain.

## CHRONIC PAIN ASSOCIATED WITH SYMPATHETIC DYSFUNCTION

Anatomists have disagreed as to whether the sympathetic nerves contain pain fibers. If such fibers exist, they do not have the same anatomic features as the somatic nerve pain fibers. There are, however, a number of painful conditions in which activity within the sympathetic system seems to contribute to the syndrome (e.g., causalgia, reflex sympathetic dystrophy, the pain and suffering syndrome, certain types of visceral pain), and electrical stimulation of a sympathetic ganglion in an awake human during a surgical operation causes pain.

### Causalgia

The cause of the severe burning pain that sometimes follows a partial injury to a peripheral nerve (termed *causalgia* by Weir Mitchell) is unknown, although its clinical features suggest that it is related to a dysfunction of the autonomic nervous system. The key to its successful treatment is early recognition of the syndrome and early treatment. A sympathetic block should be performed, and if this relieves the pain, a sympathectomy will result in cure.

### Other Causalgia-like Syndromes

Other pain syndromes that originate from damage to peripheral nerves are somewhat similar to causalgia but are not relieved by sympathetic blocks. In these patients, the use of early transcutaneous electrical nerve stimulation (TENS) may be helpful, and for longer-term relief, the implantation of a peripheral nerve stimulator should be considered. The therapeutic use of electrical stimulation to relieve pain results from Melzack and Wall's gate control theory and from Wall and Sweet's demonstration that the skin supplied by a cutaneous nerve became analgesic when that nerve was stimulated electrically. Sweet used an implantable electrode system that was activated by an RF signal transmitted through the skin to a small receiver connected to an electrode placed on the damaged nerve. Sweet and Wepsic implanted such a device in a small group of patients with painful peripheral nerve injuries and obtained good relief of pain; their results have been confirmed by many other surgeons.

In 1982, Nashold et al. reported a 4- to 9-year follow-up of the use of TENS in a group of patients with a variety of painful peripheral nerve lesions, reporting a 52% rate of relief. The overall risks of this technique are minimal. It has the advantage that patients can control the electrical stimulation to the parameters that best relieve their pain.

## Reflex Sympathetic Dystrophy

Reflex sympathetic dystrophy (Sudeck's atrophy) is a rare, painful condition that involves somatic nerves and those of the sympathetic nervous system. Although there are chronic bone and skin changes similar to those of causalgia, sympathectomy does not cure the condition; psychotherapy and physical therapy, however, may be helpful. This complex pain disorder is far from understood, and treatment is unsatisfactory.

## Summary

Chronic pain continues to challenge physicians and society. It not only represents a personal tragedy for the individuals who are in pain, but also often destroys their personal and economic well-being. Although the medical community has always been the major contributor to attempts to solve this problem, it is important that insurance companies, industry, and above all, government become more involved in research efforts to improve the condition of the many people who suffer chronic pain.

## References/Further Reading

Cook WA, Kawakami Y. Commissural myelotomy. J Neurosurg 1977;47:1–6.

Foltz EL, White LE Jr. Pain "relief" by frontal cingulotomy. J Neurosurg 1962;19:89–100.

Friedman AH, Bullitt E. Dorsal root entry zone lesions in the treatment of pain following brachial plexus avulsion, spinal cord injury and herpes zoster. Appl Neurophysiol 1988;51:164–169.

Hassenbusch P, Pillay PK. Cingulotomy for treatment of cancer-related pain. In: Arbit E, ed. Management of cancer related pain. New York: Futura, 1993:297–312.

Jannetta PJ. Microvascular decompression for trigeminal neuralgia. Surg Rounds 1983;6:24–35.

Nashold BS Jr. Extensive cephalic and oral pain relieved by midbrain tractotomy. Confin Neurol 1972;34:382–388.

Nashold BS Jr, Goldner JL, Mullan JB, Bright DS. Long-term pain control by direct peripheral nerve stimulation. J Bone Joint Surg [Am] 1982;64:1–10.

Nashold BS Jr, Ostdahl RH. Dorsal root entry zone lesions for pain relief. J Neurosurg 1979;51:59–69.

Ovelmen-Levitt J. The neurophysiology of deafferentation syndromes. In: Nashold BS Jr, Ovelmen-Levitt J, eds. Deafferentation pain syndrome. Pathophysiology and treatment. New York: Raven 1991:103–124.

Sampson JH, Cashman RE, Nashold BS Jr, et al. Conus medullaris DREZ lesions for intractable pain after trauma to the conus medullaris and cauda equina. J Neurosurg 1995;82:28–34.

Sindou M, Daher A. A spinal ablation procedure for pain. In: Dubner R, Gebhart GG, Bord MR, et al., eds. Proceedings of the Fifth World Congress on Pain. Amsterdam: Elsevier, 1988:477–495.

Sindou M, Fobe JL. The role of rhizotomies and dorsal root entry zone lesions. In: Arbit E, ed. Management of cancer related pain. New York: Futura, 1993:341–368.

Stanton-Hicks M, Jänig W, Boas RA. Reflex sympathetic dystrophy, vol. 7. In: Stanton-Hicks M, Jänig W, Boas RA, eds. Current management of pain. Boston: Kluwer, 1990.

Tasker RR, Organ LW, Hawrylyshyn P. Deafferentation and causalgia. In: Bonica JJ, ed. Pain. Research publications: Association for Research in Nervous and Mental Disease, vol. 58. New York: Raven 1980:305–329.

Wall PD. The role of substantia gelatinosa as a gate control. In: Bonica JJ, ed. Pain. Research publications: Association for Research in Nervous and Mental Disease, vol. 58. New York: Raven 1980:205–231.

# Syndromes of Cranial Nerve Hyperactivity

Peter J. Jannetta

*Neurosurgery needs a few almonds in the rice pudding, and this is one of the almonds.*

(with acknowledgment to Garrison Keillor)

In this chapter, we discuss the basics of cranial nerve syndromes, presenting an historic perspective followed by a consideration first of why microvascular decompression works and then of when and why it does not work. The concept of cranial nerve vascular compression evolved slowly and painfully during the early years of neurosurgical conceptual development. It is not difficult to understand retrospectively these "fits and starts" nor the early lack of understanding of cranial nerve syndromes. Isolated observations were made, but they were usually partial and inadequate. They were reported, but they had no general context. They were not understood. They could not be synthesized. They were therefore not considered to have any great importance or to be of lasting significance. Is this an unusual occurrence? No. Are there valid reasons for this occurrence? Yes.

## Development of the Concept

The development of a concept, or sequence of observation, which is followed only after many years by synthesis and acceptance, is a normal scientific progression rather than an anomaly, and should be considered the product of two separate but related phenomena.

### PHENOMENON ONE

Phenomenon One is something that happens to ideas that are being explored in a technologically inadequate environment. It is important to understand that, to early neurosurgical observers, practically everything they saw and did was new and "a first." The first scientific descriptions of a cranial nerve affliction were probably of trigeminal neuralgia (TGN). This is a painful disorder, and it is no less painful, in another sense, to remember that until recently nothing could be done, either medically or surgically, to relieve the pain of those who suffered it. Consider TGN as an isolated phenomenon, as our forebears did. Once it became clear that TGN was a problem of cranial nerve V (C-V), rather than of C-VII, it was natural that the treatment modalities that were developed were symptomatic. Because the intact nerve appeared to be the source of the pain, the obvious solution was to interrupt it. Peripheral neurectomy, alcohol injections, and gasserian ganglionectomy were all used, although the latter was a high-morbidity procedure. Later, total and eventually differential partial section of the dorsal root of C-V via a temporal craniectomy–middle fossa exposure (the Spiller-Frazier procedure) became the operative procedure of choice.

Note that no pathologic changes were found in the nerve because, as is now known, although TGN was then considered a disease, it is really just a symptom. Nerve section, causing numbness, simply replaced one symptom (pain) with another more tolerable one (anesthesia). Nevertheless, the procedure was usually effective and relatively straightforward technically, and it had acceptable morbidity and mortality rates.

It also must be understood that in the earlier part of the twentieth century, operations in the cerebellopontine angle were dangerous. In 1925, when Dandy began his explorations of the trigeminal nerve in patients with TGN for purposes of dorsal root section, he was the only one who performed the procedure. He was isolated both intellectually and emotionally from the small power structure of American neurosurgery. His excellent technique enabled him to operate safely in the cerebellopontine angle. Due to the natural magnification of the operative field that resulted from the high myopia from which he suffered, he was able to see surgical details better than could other less myopic surgeons. Neurosurgeons who attempted the operation were daunted by an excessive complication rate, and they were unable to confirm the presence of the abnormalities (vascular, tumor, scar) that Dandy was able to see on the nerve and that he stated might be causing TGN.

It was not until 1959 that Gardner demonstrated that he also could operate safely and effectively in the cerebellopontine angle in patients with TGN. Gardner went beyond Dandy's contribution in two ways. First, he actually moved the arteries that were compressing C-V in some patients. Second, he brilliantly extrapolated the pathophysiology of TGN to hemifacial spasm (HFS) and treated some patients with HFS by mobilizing arteries away from the C-VII and C-VIII bundle. Brilliant? Yes. Accurate, complete, and convincing? No. As Dandy had sectioned C-V, Gardner deliberately traumatized C-V and C-VII in his procedure, calling it "neurolysis."

Following Dandy's work, the next development of importance involved a single case report in 1936 by Lillie and Craig. An 18-year-old woman was operated on for severe lancinating pain in her left ear. She also was deaf in that ear and was thought to have a cerebellopontine-angle tumor, causing glossopharyngeal neuralgia (GPN) of otitic origin. At operation, a hairpin-shaped artery was seen to be looping around and stretching the auditory nerve bundle, with the apex of the "hairpin" impacted into the distal part of C-IX. No tumor was found. The authors considered that the patient truly had GPN, so they freed the artery from C-IX, sectioned the nerve, and mobilized the arterial loop from the C-VIII bundle. The patient made an uneventful recovery; she was pain-free postoperatively and remained so until at least age 58, when contact with her was lost. Of further interest is that her hearing returned to normal over a 6-week period and remained so thereafter. In publishing this case, the authors described it as one of GPN caused by pressure on C-IX from the artery that they had visualized. To my knowledge, they never studied the problem again.

The erratic developmental sequence in the study of cranial nerve syndromes is representative of a common pattern of progression in scientific and medical discovery. If some of the factors that are operant in the process of discovery, and particularly in the development of neurosurgery as a surgical specialty, are scrutinized, we find that the early neurosurgeons were very conscious of being a cadre of pioneers. They dealt with the unknown, trying to clarify and solve problems that had been regarded as insoluble.

In their work, they occasionally, through necessity, maimed or killed their patients, even while valiantly doing their best to treat insurmountable problems with new methods. Surgical results were not good then. (There continue to be certain areas of neurosurgery, such as the treatment of some malignant intrinsic tumors of the CNS, in which we still have a long way to go.) Because of these poor results, patients did not want to go to neurosurgeons, and their doctors were frequently reluctant to refer them. When a consultation with a neurosurgeon was acknowledged to be inevitable, however, and the ensuing operative treatment was reasonably successful and had acceptable morbidity and mortality rates, it was generally used by all, leaving little room for innovation of or deviation from the standard methods. Dandy and Gardner, therefore, suffered because of their creativity.

## PHENOMENON TWO

The problems go deeper than those described for Phenomenon One. Phenomenon Two is that, when an idea comes before its time, in the sense that contemporary technology is not sufficient to test it to the point of proof or disproof, it lies fallow. This is what happened with the concept of vascular compression as a possible etiology of TGN (to Dandy, and subsequently to Gardner), of HFS (to Gardner), and of GPN (to Lillie and Craig). Inadequate observation, inadequate therapy, inadequate thinking—all arose from inadequate technology and, in this case, the lack of magnification. Without the use of the operating binocular microscope and audiovisual aids, the pathology was neither seen clearly nor treated properly. The definitive therapy, vascular decompression, was done either not at all (Dandy) or inadequately (Gardner), or was not pursued after a single experience (Lillie and Craig).

One can only ponder over the delay in the application of new technology. The use of the microscope in ear surgery was first described in 1923 by Nylen and was adopted by a growing number of otologists thereafter, but it did not enter clinical neurosurgical practice until the late 1950s, and even then it was not immediately reported. Such delays have occurred throughout the history of neurosurgery.

Other factors were operant as well. Investigators did not know how to interpret their findings. The Lillie and Craig patient most likely had ear pain due to vascular compression of the nervus intermedius, the same artery causing the hearing loss; we now know that the arterial compression of C-IX was too peripheral to cause GPN. Moreover, in its time, Lillie and Craig's finding seemed to be merely another "interesting case," and they never pursued it. Retrospectively, however, it is a most important one. Even in recent years, a few observers, equipped with all of the available technology, have declared themselves unable to see the pathology causing cranial nerve syndromes. It is not easy to avoid entertaining the possibility that, in some cases, ignorance and lack of ability are factors in this failing of understanding.

With the application of the surgical binocular dissecting microscope to cranial nerve surgery, and with the utilization of audiovisual equipment and the development of electrophysiologic evaluation and monitoring techniques, a corpus of studies has grown over the past several decades. Surgeons have found that the nuances of operative technique must be learned in a hands-on fashion. Microvascular decompression (MVD) can look deceptively easy in the hands of an experienced neurosurgeon, but some are uncomfortable doing the procedure. One must do it frequently to remain "slick." Surgeons who fail to learn all of the apparently petty details of the operation also fail to treat the vascular compression adequately, and have a high complication rate. In the hands of an experienced surgeon, however, who has learned the nuances of technique, MVD for the cranial nerve syndromes is safe and effective, and it offers the best quality of life with the fewest side effects of any operation for these intractable and disabling problems.

## The Common Basis of Cranial Nerve Syndromes

The hyperactive cranial nerve problems have much in common. They are each caricatures of the normal function of the cranial nerve. This means that a nerve that subserves somatic sensation develops a syndrome of paroxysmal pain; one that subserves somatic motor function presents with an abnormal movement disorder; and one with a special sensory function, such as hearing and balance, is susceptible to tinnitus and vertigo/disequilibrium. One can clearly see the origins of TGN, HFS, GPN, Meniere's disease and its subgroups, disabling positional vertigo, and spasmodic torticollis. Arteriovenous hyperactive dysfunction due to left lateral medullary compression also exists: It is neurogenic hypertension.

All of the nerves mentioned in this chapter are located in the pons and the medulla and are normally surrounded by arteries and veins. In the genesis of the cranial nerve syndromes, two processes are operant, both the result of normal aging. The first is arterial elongation and looping around the base of the brain. This begins early and is presumably biased by genetic predisposition and living habits (e.g., diet, activity, use of tobacco). We also presumably inherit the patterns of our parents' blood vessels (which start out having a certain diameter, length, and proximity to the cranial nerves), just as we inherit the shape of their noses, chins, mouths, and so on. Second, as people age, their brains sag caudally within the posterior fossa, augmenting the amount of neurovascular contact (which is the sine qua non of the condition) and so easily predisposing to vascular compression.

The possibility of vascular compression of nonfascicular neural tissue is the common denominator of cranial nerve syndromes. Pulsatile vascular compression, arterial or venous, is the culprit. The location of the compression has clear clinical-pathologic correlations. For example, rostral compression of C-V causes TGN affecting its third division; caudal compression of the same nerve causes ophthalmic-division neuralgia; and so forth. It is not yet understood why neurovascular compression is occasionally asymptomatic. Is there a predisposition in certain myelin? Is a prior change in myelin necessary before symptoms occur? Clinicians frequently see patients who have asymptomatic gallstones or disc disease.

It is not necessary that the offending blood vessels be of large diameter to cause symptoms. We have seen TGN, HFS, vertigo, and tinnitus caused by very small arterioles and venules (50–100 μm in diameter). These have been identified and treated successfully, both alone and in conjunction with larger vessels, in patients with intractable symptoms who had never improved after previous MVD of larger vessels, as well as in patients with recurrent symptoms following an initially satisfactory period of pain relief.

## Why Microvascular Decompression Works

As discussed, pulsatile vascular compression of nonfascicular nervous tissue, such as is found in proximal lower cranial nerves and adjacent pons and medulla, is the common denominator in the hyperactive cranial nerve symptoms. Careful examination of a specimen of the affected nerve in such a case reveals both morphologic and physiologic abnormalities. Morphologically, the majority of the nerve appears to be normal; this is evident when using both light microscopy and the electron microscope. Scattered throughout the normal nerve fascicles and myelin, however, are abnormalities of both axis cylinders and myelin. These consist of hypermyelination, hypomyelination, and large whorls of aberrant myelin. Defects of myelin are found with "missing" axis cylinders, which may possibly represent short circuits. Axis cylinders that are dying and others that are reforming also are seen. All of these electron microscopic changes have been identified in cranial nerves V, VII, IX, and X of patients who suffered from the previously noted syndromes. Some evidence exists that such changes also may be found in the aged in the absence of symptoms of hyperactivity.

Electrophysiologic evidence of abnormalities of nerve conduction in patients with hyperactive cranial nerve symptoms abounds in the literature. Trigeminal evoked potentials are abnormal in 86% of patients with TGN and generally revert to normal after MVD. Careful sensory examination of the face shows abnormality in 25% to 30% of patients with TGN in the area of their pain, and this commonly reverts to normal after MVD. In HFS, strong evidence exists for both ephaptic transmission and facial motor nucleus dysfunction. In patients with C-VIII syndromes, clinical physiologic abnormalities consist of hearing loss and tinnitus, brainstem evoked potential abnormalities, and abnormal vestibular function tests. These also are reversible following successful MVD.

One must consider that neural tissue—when stimulated by pulsatile vascular compression—shows evidence of hyperactive dysfunction, with a dynamic correlate seen both morphologically and physiologically. It is not known if the electron microscopic changes are reversible, but the physiologic changes definitely are. Microvascular decompression works by getting rid of the primary cause (vascular compression) and allowing the secondary neural abnormalities caused by the vascular compression to heal over time. This explains why many of the symptoms and abnormalities associated with hyperactive cranial nerve dysfunction return to normal very slowly, taking weeks, months, or (in the case of tinnitus and spasmodic torticollis) even years.

## Why Does Microvascular Decompression Fail?

Failure of MVD is of two types: (1) a lack of relief and (2) relief followed by recurrent symptoms. Lack of relief implies that the vascular compression of the nerve has not been completely relieved. Reoperation is indicated. If symptoms are lessened but not gone, one can adopt an expectant policy, because, as previously indicated, some symptoms take time to ameliorate and the patient has a good chance of improving spontaneously.

Recurrence of symptoms implies that a blood vessel has moved back into contact with the nerve. Early recurrence after MVD for TGN is most commonly seen in patients who had pontine surface veins causing the TGN, or contributing to it. These veins tend to recanalize or collateralize over 3 to 6 months. Late recurrences are, fortunately, uncommon; for instance, only 1 in 200 TGN patients experiences a recurrence of pain, once 3 years have elapsed since MVD. The cause of such late recurrences is continuing elongation of the arteries at the base of the brain, combined with brain "sag," thus giving rise to new vascular compression. Scarring alone does not appear to cause recurrence but may bring the nerve into contact with a blood vessel by traction.

In summary, scientific progress in the development of new concepts and paradigms grew from an apparently disorderly, but probably orderly sequence of testing, dependent to a large degree on the state of contemporary technology, before becoming generally accepted. The sequence is difficult but necessary, slow but sure, and always worth it.

## References/Further Reading

Adams CB, Kaye AH, Teddy PJ. The treatment of trigeminal neuralgia by posterior fossa neurosurgery. J Neurol Neurosurg Psychiatry 1982;45:1020–1026.

Bennett MH, Jannetta PJ. Trigeminal evoked potentials in humans. Electroencephalogr Clin Neurophysiol 1980;48:517–526.

Dandy WE. Section of the sensory root of the trigeminal nerve at the pons: preliminary report of the operative procedure. Bull Johns Hopkins Hosp 1925;36:105–106.

Dandy WE. An operation for the cure of tic douloureux. Partial section of the sensory root at the pons. Arch Surg 1929;18:687–734.

Dandy WE. The treatment of trigeminal neuralgia by the cerebellar route. Ann Surg 1932; 96:787–795.

Gardner WJ, Miklos MV. Response of trigeminal neuralgia to "decompression" of sensory root: discussion of cause of trigeminal neuralgia. JAMA 1959;170:1773–1776.

Gardner WJ. Concerning the mechanism of trigeminal neuralgia and hemifacial spasm. J Neurosurg 1962;19:947–958.

Jannetta PJ. Arterial compression of the trigeminal nerve at the pons in patients with trigeminal neuralgia. J Neurosurg 1967;26:159–162.

Jannetta PJ. Observations on the etiology of trigeminal neuralgia, hemifacial spasm, acoustic nerve dysfunction and glossopharyngeal neuralgia. Definitive microsurgical treatment and results in 117 patients. Neurochirurgia (Stuttg) 1977;20:145–154.

Jannetta PJ. Microsurgery of cranial nerve cross-compression. Clin Neurosurg 1979;26:607–615.

Jannetta PJ. Cranial nerve vascular compression syndromes (other than tic douloureux and hemifacial spasm). Clin Neurosurg 1981;28:445–456.

Jho HD, Jannetta PJ. Microvascular decompression for spasmodic torticollis. Acta Neurochir (Wien) 1995;134:21–26.

Kurze, T. Microtechniques in neurological surgery. Clin Neurosurg 1964;2:128–137.

Lillie HJ, Craig W. Anomalous vascular lesion in cerebellopontine angle. Arch Otolaryngol 1936;23:642–645.

Nylen CO. The microscope in aural surgery, its first use and later developments. Acta Otolaryng (Stockholm), 1954;suppl. 116:226–249.

# 8 Advances in Epilepsy

R. Eugene Ramsay
Bryan Philbrook

During the past decade, significant advances in our understanding and management of epilepsy have been realized in several different areas. These include expanded use of EEG/video monitoring, new medications that function differently from older agents, and an expanded knowledge of the mechanisms underlying seizures, which allows us to more logically approach the pharmacologic treatment of epilepsy. This chapter presents both basic and clinical aspects of seizures and epilepsy that can be helpful to physicians involved in patient management.

## Mechanisms of Epilepsies

A seizure is the subjective and objective result of abnormal, uncontrolled discharge of cortical neurons. Epilepsy is a syndrome of recurrent seizures, the primary etiology or pathology of which resides within the CNS. It is of prime importance to distinguish not only the type of seizure but also the type of epilepsy that is present. The presently used classification of seizures and the epilepsies was drafted by the International League Against Epilepsy.

Clinically, the epilepsies have been divided into two basic types: 1) partial onset and 2) generalized seizures. This is an important distinction because the mechanisms of epileptogenesis in this dichotomy are probably dissimilar, although surely not unique. The intricacies of the cellular mechanisms underlying the various seizure types are not completely known, but certain similarities have been observed. A discussion of epileptogenic mechanisms is necessary for a basic understanding of the pharmacologic approach to treatment.

### BASIC PRINCIPLES OF PARTIAL SEIZURES

At least two theories have been proposed to explain epileptogenesis at the cellular level. One model embodies the concept of the *epileptic neuron*. According to this construct, the neuronal membranes within the epileptic focus are altered, allowing abnormal spike generation, which ultimately serves as a pacemaker of the focus. The second theory assumes that cellular structure is normal, but the *epileptic aggregate* is formed by a neuronal pool with abnormal circuitry. Proponents of this theory argue that only a group of neurons acting together is capable of self-sustained rhythmic activity.

The thumbprint of epilepsy, the paroxysmal depolarizing shift (PDS), was first described by Matsumoto and Ajmone-Marsan. Intracellular recordings demonstrate abrupt-onset, long-duration, large-amplitude depolarizations with a tendency to superimposed bursts of action potentials. The PDS ends with a slow repolarizing deflection and is typically followed by a hyperpolarization that is even slower to return to baseline. Increased synaptic drive mediated by recurrent excitatory interneurons is probably responsible for the PDS, the extracellular correlate of which is the temporally related *cortical paroxysm* or spike. The EEG records summated excitatory postsynaptic potentials (EPSP) and inhibitory postsynaptic potentials (IPSP). Many features of the PDS are identical to characteristics of the EPSP. Similarly, the cortical spike is the result of many neuronal transmembrane potentials (PDSs) within a focus and is not simply the summation the action potentials. Theoretically, all cortical structures have the necessary properties to generate epileptiform activity, but sub-populations (e.g., hippocampus) may have a greater propensity due to structural arrangements favoring recurrent excitation.

The theories of both the epileptic neuron and the epileptic aggregate are credible. The existence of two mechanisms may be the basis for the difference between partial and generalized epilepsies. It is also conceivable that in the generation of interictal discharges and the eventual evolution of a seizure, both play a role in which one relies on the other. The elemental component of a seizure is the interictal discharge, but the factors producing interictal discharge and ictal event may be different.

The development of the interictal discharge relies on the epileptic substrate's susceptibility to excitation and ability to induce synchronization. These properties are mediated by both excitatory and inhibitory activities, but in general the evolution of epileptic phenomena is determined by a combination of three factors: excitation of the neuron, loss of inhibition, and synchronization.

## Excitation

Intrinsic neuronal membrane properties permitting the generation of bursts of action potentials must be present for excitation to occur. These phenomena are primarily mediated by ion conductances, especially inward flux of sodium and calcium and outward flux of potassium. Endogenous forces, such as a genetic predisposition, and environmental factors, such as trauma or ischemia, can perturb the normal organization of the neuronal membranes, converting non-bursting neurons to potentially epileptogenic populations. Certain populations of cells possess this bursting quality endogenously (e.g., specific neurons within the hippocampus and neocortex), while other populations may take it on only under pathologic conditions.

The cell properties and cellular organization of the hippocampus and neocortex make them particularly sensitive to the development of epileptic discharges. Direct excitatory connections between pyramidal cells in the neocortex and specific subcortical and interneuronal connections promote recurrent excitatory loops. Therefore, both local and more remote influences enhance excitation. Repeated excitation may decrease the response of inhibitory neurons, further enhancing excitation. The major excitatory neurotransmitter of the CNS is glutamate, but aspartate also plays an excitatory role. Electrically fired (ephaptic) synapses also have been hypothesized.

## Inhibition

Inhibition exists in feed-back and feed-forward arrangements. Afferent axons stimulate both cells in the excitatory path and inhibitory interneurons. Inhibitory impulses are conducted rapidly to target cells, terminating excitation downstream, thus facilitating feed-forward inhibition. Feed-back inhibition is also mediated by interneurons. The afferent volley is conducted by collaterals to interneurons, which then project back, halting further activity. The principal inhibitory neurotransmitter in the brain is gamma-aminobutyric acid (GABA). Loss of inhibitory mechanisms contributes to epileptogenesis, and there is evidence that inhibitory interneurons may be especially susceptible to injury.

## Synchronization

Synchronization must be coupled to excitation for the epileptic event to materialize. The predominate mechanism facilitating synchronization appears to be synaptic excitation. A neuron with burst generator capability and feed-forward connections could serve to synchronously drive multiple neurons in the focus (the epileptic neuron influencing the epileptic aggregate). Thus, divergent afferents serve to drive a larger neuronal pool and synchronize the discharge. For example, a single excitatory afferent synapsing within the thalamocortical relay nucleus can have diffuse influences across the cortex. Presynaptic alterations in ion conductances also may have a role in synchronization, and calcium influx may enhance release of neurotransmitters, provoking the PDS.

## Interictal-Ictal Transition

An examination of cellular events progressing from the occurrence of the PDS to an actual seizure reveals complete loss of after-hyperpolarization. Furthermore, the PDS is produced more frequently, its duration is often longer, and its amplitude is larger. After-depolarizations develop following the PDS, and a relative state of constant depolarization ensues with regular small potential shifts that oscillate rhythmically. These phenomena correlate with the tonic phase of a seizure. During the clonic phase, the oscillations become less frequent, more irregular, and of greater amplitude. Long-lasting hyperpolarizations, evolving during the clonic phase of a seizure, may play a role in seizure arrest.

Conditions that favor the loss of intrinsic control mechanisms, prolonged depolarizations, and synchronization promote the transition from an interictal event to a seizure. The factors that are required for interictal discharges also may encourage propagation to an ictus. Changes in the concentrations of ions in the extracellular and intracellular environments are thought to influence the burst generators and synchronization. Cholinergic activity also may play a role in the development of a seizure by inducing transient slow depolarizations and increases in membrane resistance, and by promoting burst discharges in certain cell populations. Discharges may be propagated across vast cerebral areas due to the rich connections between excitable cells via axonal arborizations. Ultimately, the clinical manifestations of the seizure depend on the cerebral area involved.

### Nosology of Partial Seizures

A partial seizure begins in one area of the brain, gradually spreading to adjacent structures, and often generalizing to involve the entire brain. The mesial part of the temporal lobe is sensitive to injury and is most prone to develop an epileptic focus. Seizures originating from the temporal lobe were originally called *temporal lobe* or *psychomotor* seizures. In 1981 and 1985, classifications of seizures and epilepsies were proposed which used the terms *simple* and *complex partial seizures*. Simple partial seizures are those in which the patient recalls all the events that have occurred, while during a complex partial seizure (CPS), the patient is amnestic for these. The change in terminology was prompted by knowledge that not all seizures with hallucinations and disturbed cognitive perceptions are of temporal lobe origin, and that loss of memory requires bilateral dysfunction, whether frontal or temporal.

Seizures of frontal lobe origin are quite different than those involving the temporal structures (Table 8.1). Mesial temporal seizures typically begin with an aura (e.g., rising abdominal sensation, feeling of fear, or olfactory hallucination), which lasts 15 to 30 seconds (Table 8.2). Following this, memory is disrupted and the patient has no recall of what subsequently happens. Ongoing activity ceases and the patient appears to stare off into the distance. Automatisms frequently occur at this stage and may consist of repetitive rubbing of the hands, movements of the mouth, swallowing, talking, or ambulating. After 1 to 2 minutes, the seizure ceases, and in the immediate postictal period, the patient may be confused. Memory and orientation usually return to normal within 5 to 10 minutes. Seizures from the lateral temporal cortex often begin with auras consisting of elemental auditory hallucinations (hearing a buzzing sound or noise). Consciousness may be retained longer, and ictal motor activity may include bilateral clonic facial and hand movements. Later, consciousness is disturbed and automatism may be noted. Although the beginning symptoms

**Table 8.1**  Seizure Characteristics by Location of Focus

|  | Temporal | Frontal |
|---|---|---|
| Duration | 1–2 min | 15–30 sec |
| Frequency | 1/wk | >1/day |
| Aura | Olfactory | Nonspecific |
|  | Fear |  |
|  | Abdominal sensation |  |
| Automatism | Simplistic | Complex/bizarre |
|  | Reactive | Reactive |
|  | Stereotyped | Less stereotyped |
| Verbalization | Simple | Complex/explicative |
| Consciousness | Lost | Preserved |
| Postictal | Confused | No deficit |
| Occurrence | Any time | Often nocturnal |
| Precipitates | Multiple | Emotional stress |
| Response to AEDs | Good | Fair |

may differ, the later ictal and postictal findings are similar to seizures of mesial temporal onset.

Frontal lobe seizures often have a bizarre appearance; patients are frequently considered to have pseudoseizures. The seizures may be very brief, containing complex bizarre motor acts (even manipulations of the genitalia), and occur more frequently at times of emotional stress (when out in public, on a date, arguing with a spouse, etc.). The motor movements are more complex and have an appearance that suggests they are volitional. As an example, one patient was observed to "awaken" from sleep, sit up in bed, grab his pillow, and strike it on the bed and floor. Verbalizations are common and also may be complex, thus appearing to be volitional. The same patient was then noted to lay back in bed and sing two lines of "He's Got the Whole World in His Hands." This is in contrast to the relatively simple automatism (rubbing of the hands) or verbalization (e.g., "I'm OK, I'm OK, I'm OK. . . .") associated with temporal lobe CPS.

Whereas during a temporal lobe CPS, the patient does not recall the repetitive or automatistic behavior, seizures originating in the frontal motor association area produce complex posturing while the patient retains consciousness, although clouded or disturbed consciousness may occur (Fig. 8.1). Posturing with frontal seizures may involve one, two, three, or all limbs and involves contortions rather than the tonic extension or flexion seen when the motor cortex is involved. The overall pattern of frontal seizures is stereotyped but may have more intraseizure variability than temporal lobe seizures. As a result, many seizures may need to be studied to detect the overall stereotyped behavior and thus establish the diagnosis of epilepsy.

**Table 8.2**   Clinical Manifestations of Seizure by Location of Focus

**Temporal**
  Mesial (hippocampus/amgydala)
    Olfactory hallucination, feeling of fear, rising
      abdominal sensation. Aura (15–30 sec)
    Disturbance of consciousness
    Automatisms
  Lateral neocortex
    Aphasia
    Elemental auditory hallucination
    Longer aura
    Disturbance of consciousness
    Automatisms
**Frontal**
  Lateral mid (pre-motor area)
    Contralateral head and eye deviation
    Fencing posture
    Secondarily generalized tonic-clonic
  Lateral posterior
    Focal motor/sensory
    Jacksonian march
    Secondarily generalized tonic-clonic
  Midline, posterior (motor association area)
    Brief (5–10 sec)
    Nocturnal
    Clusters, 10–30/day
    Little or no disturbance of consciousness
    Elemental vocalizations
    Four-limb posturing
  Midline, central (cingulate)
    Clouded consciousness
    Duration 15–60 sec
    Complex automatisms
    Axial movements
    Head side to side
    Nocturnal
    No postictal symptoms

  Midline, anterior (frontal pole)
    Autonomic signs
      Incontinence
      Tachycardia
      Urge to urinate, or incontinence
    Abrupt loss of consciousness
    Motionless stare
    Secondary generalization
  Orbital
    Clouded consciousness
    Autonomic signs
      Salivation, excessive
      Tachycardia
      Flushing
    Olfactory hallucination
    Secondary generalization
**Parietal**
    Jacksonian or hemisensory sensation
    Speech arrest
    Disturbance of consciousness
    Automatisms
    Duration 1–2 min
**Occipital**
    Enlarging ball of light
    Disturbance of consciousness
    Automatisms
    Duration 1–2 min

The duration of frontal seizures varies from a few seconds (posterior medial frontal cortex) to minutes (orbital frontal cortex). Symptoms arising from the frontal polar and orbital cortex may include excessive salivation, feeling hot, an urge to urinate, and speech arrest. Salivation may be seen with temporal lobe seizures but is often more profuse with orbital frontal seizures. Consciousness is usually clouded but not completely disrupted. Patients often report that they can hear and understand what is said to them but cannot

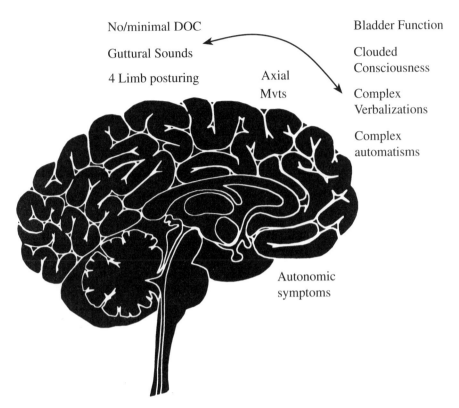

No/minimal DOC

Guttural Sounds

4 Limb posturing

Axial Mvts

Bladder Function

Clouded Consciousness

Complex Verbalizations

Complex automatisms

Autonomic symptoms

**Figure 8.1**

The characteristics of mesial frontal seizures gradually change, depending on the relative posterior or anterior location of the focus. Posterior mesial seizures are brief, with posturing, and become longer and with more complex motor and verbal patterns on involvement of more anterior areas. Autonomic function becomes evident with more anterior, polar, and orbital cortical involvement.

respond. Nocturnal seizures are frequent and, because of the complex behavior that is often present, may be difficult to differentiate from parasomnia.

A frontal lobe focus may be deep within the midline fissure, involving the motor association area or cingulate gyrus, or in the orbital cortex. These areas are remote to the electrodes used in routine EEG; hence, interictal or even ictal activity may not be seen with surface recordings. Recording from the midline (Fz and Cz) electrodes will sometimes reveal the build-up of rhythmic activity associated with a frontal seizure. Making prolonged recordings after discontinuation of anticonvulsant medications, using special montages, and having a high index of suspicion may be the only ways to diagnose some patients with seizures of frontal lobe origin.

Parietal seizures may manifest by sensory phenomena (tingling or numbness), speech arrest, or a perceived distortion of body parts (see Table 8.1). Scintillating scotoma or an enlarging light or fireball is often associated with occipital-onset seizures. Patients with a focus in the parietal occipital juncture may afterward report complex and very colorful imagery, yet at the time speak only of vague shadows. One patient always reported seeing cartoon characters at seizure onset. An aura occurred during examination, and he said he perceived only formless and colorless shadows, yet after the seizure, he again said he had seen the cartoon characters. Both the occipital and parietal lobes have rich connections into the temporal structures. Seizure activity will spread quickly from occipital or parietal areas to the temporal lobe, and the ictal manifestations are then similar to those of temporal lobe onset.

## BASIC PRINCIPLES OF PRIMARY GENERALIZED SEIZURES

The primary generalized epilepsies are characterized by interictal and ictal discharges that occur diffusely over both hemispheres, although there is usually a frontal predominance to the activity. This categorization includes primary generalized convulsive (tonic, tonic-clonic, clonic-tonic-clonic), nonconvulsive (absence and atonic), and myoclonic seizures. Except during the latter, and perhaps also during atonic spells, consciousness is impaired as epileptic activity occurs diffusely across the cortices.

Considering the diverse clinical manifestations of the generalized epilepsies, the underlying mechanisms are probably not exactly the same. However, the basic requirements of an excitable cellular membrane capable of burst potentials, impaired inhibition, and processes leading to synchronization and spread are necessary for epileptogenesis.

Generalized convulsions often begin on EEG as a rhythmic low-voltage, fast activity that evolves to high-voltage spike or polyspike discharges, these two phases correlating clinically with the tonic followed by the clonic phase of a convulsive seizure. These findings also have been explained by the experimental models of the PDS described earlier, and in general are thought to be the results of disinhibition and enhanced excitation. The mechanisms by which the entire cortex is activated have not been elucidated beyond doubt, but recent theories on generalized absence seizures may provide some insight into these processes.

The EEG signature of absence epilepsy is the bilaterally synchronous three-per-second spike and wave discharges that occur spontaneously out of a normal background and return immediately to normal background activity after the paroxysm. These correlate clinically with an alteration in consciousness without pre- or postictal symptoms.

The pathogenesis is thought to be an anomalous pacemaker within the brainstem or diencephalon, on the basis that there is an anatomic substrate with diffuse cortical projections. Experimental animal models have produced similar epileptiform activity by direct stimulation of the intralaminar nuclei of the thalamus, midbrain reticular formation, or both. Subsequent models have suggested that aberrant thalamocortical rhythms underlie the generation of the spike and wave activity. Neurons in the thalamus can shift between oscillatory and tonic firing modes, and level of arousal depends on the mode of firing.

This neuronal function is thought to be related to low-threshold calcium currents. Some investigators propose that the dysfunction present in absence epilepsy is related to mechanisms mediating the formation of spindles and the recruiting response. Ethosuximide (ESM) and trimethadione, which are antiepileptic drugs (AEDs) effective in absence epilepsy, decrease low-threshold calcium currents, and probably exert an effect by modulating the ability to accomplish the shift between bursting and tonic activities. The neuronal circuits involved in the spike and wave discharges have been proposed to be the same as those involved in generation of normal sleep, which may explain why some epilepsies occur most frequently during sleep.

In contrast to theories of generalized tonic-clonic (GTC) activity, in which a failure of inhibitory mechanisms or enhanced excitation has been suggested, generalized absence seizures are likely to be mediated by enhanced inhibition and hypersynchronization; the observation that enhanced GABA inhibition potentiates absence spells is of particular interest. Benzodiazepines and gamma-vinyl GABA (vigabatrin) exert an antiepileptic effect by enhancing inhibition.

## Nosology of Primary Generalized Seizures

The principal difference between generalized and partial epilepsies is in the *onset* of seizure activity. Partial seizures begin in a small cortical area and often produce symptoms at seizure onset. Aberrant thalamocortical activity underlies the generalized epilepsies. The thalamus has widespread cortical connections, especially to the frontal lobes. An abnormal discharge within these circuits disrupts attention and memory at the onset of the seizure. The patient feels no aura, although prior to development of a GTC seizure, short-duration spike wave discharges may be evident on the EEG and may be perceived by some patients as a warning; this prodrome is nonspecific and distinct from the more well-formed auras of partial onset seizures. After their onset, the clinical manifestations of secondary generalized and primary generalized seizures are similar. Convulsive seizures have tonic and/or clonic phases, which may be helpful in classifying the epilepsy. Generalized tonic seizures are seen more often with Lennox-Gastaut syndrome (secondarily generalized epilepsy) and in frontal lobe epilepsy, while clonic-tonic-clonic seizures are much more frequently seen in the primary generalized epilepsies.

Focal and generalized nonconvulsive seizures are usually quite distinct clinically. Absence (petit mal) seizures have no aura, last 5 to 15 seconds, and may include eye blinking or mild myoclonus. They have no postictal symptoms and have a generalized spike-wave pattern on EEG. Complex partial seizures have a characteristic aura, last 1 to 2 minutes, and produce postictal confusion. While blinking of the eyes or simple movements may be seen during a petit mal seizure, the automatisms of the partial epilepsies are more complex.

The only difficulty in clinically distinguishing absence and complex partial attacks occurs with seizures originating in the mesial frontal cortex. These may be very brief, and the EEG may show a bilateral epileptiform pattern. A well-formed, regular, single spike-wave EEG pattern is usually seen with the primary generalized seizures, while those of partial onset usually have an irregular or multiple spike-wave patterns.

## Medical Therapy of the Epilepsies

### ESTABLISHED DRUGS: EFFICACY IN PARTIAL SEIZURES

Many studies have evaluated the efficacy of various AEDs in partial seizures: A two-center study on monotherapy in previously untreated patients with epilepsy was conducted in England, sponsored by the Medical Research Council. A total of 410 patients was randomized to receive either carbamazepine (CBZ), phenytoin (PHT), phenobarbital (PB), or valproic acid (VPA). At 2 years of follow-up, 336 (82%) of the patients had achieved a 6-month seizure remission, and 246 (60%) had attained 1-year remission. No difference in efficacy was found between CBZ, PHT, PB, and VPA in the control of either partial or generalized seizures. Pediatric patients had a disproportionate incidence of adverse effects with PB.

In the United States, two large sequential, randomized double-blind trials on patients with only partial onset epilepsy have been completed. Patients were stratified by predominant seizure type into those experiencing mainly CPS and those who had predominantly generalized convulsive seizures. More patients continued to take CBZ and PHT than primidone (PRM) and PB. No difference in control of GTC seizures was found between CBZ and VPA, but by several measures (percent seizure free, percent continuing on therapy, dropout rate, etc.), CBZ was found to be better than VPA in the control of partial seizures. The Medical Research Center study did not show this difference, likely because of the smaller number of patients. The conclusions from these studies are that CBZ, PHT, and VPA are drugs of equal effectiveness in secondarily generalized tonic-clonic seizures. In CPSs, CBZ and PHT are the drugs of choice, but VPA has a significant anticonvulsant effect and should be considered as a first alternative. Nowadays, PRM is seldom used for routine therapy for partial or generalized seizures.

### ESTABLISHED DRUGS: EFFICACY IN GENERALIZED EPILEPSIES

The major AEDs used in the treatment of primary generalized nonconvulsive (absence) seizures have been ESM and VPA. Controlled trials of VPA versus

ESM in absence seizures demonstrated that 50% to 80% of patients were seizure-free with either drug. Overall, no difference in efficacy was evident, but consideration must be given to the side-effect profile. A high incidence of hepatic dysfunction was found with the use of VPA in children under the age of 2 years, so VPA should be avoided in this age group unless no alternative exists.

Valproic acid is usually considered the drug of choice in primary generalized epilepsies. This is based predominantly on clinical experience without the support of controlled clinical trials. For example, VPA is considered to be the drug of first choice in juvenile myoclonic epilepsy, yet no comparative studies have been conducted. Only one study was found, which included only patients with primary generalized convulsive seizures. At 6 months, more patients were seizure-free on valproate (64%) than on phenytoin (54%), but the difference was not statistically significant. Although CBZ is effective in secondarily generalized convulsive seizures, it has not been studied in the primary epilepsies. Because myoclonus may be seen with higher levels of CBZ, its efficacy in primary GTC seizures remains undefined.

## NEW ANTI-EPILEPTIC DRUGS: EFFICACY, PHARMACOKINETICS, AND SAFETY

### Felbamate

Felbamate (FBM) was the first of a new line of AEDs to be approved in the United States by the Food and Drug Administration. It is a broad-spectrum anticonvulsant and very effective in both generalized and partial onset epilepsies. By July 1994, after approximately 100,000 patients had been exposed to FBM, a significant incidence of bone marrow suppression and chemical hepatitis became evident. Of the total reported occurrences of aplastic anemia (n = 32) in the series, fatalities attributed to FBM occurred in 10 patients. Exposure time ranged from 23 to 339 days, and the dosage varied. Of 20 patient deaths from liver failure, only 6 (30%) were felt to be secondary to FBM. From the available information, the overall risk of fatality with FBM has been estimated to be somewhere between 1 in 8000 to 1 in 12,500 patients treated. As a result, FBM therapy was discontinued on most patients, but it should still be considered in patients with frequent seizures that are medically refractory. In some individuals, the risk of serious adverse drug effects may be less than the risk of the epilepsy.

### Gabapentin

Gabapentin (GPN) shows significant promise in the treatment of patients with partial and secondary GTC seizures. Three large randomized, multicenter, double-blind, placebo-controlled, parallel-group clinical trials have established the efficacy and safety of GPN as add-on therapy in patients with refractory partial seizures. Overall, the rate of response (defined as a 50% reduction in seizures from baseline) ranged from 17.6% to 28.0% in patients receiving

600–1800 mg daily. The median seizure frequency decreased by 17.8% to 31.9% in patients treated with GPN, compared with 0.3% to 12.5% in patients receiving placebo. A dose-response relationship exists for simple partial, complex partial, and GTC seizures; better control was evident with higher doses. Monotherapy studies of GPN are underway; in open clinical use, good seizure control has been reported. The initial recommended dosages were 900–1800 mg/day. Since then, much higher doses (2400–6000 mg/day) have been used, particularly in refractory patients, and improved efficacy has been reported.

Gabapentin (Neurontin, Parke-Davis, Morris Plains, NJ) was designed to be a structural analogue of GABA. It is a small molecule, recognized by the body as an amino acid. Absorption from the GI tract and transport into the brain and neurons is dependent on an active L-amino acid transport system. Time to maximum plasma concentration (Cmax) with oral dosing is 2 to 3 hours, and the plasma half-life ($T_{1/2}$) ranges from 5 to 7 hours. Gapabentin has a linear relationship of dose and Cmax up to dosages of 600 mg, and its bioavailability is approximately 60%. In open use of GPN, doses above 4800 mg/day have yielded progressively higher plasma levels.

Gabapentin elimination and $T_{1/2}$ are directly related to renal function as measured by creatinine clearance, an estimate of which ($C_{cl}$) can be calculated by the following equation:

$$CL_{Cr} = \frac{(140 - age) \cdot Wt}{72 \cdot S_{Cr}}$$

where weight is in kilograms and $S_{Cr}$ is the serum creatinine. As renal function declines with age, the total dose needed will be lower in patients over the age of 60. The half-life of GPN also becomes longer, so a twice-daily dosage is sufficient.

Because GPN is not metabolized in the liver, it does not induce or inhibit the mixed-function oxidase enzymes. Thus, GPN does not affect the pharmacokinetics of CBZ, CBZ-epoxide, PHT, PB, VPA, FBM, or lamotrigine (LTG). Gabapentin also is not protein-bound and has no effect on the binding of other drugs. Thus, GPN lacks any pharmacokinetic interactions with AEDs or other drugs (e.g., coumadin).

Gabapentin has been found to be safe. The adverse effects most commonly reported have been fatigue, nausea, somnolence, dizziness, slurred speech, and unsteady gait, but they have been mild. Side effects that do occur usually abate within 2 weeks. A sense of well-being was reported in some patients and volunteers given GPN. No changes in hematologic or biochemical parameters have been reported. As of August 1995, more than 100,000 people had been treated without any serious or life-threatening side effects.

Because of its unique pharmacokinetic properties, GPN may be useful in special patients. Lack of hepatic and hematopoietic changes may make it preferable for use in cancer patients receiving chemotherapeutic agents. The absence of drug-drug interactions and protein binding make GPN potentially attractive for use in patients with renal disease or porphyria, and in the elderly.

## Lamotrigine

Numerous clinical studies with LTG have established its efficacy and safety in patients with simple or complex partial seizures, with or without secondary generalized seizures. Doses of 300 mg/day and 500 mg/day decreased the frequency of seizures by 20% and 36%, respectively, while a placebo showed an 8% reduction. One monotherapy study, presented in abstract, found equal seizure control attained with 200 mg/day LTG versus 600 mg/day CBZ in patients with new onset seizures. The effective dose depended on the concurrent AEDs: 100–250 mg/day with VPA, 200–500 mg/day with monotherapy, and 400–1000 mg/day with inducing drugs (e.g., CBZ, PHT, and PB).

Lamotrigine (Lamictal) was synthesized in the Burroughs-Wellcome laboratories in England and was derived from a line of drugs that inhibit dihydrofolate reductase. Its development was based on the observation that several anticonvulsants (e.g., PHT) exhibited antifolate activity. It is a relatively weak antifolate compound but has a good anticonvulsant effect. Lamotrigine is rapidly and completely absorbed following oral administration. Dose does not affect absorption, as the Cmax increases proportionally with doses. Protein binding is approximately 55%, so significant binding interactions with other drugs is unlikely.

Lamotrigine undergoes hepatic metabolism. Autoinduction does not occur. Used in monotherapy, the $T_{1/2}$ ranges from 20 to 30 hours. Polytherapy with either CBZ or PHT potentiates the metabolism of LTG and reduces the $T_{1/2}$ to 14 to 15 hours. Valproic acid, however, inhibits LTG metabolism, increasing its $T_{1/2}$ to 50 hours. The dose of LTG used should therefore depend on the concurrent medications that the patient is taking. On the other hand, LTG does not affect the metabolism and plasma levels of CBZ, PHT, PRM, PB, or VPA. Epoxide hydrolase activity is apparently inhibited by LTG, because carbamazepine epoxide levels increase by 45%. In some patients, the higher epoxide levels may result in improved seizure control and/or greater clinical toxicity.

Lamotrigine was approved for market in Europe in 1993 and in December of 1994 in the United States. More than 110,000 patients have been treated with LTG worldwide. The most common side effects were CNS-related symptoms, including drowsiness, dizziness, headaches, unsteady gait, tremor, and nausea. These side effects were usually very mild, lasting 1 to 2 weeks, following which tolerance developed. The most frequent and significant cause for patients to discontinue its use was rash (2.3%), and some cases of Stevens-Johnson syndrome were reported. The incidence of rash was related to initial dose and the rate at which the dose was increased. The highest rate occurred in patients who were also taking VPA.

Patients should be started on low doses, which are then increased over a 4- to 8-week period to the projected chronic dose. Adults taking VPA should be started on 25 mg LTG every other day or weeks and then increased to 25 mg/day. Subsequent increases should be by 25-mg increments on a weekly basis. Patients on no AEDs and those on an enzyme inducer can be started on 25 or 50 mg/day, respectively, and subsequently increased by 25 mg on a weekly basis.

## MECHANISM OF THE ANTICONVULSANTS

Various membrane, neurotransmitter, receptor, and cellular mechanisms can reduce neuronal excitability, and an overall anticonvulsant effect can be produced by a reduction in excitatory mechanisms or an enhancement of inhibitory mechanisms. Evidence suggests that the current AEDs exert their anticonvulsant effect by 1) blocking high-frequency repetitive firing, 2) enhancing GABA-mediated synaptic transmission, or 3) having an effect on receptor-mediated ion transport.

Following membrane depolarization, most neurons respond with a train of action potentials (APs). Some AEDs affect the number and amplitude of the ADs within a burst. The initial AP is unaffected, but subsequent ones are progressively reduced in amplitude and have a slower rate of rise and fall. Finally, APs fail to be generated at all. Carbamazepine, PHT, and VPA reduce the sustained, repetitive, high-frequency firing, apparently as a result of a change in sodium ($Na^+$) channel activity.

An $Na^+$ channel exists in one of three configurations: a resting or rousable state, an open or conducting state, and an inactive or unrousable state. Membrane depolarization results in opening of some of the $Na^+$ channels, following which they become inactive. As the membrane repolarizes, the channels return to the resting state. The proportion of channels that open is dependent on the degree of depolarization. In the resting state, all the channels are closed and the AEDs do not bind to the $Na^+$ channel. Carbamazepine, PHT, and VPA bind to the inactive form of the channel and slow its return to the active state. In the presence of these AEDs and with each depolarization, an increasing number of the $Na^+$ channels enter and remain in the inactive state. A progressively smaller number of channels opens with subsequent membrane depolarizations, thus inhibiting the generation of APs. This explains the voltage-dependent and use-dependent effect of these AEDs and why little or no effect is seen in cells that are functioning normally. Both CBZ and PHT, but not VPA, inhibit the action of batrachotoxin, which binds to the $Na^+$ channel, stimulating $Na^+$ influx. Thus, VPA, while affecting the $Na^+$ channel, appears to bind to a different site than CBZ and PHT. This may explain the empiric clinical observations of a good synergistic effect of combining VPA with either CBZ or PHT.

Pharmacologically, two types of GABA receptors have been defined. The $GABA_a$ receptor is activated by GABA and GABA agonists (e.g., muscimol), enhanced by the benzodiazepines and barbiturates and inhibited by picrotoxin (PIC). Activation of the $GABA_a$ receptor opens the $Cl^-$ channel and increases $Cl^-$ conductance, allowing the $Cl^-$ ion to flow down its electrochemical gradient. In most neurons, this hyperpolarizes the cell, producing an IPSP. Picrotoxin binds to the $Cl^-$ channel noncompetitvely, reducing the time the channel remains open, limiting hyperpolarization and the resulting inhibitory effect. The benzodiazepines and barbiturates bind to adjacent sites on the $GABA_a$ receptor and enhance the opening of the $Cl^-$ channel.

Potential AEDs are identified by using a battery of tests that includes maximal electroshock (MES), subcutaneous pentylenetetrazol (PTZ) (scMET; metrazol), subcutaneous bicuculline (scBIC), PIC, and strychnine (Strych). Minimal threshold tests identify substances that elevate seizure threshold, and supramaximal tests select compounds that prevent seizure spread. Alteration and reduction of the hind limb extension of the MES have been felt to be predictive of anticonvulsive effect in partial and generalized convulsive seizures.

Carbamazepine, PHT, and VPA, which modulate sodium and potassium membrane conductances, have been very effective in the MES model. Pentyl-enetetrazol has been used as a model for anti-absence drugs. Bicuculline acts as an antagonist to the $GABA_a$ but not the $GABA_b$ receptor, while Strych appears to work as a glycine antagonist.

The models are not uniformly predictive of clinical effectiveness. For example, VPA and PB are effective in all models, while clinically VPA is much more effective than PB in the primary generalized epilepsies. These models also do not test for the development of tolerance. In humans, the benzodiazepines are very potent AEDs acutely, but because tolerance develops, they are very poor AEDs when used chronically. The spectrum of effects for the experimental anticonvulsants in these models of epilepsy is different than what we see with the presently available AEDs. This suggests that the new AEDs may employ new mechanisms of action.

### Felbamate

Felbamate blocks sustained repetitive firing of neurons by affecting voltage-dependent $Na^+$ channels. It also blocks convulsions secondary to the voltage-dependent potassium-channel antagonist, 4-aminopyridine. Felbamate inhibits both N-methyl-D-aspastate- (NMDA) and quisqualate-induced seizures, and one study found that the binding of MK-801 in rat synaptosomes is inhibited. There is no effect on the GABA or benzodiazepine receptors, adenosine uptake, or carbonic anhydrase. At present, the principal mechanism by which FBM exerts its anticonvulsant action has not been determined.

### Gabapentin

Gabapentin is most effective in MES-induced seizures, suggesting a similarity to CBZ and PHT. Gabapentin, however, appears to have a different mechanism of action. It was found to significantly reduce the $ED_{50}$ (dose that prevents seizures in half of animals) of PHT, CBZ, VPA, and PRM in a mice electroshock model of seizures. Gabapentin does not affect ligand binding to $GABA_a$, $GABA_b$, benzodiazepine, glutamate, glycine, or NMDA receptors, but it binds to a specific site in the CNS that is associated with the L-amino transport system, and it competes with other L-amino acids for transport into glia and neurons. Gabapentin also reduces glutamate, enhances GABA formation, and causes a use-dependent increase in the nonsynaptic release of GABA. These mechanisms would all enhance inhibition and reduce excitation, resulting in an anticonvulsant effect, but much more work must be done to define fully the primary mechanism of action of GPN.

### Lamotrigine

Lamotrigine was derived from a line of drugs that inhibit dihydrofolate reductase. It was found to be a relatively weak antifolate compound but has a good anticonvulsant effect in MES and PTZ models of epilepsy. It also abolished hind-limb extension to PIC and bicuculline. The overall profile of activity was similar to that of phenytoin. Lamotrigine stabilizes neuronal membranes and inhibits sustained repetitive firing by acting on voltage-sensitive $Na^+$ channels, similar to CBZ and PHT. It inhibits veratrine-evoked release of endogenous glutamate, this being apparently secondary to its effect on the $Na^+$ channel; it has no effect on release of GABA, acetylcholine, noradrenaline, or dopamine. Clinical experience has shown that LTG is very effective in the primary generalized epilepsies (absence and tonic-clonic seizures). Drugs having a use-dependent $Na^+$ channel effect are not effective in this type of epilepsy, so LTG must have other undefined effects within the brain; it will likely be shown to have more than one significant mechanism of action.

### Emerging Anti-Epileptic Drugs

As of 1995, New Drug applications have been submitted in the United States, and are likely to be approved, for vigabatrin (VGB, Sabril, R.W. Johnson Pharm. Research Inst., Spring House, PA), tiagabine (TGB, Hoechst Marion Roussel Inc., Kansas City, MO), and topirimate (TPM; Topimax). Their mechanism(s) of action will therefore be reviewed. Vigabatrin and TGB were both designed to enhance GABAnergic synaptic transmission. Vigabatrin is a deriva-tive of GABA; addition of a vinyl group allows passage through the blood-brain barrier. GABA-transaminase (GABA-T) is present in neurons and glia to break down GABA into succinate semialdehyde, an inactive compound, but the vinyl group on VGB forms an irreversible bond with GABA-T, rendering it inactive so that an increase in parenchymal concentration of GABA results.

Topirimate has a different spectrum of action. It blocks use-dependent, sustained repetitive firing presumably by affecting sodium channels as do PHT and CBZ. Topirimate does not affect ligand binding to $GABA_a$, $GABA_b$, benzodiazepine, glutamate, glycine, or NMDA receptors. It does not block the NMDA receptor but reduces the response of kainate at the same receptor complex. Although TPM does not bind to the $GABA_a$ receptor, the GABA response is enhanced. A carbonic anhydrase inhibition also has been demon-strated that is much weaker than with acetazolamide (Diamox, Lederle Labora-tories Div., Pearl River, NY). The principal mechanism of action of TPM is still not clearly defined.

## RATIONAL SELECTION OF THE ESTABLISHED AND NEW ANTI-EPILEPTIC DRUGS

### Selection by Mechanism

Classically, the selection of an anticonvulsant was based on efficacy, side-effect profile, pharmacokinetics, and cost. With the introduction of AEDs, which appear to have new ways of affecting the brain, we must consider whether the mechanism is important in drug selection. This appears to be most pertinent at present for patients with partial seizures. Valproic acid and ESM, the AEDs that work on generalized absence seizures (petit mal), both seem to modulate calcium entry into thalamic neurons by calcium-T channels. No other mecha-nism has been generally accepted as likely playing a role in the control of these seizures. In clinical trials, ESM and VPA are equally effective for patients suffering from absence seizures, and fortunately, most of them are well controlled with either drug, because we are not yet able to propose a rational selection of drugs for polytherapy.

At present, three established and three new AEDs are effective for partial onset seizures. The relative effectiveness of the established and the new AEDs remains to be defined. In the selection of the first drug in new onset epilepsy, besides efficacy, factors to be considered include safety, pharmacokinetics, and cost. Monotherapy should be pursued initially to minimize drug interactions, synergetic toxicity, and cost.

In the refractory patient, the question arises whether to combine drugs with the same, complementary, or totally different mechanisms. Although not proven in any fashion, the combination of drugs with the same mechanism is felt by many clinicians to be illogical. Thus, the combination of CBZ and PHT, both of which are $Na^+$ channel blockers, may not be as effective as combining PHT or CBZ with GPN. In uncontrolled clinical experience, combining drugs with different mechanisms appears to be very effective, but we must await the results of controlled trials.

### Selection by Type of Epilepsy

Clinical experience must continue to guide us in our selection of therapeutic agents. The partial and generalized epilepsies do not respond in the same fashion to the anticonvulsants. All of the first-line AEDs (CBZ, PHT, and VPA) are effective in the partial epilepsies, particularly in secondarily generalized seizures. Seizure activity that remains localized or spreads to only one region (simple and complex partial) is somewhat more effectively treated with CBZ or PHT. Gabapentin and LTG also are effective in partial seizures.

Very little information is available on the comparative efficacy of the old and new AEDs, but the new AEDs are useful and effective. Thus, it is reasonable to try any of them to treat focal onset seizures. The principal reasons to first select CBZ, PHT, or VPA are their lower cost and the wider experience with their use.

Perhaps most important is to identify the patients with primary generalized epilepsies and particularly syndromes such as juvenile myoclonic epilepsy. In these epilepsies, VPA is the agent of first choice, but clinical experience now suggests that LTG also will play a significant role in their treatment.

## References/Further Reading

Ajomone Marsan C. Electrographic aspects of 'epileptic' neronal aggregates. Epilepsia 1961;2: 22–38.

Ayala GF, Dichter M, Gumnit RJ, et al. Genesis of epileptic interictal spikes. New knowledge of cortical feedback systems suggests a neurophysiological explanation of brief paroxysms. Brain Res 1973;52:1–17.

Ayala GF, Matsumoto H, Gumnit RJ. Excitable changes and inhibitory mechanisms in neocortical neurons during seizures. J Neurophysiol 1970;33:73–85.

Blume WT, Whiting SE, Girvin JP. Epilepsy surgery in the posterior cortex. Ann Neurol 1991; 29:638–645.

Chang CN, Ojemann LM, Ojemann GA, Lettich E. Seizures of fronto-orbital origin: a proven case. Epilepsia 1991;32:487–491.

Commission on Classification and Terminology of the International League Against Epilepsy, The proposal for revised clinical and electroencephalographic classification of epileptic seizures. Epilepsia 1981;22:489–501.

Commission on Classification and Terminology of the International League Against Epilepsy, The Proposal for classification of epilepsies and epileptic syndromes. Epilepsia 1985;26: 268–278.

de Silva M, McArdle B, McGowan M, et al. Mono Rx for newly diagnosed childhood epilepsy: a comparative trial and prognostic evaluation. Epilepsia 1989;30:662.

Dichter M, Spencer WA. Penicillin-induced interictal discharges from cat hippocampus. I. Characteristics and topographical features. Neurophysiol 1969;32:649–662.

Dichter M, Spencer WA. Penicillin-induced interictal discharges from cat hippocampus. II. Mechanisms underlying origin and restriction. J Neurophysiol 1969;32: 663–687.

Engel J Jr. Experimental animal models of epilepsy: classification and relevance to human epileptic phenomena. In: Avanzini G, Engel J Jr, Fariello R, Heinemann U, eds. Neurotransmitters in epilepsy. New York: Elsevier, 1992.

Engle J Jr. Part II: Pathophysiology. In: Seizures and epilepsy. Davis, 1989.

Gastaut H, Zifkin BG. Ictal visual hallucinations of numerals. Neurology 1984;34:950–953.

Gloor P. Neurophysiological bases of generalized seizures termed centrencephalic. In: Gastaut H, Jasper H, Bancaud J, Waltregny A, eds. The physiopathogenesis of the epilepsies. Charles C. Thomas, 1969.

Heller AJ, Chesterman P, Elwes RDC, et al. Monotherapy for newly diagnosed adults with epilepsy: a comparative trial and prognostic evaluation. Epilepsia 1989;30:648.

MacDonald RL. Antiepileptic drug action. Epilepsia 1989;30(Suppl 1):S19–S28.

Maldonado HM, Delgado-Escueta AV, Walsh FO, et al. Complex partial seizures of hippocampal and amygdalar origin. Epilepsia 1988;29:420–433.

Matsumoto H, Ajmone Marsan C. Cortical cellular phenomena in experimental epilepsy: interictal manifestations. Exp Neurol 1964;9:286–304.

Matsumoto H, Ajmone Marsan C. Cortical cellular phenomena in experimental epilepsy: ictal manifestations. Exp Neurol 1964;9:305–326.

Mattson RH, Cramer JA, Collins JF, et al. Valproate for treatment of partial and secondarily generalized tonic-clonic seizures in adults: a comparison with carbamazepine. N Engl J Med 1992.

Mattson RH, Cramer JA, Collins JF, et al. Comparison of carbamazepine, phenobarbital, phenytoin, and primidone in partial and secondarily generalized tonic-clonic seizures. N Engl J Med 1985;313:145–151.

Penry JK. Diagnosis and treatment of absence seizures. Cleve Clin Q, 1984;77:283–286.

Penry JK, Porter RJ, Dreifuss FE. Simultaneous recording of absence seizures with video-tape and electroencephalograpy. A study of 374 seizures in 48 patients. Brain 1975;98: 427–440.

Prince DA. Electrophysiology of 'epileptic neurons.' Electroencephlogr Clin Neurophysiol 1967; 23:83–84.

Prince DA, Conners BW. Mechanisms of interictal epileptogenesis. In: Delgado-Escueta AV, Ward AA Jr, Woodbury DM, Porter RJ, eds. Advances in neurology, vol. 44. New York: Raven, 1986.

Prince DA, Connors BW, Benardo LS. Mechanisms underlying interictal-ictal transitions. In: Delgado-Escueta AV, Wasterlain CG, Treiman DM, Porter RJ, eds. Advances in neurology, vol. 34. New York: Raven, 1983.

Quesney LF. Clinical and EEG features of complex partial seizures of temporal lobe origin. Epilepsia 1986;27(Suppl 2):S27–S45.

Ramsay RE, Cohen A, Brown MC. Coexisting epilepsy and non-epileptic seizures. In: Rowan AJ, Gates JR, eds. Non-epileptic seizures. Boston: Butterworth-Heinemann, 1993: 47–54.

Ramsay RE, BJ Wilder, JV Murphy, et al. Efficacy and safety of valproic acid vs phenytoin as sole therapy for newly diagnosed primary generalized tonic-clonic seizures. J Epilepsy 1992;5:55–60.

Rasmussen T. Characteristics of a pure culture of frontal lobe epilepsy. Epilepsia 1983;24: 482–493.

Riggio S. Frontal lobe epilepsy: clinical syndromes and presurgical evaluation. J Epilepsy 1995; 5:178–189.

Sato S, White BG, Penry JK, et al. Valproic acid versus ethosuximide in the treatment of absence seizures. Neurology 1982;32:157–163.

Schmidt RP, Thomas LB, Ward AA. The hyperexcitable neuron. Microelectrode studies of chronic epileptic foci in monkeys. J Neurophysiol 1959;22:285–296.

Snead OC III. Basic mechanisms of generalized absence seizures. Ann Neurol 1995;37:146–157.

Waterman K, Purves SJ, Kosaka B, et al. An epileptic syndrome caused by mesial frontal lobe seizure foci. Neurology 1987;37:577–582.

Williamson PD, Spencer SS. Clinical and EEG features of complex partial seizures of extratemporal origin. Epilepsia 1986;27(Suppl 2):S46–S63.

# 9 Neuromuscular Disease

Karl E. Misulis

There have been many advances in our knowledge of neuromuscular disease within the past decade. Interest and research has focused especially on immune-mediated and metabolic disorders. Many disorders, long thought to be idiopathic, have been found to be inflammatory and presumably immune-mediated.

## Peripheral Neuropathy

Identification of the immune-mediated neuropathies is critically important because of the potential for improvement with treatment. Neuropathy associated with HIV has also received a great deal of attention recently, although it is usually in a setting of other clinical signs of infection.

### IMMUNE-MEDIATED NEUROPATHIES

The pathophysiologic basis of immune-mediated neuropathies is not completely understood. Classical thinking is that a novel antigen appears in the body. This may be an infectious agent or tumor. The immune system mounts an attack against the antigen, hopefully eliminating the offending agent. During or after the exposure, the immune mediators attack the neuron or myelin sheath. Damage to the sheath causes demyelination. Damage to the neuron cell body results in a neuronal or axonal degeneration.

#### Guillain-Barré Syndrome
Guillain-Barré syndrome (GBS), also known as acute immune-mediated demyelinating radiculoneuropathy, should be suspected in patients who present with weakness and areflexia. Diagnosis is supported by nerve conduction studies, which show slowing of motor and sensory conductions, especially in the proximal segments, indicating a demyelinating neuropathy. Examination of the CSF shows increased protein with few leukocytes.

*Pathophysiology*   Details of the pathogenesis of the neuropathy remain elusive. We know that antibodies attack the myelin sheath, resulting in demyelination of peripheral nerve. The origin of the antibodies is not well known, but infection has been linked frequently with GBS and results in the mounting of an immune response. While the attack is focused on the infection, shared immunologic appearance results in peripheral nerve being attacked as well. Associations with respiratory infections, influenza, hepatitis, HIV infection, meningococcal meningitis, and now *Campylobacter jejuni* have all been suspected. The latter is of most interest recently because it is believed that peripheral nerve myelin and *C. jejuni* share epitopes, but the diversity of the associations suggests that the epitopes are shared with numerous other organisms, so identification of one specific antigen with pathogenesis of GBS will be impossible.

The attack on peripheral nerve appears to involve both cellular and humoral immune mechanisms. Autoreactive T-lymphocytes specific for the P0 and P2 antigens on myelin mediate a cellular attack, while circulating antibodies to these same antigens, as well as to glycolipids and glycoproteins, make up the humoral response.

*Treatment*   Not long ago, no specific treatments were known to alter the course of GBS. Supportive care resulted in survival of most patients; those who did not succumb to complications would eventually improve. Plasma exchange was used extensively, and its efficacy is well founded on randomized studies. Recently, human immune globulin (HIG) has been used for treatment of GBS, although its mechanism is unknown. Corticosteroids have not been found to be helpful and are considered contraindicated by most clinicians.

Clinicians fall into two camps regarding treatment of GBS. The first uses plasma exchange because it has been shown effective in multiple trials and has been studied in a larger number of patients than has HIG. On the other hand, plasma exchange is more invasive, usually requiring central lines, and is associated with fluid shifts that can be difficult to manage. The second camp favors HIG because it is generally better tolerated by the patient. While allergic reactions, renal failure, and viral transmission have been reported with HIG, the relative frequency of these adverse effects is low. Human immune globulin would be used routinely if more data supported its efficacy.

*Related Disorders*   Occasional patients with otherwise typical GBS are shown on electromyogram (EMG) to have an axonal neuropathy rather than, predominantly, demyelination. In the acute phase, they show reduced numbers of units on EMG; subsequently, fibrillation potentials are seen, and motor unit reorganization gives rise to polyphasic potentials. These patients are more likely to have preserved though depressed tendon reflexes. They also are less likely to respond to treatment. They often improve, but at a slower rate than their counterparts with demyelinating neuropathy.

Miller Fisher syndrome is the clinical triad of ataxia, areflexia, and ophthalmoplegia. Patients with this syndrome are less likely to have limb weakness and ventilatory failure than are patients with GBS. They have an IgG antibody to GQ1b, as do GBS patients with ophthalmoplegia, whereas GBS patients without ophthalmoplegia do not. The exact role of this ganglioside in the pathogenesis of Miller Fisher syndrome is not completely understood, but GQ1b is much more prominent in oculomotor nerves than in other nerves. Measurement of these antibodies is of interest, but it does not assist with management of the patient.

## Chronic Inflammatory Demyelinating Polyradiculoneuropathy

Chronic inflammatory demyelinating polyradiculoneuropathy (CIDP) is sometimes deemed a chronic form of GBS, but the two entities are clinically quite different in presentation and treatment. Patients present with a symmetric sensorimotor polyneuropathy with involvement of proximal and distal muscles. Sensory symptoms include pain and loss of large-fiber modalities (vibration and proprioception) in some patients. The neuropathy is usually slowly progressive, though stepwise progression and a relapsing-remitting pattern is occasionally seen.

*Diagnosis*    Diagnosis of CIDP should be suspected when a patient presents with a chronic sensorimotor demyelinating neuropathy. Nerve conduction studies show slowing, with dispersion of the compound action potentials of both motor and sensory nerves. Conduction block may be seen, but the sensory symptoms and abnormalities on sensory nerve conduction studies differentiate CIDP from multifocal motor neuropathy (MMN). An electromyogram shows denervation with signs of ongoing reinnervation in some patients.

Cerebrospinal fluid shows elevated protein levels with little or no increase in WBC. A nerve biopsy may show segmental demyelination and a mononuclear infiltrate in the nerve, but the findings are often equivocal.

*Treatment*    Treatment of CIDP has included corticosteroids, immunosuppressants, plasma exchange, and HIG. Human immune globulin has received a great deal of attention recently, though expense keeps this from being first-line therapy for most patients; a multicenter study examining its efficacy in CIDP is under way.

## Multifocal Motor Neuropathy

Multifocal motor neuropathy is characterized by muscle weakness and wasting without sensory symptoms. While other immune-mediated neuropathies are mainly considered in the diagnosis of *peripheral sensorimotor neuropathy*, MMN is considered in the differential diagnosis of *motoneuron disease*. Clinically, MMN resembles progressive muscular atrophy, a degeneration affecting the lower motoneurons with sparing of upper motoneurons. Patients typically have asymmetric weakness, cramps in various muscles, and fasciculations, but they typically are younger than patients with amyotrophic lateral sclerosis (ALS) or progressive muscular atrophy, with an age range of 25 to 58 years. Bulbar symptoms and signs are less prominent than in patients with other motoneuron diseases.

***Diagnosis***   Nerve conduction studies show slowing in proximal segments of the nerves. This is usually manifest as increased F-wave latencies. In addition, conduction in the motor nerves shows segmental conduction block, character-ized by slowed conduction of motor axons through a region of the nerve. The compound motor action potential (CMAP) is dispersed and of low amplitude when the nerve is stimulated above the site of conduction block. Sensory conduction is normal, even through the sites of motor nerve conduction block.

Electromyography shows mild denervation with signs of reinnervation. The signs of demyelination on nerve conduction studies are much more promi-nent than the signs of axonal degeneration.

***Immunology***   Immunopathogenesis of MMN is suspected on the basis of some patients having a gammopathy, usually IgM. The IgM antibodies were directed against gangliosides, which share a common carbohydrate section. Not all patients with anti-ganglioside antibodies have conduction block; some have a lower motoneuron syndrome without this feature. Not all patients with MMN have elevated anti-GM1 antibodies; approximately 70% of patients with MMN have high titers of the anti-GM1 antibody.

***Treatment***   Multifocal motor neuropathy is particularly attractive as a candi-date for immune suppression because of the identification of abnormal IgM protein. Cyclophosphamide has been effective for the treatment of patients with MMN but is less effective for patients with anti-ganglioside–mediated motor axonopathy without conduction block. In patients with conduction block, the presence of anti-GM1 antibodies has not been shown to predict response to treatment.

Treatment with corticosteroids has been disappointing, with a small and transient response. Recent evidence suggests that intravenous HIG produces a good response, though most patients had to be re-treated after several months. A commonly used protocol is treatment with HIG at a dose of 0.4 g/kg/day for 5 days.

## Other Immune-Mediated Neuropathies

Peripheral neuropathy associated with anti-myelin–associated glycoprotein (anti-MAG) antibodies presents as a slowly progressive sensorimotor neuropa-thy, more prominent distally. Motor symptoms develop later than the sensory symptoms, which include paresthesias and sensory ataxia. Bulbar and autonomic functions are spared. Because of the prominent sensory involvement, reflexes are markedly reduced or absent.

Nerve conduction velocity (NCV) studies and EMG show a mainly demyelinating neuropathy, but a mixed axonal and demyelinating pattern is often seen. Cerebrospinal fluid is normal, except for elevated protein. The anti-MAG antibody is an IgM protein, often associated with a monoclonal gammopathy. The etiology of the gammopathy is usually unknown, though some patients have lymphoma, Waldenstrom's disease, or chronic lymphocytic leukemia.

Studies are under way to determine the best treatment for these immune-mediated neuropathies. Possible modalities include corticosteroids and immunosuppressants such as cyclophosphamide and azathioprine. Human immune globulin also has been used. If the patient has predominantly sensory symptoms, symptomatic treatment of the neuropathic pain is usually all that is necessary. Patients with motor deficit probably should be treated with plasma exchange, followed by prednisone (80–100 mg/day) or other immunosuppressants.

Although laboratories market antibody assays, there is seldom reason to use them in routine clinical practice. Multifocal motor neuropathy is diagnosed by clinical presentation and demonstration of partial conduction block on NCV studies. The presence of anti-GM1 antibodies is of interest, but does not appear to identify patients who will respond to treatment; treatment is warranted regardless of whether anti-GM1 antibodies are found.

Neurologists evaluate many patients with neuropathies affecting predominantly sensory nerves, of whom some have anti-sulfatide antibodies; this acidic glycosphingolipid is a major constituent of myelin, much higher in concentration than any ganglioside. Unlike anti-GQ1b antibodies, the anti-sulfatide antibodies have not been associated with a consistent clinical presentation, which casts doubt on the role of sulfatides in sensory neuropathy.

## HIV-Associated Neuropathy

HIV is associated with several types of neuropathy, especially those that have been attributed to immune-mediated attack. Autoimmune demyelinating polyradiculoneuropathy (AIDP) and CIDP are the most recognized. Neuropathy in AIDS patients is underdiagnosed because of the magnitude of the patients' other medical problems.

### Guillain-Barré Syndrome

The relationship of GBS in AIDS to that in patients without AIDS is unclear. One plausible theory suggests that cytokines promote movement of activated lymphocytes and macrophages and induce major histocompatibility complex antigens on Schwann cells, exposing new antigens on the myelin sheath. These antigens do not have self-tolerance, so an immune response is mounted against them.

Patients present with ascending neuropathy with both motor and sensory involvement, though motor symptoms and signs predominate in most instances. Nerve conduction studies show conduction block, reduced NCVs, and prolongation of F-wave responses, all findings similar to GBS unassociated with HIV. Guillain-Barré syndrome with HIV, however, often shows a mild lymphocytic pleocytosis in the CSF.

Guillaine-Barré syndrome typically develops early in HIV infection and is occasionally associated with initial infection. Neuropathy that mimics GBS appears in late HIV infection and often is associated with cytomegalovirus (CMV) infection.

### Chronic Inflammatory Demyelinating Polyradiculoneuropathy
The pathophysiology of CIDP is similar to that described for GBS in HIV infection. The clinical presentation is very similar to that seen in patients without HIV infection: slowly progressive sensory and motor loss. Nerve conduction velocity studies show demyelinating neuropathy with partial conduction block. Axonal changes are occasionally present. Examination of CSF shows increased protein. In other respects, CIDP associated with HIV is similar to CIDP in HIV-negative patients.

### Mononeuropathy Multiplex
Mononeuropathy multiplex is encountered in relatively few clinical conditions, with most patients having diabetes and lesser numbers having vasculitis and leprosy. Patients with HIV may develop a multiplex neuropathy on the basis of secondary disorders, including AIDPs and plexopathies, vasculitis, varicella-zoster eruptions, and lymphomatous meningitis. Mononeuropathy secondary to vasculitis may be associated with deposition of immune complexes involving HIV antigens and associated antibodies.

### Painful, Predominantly Sensory Neuropathy
Unlike many of the other immune-mediated neuropathies in HIV infection, painful sensory neuropathy develops later in the course of the illness. Patients present with painful dysesthesias in a stocking-and-glove distribution. Motor symptoms and signs may develop, but these findings are mild.

Nerve conductions are usually normal, but conduction velocities are occasionally slightly slowed. An EMG shows axonal degeneration, and the slowed conduction, when present, is due to secondary demyelination.

The pathophysiology of this neuropathy is not completely understood. Autopsy findings show inflammatory infiltrates in the dorsal root ganglia, suggesting that this is an immune-mediated neuronal degeneration. HIV-induced production of cytokines may play a role in the immune response, as may the drugs administered to patients with HIV, including especially antibiotics and chemotherapeutic agents.

## Cytomegalovirus-Associated Neuropathy

Neuropathy associated with CMV infection develops late in the course of AIDS, and is characterized by either a polyradiculomyelopathy or multifocal neuropathy. Polyradiculomyeloneuropathy presents with ascending weakness and paresthesias, often associated with corticospinal tract signs and sphincter dysfunction. Reflexes are depressed or absent in the legs, although plantar responses are often extensor. Laboratory studies are remarkable for EMG findings of axonal neuropathy with secondary demyelination. The CSF shows a polymorphonuclear pleocytosis with markedly elevated protein. Autopsy shows extensive inflammatory neuropathy with necrotizing myelopathy. Some patients respond to ganciclovir.

Multifocal neuropathy presents with patchy sensory symptoms, including paresthesias, which may be painful. Motor symptoms develop subsequent to the sensory symptoms. Laboratory findings include EMG findings of axonal neuropathy with secondary demyelination. Compound motor and sensory nerve action potentials are of low amplitude, indicating partial conduction block.

## IMMUNOLOGY OF MOTONEURON DISEASE

Neurodegenerative diseases such as Alzheimer's, Parkinson's, and ALS share the common feature of degeneration of selected nerve cells; the specific clinical syndrome is dependent on which neurons are affected. The final pathway of neuronal degeneration appears to be loss of membrane integrity, influx of ions (especially calcium), and activation of proteases and phospholipases within the cells. In some conditions of neuronal and muscular degeneration, there are signs of immune activation, including inflammation and products of complement activation. Therefore, it is thought that the immune system may play a role in mediating or initiating the neuronal damage.

Amyotrophic lateral sclerosis is characterized by degeneration of upper and lower motoneurons, resulting in weakness and wasting of muscles. The upper extremities are often affected first, with the intrinsic muscles of the hands prominantly affected. Initial asymmetry ultimately gives way to generalized weakness involving axial, appendicular, and bulbar muscles.

In approximately 50% of patients with ALS, IgM anti-GM1 antibodies are present, but this is not a specific finding, because other autoimmune disorders are associated with modest elevations. Approximately 10% of patients have *high* titers of IgM anti-GM1 antibodies, however, and in these patients the suspicion of immune mediation is more secure.

Many ALS patients have antibodies to voltage-gated calcium channels, which differ from the antibodies found in myasthenic syndrome. Antibodies in ALS bind to the alpha-1 subunit of the calcium channel, whereas antibodies in myasthenic syndrome bind to the alpha-1 and beta subunits of the calcium channel. Any role of anti–calcium channel antibodies in the pathogenesis of ALS is unknown.

Unfortunately, studies of the results of immunosuppression in ALS have been disappointing; while immunosuppression may be effective for selected patients with pure, lower motoneuron syndromes or conduction block, it is ineffective for typical ALS with upper and lower motoneuron involvement.

It does not appear that anti-ganglioside antibody testing is helpful in selecting patients for treatment, so it should not be performed in routine practice for patients with suspected ALS or primary lateral sclerosis (PLS). Clinical and electrophysiologic data must rule out MMN, however, because of the response to treatment of this disease. Part of the difficulty in interpreting anti-GM1 antibodies is that 10% of normal individuals have levels of these that are in the same range as that of patients with motoneuron disease.

# Myopathy

## MITOCHONDRIAL DISORDERS

The modern biochemistry of mitochondrial disorders has been pioneered especially by Darryl DeVivo and Salvatore DiMauro. This is an expanding area of medicine that will receive increasing attention in the next decade. Identification and characterization of mitochondrial disorders has been the predominant advance. It is hoped that increasing understanding of pathophysiology will allow for design of effective treatments.

The diagnosis of mitochondrial disorders depends on clinical suspicion followed by appropriate laboratory testing. Patients are typically thought to have a multisystem disorder with both central and peripheral nervous system involvement. Symptoms include developmental delay, mental retardation, ataxia, hearing loss, ocular motor deficit, visual loss, and weakness. Signs include microcephaly, intellectual impairment, peripheral neuropathy, optic atrophy, ptosis and/or ophthalmoplegia, and indications of cardiomyopathy. Other organs, including the liver, kidney, skin, and endocrine glands, may be affected.

Inheritance of mitochondrial disorders has classically been described as from mother to male and female children, since the mitochondria all are derived from those in the egg. This form of inheritance can be confused with X-linked inheritance. Many mitochondrial disorders, however, are due to disorders of nuclear DNA, affecting mitochondrial function.

### Substrate Transport

Substrate is transported across the plasma and mitochondrial membranes. Carnitine is actively transported across the plasma membrane and then participates in transport of long-chain fatty acids (LCFA) as carnitine esters through the mitochondrial membrane. It is then recycled for use in transport again. The first step is conversion of the fatty acid to a CoA derivative by fatty acyl CoA synthase. Carnitine acyltransferase I forms O-acyl carnitine by transferring an acyl group from the cytoplasmic CoA. This group is transferred across the mitochondrial membrane and then bound to another CoA by carnitine acyltransferase II. Defects of carnitine transport result in cardiomyopathy, hypoketotic hypoglycemia, and weakness with hypotonia. Patients improve with carnitine supplementation.

Carnitine palmitoyl transferase-1 (CPT-1) is important for formation of acylcarnitine from acyl CoA plus carnitine. Deficiency of this enzyme produces nonketotic hypoglycemia in infancy, with hepatomegaly and hypertriglyceridemia.

Carnitine palmitoyl transferase-2 is responsible for conversion of acylcarnitine back to acyl CoA and carnitine after it has passed through the inner mitochondrial membrane. Deficiency of CPT-2 presents as a myopathy with cramps and myoglobinuria. Adult onset is typically benign, but onset in infancy is a severe and ultimately lethal condition, characterized by multi-organ involvement.

## Substrate Utilization

Substrate utilization implies metabolism of the molecules transported to the inside of the mitochondria. During the process of oxidation, electrons are given off, with dehydrogenation of the substrate. The dehydrogenation converts a primary acceptor, such as nicotinamide adenine dinucleotide (NAD), to NADH which donates a pair of electrons to the electron transport chain. The electrons are passed down the chain until they react with oxygen and protons to produce water. Free energy of the electrons is lost during their passage down the transport chain, some by being captured as adenosine triphosphate (ATP) (formed from adenosine diphosphate [ADP]) and the rest as heat. The electron transport chain is the body's greatest requirement for oxygen. The ATP, of course, is used for energy elsewhere. The electron transport chain is called the respiratory chain because it requires oxygen. The term *oxidative phosphorylation* refers to the production of ATP from ADP.

*Pyruvate Utilization*    Pyruvate can follow three basic pathways of metabolism. First, the most classic is dehydrogenation to form acetyl CoA, which enters the citric acid cycle. Second, pyruvate can be transformed to lactate during anaerobic conditions, producing $NAD^+$, which is a required cofactor for glyceraldehyde-3-phosphate dehydrogenase to maintain glycolysis. Third, pyruvate carboxylase catalyzes the carboxylation of pyruvate to oxaloacetate; this reaction is especially important for gluconeogenesis and for replenishment of cofactors of the citric acid cycle.

Deficiency of pyruvate carboxylase presents in infants with failure to thrive, hypotonia, psychomotor retardation, and metabolic acidosis. Complete absence of the enzyme is rapidly fatal in infancy, but patients with some residual enzyme activity survive until early childhood.

Pyruvate cannot be dehydrogenated directly: It does not have a structure that is susceptible to a dehydrogenase. The pyruvate dehydrogenase complex consists of three enzymes, the first of which is pyruvate decarboxylase, and employs five cofactors. The final product of pyruvate dehydrogenase complex is acetyl CoA, which enters the citric acid cycle for further metabolism. Deficiency of pyruvate dehydrogenase, like many other defects of mitochondrial metabolism, produces failure to thrive, hypotonia, and psychomotor retardation. There are also, however, a variety of dysmorphic features, including broad nasal bridge, micrognanthia, low-set ears, short arms and fingers, short stature, simian creases, and hypospadias. Other neurologic findings include ptosis, other cranial nerve palsies, and dysphagia. Infantile patients often resemble sufferers of Leigh's syndrome.

While treatment of mitochondrial disorders has been unsatisfactory, thiamine supplementation has been tried in many, and a high-fat diet may benefit patients with pyruvate dehydrogenase deficiency by providing an alternative source of acetyl CoA through beta-oxidation of fatty acids rather than by metabolism of pyruvate. Lipoic acid and l-carnitine supplementation also has seemed helpful in some patients.

*Fatty Acid Utilization*    Beta-oxidation of fatty acids results in generation of acetyl CoA, which is then able to enter the citric acid cycle. Successive 2-carbon segments are removed from the fatty acid chain, resulting in shorter chains that re-enter the loop until metabolized to acetyl CoA. Beta-oxidation occurs in the mitochondria, so carnitine is used to facilitate transport of LCFA. Medium-chain fatty acids pass directly through the inner mitochondrial membrane. Fatty acids with an even number of carbons are completely metabolized to acetyl CoA; however, fatty acids with an uneven number of carbons terminate beta-oxidation as propionyl CoA; this is carboxylated to methylmalonyl CoA; then the carbon atoms are rearranged to form succinyl CoA, which can enter the citric acid cycle.

Ten different conditions involving various sites along the pathway of beta-oxidation have been described. It is impossible to definitively separate the syndromes based on clinical presentation alone. Medium-chain acyl-CoA dehydrogenase (MCAD) deficiency accounts for half of the cases of defects in beta-oxidation. Patients present in infancy or early childhood with episodic vomiting, encephalopathy, and hypotonia. Attacks are precipitated by fasting or infection; between attacks, patients are essentially normal.

Other defects of beta-oxidation are characterized by symptoms and signs similar to MCAD deficiency, but in addition may include limb weakness, which is not found with MCAD deficiency. Other patients may develop cardiac defects.

## Citric Acid Cycle

The citric acid cycle is also known as Kreb's cycle or tricarboxylic acid (TCA) cycle and is responsible for metabolism of acetyl CoA derived from beta-oxidation of fatty acids or pyruvate, or metabolism of pyruvate directly. Two of the products of the citric acid cycle are NADH and $FADH_2$, which are then metabolized to ATP by the respiratory chain; its other major product is carbon dioxide. Neurologic abnormalities have been associated with several defects of the cycle, but in most, the neurologic difficulties are part of a multisystem disorder.

Alpha-ketoglutarate dehydrogenase catalyzes the formation of succinyl CoA from alpha-ketoglutarate. In the process, NADH is produced from $NAD^+$. This is the second NADH produced in the citric acid cycle. Deficiency of alpha-ketoglutarate dehydrogenase presents in childhood with extrapyramidal signs and results in death in early childhood.

Fumarase is responsible for formation of l-malate from fumarate. This is a freely reversible conversion and is a simple hydration, not generating any reduced cofactors that can enter the electron transport chain. Fumarase deficiency causes progressive encephalopathy in infancy, with fumaric acidemia.

Succinate dehydrogenase is important not only for the citric acid cycle, but also in the respiratory chain. In the citric acid cycle, it catalyzes the formation of fumarate from succinate, with the formation of $FADH_2$ from flavin adenine dinucleotide (FAD), which can in turn enter the electron transport chain. Deficiency of succinate dehydrogenase results in progressive encephalopathy and myopathy.

## Oxidative Phosphorylation

Oxidative phosphorylation refers to the formation of high-energy phosphates that are used for specific functions in the cells. For example, ATP is made from the oxidative phosphorylation of ADP. In the process of phosphorylating ADP, the cofactors are oxidized, producing acceptors that are reused by glycolysis and the citric acid cycle.

Luft's syndrome is the only described defect of oxidative phosphorylation. Case reports of the few patients with this condition presented with a syndrome of hypermetabolism that could not be explained by hyperthyroidism. Heat intolerance, exercise intolerance, resting tachycardia plus excessive perspiration, polydipsia, and polyphagia were present. Death occurred in middle age.

## Electron Transport Chain

The electron transport chain consists of five complexes that are responsible for generation of ATP from the oxidized intermediaries of metabolism. It passes electrons from a higher energy state to a lower energy state, generating ATP in the process and regenerating the cofactors for glycolysis and the citric acid cycle. The typical presentation of electron transport chain defects is either a myopathy or a multisystem disorder characterized by weakness in addition to signs such as encephalopathy, developmental delay, and seizures.

Complex I deficiency can present as either a myopathy or a systemic disorder affecting multiple organ systems. When it presents as a myopathy, only skeletal muscle is affected in most patients; weakness and easy fatigability are the predominant symptoms. Patients with multisystem disease may present early in life with encephalopathy with hypotonia and developmental delay, and may die relatively early. Others can present with mitochondrial encephalopathy lactic acidosis and stroke. No large studies have demonstrated efficacy of any treatment protocol.

Defects of complex II are rarely identified, but the few known cases have presented as encephalomyopathy. Defects of complex III produce weakness in addition to other organ involvement. Some patients have a multisystem disorder, while others have a relatively pure myopathy. Some present in childhood with only skeletal muscle affected, whereas others present in infancy with cardiomyopathy.

Leigh's syndrome can be associated with complex IV deficiency as well as pyruvate dehydrogenase deficiency. Complex IV deficiency–associated Leigh's syndrome presents later than that associated with pyruvate dehydrogenase deficiency, and peripheral neuropathy is common; with pyruvate dehydrogenase deficiency, patients present in infancy with seizures and apnea.

Complex V is the mitochondrial ATP synthase. Only two patients have been described with a defect in this enzyme; both presented with weakness, but one had a myopathy with ragged red fibers and the other had neuropathy with involvement of other organs.

## Neuromuscular Transmission Defects

Most of the advances in neuromuscular transmission defects have been in understanding their pathophysiology and in diagnosis. The normal neuromuscular junction transmits with near-perfect synaptic security; that is, for every motor axon action potential, virtually every muscle fiber innervated by that axon will have an action potential. Neuromuscular transmission defects result in loss of this security.

### Myasthenia Gravis

Antibodies to the acetylcholine receptor bind to the receptor and cause receptor degeneration. A clinical tie has been established between the thymus and myasthenia gravis. Because the thymus contains myoid cells, one theory is that the antibodies are to the acetylcholine receptors on the intrinsic myoid cells.

Diagnosis of myasthenia gravis depends on clinical history, examination, and supportive laboratory tests. A typical history would be progressive weakness, fatigue, ptosis, diplopia, and dysarthria or dysphagia. Findings on examination include extraocular motor defects that cannot be explained by a single cranial nerve lesion and proximal weakness.

Myasthenia is traditionally divided into ocular and generalized forms; however, the majority of patients with purely ocular myasthenia will eventually progress to generalized myasthenia, although this progression may take several years.

Diagnosis is supported by a positive edrophonium (Tensilon) test, but sensitivity and specificity are less than perfect. Unless there is a definite endpoint, the benefit of edrophonium injection may be difficult to identify. Also, other disorders, such as partial denervation, can benefit from acetylcholinesterase blockade, resulting in a misinterpretation of the test. Other supporting evidence of myasthenia comes from electrophysiologic and serologic tests.

Electrophysiologic techniques for identification of myasthenia gravis include repetitive stimulation and single-fiber EMG. Electrical stimulation of the motor nerve produces activation of the muscle fibers, which is recorded by surface electrodes over the muscle. In normal subjects, repetitive stimulation at 3 per second produces muscle action potentials that are of similar amplitude and duration. In patients with myasthenia, with repetitive stimulation some muscle fibers fail to be activated, producing a decremental response. A decrement of more than 10% between first and fifth pulses is considered abnormal. Although some patients may exhibit the decremental response during a baseline recording, others require 1 minute of exercise of the tested muscle for the decrement to be obvious. Proximal muscles are more likely to exhibit an abnormal response, but none have universal sensitivity. False-negative results range from 17% to 45% in generalized myasthenia, depending on the severity of symptoms.

Single-fiber EMG has been discussed in the neurophysiologic literature for years, and is still widely employed in academic centers, but its use in routine clinical practice has declined. Single-fiber EMG is abnormal in 94% of generalized myasthenia and 80% of ocular myasthenia. Unfortunately, as with repetitive stimulation, changes in single-fiber EMG are not specific for myasthenia. Denervating diseases and myopathies also may show increased blocking and may jitter on single-fiber EMG.

Antibody testing is routine for the evaluation of patients with myasthenia gravis. Anti-acetylcholine receptor antibodies are present in approximately 85% of patients with generalized myasthenia, but in only 50% of patients with pure ocular myasthenia. Children also are often antibody-negative. A commonly used antibody panel reports binding, blocking, and modulating antibodies. Binding antibodies are used most often and represent the binding of bungarotoxin to the acetylcholine receptor. Blocking antibodies are positive in some patients who do not show binding antibodies; blocking antibodies reduce the binding of bungarotoxin to acetylcholine receptors, while modulating antibodies reduce the density of acetylcholine receptors on cultured myotubes.

Anti-striated muscle antibody testing is often combined with anti-acetylcholine receptor antibody testing. Anti-striated muscle antibodies are positive in approximately 84% of patients with thymoma. However, only 17% of patients with positive antibodies have thymoma, so antibodies cannot be used to screen definitively for it. The clinical utility of these antibodies, therefore, is relatively low.

Traditional treatment of patients with myasthenia gravis includes corticosteroids and acetylcholinesterase inhibitors. (Conventional therapy is not covered in this chapter.) Many clinicians have become more aggressive about the use of immunosuppressants, including cyclophosphamide, azathioprine, and cyclosporine. Azathioprine (Imuran) is usually used for patients who do not respond to prednisone or who relapse during prednisone therapy. Plasmapheresis is used for patients who have rapid worsening of weakness due to myasthenia (i.e., myasthenic crisis). Plasmapheresis also is used in patients prior to thymectomy. Unfortunately, the effects of plasmapheresis are not long-lasting, and its use must be complemented by other treatments, whether prednisone or immunosuppressants.

Human immune globulin has been used for myasthenia gravis, just as it has for patients with other autoimmune neurologic disorders. The response rate has been established in only a relatively small number of patients. Typical dosing is the same as discussed previously, 0.4 mg/kg/day for 5 days. Alternative dosing also is being used, but no clear guidelines exist on indications for use in myasthenia or on dosing.

Thymectomy is commonly performed, but data indicating efficacy are conflicting. Current belief is that thymectomy is indicated for all patients with thymoma and for selected patients with active disease who are under 60 years of age. The beneficial effects of thymectomy may not be realized for years. However, the rate of complete remission of myasthenia gravis is so low that thymectomy is justified.

## MYASTHENIC SYNDROME

The Lambert-Eaton myasthenic syndrome is characterized by weakness and fatigability, which affect predominantly lower extremity muscles, especially proximally. Some patients may have ptosis, but prominent bulbar symptoms are uncommon. Autonomic symptoms include dry mouth, decreased lacrimation, and impotence.

Approximately two-thirds of patients with myasthenic syndrome have cancer; small-cell lung cancer is the most commonly associated tumor. Up to 80% of patients with myasthenic syndrome have IgG antibodies to voltage-gated calcium channels; the proportion does not differ between patients with and without cancer. The antibodies are thought to be related to a 58-kD synaptic vesicle protein, synaptogmin. In rats, antibodies made to synaptogmin produce electrophysiologic findings similar to myasthenic syndrome; a similar mechanism may be in effect in humans. It is thought that all myasthenic syndrome patients have antibodies that impair calcium movement across the cell membranes, but that there are limitations in the sensitivity of the antibody assay, which unfortunately, is not routinely available. Diagnosis therefore depends on clinical features and electrophysiologic abnormalities.

Nerve conduction velocity studies are normal except for low-amplitude CMAPs. With exercise, the amplitude of the CMAP is increased because of increased mobilization of transmitter; more calcium is available for mediation of transmitter release. Repetitive stimulation at high rates (20–50 Hz) results in an incremental response, with successive CMAPs being larger than the first. At low rates of stimulation (2–5 Hz), however, there is a decremental response. This is because the slow rate of repetitive stimulation does not allow for build-up of calcium in the presynaptic terminal. Hence some terminals fail to transmit subsequent stimuli. An increment in CMAP amplitude and rectified-integrated amplitude after exercise is more supportive of a diagnosis of myasthenic syndrome than other neurophysiologic tests. Future research will almost certainly focus on better diagnosis of this rare condition; unless it is looked for by careful clinical inspection and detailed neurophysiologic study, the diagnosis will be missed.

Treatment of myasthenic syndrome has typically been treatment of the underlying tumor in patients with associated malignancy. Even if this is successful, there is no guarantee that the myasthenic syndrome symptoms will improve. Guanidine has been used to augment the release of acetylcholine, but it is poorly tolerated because of numerous systemic adverse effects. 3,4-Diaminopyridine prolongs the duration of the presynaptic action potential, thereby allowing for more calcium entry into the muscle fibers. This has the same effect as exercising or high-frequency repetitive stimulation, in that terminal calcium is increased. Although diaminopyridine is probably the most effective agent, it is not available for routine use, so patients considered for treatment must be part of an ongoing study of the drug at an institution.

Other mechanisms of immune suppression have been used for patients with myasthenic syndrome. Plasmapheresis has only a limited effect, requiring use of other immune suppressants to maintain the improvement. None of the immune suppressants has been studied systematically in myasthenic syndrome, undoubtedly because few patients with this condition are encountered. Human immune globulin has been tried in a few patients with myasthenic syndrome, with conflicting results; a larger trial is clearly needed.

## Summary

During the past decade, advances in neuromuscular disease have been mainly in identification of metabolic and genetic diseases, improved neurophysiologic and biochemical diagnostic skills, and increased awareness of autoimmune neuromuscular disease. It is hoped that, in the near future, better guidelines for treatment of neuromuscular autoimmune diseases will be developed, and methods to better treat mitochondrial disorders and other metabolic defects of nerve and muscle will be discovered.

## References/Further Reading

Azulay J-P, Blin O, Pouget J. Intravenous immunoglobulin treatment in patients with motor neuron syndromes associated with anti-GM1 antibodies: a double-blind, placebo-controlled study. Neurology 1994;44:429–432.

Chaudhry V, Corse AM, et al. Multifocal motor neuropathy: response to human immune globulin. Ann Neurol 1993;30:397–401.

Chaudhry V, Corse AM, et al. Multifocal motor neuropathy: electrodiagnostic features. Muscle Nerve 1994;17:198–205.

Guillain-Barré Study Group. Plasmapheresis and acute Guillain-Barré syndrome. Neurology 1985;35:1096–1104.

Hartung HP, Pollard JD, Harvey GK, Toyka KV. Immunopathogenesis and treatment of the Guillain-Barré syndrome. Part I. Muscle Nerve 1995;18:137–153.

Hartung HP, Pollard JD, Harvey GK, Toyka KV. Immunopathogenesis and treatment of the Guillain-Barré syndrome. Part II. Muscle Nerve 1995;18:154–164.

Limburg PC, The TH, Hummel-Tappel E, et al. Anti-acetylcholine receptor antibodies in myasthenia gravis. Part 1. Relation to clinical parameters in 250 patients. J Neurol Sci 1983;58:357.

Nobile-Orazio E, Meucci N, Barbieri S, et al. High dose intravenous immunoglobulin therapy in multifocal motor neuropathy. Neurology 1993;43:537–544.

Oh SJ, Kim DE, Kuruoglu R, et al. Diagnostic sensitivity of the laboratory tests in myasthenia gravis. Muscle Nerve 1993;15:94–100.

Pestronk A, Cornblath DR, Ilyas AA, et al. A treatable multifocal motor neuropathy with antibodies to GM1 ganglioside. Ann Neurol 1993;27:316–326.

Rowland LP, Defenini R, Sherman W, et al. Macroglobulinemia with peripheral neuropathy simulating motor neuron disease. Ann Neurol 1982;11:532–536.

Stricker BH, van der Klauw MM, Ottervanger JP, van der Meche FG. A case-control study of drugs and other determinants as potential causes of Guillain-Barré syndrome. J Clin Epidemiol 1994;47:1203–1210.

van der Meche FGA, Schmidt PIM. Dutch Guillain Barré Study Group: a randomized trial comparing intravenous immune globulin and plasma exchange in Guillain-Barré syndrome. N Engl J Med 1992;326:1123–1129.

# DEGENERATIVE DISEASES

# 10 Neurodegenerative Diseases

James R. Burke
Jeffrey M. Vance
Mark J. Alberts

Neurodegenerative diseases (NDGDs) are more notable for differences in their clinical manifestations and mechanisms than their similarities, but they share some common features. One of these is the marked specificity of target cell destruction, a second is the onset of symptoms after a period of apparent normality, and a third is inexorable progression. The characteristic delay in onset of symptoms of NDGD until years after birth and the specificity of pathology led to many hypotheses about the etiology of neuronal degeneration, including a causative role for genetic mutations, exposure to organic or inorganic toxins, immune-mediated damage, excitotoxicity, loss of trophic factors, and programmed cell death.

## Etiology of Neurodegenerative Diseases

### GENETICS

In the assessment of a patient with an unidentified, progressive, neurologic syndrome, the family history is essential not only in diseases that are known to be genetic, like Huntington's disease (HD), but also in diseases generally thought to be sporadic, such as amyotrophic lateral sclerosis (ALS). In ALS, study of patients with an autosomal dominant form (5% of total ALS cases) led to the identification of causative mutations in the gene for superoxide dismutase. Genetic analysis also led to identification of different forms of NDGD that share a common clinical and neuropathologic phenotype: Alzheimer's disease (AD) was recently shown to be linked to at least four distinct genetic loci. Neurodegenerative diseases also have been shown to be the result of a new class of mutations: expansion of trinucleotide repeats.

Mutations in the nuclear genome are not the only cause of NDGD. Mitochondrial DNA mutations are found in Leber's optic atrophy and deletions of portions of the mitochondrial genome are found in chronic progressive external ophthalmoplegia. Prion diseases, such as Creutzfeldt-Jakob disease

(CJD), can be the result of either infection with a nonnucleic acid–containing proteinaceous particle or inherited or spontaneous mutation of the host prion gene. In addition, in inherited prion diseases, the same mutation can result in different clinical and neuropathologic findings. Familial CJD and fatal familial insomnia (FFI) are associated with the same mutation in the prion gene at codon 178 but have different clinical presentations.

## ENVIRONMENTAL TOXINS

Many NDGDs occur sporadically and do not appear to have a genetic component. Some may be caused by environmental toxins. A dramatic example of geographic clustering is seen on the island of Guam. The Chamorro residents of this western Pacific island, until recently, had a markedly increased rate of ALS-parkinsonism-dementia complex. Some studies implicated cycad flour, which was a major dietary constituent and contains an amino acid that could result in excitotoxicity. Later work has tempered the enthusiasm for this explanation, but an environmental cause remains likely. Other organic compounds, such as 1-methyl-4-phenyl-1,2,3,6-tetrahydropyridine (MPTP), cause parkinsonism, and such inorganic toxins as aluminum and iron have been proposed to play a role in the development of AD and of Hallevorden-Spatz and Parkinson's diseases, respectively.

## THE IMMUNE SYSTEM IN NEURODEGENERATION

The pathology of some NDGDs is notable for signs of inflammation, including complement deposition and the presence of microglia and reactive astrocytes. Microglial infiltrates are seen in AD and Parkinson's disease (PD), but it is unknown whether the affected cells are responsible for pathogenesis, are serving to slow disease progression, or are an epiphenomenon. An immune pathogenesis is supported by a slowing of AD progression by indomethacin, a nonsteroidal anti-inflammatory agent (NSAIA), and by twin studies showing an inverse correlation between treatment with NSAIAs or steroids and AD. Unfortunately, conclusions from these data must await replication and clarification.

## EXCITOTOXICITY

Excitotoxicity is the phenomenon whereby excessive stimulation of neurons results in cell death; it can result from generation of free radicals, or excessive release of glutamate or calcium, and may be a final common pathway for neuronal cell death (see Chapter 1).

## GROWTH FACTORS

Nerve growth factors are candidates likely to play a role in NDGD because they have marked specificity of action. One of them is critical to the development of sympathetic and sensory neurons of the peripheral nervous system, as well as forebrain cholinergic neurons, which are severely affected in AD. Depletion of growth factors may cause neurodegeneration in PD, AD, and ALS, but evidence is lacking. Even if the primary cause of these disorders is not loss of trophic factors, supplementation may result in decreased cell loss and slowing of disease progression.

## PROGRAMMED CELL DEATH

Approximately half of the neurons generated in the mammalian CNS die before adulthood. This preprogrammed death is the result of environmental cues and activation of a genetic program. Aberrant regulation of the cell death program may be involved in the etiology of NDGD. Similarities between programmed cell death and neurodegeneration include marked target selectivity, latency of activation, and the occurrence of both familial and spontaneous disorders.

# Alzheimer's Disease

## CLINICAL

The onset of AD is typically gradual, with insidious progression of memory problems, primarily difficulty with short-term memory. Later, other spheres of cognition are affected, the patients lose the ability to care for themselves, and they eventually become totally dependent on others. Ultimately, they become bedridden, unable to eat, and often immobile. Death is usually due to intercurrent infection, cardiac arrest, or malnutrition. The average time from disease onset to death is 7 to 10 years.

## PATHOLOGY

In AD, the brain shows some degree of cortical atrophy with neuronal loss, particularly in layers III and V of the cortex. Loss of cholinergic neurons projecting from the nucleus basalis of Meynert (nbM) and other forebrain areas is invariably present; in the nbM, it amounts to an overall neuronal loss of 15% to 18%, with 70% of larger neurons from some regions.

Two cellular pathologic findings considered hallmarks of AD are neurofibrillary tangles (NFTs) and senile plaques (SPs). A major constituent of the NFT is the paired helical filament (PHF), composed virtually entirely of the microtubule-associated protein tau. Tau enhances tubulin polymerization and is regulated by the extent and site of phosphorylation. When tau is hyperphosphorylated, its ability to bind microtubules is reduced by approximately 90%; it forms PHFs instead.

Senile plaques are spherical, extracellular lesions with a central deposit of beta-amyloid (also called A), which is a constituent of the larger amyloid precursor protein (APP). Two types of SPs have been described: diffuse and classic (or mature). Diffuse SPs are probably precursors of the classic plaque. The highest concentration of SPs occurs in the hippocampus, amygdala, and cerebral cortex. A can also be deposited in the walls of cerebral and leptomeningeal vessels, producing an amyloid angiopathy, but this occurs neither invariably nor exclusively in AD.

Neither NFTs nor SPs are unique or specific for AD, but they are seen in other degenerative disorders such as Down syndrome, the parkinsonian-dementia complex of Guam, subacute sclerosing panencephalitis, and normal aging. Synaptic loss correlates better with severity of dementia than does the number of NFTs or SPs.

## ETIOLOGY

Recent research into the etiology of AD has centered around genetic factors and the role of beta-amyloid, there being relatively little evidence to support other etiologies, whereas the evidence supporting a genetic etiology is, in many cases, convincing.

The exact percentage of AD cases that are familial or due to genetic factors remains unclear. Once thought to be rare, familial AD (FAD) is now recognized to be common. From our experience in a Memory Disorders Clinic, approximately 50% of all AD patients have one or more first-degree relatives with AD or an unspecified dementing illness. Considering incomplete ascertainment and misdiagnosis, a 50% prevalence of FAD may be conservative.

In the early genetic studies, FAD was divided into early-onset (<60 or 65 years) and late-onset (> 60 or 65 years) cases. As seen in Table 10.1, early-

**Table 10.1**   Genetic Loci for Familial Alzheimer's Disease

| Chromosome | Gene | Clinical Features | Frequency |
|---|---|---|---|
| 14q24.3 | Unk | Early-onset, rapid progression, myoclonus, aphasia | Infrequent |
| 19q13 | ApoE | Late-onset, slowly progressive | Common |
| 21q21.3 | APP | Early-onset | Rare |
| ? | Unk | Volga-German ancestry, mean age 58 yr | Rare |
| ? | Unk | Late-onset, slowly progressive | Common |

Early-onset typically refers to a mean age of onset less than 60 to 65 years; late-onset typically refers to a mean age of onset of greater than 65 years.
Unk, unknown; ApoE, apolipoprotein E.

onset FAD has been linked to several chromosomes (and genes in some cases), including chromosomes 21 and 14. Late-onset FAD has been linked to chromosome 19, but there are likely to be other loci involved in this disease. Another variety of early-onset FAD, the Volga-German kindred, appears not to be linked to any of the loci described to date.

Among the candidate genes and proteins studied in AD, beta-amyloid has been the focus of much attention; its parent molecule is APP, which maps to the proximal long arm of chromosome 21. It is the pathogenic gene for at least some rare cases of FAD. A variety of mutations in APP can produce early-onset FAD. The first described of these occurs at codon 717 of APP, just beyond the C-terminus of the A moiety, and is a single-base point mutation; a 2-base mutation occurs proximal to the N-terminus of the A moiety. The known APP/FAD mutations are summarized in Table 10.2. The APP mutations are rare, and probably account for fewer than 10% of early-onset FAD cases and less than 0.5% of all AD. No mutations in the APP gene have been found in cases of sporadic AD.

The pathophysiology of the APP mutations is unclear. Although these mutations may alter APP processing and cause the increased release of $A\beta$ peptide and increased deposition or decreased metabolism of $APP/A,\beta$ no definitive proof of any specific mechanism has emerged. Although $A\beta$ may be toxic to neurons, recent studies have failed to document this; in fact, $A\beta$ may actually promote neuronal outgrowth. More work is needed before the pathophysiology of $A\beta$ secretion, processing, and deposition in patients with the $APP_{717}$ mutations can be fully understood.

The gene on chromosome 14 that is responsible for some cases of early-onset FAD remains unidentified. Recent clinical studies of the FAD pedigrees linked to chromosome 14 have shown intriguing differences in both the clinical and pathologic manifestations of this type of AD. These patients had a mean age of onset of 36 years, and aphasia and myoclonus became prominent early in the course of the disease, suggesting that chromosome 14–linked FAD may differ from other varieties of FAD.

**Table 10.2**  Mutations in the Amyloid Precursor Protein Known to Cause Familial Alzheimer's Disease

| Clinical Features | Codon(s) | Mutation |
|---|---|---|
| Dementia, ICH | 692 | C to G transversion |
| Dementia | 670/671 | G to T and A to C |
| Dementia | 717 | Three different single-base changes |

ICH, intracerebral hemorrhage.
SOURCE: Adapted by permission from Schellenberg G, Bird T, Wijsman E, et al. Genetic linkage evidence for a familial Alzheimer's disease locus. *Science* 1992;258:668–771.

## APOLIPOPROTEIN E

Perhaps no recent discovery has affected our understanding of AD pathogenesis more than the discovery that apolipoprotein E is a major determinant of AD risk and age of onset. To avoid confusion, the causative gene is designated as APOE, and the protein, apoE.

APOE has three major isoforms, or alleles—E2, E3, and E4—which vary by one amino acid and have different biophysical properties. ApoE binds to the A protein; the avidity of binding is isoform-specific, with apoE4 binding more rapidly than apoE3. This was significant because earlier studies demonstrated genetic linkage of late-onset FAD to the proximal long arm of chromosome 19, in the region of the APOE gene. The allele frequency of APOE 4 is markedly increased in patients with late-onset FAD (0.40), compared with controls (0.09–0.17), both in familial and in sporadic late-onset AD. Figure 10.1 shows the effect of APOE genotype on age of disease onset in individuals presenting to a memory disorders clinic. The average age of onset increases from 70 years in individuals homozygous for E4, to 85 years in individuals with the genotype E3/E3. The E2 allele is protective; individuals with the E2 allele have a reduced risk of developing AD and a much later age of onset. This may have significant applications in terms of designing therapeutic agents.

The exact mechanism by which the apoE genotype affects the development and age of onset for AD is a subject of some debate and speculation. Strittmatter and colleagues have developed a hypothesis for AD causation based on isoform-specific interactions of apoE and tau. In vitro, tau irreversibly binds to apoE3 but not to apoE4. Tau bound by apoE3 may not become hyperphosphorylated, remaining available to bind to microtubules and thereby stabilize them. Because apoE4 does not readily bind to tau, unbound tau can become hyperphosphorylated and form PHFs. The loss of tau bound to

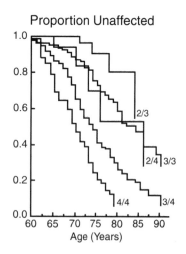

**Figure 10.1**

Probability of remaining unaffected by Alzheimer's disease in relation to the apolipoprotein E genotype. This illustration shows age of onset for subjects with indicated APOE genotypes. Onset curves were estimated by Kaplan-Meier product limit distributions. Data are limited to late-onset AD and should not be generalized to all populations at risk. (Courtesy of Dr. Allen Roses, Duke University Medical Center).

microtubules causes destabilization of the microtubules; this results in impaired axonal transport, eventually leading to cell death.

Selkoe, however, has pointed out that (1) neurons do not synthesize apoE, (2) clinical AD and its related pathology can be seen in APOE3/E3 individuals, and (3) the hypothesis does not account for A deposition. However, (1) in the brain, apoE is synthesized in glial cells; recent studies by Han et al. have shown apoE within brain neurons, perhaps resulting from the binding and internalization of the apoE synthesized by neuroglia. (2) Although AD (both familial and sporadic) certainly occurs in individuals with the E3/E3 APOE genotype, this does not negate the role of apoE4 in AD: E3/E3 individuals with AD may have other *causative* genetic factors, apart from the apoE locus; APOE appears to be a *susceptibility* locus, altering the age of onset and probability of developing the disease. (3) The significance of A deposition in AD pathogenesis remains controversial. Even cases with the APP$_{717}$ mutation show marked variability in the density and distribution of A plaques; A deposition, therefore, may not be the gold standard by which explanations of disease expression should be measured.

These recent genetic findings have greatly advanced our understanding of the molecular pathogenesis of AD. Genetic markers are now available that identify populations at risk for AD, though we do not encourage the use of those currently available for individual genetic counseling or risk assessment; and new approaches for the development of prophylactic and therapeutic agents are now possible. Families and at-risk individuals can be screened for the various APP mutations. Such screening might be useful for detailed genetic studies and in therapeutic trials that aim to prevent disease expression. The use of the APOE genotype for the diagnosis of AD is an evolving area. Genotype data could be used to stratify patients in treatment trials and epidemiologic studies, since they provide a clear indication of age of onset. The therapeutic implications of APOE genotype are more profound. Agents that mimic the effects of apoE2 or apoE3 could greatly retard disease expression for some at-risk individuals with the E3/E4 or E4/E4 genotype.

## Trinucleotide Repeat Diseases

Prior to 1990, only a few types of mutations were known to cause inherited disease, the most well known being point mutations (substitution of one base for another). If such a mutation produces a change in the amino acid coded by that codon, an aberrant protein may be produced, the expression of which results in disease. Less common mutations are modifications of the same theme: deletions, duplications, and inversions. All are rare events, and once introduced into genomic DNA, are stably transmitted to subsequent generations. The newly discovered dynamic mutations, trinucleotide repeat expansions, do not follow these rules.

Trinucleotide repeats (e.g., CAG CAG CAG . . .) are normal nucleotide sequences found in some genes. Disease occurs when the number of repeats becomes too large. Alleles with a small number of repeats are generally stable and inherited in traditional mendelian fashion. On rare occasions, small repeats expand, generating a "premutation" allele not associated with disease. Premutation alleles are much more unstable then the common, smaller repeat alleles. They can, in a single generation, jump to a much larger size and produce disease. Premutation alleles also can vary in size from generation to generation.

Nine disorders are known to be due to trinucleotide expansions (Table 10.3). Trinucleotide repeats can be divided into two primary groups: The expressed CAG repeat diseases, which result from expression of the CAG as a polyglutamine tract in the disease protein, are NDGDs. In the non-expressed repeat diseases (myotonic dystrophy [DM] and fragile X [FRAX] syndromes), the trinucleotide repeat occurs within a noncoding region of DNA. These are all neurologic disorders and manifest the phenomenon of anticipation; that is, disease severity increases in succeeding generations. Because repeat size is inversely correlated with age of disease onset, the demonstration of rapid expansion of repeats from generation to generation provides an explanation for anticipation.

## CAG NEURODEGENERATIVE DISORDERS

The CAG neurodegenerate (CAG-NDG) disorders group includes HD, spinal cerebellar ataxia type 1 (SCA1), dentatorubral pallidoluysian atrophy (DRPLA), the allelic Haw River syndrome (HRS), Machado-Joseph disease (MJD), and Kennedy's disease, or spinobulbar muscular atrophy (SBMA). All but SBMA are autosomal-dominant diseases with variable ages of onset; they present with a variable combination of movement abnormalities, dementia, and ataxia. The pathologic findings overlap, but all have distinctive features.

**Table 10.3** Trinucleotide Repeat Disorders

| Disorder | Repeat Sequence | Expression | Expanded Range |
|---|---|---|---|
| Huntington | CAG | Polyglutamine | ?38–121 |
| SCA1 | CAG | Polyglutamine | >43 |
| SBMA | CAG | Polyglutamine | 40–62 |
| HRS | CAG | Polyglutamine | >49 |
| DRPLA | CAG | Polyglutamine | 49–75 |
| MJD | CAG | Polyglutamine | 68–79 |
| DM | CTG/CAG | Not expressed | 100–2000 |
| FRAXA | CGG | Not expressed | 200–2000 |
| FRAXE | CGG/CCG | ? | 200–1000 |
| FRAXF | CGG/CCG | ? | 300–500 |

The age of onset of the trinucleotide repeat diseases is inversely correlated with increasing size of expansion, but only 50% of the variability of age of onset in HD can be explained by the repeat length. The factors responsible for the remaining variation are unknown. Similarly, variability in clinical expression is not correlated with repeat length, so other unidentified factors must be important.

The CAG-NDG disorders share several striking similarities. The range of abnormal repeats in CAG-NDG is relatively small (< 160), contrasting with the large number of repeats seen in DM and FRAX. The CAG-NDG disease repeats also have limited generational instability, compared with DM and FRAX. It is not clear, however, at what time and in what tissue the instability occurs. Trinucleotide expansion is more likely transmitted from an affected father, so one large source of this expansion must occur in sperm.

The CAG repeat in all of these diseases is in the expressed portion of the gene, and codes for the amino acid glutamine. The mechanism by which expansion of a polyglutamine stretch in a protein causes neurodegeneration is unknown, but several mechanisms have been suggested: Slow toxic effects of abnormally large polyglutamine cleavage products may disturb the intracellular and mitochondrial milieu; polyglutamine mimics polyamines and causes excitotoxicity by stimulation of glutamate receptors; and formation of tetrahelical structures by the amino acid repeat causes cross-links that ultimately result in neuronal death. Because of their long life and inability to divide, neurons may be particularly susceptible to the long-term toxic effects of polyglutamine. Perutz et al. demonstrated that polyglutamine forms beta-sheets and suggested that the beta-sheet structure may function as a binding mechanism that causes protein precipitation or altered transcriptional regulation.

How these diseases cause specific targeting of neuronal degeneration is unknown. Differential expression of the disease protein does not explain the neuropathologic specificity, because transcriptional analysis by Northern blot has shown that messenger RNA of the trinucleotide repeat disease genes is expressed in virtually all tissues studied, as also is huntingtin, the protein product of the HD's gene. The mechanism of how an expanded glutamine repeat causes neurodegeneration may become apparent when the normal function of these genes is discovered.

## Dentatorubral Pallidoluysian Atrophy and Haw River Syndrome
Dentatorubral pallidoluysian atrophy is an autosomal-dominant disorder occurring almost exclusively in Japan. It is associated with progressive myoclonus, dementia, epilepsy, ataxia, and chorea. DRPLA is due to expansion of a CAG trinucleotide repeat in a gene on chromosome 12p. Screening of similar cases in Great Britain has demonstrated that DRPLA is also found in non-Japanese individuals. Haw River syndrome, a similar but not identical NDGD found in

an African-American family living near the Haw River in central North Carolina, shares the same molecular CAG expansion as DRPLA.

Dentatorubral pallidoluysian atrophy is unique among the trinucleotide repeat diseases in that it has a markedly restricted geographic distribution. The large difference in prevalence between the Japanese and other populations may be due to the greater prevalence (7%–10%) of large, premutation expansions (also known as intermediate alleles) in the normal Japanese population, compared with greater than 1% of Caucasians and African-Americans. A decreased frequency in premutation-size alleles also has been noted in the HD gene in Japan, where HD is rare, and in the DM gene in black South Africans, where DM is reported to be nonexistent. Premutation allele frequency is therefore a likely explanation for the varying incidence of these diseases.

### THE NON-EXPRESSED TRINUCLEOTIDE REPEAT DISORDERS

Fragile X syndrome (FRAX types A, E, and F), which is the most common form of inherited mental retardation and was the original disorder identified as the result of a trinucleotide repeat expansion, and DM are members of the non-expressed (non-CAG) trinucleotide disorders group. Both diseases exhibit intergenerational instability and can form very large repeats. The repeat lies in the 5' untranslated region of the protein FMR1, which is thought to bind RNA. In affected individuals, the protein is not transcribed, due to the expansion of the CGG repeat and the hypermethylation of cytosine residues.

Myotonic dystrophy is the most common muscular dystrophy in adults, and is due to an expanded CTG repeat in the 3' untranslated region of a protein kinase gene. Although its primary expression is in neuronal tissue, it has, unlike the other trinucleotide repeats, significant disease expression in non-neuronal tissues: DM patients commonly develop cataracts, insulin-dependent diabetes, gastrointestinal abnormalities, and cardiac arrhythmias, along with myotonia, muscle weakness, atrophy, and personality changes. Intergenerational instability and somatic variability contribute to the wide clinical variability seen in the disorder. Conflicting results have been reported in measuring the transcription of the DM gene. Most recent studies, however, suggest that although the CTG is in the untranslated portion of the gene, the expression of the DM gene is normal. The mechanism that leads to disease is not known; the CTG expansion may bind nucleosomes, increasing repression of transcription.

## Parkinson's Disease

The role of genetics in the pathogenesis of PD is unknown. Environmental factors may play a role in the causation of PD, but no specific toxin has been identified in patients with idiopathic PD.

## CLINICAL AND PATHOLOGIC FEATURES

The clinical diagnosis of PD is based on a classic symptom complex that includes bradykinesia, rest tremor, motor rigidity, cogwheeling, and postural instability. Although the clinical diagnosis of PD is often straightforward, many diseases mimic its symptoms and signs. In an autopsy study of 100 patients given the diagnosis of PD by neurologists, Hughes found that the clinical diagnosis was erroneous in 20 (20%). Diseases misdiagnosed as idiopathic PD include progressive supranuclear palsy, multisystem atrophy, Alzheimer's-type pathology, AD, and basal ganglia vascular disease.

The neuropathologic diagnosis of PD requires the presence of Lewy bodies. Classic Lewy bodies are eosinophilic intraneuronal inclusions found in the substantia nigra, locus ceruleus, and dorsal motor nucleus of the vagus; they are also found in normal aging and are present at autopsy in up to 10% of asymptomatic elderly people. Another pathologic hallmark of PD is loss of dopamine-containing neurons, which leads to depigmentation of the substantia nigra and locus ceruleus. There is also loss of neurons containing norepinephrine, serotonin, gamma-aminobutyric acid, and acetylcholine.

Clinical signs and symptoms of PD result when fewer than 50% of the substantia nigra pars compacta neurons are destroyed and 80% of striatal dopamine is lost. Substantia nigra neurons also are lost as a result of aging, but the pattern of cell loss within the substantia nigra is different in aging and PD. Signs of inflammation such as microglial infiltration are present in PD, but not in normal aging; and bradykinesia of the elderly does not appear to be the result of an insufficient supply of dopamine.

## EPIDEMIOLOGY

Parkinson's disease is one of the most common NDGDs. The average age of onset is approximately 60 years, and its incidence increases until age 75 and then declines.

Many studies have examined the role of genetics and the environment in the genesis of PD. Monozygotic twins are no more likely than dizygotic twins to be concordant for PD, but there are rare individual families with multiple, affected individuals, which suggests a genetic etiology. The genetic data are consistent with the hypothesis that inheritance of PD as a single gene trait is uncommon, but polygenic inheritance or a gene with low penetrance cannot be eliminated.

Epidemiologic studies have been successful in identifying several factors associated with development of PD. Individuals who do not smoke have an increased risk of PD. The mechanism by which cigarette smoking protects against the onset of PD is unknown. The initiation of smoking following onset of PD, however, does not protect against progression of the disease. It is possible that absence of smoking is an epiphenomenon, and the true risk factor is unknown. Another risk factor for PD is rural living; again, the exact correlate is only speculative.

## ETIOLOGY

The etiology of PD is unknown, but hypotheses span the spectrum from gene defects to exposure to environmental toxins. The hypothesis that environmental factors are important in the development of PD was strengthened when a chemist, attempting to make an illicit synthetic opiate, inadvertently synthesized a potent specific neurotoxin, MPTP. Injection of MPTP results in a permanent disease that is clinically indistinguishable from PD.

The mechanism by which MPTP causes parkinsonism has been studied intensely. MPTP is a lipophilic molecule that rapidly enters the brain. There it is oxidized by extraneuronal monoamine oxidase B (MAO-B) to an unstable form, $MPDP^+$, which breaks down to $MPP^+$. $MPP^+$ is transported by neuronal dopamine-uptake systems and accumulates inside neurons; it is presumed to be the agent responsible for neuronal death. Once inside neurons, $MPP^+$ is sequestered in mitochondria, where it interferes with energy metabolism by inhibiting nicotinamide adenine dinucleotide (NAD)-linked oxidation in complex I. Inhibition of complex I may lead to an increase in free radicals from the respiratory chain, which then cause extensive cell damage and ultimately neuronal cell death.

The finding that $MPP^+$ inhibits mitochondrial complex I led several groups to examine mitochondrial function in the brains of patients with PD. As predicted by the MPTP model, decreased complex I activity has been found in the substantia nigra of patients with PD, but this finding is not universal.

Data from the MPTP model implicate free radicals as a cause of neurodegeneration in PD. Normal oxidation of dopamine has been proposed to generate free radicals, and markers of free radical production are increased in the substantia nigra of patients with PD, while glutathione, an antioxidant, is decreased. Unfortunately, supplementation of antioxidant defenses with vitamin E in patients with mild symptoms of PD does not slow disease progression. A trial of the MAO-B inhibitor deprenyl in early PD was also undertaken, because MAO-B is implicated in oxidation of MPTP to $MPP^+$, oxidation of dopamine, and free radical generation. Treatment of mildly symptomatic PD patients with deprenyl (selegiline) resulted in a significant delay of disease progression until L-dopa therapy was necessary. It remains unclear, however, whether this was due to its effect on inhibiting the breakdown of dopamine or occurred via a neuroprotective mechanism. Figure 10.2 shows a model of PD pathogenesis, incorporating a role for free-radical generation and decreased antioxidant defenses.

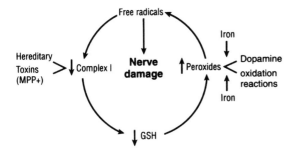

**Figure 10.2**
A proposed model for the pathogenesis of Parkinson's disease. This illustration shows how a deficiency in mitochondrial complex I, a deficiency in glutathione, or an increase in peroxide formation can lead to a cycle resulting in free radical formation and neuronal damage. This model takes into account that different or even multiple etiologic factors might contribute to the pathogenesis of PD. (Reproduced by permission from Olanow CW, Calne D. Does selegiline monotherapy in Parkinson's disease act by symptomatic or protective mechanisms? Neurology 1991;42(Suppl 4):13–26. Copyright 1991, Advanstar Communications.)

The possible role of free radicals and MPTP in PD suggests an explanation for the protective effect of cigarette smoking. Carbon monoxide can act as a reducing agent, thereby reducing the oxidative stress that has been proposed to lead to nigral neuronal death. In addition, MAO-B levels in platelets from smokers are 75% of those from nonsmokers. If the development of PD requires activation of an MPTP-like neurotoxin by MAO-B oxidation, then the lower MAO-B levels in smokers may explain their decreased incidence of PD.

Given the epidemiologic association of PD with rural living and absence of smoking, why do not all individuals with these risk factors develop disease? In fact, PD is uncommon, even among people at increased risk. The ability to metabolize exogenously administered compounds is a genetic trait, and patients with PD have an altered ability to deal with sulfur-containing compounds and show abnormalities of pyridine metabolism; MPTP is a pyridine derivative. Further studies on metabolism in PD patients clearly need to be done.

Epidemiologic, biochemical, and genetic studies suggest a multi-step mechanism for the pathogenesis of PD. Environmental exposure to a weak neurotoxin may have to be coupled with impaired metabolic pathways to result in PD. A persistent area of inquiry involves the existence or identity of a putative neurotoxin(s). If a toxin is identified, prophylaxis may then be possible by reducing exposure to it or supplementing host defenses to prevent or delay neurodegeneration.

## Conclusion

Progress in understanding NDGD has been substantial, but fundamental questions remain. Why are selected neuronal groups vulnerable to degeneration, while adjacent areas are spared? What accounts for the sometimes decades-long delay between birth and onset of symptoms? Does specific neuronal pathology imply a single etiology, or do multiple factors combine into a final common pathway? A detailed understanding of pathogenesis will be required to answer these questions. If the progress made in learning about NDGD in the past 10 years is any indication, many of the answers may soon be available.

## References/Further Reading

Agid Y, Javoy-Agid F, Ruberg M. Biochemistry of neurotransmitters in Parkinson's disease. In: Marsden CD, Fahn S, eds. Movement disorders 2. London: Butterworths, 1987:166–230.

Amouyel P, Brousseau T, Fruchart J-C, Dallongeville J. Apolipoprotein E- 4 allele and Alzheimer's disease. Lancet 1993;342:1309.

Appel SH. A unifying hypothesis for the cause of amyotrophic lateral sclerosis, parkinsonism and Alzheimer's disease. Ann Neurol 1991;10:499–505.

Bernheimer H, Birkmayer W, Hornykiewicz O, et al. Brain dopamine and the syndromes of Parkinson and Huntington: clinical, morphological and neurochemical correlations. J Neurol Sci 1973;20:415–455.

Bird TD. Familial Alzheimers disease. Ann Neurol 1994;36:335–336.

Borgaonkar DS, Schmidt LC, Martin SE, et al. Linkage of late-onset Alzheimer's disease with apolipoprotein E type 4 on chromosome 19. Lancet 1993;342:625.

Bramblett GT, Goedert M, Jakes R, et al. Abnormal tau phosphorylation at ser396 in Alzheimer's disease recapitulates development and contributes to reduced microtubule binding. Neuron 1993;10:1089–1099.

Breitner JCS, Gau BA, Welsh KA, et al. Inverse association of anti-inflammatory treatments and Alzheimer's disease: initial results of a co-twin study. Neurology 1994;44:227–232.

Corder EH, Saunders AM, Strittmatter WJ, et al. Gene dose of apolipoprotein E type 4 allele and the risk of Alzheimer's disease in late onset families. Science 1993;261:921–923.

Corder EH, Saunders AM, Risch NJ. Protective effect of apolipoprotein E type 2 allele for late onset Alzheimer disease. Nat Genetics 1994;7:180–184.

Fearnley JM, Lees AJ. Aging and Parkinson's disease: substantia nigra regional selectivity. Brain 1991;114:2283–2301.

Goedert M. Tau protein and the neurofibrillary pathology of Alzheimer's disease. Trends Neurosci 1993;16:460–465.

Haltia M, Viitanen M, Sulkava R, et al. Chromosome 14-encoded Alzheimers disease—genetic and clinicopathological description. Ann Neurol 1994;36:362–367.

Han SH, Einstein G, Weisgraber KH, et al. Apolipoprotein E is localized to the cytoplasm of human cortical neurons—a light and electron microscopic study. J Neuropathol Exp Neurol 1994;53:535–544.

Hardy J, Higgins G. Alzheimer's disease: the amyloid cascade hypothesis. Science 1992;256:184–185.

Hefti F. Growth factors and neurodegeneration. In: Calne DB, ed. Neurodegenerative diseases. Philadelphia: Saunders, 1994:177–194.

Hefti F, Hartikka J, Knusel B. Function of neurotrophic factors in the adult and aging brain and their possible use in the treatment of neurodegenerative diseases. Neurobiol Aging 1989;10:515–533.

Hughes AJ, Daniel SE, Kilford L, Lees AJ. Accuracy of clinical diagnosis of idiopathic Parkinson's disease: a clinico-pathological study of 100 cases. J Neurol Neurosurg Psychiatry 1992;55:181–184.

Huntington's Disease Collaborative Research Group. A novel gene containing a trinucleotide repeat that is expanded and unstable on Huntington's disease chromosomes. Cell 1993;72:971–983.

Jenner P. Oxidative damage in neurodegenerative disease. Lancet 1994;344:796–798.

Koller WC, Vetere-Overfield B, Gray C, et al. Environmental risk factors in Parkinson's disease. Neurology 1990;40:1218–1221.

Kosik K. Tau protein and Alzheimer's disease. Curr Opinion Cell Biol 1990;2:101–104.

Marsden CD. Parkinson's disease in twins. J Neurol Neurosurg Psychiatry 1987;50:105–106.

Martila RJ. Epidemiology. In: Koller W, ed. Handbook of Parkinson's disease. New York: Marcel Dekker, 1987:50–69.

McGeer P, McGeer E, Suzuki J, et al. Aging, Alzheimer's disease and the cholinergic system of the basal forebrain. Neurology 1984;34:741–745.

McGeer PL, McGeer EG, Suzuki JS. Aging and extrapyramidal function. Arch Neurol 1977; 34:33–35.

Olanow CW. A radical hypothesis for neurodegeneration. Trends Neurosci 1993;16:439–444.

Pericak-Vance MA, Bebout JL, Gaskell PC Jr, et al. Linkage studies in familial Alzheimer disease: evidence for chromosome 19 linkage. Am J Hum Genet 1991;48:1034–1050.

Perutz MF, Johnson T, Suzuki M, Finch JT. Glutamine repeats as polar zippers: their possible role in inherited neurodegenerative diseases. Proc Natl Acad Sci USA 1994;91:5355–5358.

Rajput AH. Clinical features and natural history of Parkinson's disease (special consideration of aging). In: Calne DB, ed. Neurodegenerative diseases. Philadelphia: Saunders, 1994: 555–572.

Rogers J, Kirby LC, Hempelman SR, et al. Clinical trial of indomethacin in Alzheimer's disease. Neurology 1993;43:1609–1611.

Rosen DR, Siddique T, Patterson N, et al. Mutations in Cu/Zn superoxide dismutase gene are associated with familial amyotrophic lateral sclerosis. Nature 1993;362:59–62.

Roses A. Apolipoprotein E affects the rate of Alzheimer disease expression: beta amyloid burden is a secondary consequence dependent on APOE genotype and duration of disease. J Neuropathol Exp Neurol 1994;53:429–437.

Roses AD, Pericak-Vance MA, Clark CM, et al. Linkage studies of late-onset familial Alzheimer's disease. Adv Neurol 1990;51:185–196.

Saunders AM, Schmader K, Breitner JCS, et al. Apolipoprotein E 4 allele distributions in late-onset Alzheimer's disease and in other amyloid-forming diseases. Lancet 1993;342: 710–711.

Saunders AM, Strittmatter WJ, Schmechel D, et al. Association of apolipoprotein E allele epsilon 4 with late-onset familial and sporadic Alzheimer's disease. Neurology 1993;43: 1467–1472.

Schmechel DE, Saunders AM, Strittmatter WJ, et al. Increased amyloid-peptide deposition in cerebral cortex as a consequence of apolipoprotein E genotype in late-onset Alzheimer disease. Proc Natl Acad Sci USA 1993;90:9649–9653.

Schoenberg BS. Descriptive epidemiology of Parkinson's disease: disease distribution and hypothesis formulation. In: Yahr, Bergman, eds. Parkinson's disease. Advances in Neurology, Volume 45. New York: Raven Press, 1986;277–284.

Schoenberg BS, Anderson DW, Haerer AF. Prevalence of Parkinson's disease in the biracial population of Copiah county, Mississippi. Neurology 1985;35:841–845.

Selkoe D. Alzheimer's disease: A central role for amyloid. J Neuropathol Exp Neurol 1994;53: 438–447.

Spencer PS, Allen RG, Kisby GE, et al. Excitotoxic disorders. Science 1990;248:144.

St. George-Hyslop P, Haines J, Rogaev E, et al. Genetic evidence for a novel familial Alzheimer's disease locus on chromosome 14. Nature Genet 1992;2:330–334.

St. George-Hyslop P, Myers R, Haines J, et al. Familial Alzheimer's disease: progress and problems. Neurobiol Aging 1989;10:417–425.

Strittmatter WJ, Weisgraber KH, Goedert M et al. Hypothesis: microtubule instability and paired helical filament formation in the Alzheimer disease brain are related to apolipoprotein E genotype. Exp Neurol 1994;125:163–171.

Strittmatter WJ, Saunders AM, Schmechel D, et al. Apolipoprotein E: high-avidity binding to beta-amyloid and increased frequency of type 4 allele in late-onset familial Alzheimer disease. Proc Natl Acad Sci USA 1993;90:1977–1981.

Tanner CM, Langston JW. Do environmental toxins cause Parkinson's disease? A critical review. Neurology 1990;40(Suppl 3):17–30.

Yoshizawa T, Yamakawa-Kobayashi K, Komatsuzaki Y, et al. Dose-dependent association of apolipoprotein E allele e4 with late-onset sporadic Alzheimer's disease. Ann Neurol 1994; 36:656–659.

Ward CD, Duvoisin RC, Ince SE, et al. Parkinson's disease in 65 pairs of twins and in a set of quadruplets. Neurology 1983;33:815–824.

# Movement Disorders

John B. Penney, Jr.

This chapter reviews the anatomy and functions of the basal ganglia, particularly with regard to their control of movement. The topics covered include input and output connections of the various nuclei; details of the internal anatomy of the striatum; the organization of several feedback loops, and how these feedback loops are thought to function; how dysfunction of these feedback loops results in the various disease states associated with basal ganglia disease; the pathogenic mechanisms that are thought to be involved in Huntington's disease and Parkinson's disease; and the pathophysiology of parkinsonian dose-response fluctuations.

## Basal Ganglia Input and Output Connections

The basal ganglia are a collection of gray matter nuclei located deep within the cerebral hemispheres. Although the amygdala, claustrum hippocampus, and thalamus are, strictly speaking, gray matter nuclei within the hemispheres, their connections and functions are different from the nuclei that are commonly considered part of the basal ganglia. The major input nuclei for the basal ganglia are the caudate nucleus, the putamen, and the nucleus accumbens. These three nuclei develop from the same embryonic primordium, are contiguous with each other, and have the same internal histologic features. Therefore, together they are considered one histologic entity, the striatum.

The striatal structures (caudate, putamen, and nucleus accumbens) receive somatotopic input from the cerebral cortex as well as input from the midline (centromedian and parafascicular) nuclei of the thalamus, and from the dopamine neurons found in the substantia nigra pars compacta and ventral tegmental area. The other basal ganglia structure that receives somatotopic cortical input is the subthalamic nucleus. The cortical input to the striatum is known to use glutamate as its neurotransmitter and to be excitatory on basal ganglia neurons. The thalamic input to striatum also is excitatory and is probably glutamatergic.

The substantia nigra pars compacta itself receives some input from the cortex, input from the other basal ganglia structures, and ascending input from the acetylcholine neurons of the pedunculopontine nucleus of the brainstem.

There is input from the brainstem to the other basal ganglia structures as well, particularly norepinephrine input to the striatum from the locus ceruleus, serotonin input to the globus pallidus and substantia nigra pars reticulata from the raphe nuclei, and noncholinergic pedunculopontine input to the globus pallidus. (These pathways are not shown in Figure 11.1).

Each striatal area has two parallel output pathways (Figs. 11.1 and 11.2). One goes to the external segment of the globus pallidus. The external pallidal projecting neurons use the inhibitory amino acid neurotransmitter gamma-

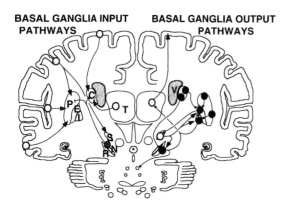

**BASAL GANGLIA INPUT PATHWAYS**

**BASAL GANGLIA OUTPUT PATHWAYS**

**Figure 11.1**

Basal ganglia input and output pathways. Input pathways are shown on the left and output pathways on the right. Excitatory pathways are shown by an open circle representing the neuron cell body and an arrowhead representing the end of the pathway. Inhibitory pathways are shown by a closed circle representing the neuron cell body and a dot representing the end of the pathway. C, caudate nucleus; E, external segment of globus pallidus; I, internal segment of globus pallidus; N, substantia nigra pars compacta; P, putamen; R, substantia nigra pars reticulata; S, subthalamic nucleus; T, thalamus; V, ventricle.

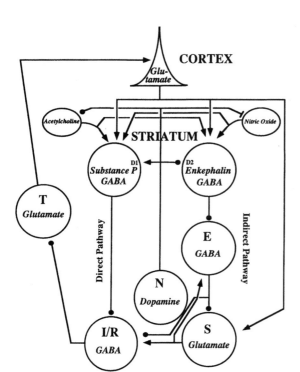

**Figure 11.2**

Schematic diagram of basal ganglia neuronal connections and the neurotransmitters used by each of the cell types. The neurotransmitter of each cell type is printed in italic style over each "cell body." Arrowheads and dots at the end of pathways reflect excitatory and inhibitory connections, as in Figure 11.1. "Striatum" refers to the histologically and connectionally similar caudate nucleus, putamen, and nucleus accumbens. In addition, "D1" and "D2" show the location of dopamine D1 and D2 receptors on striatal substance P and enkephalin neurons, respectively. Abbreviations as in Figure 11.1.

aminobutyric acid (GABA) and the peptide, enkephalin, as their neurotransmitters. The other output pathway goes to either the internal segment of the globus pallidus or to the substantia nigra pars reticulata. These neurons use GABA and the peptides, substance-P and dynorphin.

The external segment of the globus pallidus consists of GABAergic neurons, which generate action potentials at a high rate and in a bursting pattern of firing and which send their output to the subthalamic nucleus and the internal segment of the globus pallidus and the substantia nigra pars reticulata.

The subthalamic nucleus consists of glutamatergic neurons, which, under ordinary circumstances, have a relatively high firing rate. The subthalamic nucleus output goes back to the external segment of the globus pallidus and also goes onward to the internal segment of the globus pallidus and to the substantia nigra pars reticulata. The relationship between the pallidal segments and the subthalamic nucleus is somatotopic and forms a feedback loop. External pallidal cells receive input from the same part of the subthalamic nucleus to which they send output, and project to the same part of the internal segment and substantia nigra as the subthalamic neurons to which they project.

The internal segment of the globus pallidus and the substantia nigra pars reticulata are the main output centers for the basal ganglia. Like the striatal structures, the internal globus pallidus and the substantia nigra pars reticulata share the same developmental primordium and the same internal histologic features. Both consist of large GABAergic neurons that have a high intrinsic firing frequency. The somatotopic inputs to the two structures also suggest that these are one developmental structure that was split into two parts by the growth of the internal capsule. In motor portions, the lateral part of the internal segment receives input from distal limb structures, while its medial part receives information about proximal limbs. The lateral part of the pars reticulata receives input about trunk structures. The more medial parts of the pars reticulata are concerned with eye movements and internal states. Both the internal segment and the pars reticulata send their main outputs to the ventral lateral complex of the thalamus. Both nuclei also send descending projections into the brainstem, noncholinergic motor control areas adjacent to the pedunculopontine nucleus.

The pars reticulata has a number of connections that the internal segment does not. The neurons of the pars reticulata receive dopaminergic input from the dendrites of pars compacta neurons, while there are only a few dopamine axons projecting to the internal segment. The pars reticulata has additional output pathways as well. One is an eye movement control pathway that goes to the superior colliculus. There also are projections from the pars reticulata to the mediodorsal nucleus of the thalamus.

The segregation of inputs to the striatum involve more than simply the somatotopic organization of motor and sensory projections to the putamen. The temporal, parietal, and occipital cortices all send their projections to the body and tail of the caudate nucleus. Primary sensory, primary motor, and premotor areas send their projections to the putamen. The loss of dopamine in Parkinson's disease is primarily in the projections to the putamen. This is why the symptoms of Parkinson's disease are primarily motor. Cortical area 8, which controls eye movement, sends a prominent projection into the anterior caudate, which in turn projects to the mid parts of the substantia nigra pars reticulata, which in turn project to the superior colliculus. This is a motor control pathway for eye movements that is an alternative to the direct fronto-pontine projection. It is dysfunction of this pathway that probably causes the eye movement abnormalities that are seen in Parkinson's and Huntington's diseases.

The lateral prefrontal cortex projects onto the lateral dorsal head of the caudate nucleus. It is this area of cortex that is responsible for organization and planning of future activities as well as the ability to perform well on neuropsychological tests that measure the ability to "shift sets," such as the Wisconsin card sort test. Loss of the dorsal part of the head of the caudate nucleus is prominent in Huntington's disease, and this is presumably why these patients have difficulty with organization and on performance of tests such as the Wisconsin card sort. Abnormalities of these tests can be detected in Parkinson's disease, but these are rarely correlated with clinically significant behavioral observations in parkinsonism.

The medial part of the frontal cortex, particularly the cingulate gyrus, projects to the ventral and medial parts of the head of the caudate nucleus. Lesions of this cortex result in patient apathy and lack of initiative, even akinetic mutism, and the damage to this region in Huntington's disease may be the cause of apathy and decreased initiative.

The orbitofrontal cortex projects to the ventral caudate nucleus and nucleus accumbens. Lesions of this cortex often result in irritability and a labile affect. Irritability and lability also can be part of the symptoms of Huntington's disease. Lesions of either the orbitofrontal or cingulate cortices have been associated with obsessive-compulsive disorder, which is occasionally seen in Huntington's disease and is common in Tourette's syndrome. Other non-neocortical structures, such as the amygdala and the hippocampus, also have projections that go to the nucleus accumbens.

Thus, each cortical region has its own region of striatum to which it projects. This separation of pathways is maintained in the striatopallidal, pallid-othalamic, and thalamocortical projections so that the corticostriatopallidotha-lamocortical loop is a closed loop that feeds information back to the same area of cortex from which it left.

The somatotopic areas within the striatum are not single, but multiple. Thus, the hand area of cerebral cortex sends projections to multiple discrete areas of the striatum (Fig. 11.3). These discrete striatal hand areas are thought to each integrate their information with that of adjacent striatal areas that receive input from other cortical regions. This integrating activity of the basal

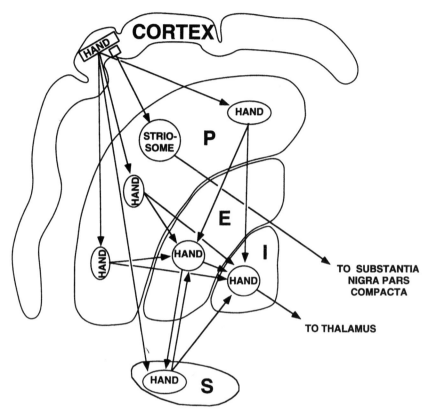

**Figure 11.3**
Divergence and convergence in projections through the striatum. Each cortical region, such as the hand region of the motor cortex, projects to multiple regions within the matrix of the striatum. These projections come from neurons in layer III and upper layer V of the cortex. A separate projection goes from deep-layer V and layer VI of the cortex to the striosomal compartment of the striatum. The matrix projections from each of the striatal hand areas converge on a single hand region in both external and internal globus pallidus. The pallidal regions are, in turn, reciprocally connected through the hand region of the subthalamic nucleus. Abbreviations as in Figure 11.1.

ganglia is further enhanced in the globus pallidus, because the multiple regions of the striatum that are processing information about a single somatotopic area of body all project to the same region of globus pallidus. Thus, hand information leaves the motor cortex and goes to multiple regions of the striatum where it is integrated with information from other cortical regions, but all the striatal hand areas project to a single hand region of the globus pallidus.

All of the cortical striatal projections described are somatotopic and come from neurons that lie in the upper part of layer 5 and the lower parts of layer 3 of the cortex. In addition, there is another corticostriatal projection that comes from layer 6 and deep parts of layer 5. This projection comes from all parts of the cortex, but particularly from prefrontal regions, and projects to a separate histologic part of the striatum called the *striosomes*, or *patches*.

Striosomes are histologically different from the surrounding striatal matrix. They have lower levels of the enzyme acetylcholinesterase and receive denser dopaminergic innervation than the matrix. Striosomes receive this dense innervation from the dorsal tier of the substantia nigra pars compacta dopamine neurons, while matrix receives its dopamine input from the ventral tier of the substantia nigra pars compacta and (in the nucleus accumbens) from the ventral tegmental area. The output of the striosomes does not go to the external or internal segments of the globus pallidus or to the substantia nigra pars reticulata. Instead, this output goes to the dopamine neurons of the substantia nigra pars compacta. This deep cortex to striosome to substantia nigra pars compacta pathway is a way for cortical information to reach the dopamine neurons of the substantia nigra.

## Details of Striatal Anatomy

There are several different types of striatal neurons. The vast majority (approximately 90%) of striatal neurons are medium-sized (approximately 40 μm in diameter) and have dendrites that are studded with numerous spines. These medium-sized spiny neurons are the striatal output neurons. There are two main classes of spiny neurons (see Fig. 11.2). The neurons that project to the external segment of the globus pallidus use GABA and enkephalin as their transmitters. These neurons are enriched in dopamine receptors with a D2 pharmacology, as well as having adenosine A2a receptors. These neurons, because of their D2 receptors, are largely inhibited by dopamine. The spiny neurons that project to the internal globus pallidus and substantia nigra pars reticulata use GABA and substance P as their neurotransmitters and are enriched in dopamine receptors of the D1 pharmacology. These neurons are excited by dopamine.

Both types of striatal neurons have voltage-gated potassium channels in their membranes, which tend to maintain the membrane's voltage at a constant level. These membranes are usually hyperpolarized, and thus striatal spiny neurons are usually silent. If a sufficient excitatory input arrives simultaneously at many places on the spiny neuron's dendrites, the voltage will shift to a depolarized state that will be maintained for a second or so, during which time the neuron may fire a burst of action potentials. Thus, single-cell recordings of these neurons find that they are usually silent and occasionally fire a brief burst of action potentials that are correlated with the somatotopic function of that particular region of the striatum.

The other 10% of striatal neurons consists of interneurons, which do not have long external projections. The largest striatal neurons (about 100 μm in diameter) use acetylcholine as their neurotransmitter. These neurons have prominent dopamine D2 receptors on them and send excitatory outputs to the striatal spiny neurons. There are also GABA interneurons, which contain the peptide parvalbumin. Finally, there are the interneurons, which contain nitric oxide, a prominent local circuit neurotransmitter. These neurons also contain somatostatin and neuropeptide-Y. The role of the GABA and nitric oxide interneurons in governing striatal output is currently unclear.

## Feedback Loops Through the Basal Ganglia and How They Might Function

Figure 11.2 shows the organization of the basal ganglia feedback loop that has been postulated to play an important role in controlling motor function. Inputs from multiple areas of cortex, but particularly from motor, sensory, premotor, and supplementary motor cortices, go to the striatum. In addition, there is direct cortical input to the subthalamic nucleus. Within the striatum, cortical input makes excitatory connections with the interneurons as well as the two types of output neurons. Striatal output from the GABA/substance P neurons that project to the internal segment of the globus pallidus and pars reticulata of the substantia nigra has been called the direct striatal output pathway. Output from the enkephalin/GABA neurons goes first to the external segment of the globus pallidus, where it inhibits these neurons.

The external pallidal neurons in turn project to the subthalamic nucleus neurons, which are excitatory on internal pallidal and pars reticulata neurons. This striatal to external pallidal to subthalamic to internal pallidal pathway has been called the *indirect striatal output pathway*. All striatal output ultimately ends up in the internal segment of the globus pallidus and substantia nigra pars reticulata, which in turn projects to the ventral anterior and ventral lateral nuclei of the thalamus. The ventral thalamic complex that receives input from basal ganglia in turn projects to the cortex, with each cortical region receiving basal ganglia output that originated in the same region of cortex.

The simplest feedback loop through the basal ganglia is a negative feedback loop involving the subthalamic nucleus (Fig. 11.4). Excitatory input from the cortex goes to the subthalamic nucleus, which in turn excites the internal pallidum, which sends a strong inhibitory signal to the ventral thalamus. The ventral thalamus, therefore, fails to excite the cortex. Thus, output from the cortex will tend to inhibit itself.

This negative feedback loop is not absolute in shutting down cortical activity, because embedded within it is another negative feedback loop (Fig. 11.5). The subthalamic nucleus output neurons excite the inhibitory GABAergic neurons of the external segment of the globus pallidus, which in turn

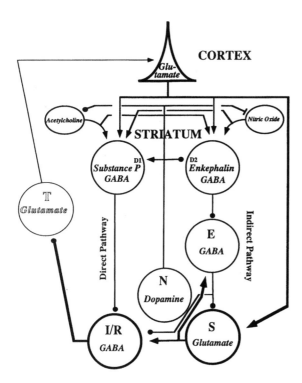

**Figure 11.4**
The subthalamic negative feedback loop's effect on the activity of basal ganglia neurons. Relative activities of different pathways are shown by the thickness of the connection. Abbreviations and symbols as in Figures 11.1 and 11.2.

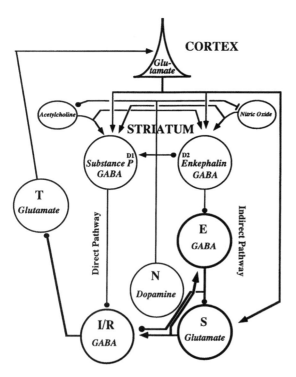

**Figure 11.5**
The modified subthalamic loop's effect on the activity of basal ganglia neurons. Relative activities of different pathways are shown by the thickness of the connection. Abbreviations and symbols as in Figures 11.1 and 11.2.

inhibit the subthalamic nucleus. Recordings from internal pallidal cells show that cortical stimulation produces an excitatory postsynaptic potential due to excitatory input from subthalamus, followed immediately by an inhibitory postsynaptic potential, due to inhibitory lateral pallidal output that has been stimulated by subthalamic activity. Thus, subthalamic nucleus activity tends to dampen itself to some extent. This leaves the internal pallidum susceptible to influences from the striatum.

The direct striatal output pathway (Fig. 11.6) provides an inhibitory input to the internal pallidal neurons. Thus, activity of the direct pathway supports an ongoing activity that the striatum considers appropriate by inhibiting the internal pallidal neurons that would shut the activity down.

Activity in the indirect striatal output pathway (Fig. 11.7) has the opposite effect. The striatal GABA enkephalin neurons inhibit the external pallidal neurons, which ordinarily dampen subthalamic activity. Thus, subthalamic activity is heightened, causing a heightening of internal pallidal output, thus shutting down the thalamus and cortex. Activity in the indirect pathway can be used to stop an ongoing movement that has become inappropriate and to enable the cortex to program a new movement.

The dopamine input to the striatum from the substantia nigra pars compacta (Fig. 11.8) has opposite effects on the two striatal output pathways.

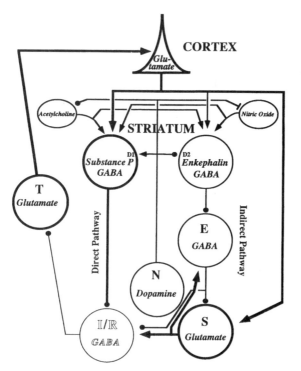

**Figure 11.6**
The direct pathway "GO" loop's effect on the activity of basal ganglia neurons. Relative activities of different pathways are shown by the thickness of the connection. Abbreviations and symbols as in Figures 11.1 and 11.2.

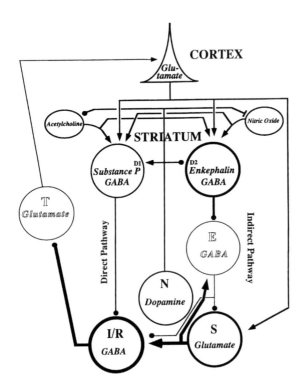

**Figure 11.7**
The indirect pathway "STOP" loop's effect on the activity of basal ganglia neurons. Relative activities of different pathways are shown by the thickness of the connection. Abbreviations and symbols as in Figures 11.1 and 11.2.

Dopamine input excites the direct striatal output pathway via dopamine D1 receptors on GABA/substance-P neurons while it inhibits the indirect striatal output pathway via dopamine D2 receptors on GABA enkephalin neurons. Both actions of dopamine tend to reinforce ongoing activity by inhibiting internal pallidal output directly through the striatal substance P/GABA neurons and by freeing the external pallidal neurons from striatal inhibition through the indirect pathway, thus leading to a dampening of subthalamic activity. This physiologic action of dopamine is consistent with its proposed action in the brain's reward mechanisms. Dopamine reinforces activities that are being rewarded. Dopamine's rewarding actions may explain why dopaminergic drugs such as cocaine and amphetamines are drugs of abuse.

## Dysfunction of Basal Ganglia Feedback Loops in Disease States

The simplest of the movement disorders encompassed by this model is hemiballism (Fig. 11.9). This is a unilateral disorder caused by an infarct in the subthalamic nucleus. Loss of the subthalamic nucleus means that internal pallidal neurons cannot be excited. Thus, they cannot increase their inhibition of the thalamus to shut down an inappropriate activity. This results in the

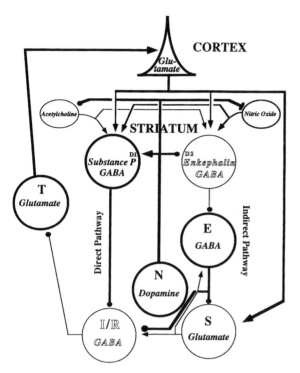

**Figure 11.8**
Dopamine effects on the activity of basal ganglia neurons. Relative activities of different pathways are shown by the thickness of the connection. Abbreviations and symbols as in Figures 11.1 and 11.2.

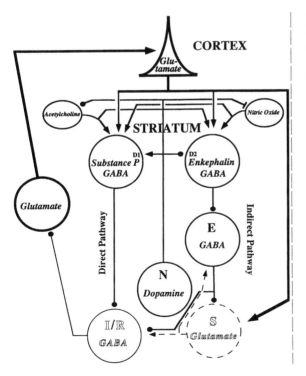

**Figure 11.9**
Hemiballism's effect on the activity of basal ganglia neurons. Destroyed neurons and connections are shown by dotted lines. Relative activities of different pathways are shown by the thickness of the connection. Abbreviations and symbols as in Figures 11.1 and 11.2.

large amplitude, violent, inappropriate, extra movements of hemiballism. Activation of the indirect pathway has very little effect, because the indirect pathway only inhibits the external pallidal neurons that are already underactive due to their loss of subthalamic input. Nevertheless, this minimal effect may be why dopamine D2 receptor antagonists are of some benefit in hemiballism.

Parkinson's disease is the opposite of hemiballism (Fig. 11.10). In this disease, the pars compacta dopamine neurons degenerate. This produces a prominent dopamine loss in the motor areas of the putamen. Parkinsonian states also can be produced by drugs that block dopamine receptors or which inhibit dopamine synthesis. The result is less excitation of the direct striatal output pathway and less inhibition of the indirect striatal output pathway. Therefore, external pallidum is over-inhibited and is unable to shut down the subthalamic nucleus. Subthalamic nucleus over-activity, in turn, drives the internal pallidum to heightened activity. This over-activity cannot be shut down by the direct striatal output pathway, which is under-functional. Thus, appropriate activity cannot be sustained, and the akinesia, bradykinesia, and rigidity of parkinsonism result.

It is this hyperactivity of the subthalamic nucleus and internal globus pallidus that explains the recently rediscovered success of pallidotomy in this

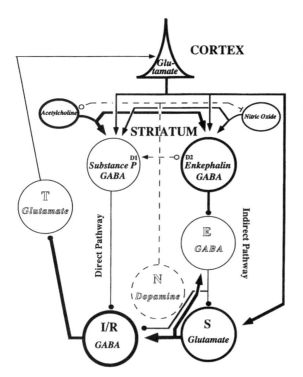

**Figure 11.10**
Parkinson's disease's effect on the activity of basal ganglia neurons. Destroyed neurons and connections are shown by dotted lines. Relative activities of different pathways are shown by the thickness of the connection. Abbreviations and symbols as in Figures 11.1 and 11.2.

disease. This operation has been found to be quite useful for the rigidity and bradykinesia of parkinsonism. It is also quite good for relieving the dyskinesias induced by dopamine agonists. These dyskinesias may be caused by hyperactivity in the direct striatal output pathway, producing inappropriate support for ongoing movements. Such support will not be present if the cortex and thalamus are forced to function on their own.

In early Huntington's disease (Fig. 11.11), the primary pathology is loss of striatal spiny neurons. The first cells to be lost are the GABA enkephalin neurons that are the origin of the indirect striatal output pathway. Thus, external pallidal neurons are free to inhibit ongoing subthalamic activity. This dampening of subthalamic activity is like a minor version of the subthalamic loss seen in hemiballism. Thus, there are inappropriate movements, (i.e., movements that occur at the wrong place and at the wrong time). These movements are lesser in amplitude than those of hemiballism and become apparent as chorea. Later in Huntington's disease (Fig. 11.12), the GABA/substance P neurons of the direct output neurons are lost as well. This results in rigidity and slowness, which become apparent late in adult-onset Huntington's disease and are prominent early in juvenile-onset, rigid Huntington's disease. The pathologic process begins in the tail of the caudate, then spreads to the head of the caudate, then the putamen, and finally to the nucleus accumbens. There

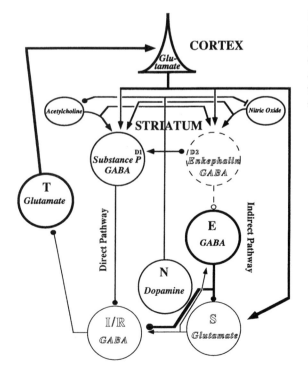

**Figure 11.11**

Early Huntington's disease's effect on the activity of basal ganglia neurons. Destroyed neurons and connections are shown by dotted lines. Relative activities of different pathways are shown by the thickness of the connection. Abbreviations and symbols as in Figures 11.1 and 11.2.

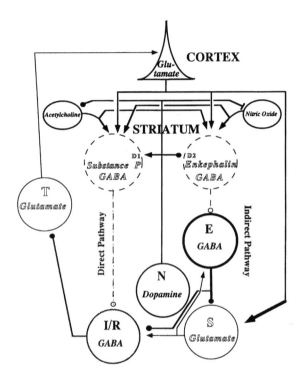

**Figure 11.12**
Late Huntington's disease's effect on the activity of basal ganglia neurons. Destroyed neurons and connections are shown by dotted lines. Relative activities of different pathways are shown by the thickness of the connection. Abbreviations and symbols as in Figures 11.1 and 11.2.

is relative sparing of the striatal interneurons, including the acetylcholine neurons and the nitric oxide/somatostatin neurons.

This lesion of both striatal output pathways also may be the origin of the dystonia that is prominent in late Huntington's disease. Internal pallidal neurons receive no input from the direct pathway, so there is nothing to sustain appropriate ongoing movements. At the same time, they are not receiving the correct information from the subthalamic nucleus. Thus, they are not in a position to shut down inappropriate movements. They probably maintain a steady firing rate on an intrinsic basis that may result in the abnormal postures seen in dystonia, because no feedback about the appropriateness or inappropriateness of ongoing activity is received by the cortex.

Pallidal lesions themselves also can result in a symptomatic dystonia, perhaps because the thalamus is disinhibited. The location of the lesions that result in spontaneous dystonias is unknown. They may result from inappropriate connections being made in the striatum that cause both the indirect and direct pathways to be activated at the same time, rather than in the reciprocal pattern that would be expected for control of normal movements.

While the model that has been presented explains many of the features of basal ganglia disease, it is clearly incomplete. There are feedback pathways that go from both segments of pallidum to the striatum. It may be the activities of these pathways that will ultimately explain why this model fails to explain the symptoms of progressive supranuclear palsy (PSP). In this disease, there are prominent lesions in the subthalamic nucleus and internal segment of the globus pallidus. Given this model, one would expect that PSP patients would have a syndrome like hemiballism with extra movements but, instead, they have a parkinsonian syndrome of rigidity and bradykinesia. While there are also brainstem lesions in PSP that may explain its symptoms, PSP may represent an inadequacy of the model.

Another shortcoming of the model is its failure to provide an obvious explanation for the tremor seen in Parkinson's disease. It may be that tremor is a property of the intrinsic firing rate of thalamic neurons, since thalamic neuron firing at the same rate as that of the tremor can be seen in parkinsonism. This model, however, would predict that thalamic neurons should be shut down by medial pallidal over-activity.

One possible explanation may be that the thalamic neurons that are hyperactive are not those in the basal ganglia output pathway, but are in fact those in the cerebellar output pathway, because the same thalamic lesion can be used to relieve both parkinsonian tremor and cerebellar outflow tract (rubral) tremor. Perhaps the rest tremor seen in Parkinson's disease is simply a reverberation of the corticocerebellar loop that is not dampened by output from the basal ganglia.

## Pathogenic Mechanisms That Might Be Involved In Huntington's Disease and Parkinson's Disease

It has recently been shown that Huntington's disease is due to an abnormally increased number of glutamine moieties in the amino terminal end of a previously unknown protein. This polyglutamine stretch is coded by multiple repeats of the DNA codon, CAG. In most persons, this protein has a stretch of about 20 glutamines in a row. In Huntington's disease, the number of glutamines is over 40. The higher the number of glutamines, the earlier the average age of onset.

In patient's with the juvenile, rigid form, more than 60 glutamine repeats are common. The number of glutamine repeats can increase during spermatogenesis; thus, a father with Huntington's disease often has children whose age of onset of the disease is younger than his. This is not true for mothers who pass the same number of glutamine repeats on to children who inherit the disease. There also are individuals in the population who have 35 to 40 glutamine repeats. If a father with this number of repeats should have an increase in repeat number to over 40 during spermatogenesis, he may produce a child with a new mutation to Huntington's disease.

Several other neurologic diseases have been described that have the same expansion of CAG repeats, with anticipation from generation to generation and multiple forms of the disease. These include spinal bulbar muscular atrophy, spinocerebellar ataxia types I and II, dentatorubral pallidoluysian atrophy, and Machado-Joseph disease. All have glutamine repeats; all have disease, if there are more than 40 repeats, and apparent, new spontaneous mutations. Other triplet repeat diseases, myotonic dystrophy and fragile X syndrome, have triplet repeats that are not in the coding region of their proteins.

How a triplet glutamine repeat produces disease is unclear. All of these diseases are autosomal-dominant diseases. In Huntington's disease, it has been shown that homozygotes do not have significantly more severe disease than do heterozygotes. This suggests that the polyglutamine repeat has conveyed some new function to the Huntington's disease protein that it did not have, rather than causing disease by making the protein dysfunctional. Somehow, this aberrant protein must be toxic to the cell.

The Huntington's disease gene is expressed in many tissues, not solely in the brain, and within the brain there is no increase of expression in the neurons that are vulnerable. Thus, it is unclear why pathology in the disease rests primarily on the medium spiny neurons of the striatum. On the other hand, animal models of Huntington's disease can be produced by systemic injection of certain inhibitors of the mitochondrial respiratory chain. Most prominent among these is 3-nitropropionic acid, an inhibitor of complex II of the mitochondrial respiratory chain. This toxin produces loss of striatal, medium spiny neurons, with sparing of striatal somatostatin and acetylcholine neurons, just like in Huntington's disease. In addition, patients with Huntington's disease have increased lactic acid in their brains, suggesting mitochondrial dysfunction. It, therefore, may be that the product of the Huntington's disease gene produces mild mitochondrial dysfunction and that this mitochondrial dysfunction first produces abnormalities in the striatum.

Consistent with this hypothesis, Leigh's syndrome, a mitochondrial disorder of children, often causes striking striatal necrosis, although other mitochondrial diseases, such as MELAS and MERRF syndromes, usually do not cause striatal damage. With better understanding of the normal and abnormal function of the Huntington's disease protein, it should become obvious why the pathology in the disease exists where it does.

The understanding of the defect in Huntington's disease is much more advanced than that of Parkinson's disease, where no known gene defects exist and where most cases do not appear to have any genetic basis. It is known that all humans lose dopamine neurons with aging. One hypothesis is that this dopamine neuron loss is due to toxic effects of dopamine itself. Dopamine is metabolized by the enzyme monoamine oxidase, with the production of free radicals. It may be that this excess load of free radicals is toxic to dopamine neurons over the long lifetime of a human. In addition, there is an excess of iron in the substantia nigra, and iron acting through the Fenton reaction can increase free radical production. Thus, Parkinson's disease may result from accelerated aging, producing increased free radicals and death of dopamine neurons. Accelerated aging alone, however, could not explain selective loss in the substantia nigra pars compacta.

## Pathophysiology of Parkinsonian Dose-Response Fluctuations

The standard therapy for Parkinson's disease is treatment with levodopa, the precursor of dopamine, along with a peripherally acting inhibitor of the dopa decarboxylase enzyme that converts dopa to dopamine. The peripheral inhibitor prevents conversion of dopa to dopamine in the bloodstream, thus preventing the nausea and hypotension that characterized treatment of parkinsonians with pure levodopa. Levodopa presumably works so effectively because it bypasses the need for the enzyme tyrosine hydroxylase, the rate-limiting step in the synthesis of dopamine. There is very little tyrosine hydroxylase in the striatum of a patient with Parkinson's disease. There is, however, sufficient dopa decarboxylase to produce significant amounts of dopamine from exogenous supplies of dopa. Most of this dopa decarboxylase is probably present in the axon terminals of surviving dopamine neurons. The symptoms of Parkinson's disease begin when approximately 70% of dopamine terminals have been lost, but in the early stages of Parkinson's disease there are still enough terminals present to carry out efficient dopa decarboxylation.

When patients with early, mild Parkinson's disease begin taking levodopa, they experience a sustained relief of symptoms from the drug, which may last for many hours after a single dose. As the disease progresses, however, patients first note that an individual levodopa dose will last only for a few hours, typically 3 to 4 hours, and that they then must take another pill. This stage is often followed by the development of dyskinesias at the peak of their plasma levodopa dose. Ultimately, many patients are left alternating between a dyskinetic state and a parkinsonian "off" state. The clinical state of the patient may fluctuate rapidly and severely, even with quite small changes in the plasma levodopa level. The plasma levodopa levels fluctuate just as much early in the course of the disease as they do late in the course, but presumably the presence of a large, although decreased, number of dopamine presynaptic terminals provides a storage mechanism through which the brain can buffer its own dopamine levels against the swings in delivery from the bloodstream.

Dopa dose-response fluctuations can be understood as an interaction between fluctuating delivery of dopa to the brain from the bloodstream and biphasic antagonist-agonist responses to dopamine. As shown in Figure 11.13, at low levels of the drug, the antagonist effects of dopamine predominate, presumably by acting on presynaptic receptors governing dopamine release. At higher levels, direct agonist effects on postsynaptic receptors predominate. The summed response of the antagonist and agonist effects has a sharp transition between the antagonist and agonist response states.

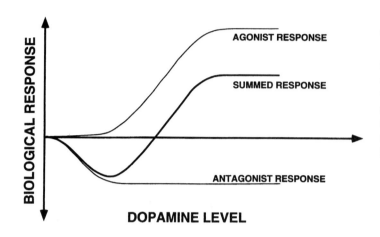

**Figure 11.13**

The biologic response to different levels of dopamine. Improvement is up; worsening is down. Antagonist responses occur at lower doses than agonist responses. Therefore, as the dopamine level increases, the summed response is antagonist at low levels and agonist at higher levels.

An idealized plasma curve of levodopa after a single intake is shown in Figure 11.14. Plasma drug levels rise fairly rapidly, then have a slow exponential decay. Combining the antagonist-agonist responses to drug levels shown in Figure 11.13 with the fluctuating plasma levels shown in Figure 11.14 produces the biologic response fluctuations shown in Figure 11.15. Immediately after a dose is taken, there is an antagonist response to the low level of levodopa, producing a more severe "off" state in the patient, followed by a rapid swing to an "on" state, which is sustained for some time, and then another rapid swing back to an "off" state, which may be, again, more severe than the state of having no dopa at all. If another dose of medication is taken at the wrong moment in the fall of the blood levels, it may produce a more severe "off" state, at least temporarily.

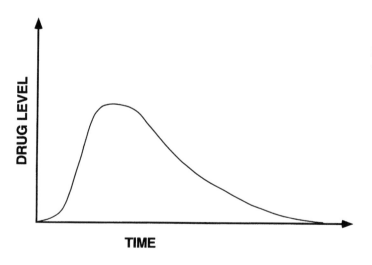

**Figure 11.14**

Idealized time course of the blood levodopa level after a single oral dose.

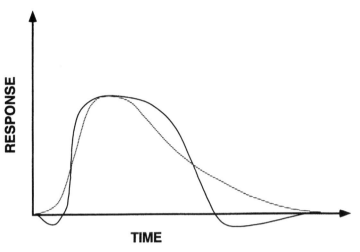

**Figure 11.15**
Idealized biologic response to a single oral dose of levodopa. The agonist and antagonist effects of the different dose levels produce a much more abruptly contoured response curve (solid line) than the actual drug level curve (dotted line).

RESPONSE

TIME

Delivery of levodopa to the brain is controlled by several factors that may produce seemingly random fluctuations in the rate of delivery. One is the pyloric valve between the stomach and the duodenum. Dopamine agonists tend to close this valve and inhibit delivery of levodopa to the intestine where it must be absorbed. The second cause of fluctuations in delivery is that levodopa is transported from the intestine to the bloodstream by the facilitated transport mechanism that is used to transport neutral amino acids. Neutral amino acids in the intestines from a protein meal will interfere with transport of levodopa to the bloodstream; thus, many patients discover that if they eat a protein meal, their subsequent levodopa pills will not work, because they are not absorbed in a timely fashion. A third, similar cause of fluctuations is that a facilitated transport mechanism also exists for neutral amino acids from the bloodstream, and it is through this mechanism that levodopa enters the brain.

These fluctuations in the delivery of levodopa to the brain, coupled with the biologic response to fluctuations in brain levodopa levels, produces two types of dyskinesia in patients. The more common is the peak-dose dyskinesia, shown in Figure 11.16. After patients with peak-dose dyskinesias take a medication, they are at first "off." They then rapidly turn on as the biologic response peaks. The peak, however, is accompanied by dyskinesias, usually of the choreoathetoid type.

The best therapeutic management strategy for dose-response fluctuations with peak-dose dyskinesias is to keep a constant plasma levodopa concentration at a level that produces an "on" state without too much, if any, dyskinesia. Sustained-release preparations of levodopa may accomplish this in patients with relatively mild dose-response fluctuations. With severe response fluctuations, however, even sustained-release capsules are insufficient. Patients with such

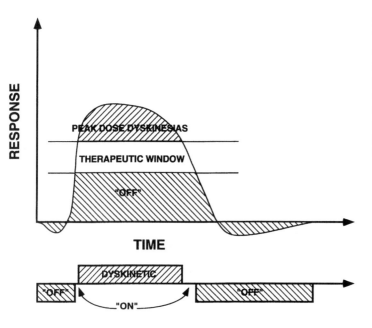

**Figure 11.16**
Typical time course of peak-dose dyskinesias. "Off" times (\\\) and dyskinesias (///) are superimposed on the biologic response curve. Typically, patients are "on" when they are dyskinetic, and may have periods of normal "on" function before and after the dyskinetic episodes.

fluctuations may have to take many small doses of medications, as often as every 1 to 2 hours. Absorption of levodopa may be accelerated if they dissolve their levodopa in a liquid form buffered with ascorbic acid rather than take it as pills. Alternatively, patients with extremely severe fluctuations may get their best response by having a duodenal pump deliver the medication.

A second type of levodopa dose-response fluctuation is the so-called biphasic dyskinesia, shown in Figure 11.17. Patients with this type develop their dyskinesias in the transition period between being "on" and being "off." They have ballistic or dystonic dyskinesias at the beginning and at the end of their dose interval. Such patients often respond best to keeping the plasma levodopa high for as long as possible during the day and by adding dopamine agonist drugs. Such patients are *not* helped by giving smaller doses more often, because they simply fluctuate more often, producing more dyskinesias. Patients with biphasic dyskinesias are much more likely to complain about the dyskinesias. These dyskinesias are frequently ballistic and frequently in the legs. Not infrequently, they are quite incapacitating to the patient, while the peak-dose dyskinesias tend to be accompanied by the "on" state and are usually better tolerated by the patient. Treatment of biphasic dyskinesias is often complicated in the evening because there may be a prolonged slow fall of medication levels accompanied by prolonged incapacitating dyskinesias.

The pathophysiologic mechanism underlying biphasic dyskinesias is unknown. Perhaps better understanding of the mechanism involved will lead to better treatment for this disabling condition.

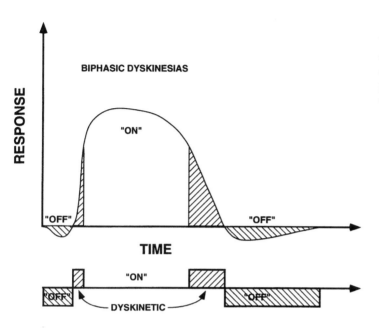

**Figure 11.17**
Typical time course of biphasic dyskinesias. "Off" times (\\\\) and dyskinesias (///) are superimposed on the biologic response curve.

## Summary

In conclusion, current understanding of the anatomy, neurotransmitters, and physiology of the basal ganglia has led to a model of basal ganglia function that can explain the many symptoms of basal ganglia disease. Furthermore, understanding of the pharmacokinetics of levodopa delivery to the brain has explained many of the dose-response fluctuations that are seen in advanced parkinsonian patients on long-term levodopa therapy. However, the basic pathogenic mechanisms that produce death of striatal neurons in Huntington's disease and of substantia nigra pars compacta dopamine neurons in Parkinson's disease remain unknown. It is hoped that future scientific advances will fill this gap in our knowledge.

## References/Further Reading

Alexander GE, DeLong NR, Strick PL. Parallel organization of functionally segregated circuits linking basal ganglia and cortex. Annu Rev Neurosci 1986;9:351–357.

Cummings JL. Frontal-subcortical circuits and human behavior. Arch Neurol 1993;50:873–880.

DeLong MR, Georgopoulos AB. Motor functions of the basal ganglia. In: Brooks VB, ed. Handbook of physiology: the nervous system, vol. 11. Washington, DC: American Physiological Society, 1981:1017–1061.

Gancher ST, Nutt JG, Woodward WR. Peripheral pharmacokinetics of levodopa in untreated stable and fluctuating parkinsonian patients. Neurology 1987;37:940–944.

Gerfen CR. The neostriatal mosaic: multiple levels of compartmental organization in the basal ganglia. Annu Rev Neurosci 1992;15:285–320.

Graybiel AM, Ragsdale CW. Biochemical anatomy of the striatum. In: Emson PC, ed. Chemical neuroanatomy. New York: Raven, 1983:427–504.

Gusella JF, Macdonald ME, Ambrose CM, Duyao MP. Molecular genetics of Huntington's disease. Arch Neurol 1993;50:1157–1163.

Hornykiewicz O, Kish S. Biochemical pathophysiology of Parkinson's disease. In: Yahr MD, Bergman KJ, eds. Parkinson's disease. New York: Raven, 1987:19–34.

Laitinen LV, Bergenheim AT, Hariz MI. Ventroposterolateral pallidotomy can abolish all parkinsonian symptoms. Stereotact Funct Neurosurg 1992;58:14–21.

Macdonald ME, Barnes ME, Srindhi J, et al. Gametic but not somatic instability of CAG repeat length in Huntington's disease. J Med Genet 1993;30:982–986.

Mink JW, Thach WT. Basal ganglia intrinsic circuits and their role in behavior. Curr Opin Neurobiol 1993;3:950–957.

Myers RH, Macdonald ME, Koroshetz WJ, et al. De novo expansion of a (CAG)N repeat in sporadic Huntington's disease. Nature Genet 1993;5:168–173.

Nutt JG. On-off phenomenon: relation to levodopa pharmacokinetics and pharmacodynamics. Ann Neurol 1987;22:535–540.

Paalzow GHM, Paalzow LK. L-dopa: how it may exacerbate parkinsonian symptoms. TIPS 1986;7:15–19.

Youdim MBH, Ben-Shachar D, Yehuda S, Riederer P. The role of iron in the basal ganglion. Adv Neurol 1990;53:155–162.

# Transplantation and the Central Nervous System

Robert J. Maciunas

The most frequent function of neurosurgery is to ablate tissue that is diseased, dysfunctional, or impinging on viable neural parenchyma; neural destructive lesions have also been employed to treat intractable pain and psychiatric disorders. Following the development of the use of prosthetics in neurosurgery, however, the vector of surgical manipulation has shifted from out to in. The developments in this field will dovetail with those of neural transplantation, as the former provides the vehicle and the latter the cargo for the coming generation of neurosurgical procedures. The smallest and finest of prosthetics in neurosurgery will undoubtedly be produced to precisely restore functionality through replacement of identifiable subcellular organelle mechanisms or synaptic connections and the targeted insertion of new genetic blueprints/instructions to redirect the expression of cellular machinery and restore deficient or deranged functions.

The first report of successful neural transplantation was in 1904 by Dr. Elizabeth H. Dunn of the University of Chicago, who demonstrated survival of immature neurons transplanted into the brain of another animal. Decades elapsed before the application of preliminary data from animal trials of transplantation propelled several waves of enthusiasm for trials of adrenal cell transplantation surgery in patients with parkinsonian movement disorders. Although this enthusiasm was dashed by the mixed results achieved by North American clinical centers, a new era of neurosurgical intervention involving tissue transplantation to the brain had begun. After some retrenchment and further basic scientific evaluation of tissue substrates and their characteristics, several centers have reported promising clinical results using fetal tissue transplants. Such human studies have prompted deeply considered discussions regarding their sociopolitical, ethical, and religious ramifications, in addition to their scientific and clinical merits.

# Parkinson's Disease

## LABORATORY INVESTIGATION

The histopathologic and biochemical abnormalities of Parkinson's disease include degeneration of dopaminergic cells in the substantia nigra and secondary depletion of dopamine in the striatum. Clinical motor manifestations are partially reversible by administering the precursor of dopamine, L-dopa.

The initial animal model of parkinsonism was produced by injecting the dopamine analogue, 6-hydroxydopamine (6-OHDA), into the nigrostriatal tract of a rodent. This caused degeneration of the dopaminergic neurons in the substantia nigra and depleted the striatum of dopamine. The rats developed a characteristic and quantifiable syndrome of turning toward the side of the cerebral lesion. Normal function could be restored by inserting grafts of fetal substantia nigra, used because of its relative resistance to mechanical trauma and anoxia, as well as its proliferative ability. The grafts lessen the rotational behavior induced by 6-OHDA, and survival of grafted cells containing catecholamines has been demonstrated histochemically. The beneficial effect is thought to be due to dopamine production by the nigral grafts.

A useful model of Parkinson's disease is produced in humans, nonhuman primates, and rodents by injection of 1-methyl-4-phenyl-1,2,3,6-tetrahydropyridine (MPTP). Fetal striatal tissue grafts produce clinical improvement in animals, and have formed the basis for clinical trials in humans exposed to MPTP through illicit drug use.

The medical use of fetal tissue is discouraged in many countries, even when produced by therapeutic abortion. Equally significant are concerns regarding potential immunologic rejection of fetal tissue grafts. These constraints led to consideration of adrenal medullary tissue as a more practical short-term solution to the procurement of a graft source with properties similar to those of fetal substantia nigra.

In the absence of corticosteroids, adrenal medullary tissue synthesizes catecholamines, especially norepinephrine and dopamine. In tissue culture, chromaffin cells transform to a neuronal phenotype. These transplanted adrenal medullary cells were shown to form nerve fibers and innervate brain tissue. Early experiments suggested that adult adrenal medullary tissue grafts were equally as effective as fetal nigral tissue in reducing rotational behavior in the 6-OHDA rodent model.

Animal models of parkinsonism have shown that a graft need not structurally replace the degenerating neurons in the substantia nigra to effectively treat parkinsonism. Rather, it seemed that a graft might ameliorate the neurologic changes by acting as a nonspecific source of dopamine. That hypothesis provided the rationale for initiating trials of adrenal medullary tissue transplantation in animals and humans.

## ADRENAL MEDULLARY TRANSPLANT CLINICAL TRIALS

The clinical basis for selection of patients for the first human transplantation study in parkinsonism, by Backlund and his colleagues in Sweden, was based on the clinical features of their disease. Patients with hypokinesia and rigidity were viewed as suffering from problems related to L-dopa treatment, and disease progressions and nerve cell replacement or reinnervation, as well as neurotransmitter replacement, were hypothesized as potential mechanisms of a graft effect.

In the Swedish group's first reported clinical trial, two severely disabled patients, aged 55 and 46 years, underwent surgery. Both were L-dopa–responsive and had on-off reactions. Two-thirds of the left adrenal gland was surgically removed and fragments of the medulla were loaded into a steel spiral carrier and transplanted stereotactically through a burr hole into the central region of the head of the right caudate nucleus. Patients were taken off medication perioperatively for 4 to 24 days. Increased levels of catecholamine metabolites found in the patients' lumbar CSF were considered to be evidence that chromaffin cells survived and released catecholamines. It was postulated that dyskinesia and paranoia in the patients during the early postoperative period resulted from catecholamine release by the transplant. The clinical results were disappointing, with neither patient showing a significant response; nevertheless, the study showed that adrenal medullary transplantation was technically feasible.

Since their initial study, Backlund's group has reported that despite only transient improvement lasting much less than a year, a total of four patients had shown no clinical deterioration since their adrenal transplant for up to 5 years. They speculated that these results raise the possibility of a transplant-related slowing of disease progression, even in the face of lack of distinct improvement.

In 1987, Lindvall et al. reported operating on two patients in a slightly different manner. The patients were 46 and 63 years old, with L-dopa–responsive stage IV disease and experienced dyskinesia with on-off responses. Adrenal medullary fragments were implanted in the central region of the right putamen. These patients were taken off their medication for 2 days at 1 week before surgery and again for 2 days after surgery. It was concluded from this study that transplantation of adrenal medullary tissue to the striatum in patients with severe Parkinson's disease can produce transient and partial restoration of motor function. No change in tremor and no chronic changes in rigidity were observed.

In Mexico in 1987, Madrazo et al. reported dramatic and persistent improvement in two patients, aged 35 and 39 years, with severe Parkinson's disease. Both patients had intolerable side effects from L-dopa, and one had become unresponsive to medication. They received no medication at the time of surgery or during the follow-up period. Madrazo's technique was radically different from that of the Swedish investigators: The fragments of adrenal medullary tissue were left in contact with ventricular CSF, being placed into a cavity created in the ependymal surface of the head of the caudate nucleus that was approached transcortically after a right frontal craniotomy. The entire right adrenal was harvested. A marked bilateral improvement was reported to have occurred gradually and continuously over the 15-month follow-up period in the absence of drug treatment. The extended period of progressive improvement suggested that the transplant was affecting the progression of disease as well as the symptoms and signs. Subsequent to this report, the same group reported on eight additional patients, with similarly remarkable results. These findings generated considerable interest and led to the initiation of multiple clinical trials.

A pilot series of 18 patients was recently evaluated at Vanderbilt University. The subjects were divided into three groups of six according to age and severity of disease. Group 1 consisted of young patients (under 50 years old) with mild to moderate disease. Group 2 consisted of young patients with moderately severe to severe disease. Group 3 consisted of older patients (greater than 60 years old) with moderately severe to severe disease. All of the patients exhibited bradykinesia, rigidity, and resting tremor. All were responsive to L-dopa. Two had on-off reactions. None had ataxia, intention tremor, pyramidal tract signs, or deficits in downward gaze.

Computed tomography–directed stereotactic craniotomies were performed to access the interface between the frontal horn of the right lateral ventricle and the dorsomedial ependymal surface of the head of the caudate. A right adrenalectomy was performed via a twelfth rib flank retroperitoneal approach. Several fragments of adrenal medullary tissue were placed into a cavity in the caudate. Much less time elapsed between harvest and implantation than in the previous studies. Medications were not altered during the first year following the transplant, although six patients were taken off L-dopa perioperatively for 7 to 26 days so that biochemical studies of their CSF might be done. No significant differences were seen between preoperative and postoperative values for CSF levels of catecholamines (dopamine, norepinephrine, epinephrine, dopa) and/or their metabolites; this suggested that dopamine was not responsible for the effects of grafting in this series.

Of the 12 younger patients, the scores of nine showed improvement at 1 year, and those of three worsened. If a difference of 10 points in the Columbia Rating Scale were required for a score to be considered distinctly changed, however, then four patients demonstrated distinct improvement and no patient distinctly deteriorated. In general, those patients with a higher baseline score showed a greater change in their scores. No consistent pattern of clinical improvement could be seen in the last six patients, all of whom were over the age of 60. Improvement was seen in stability, dexterity, gait, and speech. Only rigidity did not show improvement in any patient. Interestingly, these improvements were bilateral and affected both axial and peripheral signs, indicating that the effect was likely not an effect of the surgical procedure per se. The changes took several weeks to become apparent, and the improvement was then persistent or progressive.

With the enthusiastic early reports from Mexico and initial clinical experience from North America, investigators around the world began trials of human adrenal medullary transplantation in Parkinson's disease. Clinical data rapidly accumulated at individual institutions regarding the technique, safety, and efficacy of these procedures. The American Association of Neurological Surgeons in conjunction with the National Institute of Neurological Disorders and Stroke established the General Registry for Adrenal/Fetal Transplantation (GRAFT) Project in 1987. Analysis of data from these series from different institutions suggested a considerable variation in both the number and severity of complications encountered and the number and significance of positive clinical results. When the data were lumped together, the participating physicians reported some degree of improvement in 79% of the patients, but only 21% of patients were considered to show "significant" improvement at 12 months. Subsequently, individual clinical trials confirmed that most patients experienced only mild to moderate benefits. Another registry was established by the United Parkinson Foundation in conjunction with the Rush-Presbyterian St. Luke's group in Chicago to collect clinical efficacy data. Neural grafting did seem to improve certain objective functions, particularly the duration of "on" time periods, and this improvement in "on" time was not associated with increased dyskinesia. There was no significant improvement in mental function nor decrease in the medication required postoperatively.

In the surgical approaches used to date, adrenal medullary cells have appeared to survive poorly following implantation into the human striatum. Pathologic analysis of autopsy specimens revealed poor evidence for graft viability, even if grafts were placed in contact with the CSF.

In these initial series of neural transplantation, there were significant perioperative risks. Open microsurgical craniotomy procedures were associated with more frequent and more severe complications than stereotactically directed craniotomies or stereotactic fine-needle transplant procedures. The complication rate ranged from 5% to 40%. The conglomerate GRAFT series data on mortality and morbidity were in fact worse than those for the natural history of Parkinson's disease.

## COGRAFT CLINICAL TRIALS

Limited clinical trials have been initiated in the United States and China to assess graft viability and efficacy for cograft preparations of peripheral nerve and adrenal medullary tissue. These were supported by convincing data obtained from rat and primate models that showed cografting of the peripheral nerve enhanced the survival of adrenal medullary grafts as well as the capability for enhancing the host dopaminergic response. Various laboratories are investigating striatal infusions of nerve growth factor (NGF), a- and b-fibroblast growth factors (FGFs), and gangliosides to induce partial recovery of the nigrostriatal dopaminergic system in lesioned animal models. Based on promising work on a rat model in Sweden, Olson et al. undertook a trial to evaluate NGF infusions into the region of striatal medullary grafts. Their clinical data suggested mild to moderate improvement or stabilization in patients' functional status.

## FETAL CELL TRANSPLANTATION TRIALS

The transplantation of human fetal tissue produced behavioral improvement in rodent models, primarily by reinnervation of the host by dopaminergic fibers from the graft. There also was behavioral improvement in MPTP-lesioned primate models after transplantation, although this was correlated with sprouting of the surviving dopaminergic neurons rather than through synaptic connections from the graft. In both cases, it was felt that synaptic formation alone was not the answer because it could not explain the bilateral effects of unilateral grafting. One postulated mechanism of action for this bilateral improvement has invoked a *neurotrophic* effect.

Fetal mesencephalic allografts for human Parkinson's disease were reported in 1987 from Mexico and Sweden. Using a single donor and an open microsurgical technique, Madrazo et al. reported significant early clinical improvement in their patients. Lindvall et al. stereotactically placed transplants into the caudate and putamen bilaterally in their patients.

Three reports published simultaneously in *The New England Journal of Medicine* in 1992 indicated both clinical safety and efficacy for the stereotactic implantation of fetal ventral mesencephalic tissue into patients with parkinsonian symptoms. In each report there was a statistically significant decrease in "off" time and an increase in "on" time, less severe parkinsonian symptoms at all times, and less severe dyskinesia and drug-induced psychosis due to reduced dosages of levodopa. Surgical complications were minimal. These studies, in aggregate, included patients with idiopathic Parkinson's disease and patients with MPTP-induced parkisonism.

Spencer's group transplanted cryopreserved tissue from single fetal donors unilaterally into the right caudate nucleus in four patients. Cyclosporine was given for 6 months postoperatively. Diminished signs and symptoms of parkinsonism were observed during 18 months of evaluation. Positron emission tomography (PET) with ($^{18}$F) fluorodopa in one patient before and after surgery revealed a bilateral restoration of caudate dopamine synthesis, but continued deficits in the putamen.

Freed et al. transplanted fresh tissue from single fetal donors into the caudate and putamen in a series of seven patients with advanced idiopathic

Parkinson's disease. Two patients received unilateral grafts into the caudate and the putamen on the side contralateral to the side with worse symptoms. Five patients received bilateral grafts implanted into the putamen only. The rationale for this was that these regions showed the greatest loss of dopamine. There were no surgical complications. All patients reported improvement, according to the Activities of Daily Living Scale, when in the "on" state between 3 and 12 months after surgery. Five of the seven improved according to the Unified Disease Rating Scale, and their mean Hoehn-Yahr score improved from 3.71 to 2.50. Dyskinesia was lessened, and "off" episodes were lessened postoperatively. At 46 months postoperatively, in one patient PET indicated graft survival.

Widner and his colleagues evaluated implantation of ventral mesencephalic tissue from multiple fetal donors into the caudate and putamen bilaterally in two young adults with severe MPTP-induced parkinsonism. Both patients demonstrated sustained improvements in motor function over 24 months, increased independence, and reduced dosages of levodopa. Positron emission tomography showed marked increased bilateral striatal uptake of fluorodopa. It was postulated that the excellence of these results was due to a unique suitability of patients with MPTP-induced parkinsonism as candidates for neural grafting, because of the nonprogressive nature of their disease, and limitation of the lesion to the nigrostriatal dopaminergic system, resulting in a pure hypodopaminergic state in MPTP-induced parkinsonism.

Despite the encouraging preliminary evidence for the therapeutic efficacy of tissue transplantation in treating parkinsonism, significantly longer follow-up of large numbers of patients admitted to well-controlled trials is needed. Persistent improvement over several years would imply that the progression of the disease had been altered. If so, tissue transplantation would be the first therapy for Parkinson's disease to achieve such a result.

## Huntington's Disease

Unilateral or bilateral striatal injection of excitotoxins, including quinolinic acid, kainic acid, and ibotenic acid, provide an experimental rodent model for this Huntington's disease. There is degeneration of those neurons whose cell bodies lie in the region of the injection site, while axons of passage and nerve terminals remain intact. Fetal striatal tissue grafting before toxin injection attenuates the toxin-induced damage, and post-toxin injury grafting also lessens the degree of hyperactive behavioral abnormalities.

Madrazo's group in Mexico reported on a single Huntington's chorea patient undergoing unilateral caudate head transplantation with human fetal striatal allografts. They observed stabilization in the clinical deterioration and, possibly, neurologic improvement. Long-term follow-up has not been reported.

## Dementia

The performance of aged rats in certain learning and coordination tasks is inferior to that of young rats. Grafts of fetal septal band tissue, containing cholinergic tissue, to the hippocampus bilaterally cause marked improvement in spatial learning tasks. Grafts of fetal mesencephalic tissue, corresponding to the substantia nigra, to the striatum cause significant improvement in performance in tests of coordination. These effects appear to be both graft-specific and donor site–specific.

A model of Alzheimer's disease is created by lesioning the nucleus basalis in rats, thereby depleting cholinergic terminals throughout broad areas of the cortex and causing quantifiable memory deficits. When fetal ventral forebrain was transplanted into the cortex of these rats, histologically demonstrable increases in choline acetyltransferase around the grafts was demonstrated. Human trials have been limited to chemode administration of intraventricular acetylcholine; no tissue transplantation trials have been reported.

## Developmental Deficiencies

Suspended Purkinje cells from normal mice can be injected into the cerebellum of *pcd* mice, a strain of mice with progressive degeneration of cerebellar Purkinje cells. The transplanted cells migrate into the molecular layer and become invested with connections from the intrinsic cells of the cerebellum.

Myelination has been induced in shiverer mouse brains by transplanting suspensions of normal human or mouse brain cells. These cells migrate and produce normal, organized myelin around axons of intrinsic neurons. Even though the shiverer mouse is genetically deficient in myelin, and the extant myelin is devoid of myelin basic protein, this transplant-produced myelin does not appear to incite any inflammatory or immunologic response. No human trials involving developmental disorders have been published.

## Trauma

Destruction of the medial frontal lobe parenchyma in rats causes worsened performance in a T-maze alternation task. Transplanting fetal frontal cortex into this region resulted in significantly improved performance. Fetal cerebellar grafts did not produce improvement, perhaps suggesting that the effect is donor site–specific. No human trials of frontal lobe transplantation have been reported.

## Pain

Transplanting adrenal medullary tissue into rat periacqueductal gray matter results in a markedly attenuated response to the painful stimulus of administering nicotinic acid. This is perhaps related to the fact that nicotinic acid stimulates the release of met- and leu-enkephalin, both of which can be secreted by the transplanted adrenal medullary tissue.

## Hormonal Deficiencies

One experimental model has been the Brattleboro rat, which is genetically deficient in vasopressin and therefore demonstrates diabetes insipidus. Transplanting fetal vasopressin-containing preoptic-region neurons to the third ventricle of Brattleboro rats reverses this deficit. Other mice have been bred to be genetically deficient in gonadotropin-releasing hormone (GnRH), causing hypogonadism and sterility. When normal fetal preoptic neurons containing GnRH are transplanted into the hypothalamus of Brattleboro rats, their genitalia increase in size and they can reproduce. The entire pituitary gland itself has been transplanted to the median eminence of hypophysectomized rats, with survival of pituitary tissue and normalization of blood levels of luteinizing hormone, prolactin, and thyroxine. Although some investigators have speculated about the applicability of this technique to radical surgical therapies for patients with Cushing's disease, no clinical trials have been reported to date.

## Future Tissue Graft Considerations

Investigators have used both tissue fragments and cell suspensions for neural grafting. Theoretical considerations regarding nutrient diffusion into the center of grafts and development of a new blood supply require considerable attention before the optimal form in which to deliver cell grafts can be determined conclusively. It appears likely, however, that cell suspensions may prove superior to tissue blocks or fragments for cell survival and integration. As a side issue, tissue suspensions or fragments used during initial experiments involving neural transplantation contained both neuronal and glial cells, although the relative contribution of each cell type to the graft remains to be determined. Tissue culture techniques are proving helpful to study these issues and optimize the transplanted cellular mix.

Fetal tissue will likely supplant adult homograft tissue as a source of graft material, because of its proliferative ability as well as its resistance to trauma and ischemia. Unfortunately, it is possible that both xenografts and allografts of fetal tissue may require chronic immunosuppressive therapy to prevent their rejection. Chronic immunosuppression may add significant lifetime risk to these patients after their transplantation procedure, and they might even experience an active graft versus host rejection reaction that could injure the surrounding normal brain.

An attractive alternative to using directly harvested cells in transplants is to employ tissue culture techniques to provide a donor pool. Cells grown in tissue culture often lose their histocompatibility antigenicity and might therefore not be as prone to rejection. At the very least, such cells could be definitively tissue-typed and stored until an ideally matched recipient could be found. Also, transmission of infectious agents could be excluded before transplantation, and the stresses of emergent harvesting prior to transplantation could be obviated.

The ideal sites for tissue transplantation to the brain remain to be defined. There are some who believe that placement of grafts on the ventricular surface may be superior to intraparenchymal implantation. Comparison of the Mexican with the Swedish studies suggests that this might be true. Targeting of tissue placement will be aided by digital tomographic imaging systems, using biochemical or functional markers. The limited migration of implanted cells suggests that precise localization is important for therapeutic effect.

The brain produces trophic factors, which promote survival and proliferation of neural cells in tissue culture. This trophic activity peaks approximately 6 days after parenchymal damage. Grafting into tissue cavities prepared in the brain several days earlier enhances the survival of grafted cells. A better understanding of the factors mediating this effect may enhance the viability of neural grafts. Some investigators suggested that the sprouting is due to surgically induced trauma, with or without the transplanting of viable cells. Recent work with fetal tissue may suggest that to use viable cells is, indeed, important.

Delayed adverse effects of transplantation remain a possibility. A potential risk is proliferation or neoplastic change within the transplant. Also, the blood-brain barrier seems to be permanently disrupted at the site of neural grafting. Although, in theory, this might permit beneficial pharmacologic manipulation of graft function, the presumed protective role of this barrier would be bypassed, exposing the brain to multiple blood-borne agents, including soluble components of plasma, infectious agents, and cellular and humoral components of the immunologic system.

## Molecular Neurosurgery and Gene Therapy of Brain Tumors

The ultimate neurosurgical treatment of brain tumors will undoubtably forego removal or ablation of tissue in favor of redirecting cellular machinery or stimulating apoptosis in unsalvageable tumor cells. This will require removal and replacement of identifiable subcellular organelle mechanisms.

Techniques of genetic engineering have allowed the construction and recombination of genes and controlling elements, and of in situ transfer of genetic information, typically involving viruses. These involve the direct transfer of specific tumor-suppressor genes, of genes that encode a particular toxic product, or of genes whose products induce a new sensitivity on the part of tumor cells to exogenous agents. Many types of viruses are under investigation for these purposes, but two categories, retroviruses and herpes viruses, have received special attention for use in human CNS tumors.

Retroviruses have characteristics favorable for use in gene transfer to brain tumors; they are RNA viruses, and depend on a target cell exhibiting DNA synthesis. Retroviruses may thus have a selective advantage in targeting tumor cells because malignant gliomas are composed of an actively dividing cell population that proliferates against a mitotically inactive background of the CNS. Replication-defective retroviruses have been constructed, conferring a greater measure of safety because the virus cannot make additional infectious copies once it has transfected a host cell.

A major disadvantage of retroviruses is that their small size excludes the insertion of large genes or more than one gene. They also demonstrate a short half-life in vivo and have the potential for oncogenesis. Experimental studies have included the cerebral stereotactic inoculation of fibroblast cultures producing replication-defective retroviral vectors, resulting in transfer of the herpes simplex thymidine kinase (HSV-TK) gene to 9L and C6 gliomas, in order to confer tumor cell sensitivity to the antiviral drug ganciclovir. Subsequent systemic administration of ganciclovir has produced excellent tumoricidal responses. Implantation of the viral producer fibroblast cell lines, psi-2-BAG or NIH3T3, allows sustained local packaging and release of the retroviral vector with improved transfer and expression of the gene. However, low efficiencies (1%–10%) for gene transfer still remain a significant limitation for replication-defective retroviral vectors.

Herpes viruses, being larger than retroviruses, can accommodate the transfer of more genetic material. Because they do not require host cell DNA synthesis for gene transfer, they do not selectively target proliferating tumor cells. Antiherpetic agents exist to minimize or abort an ongoing herpes infection, providing a means for control in future clinical use. Several herpes mutants have been produced that are deficient in producing DNA replication enzymes such as thymidine kinase and DNA polymerase, therefore demonstrating growth in proliferating glioma cells but compromised growth within normal glia and neurons. Experimental studies have included use of an engineered herpes simplex virus-1 mutant (dlsptk) with a deletion at the HSV-TK gene to kill human glioma cells in vitro and in vivo. Fatal encephalitis has limited the use of inoculation with higher doses of virus, prompting a search for mutant lines with less host toxicity.

The future of what has been termed *molecular neurosurgery* remains extremely promising for providing a method of more definitively assaulting the cellular machinery of gliomas by targeting the genetic code itself for transplantation, to significant clinical benefit.

## Interactive Image-Guided Surgical Technology

Future surgical trials of tissue transplantation to the brain may require placing multiple grafts in spatial arrays distributed through an anatomically or functionally defined target, owing to the limited ability for incorporation and outgrowth by donor graft tissue. Stereotactic techniques will assist progressively more and more in this (see Chapters 22 and 24).

## Conclusion

The transpantation of whole tissue, cellular suspension, and subcellular components to alleviate CNS dysfunction is a rapidly evolving field. The promise of this field must be rigorously and systematically defined through pragmatic engineering research and development of surgical navigation systems, careful experimental studies of fundamental cellular manipulations in appropriate animal models, well-controlled pilot clinical trials, and, ultimately, large, prospectively randomized clincal trials.

Although Parkinson's disease has attracted the greatest clinical interst for neural transplantation, the surgical methodology is easily transferable to other clinical syndromes. The advances in genetic therapy of brain tumors bring transplantation down to the subcellular level of organization. Advances in digital imaging, robotics, and virtual reality computer control interfaces will provide a new generation of surgical tools to provide microscopic percision to better deliver transplanted materials to their targets. Once thought to be futuristic, these technology-intensive developments are squarely in the tradition of precise anatomic and functional localization underpinning the diagnostic and therapeutic missions of medical and surgical neurology.

## References/Further Reading

Allen GS, Burns RS, Tulipan NB, et al. Adrenal medullary transplantation to the caudate nucleus in Parkinson's disease: initial clinical results in 18 patients. Arch Neurol 1989;46:478–491.

Backlund EO, Granberg PO, Hamberger B, et al. Transplantation of adrenal medullary tissue to striatum in parkinsonism: first clinical trials. J Neurosurg 1985;62:169–173.

Bakay RAE, Allen GS, Apuzzo M, et al. Preliminary report on adrenal medullary grafting from the American Association of Neurological Surgeons Graft Project. Prog Brain Res 1990;82:581–591.

Bakay RAE, Sladek JR Jr. Fetal tissue grafting into the central nervous system: yesterday, today, and tomorrow. Neurosurgery 1993;33:645–647.

Burns RS, LeWitt PA, Ebert MH, et al. The clinical syndrome of striatal dopamine deficiency: parkinsonism induced by 1-methyl-4-phenyl-1,2,3,6-tetrahydropyridine (MPTP). N Engl J Med 1985;312:1418–1421.

Freed WJ, Cannon-Spoor HE, Krauthamer E. Intrastriatal adrenal medulla grafts in rats: long-term survival and behavioral effects. J Neurosurg 1986;65:664–670.

Galloway RL Jr, Berger MS, Bass WA, Maciunas RJ. Registered intraoperative information: electrophysiology, ultrasound, and endoscopy. In: Maciunas RJ, ed. Interactive image-guided neurosurgery. Park Ridge, IL: American Association of Neurological Surgeons, 1993:247–258.

Goetz CG, Olanow CW, Koller WC, et al. Multicenter study of autologous adrenal medullary transplantation to the corpus striatum in patients with advanced Parkinson's disease. N Engl J Med 1989;320:337–341.

Lindvall O, Backlund EO, Farde L, et al. Transplantation in Parkinson's disease: two cases of adrenal medullary grafts to the putamen. Ann Neurol 1987;22:457–468.

Maciunas RJ, ed. Interactive image-guided neurosurgery. Park Ridge, IL: American Association of Neurological Surgeons, 1993.

Maciunas RJ, Fitzpatrick JM, Galloway RL Jr, Allen GS. Beyond stereotaxy: extreme levels of application accuracy are provided by implantable fiducial markers for interactive image-guided neurosurgery. In: Maciunas RJ, ed. Interactive image-guided neurosurgery. Park Ridge, IL: American Association of Neurological Surgeons, 1993:261–270.

Madrazo I, Drucker-Colin R, Diaz V, et al. Open microsurgical autograft of adrenal medulla to the right caudate nucleus in two patients with intractable Parkinson's disease. N Engl J Med 1987;316:831–834.

Madrazo I, Drucker-Colin R, Leon V, et al. Adrenal medulla transplanted to caudate nucleus for treatment of Parkinson's disease: report of 10 cases. Surg Forum 1987;38:510–512.

Maurer CR Jr, Fitzpatrick JM. A review of medical image registration. In: Maciunas RJ, ed. Interactive image-guided neurosurgery. Park Ridge, IL: American Association of Neurological Surgeons, 1993:17–44.

Olson L, Backlund EO, Ebendal T, et al. Intraputaminal infusion of nerve growth factor to support adrenal medullany autografts in Parkinson's disease: one year follow-up of first clinical trial. Arch Neurol 1991;48:373–381.

Spencer DD, Robbins RJ, Naftolin F, et al. Unilateral transplantation of human fetal mesenceplalic tissue into the candate nucleus of patients with Parkinson's disease. N Engl J Med 1992;327:1541–1548.

Tulipan N. Brain transplants: a new approach to the therapy of neurodegenerative disease. Neurol Clin 1988;6:405–420.

Tulipan NB, Zacur HA, Allen GS. Pituitary transplantation. Part 1. Successful reconstitution of pituitary-dependent hormone levels. Neurosurgery 1985;16:331–335.

Turner DA, Kearney W. Scientific and ethical concerns in neural fetal tissue transplantation. Neurosurgery 1993;33:1031–1037.

Widner H, Tetrud J, Rehncrona S, et al. Bilateral fatal mesenceplalic grafting in two patients with parkinsonism induced by 1-methyl-4 phenyl-1,2,3,6 tetrahydropyridine. New Engl J Med 1992;327:1556–1563.

# IV INTERMEZZO

# Methods of Moral Philosophy and the Basis of Medical Ethics

**13**

A. Norman Guthkelch

The main purpose of this chapter is to address some basic problems of methodology which lie behind any consideration of the issues of medical ethics. The practical applications of ethical thought to clinical practice in the neurosciences are dealt with elsewhere, for example in Bernat's recently published *Ethical Issues in Neurology*. However, unless we are aware of the principles from which we are reasoning, we court two dangers: the first is of becoming mere ethical technicians, making decisions in response to persuasive advocacy or ingrained habit, with little idea of their justification or implications; the second is of operating within an exclusively patient-problem–oriented ethical system without heed to intra- and inter-professional relationships, family systems, economic realities, and so on.

## Moral Philosophy, Old and New

Western ethical thought began with the Greek philosophers, for whom there was an ultimate Good to which all reasonable men (note the use of the male gender!) wished to attain: It could be achieved by leading a virtuous life. Success led to personal fulfillment and to the respect and admiration of one's fellows. If everyone led a virtuous life, happiness would follow.

Such optimism was perhaps understandable when the Greek city-states were flourishing and a small male intellectual and economic elite felt they had some reason to regard themselves superior to their barbarian neighbors, at least in all of the attainments that they themselves valued. But of course, their material needs were supplied by slave labor under the direction of compliant womenfolk, whose happiness and fulfillment were secondary considerations. In any event, the politicoethical structure that the Greeks erected proved all too susceptible to the winds of change. Their ideal of an egalitarian union of civilized city-states was out of date before it had been fully formulated. Nor were the assumptions of Greek ethics beyond challenge. The peoples of the

Roman Empire, subject to the whims of uncaring emperors and their brutal subordinates, found Greek ideals unrealistic and lacking in warmth. Aristotle's ideal Greek gentleman was well educated and well mannered, but he was also a sanctimonious snob, while Plato's assertion, put into the mouth of Socrates, that evil doing is due simply to lack of education, takes no account of the strength of the emotional forces that lurk within the shadows of the human mind.

Nowadays, we recognize that ethics cannot exist in isolation from culture and belief, and that there is more than one possible methodology that we could use to examine our judgments of what is right. Before passing to the consideration of actual principles and rules, therefore, we need to take a glance at some of the work of the moral philosophers of the twentieth century.

Consideration of four fundamental questions lies at the heart of much of modern moral philosophy. 1) Given that it means something to assert that certain actions are good, what mental processes do we use to make a judgment in a given case? Can ethical statements be said to be true, or are they simply expressions of approval or disapproval? 2) If ethical statements *can* be true, do we judge them by the same sort of logical reasoning by which we arrive at a medical diagnosis, or do we use some other mental faculty? 3) In any event, is the truth of an ethical statement universal, or is it valid only for the person who makes it? 4) Finally, in judging whether a course of action is right, should we look at the actions themselves, or at their consequences?

## ARE ETHICAL STATEMENTS EVER TRUE?

This question asks whether, in making ethical judgments, we use cognition. It does *not* ask whether we ever have sufficient *information* to make such judgments with confidence (anyone who has sat on an ethical review body knows how rarely we do), but rather whether, *in principle*, a statement such as, "It would be wrong to terminate life support in that case," can ever be called true or false in the sense that the statement, "It rained today," is true or false. Cognitivist philosophers answer, "Yes," while emotivists assert that such judgments as, "Euthanasia is wrong," are neither true nor false but rather expressions of approval or disapproval (e.g., "Shame on you if you practice euthanasia!") or perhaps veiled imperatives (e.g., "Don't you dare . . . !"). Some emotivists have even held that moral philosophy should be confined to definitions and discussions of ethical terms and should avoid actual ethical pronouncements, in which case, despite their expertise in handling questions of meaning and implication, philosophers would have no place in an ethical review committee.

## HOW CAN WE KNOW WHAT IS GOOD?

The emotivist position more or less closes the door to discussion of this question. Cognitivists, however, while agreeing that there is some sense in which ethical judgments can be true or false, differ as to how the truth or falsity of these is to be known. One school would propose a definition of what is good and then by deduction or induction arrive at an appropriate judgment in each particular case. This is called the naturalist position, because it makes ethics subject to the same laws of reasoning as is Natural Science. Against this, non-naturalists argue that the idea of what is Good is unanalyzable and perceived only by an act of intuition. Both positions have their supporters; neither is unassailable, and the possibility of combining them will be discussed later. At present, we need only note that the choice is not purely academic. If we adopt an extreme naturalist position, we are in danger of believing that every ethical problem has a unique correct solution; pure intuitionism risks my asserting that, or behaving as if, my intuition is more valid than yours, for reasons that are beyond discussion.

## IF ETHICAL STATEMENTS CAN BE TRUE, FOR WHOM ARE THEY TRUE?

If there *is* some sense in which ethical statements are true, we must consider whether they are ultimately statements about ourselves, so that "X is good" means "I believe X is good (and I hope you will think so too)"—the subjectivist position—or whether they are absolute, as some objectivists would say, or at least inescapable, given the speaker's cultural and religious heritage. A possible way of blending subjective and objective elements in forming our judgments will be discussed later, when we look at Good Reasons theory.

## ON WHAT BASIS SHOULD ETHICAL DECISIONS BE MADE?

The main controversy here is whether the ethical value of an action should be assessed *non-teleologically,* usually by asking whether the action itself is right, or *teleologically,* by asking whether its consequences are good. An example of the former position is the deontologic approach to ethics associated with many religions and with Immanuel Kant, while the latter may be exemplified by the utilitarianism of the British philosophers Jeremy Bentham and John Stuart Mill.

Deontologists may appeal to religious inspiration, to obedience to law, or simply to a personal conviction of obligation (Kant). Utilitarians look toward goals such as the greatest happiness for the greatest number (Bentham) or the preservation and improvement of life; some of them believe each problem must be considered separately (act utilitarianism), while others use the principle of utility to make rules that can then be applied to classes of acts (rule utilitarianism). An important difference between the deontologic and the utilitarian positions is that whereas the deontologic position requires only that one accepts the idea of an ultimate authority, utilitarianism is meaningless unless there is someone around to evaluate consequences.

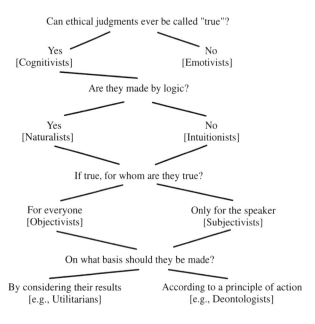

Can ethical judgments ever be called "true"?

Yes
[Cognitivists]

No
[Emotivists]

Are they made by logic?

Yes
[Naturalists]

No
[Intuitionists]

If true, for whom are they true?

For everyone
[Objectivists]

Only for the speaker
[Subjectivists]

On what basis should they be made?

By considering their results
[e.g., Utilitarians]

According to a principle of action
[e.g., Deontologists]

**Figure 13.1**
Questions underlying the formulation of ethical theories and how they may be answered. (Compare with Ashley and O'Rourke, p. 149.)

In summary, it is possible to visualize the analytical work of recent moral philosophers in at least four different dimensions, as it were: namely, in their attitudes to the roles that are played therein by cognition, naturalism, subjectivism, and teleology (Fig. 13.1). Noncognitive (emotivist) theories stand pretty much on their own, but within the other fields, although some sets of choices are more popular than others, any are logically possible. Not unexpectedly, cogent objections have been raised to all of the resulting systems. For systematic discussion of these objections, from two very different points of view, see *Healthcare Ethics: A Theological Analysis*, by Ashley and O'Rourke, and *Twentieth Century Ethics*, by Hancock.

Does all of this striving toward formal justification have any relevance to real ethical problems in all their messiness, and does it take sufficient account of the subconscious conflicts that arise within our minds as we reflect on how to behave for the best? Cannot the realities of the human condition be faced in a more personal way?

René Descartes derived his conception of reality from the fundamental proposition *cogito ergo sum*.* For Descartes, this statement was a sort of rock of intellectual certainty that was unmoved by metaphysical doubts about the *meaning* of existence. Two centuries after Descartes, however, Søren Kierkegaard used the certainty of self-awareness in a much less reassuring way. To him, it

*Literally, this Latin phrase means, "I think, therefore I am," suggesting the syllogism, "I think—whatever thinks *is*—therefore I am." It is more probable, however, that Descartes was saying, "My most certain knowledge is of my existence."

seemed that, because the unique feature of being human is what he described as our painful self-knowledge, this must also be the starting point for any consideration of what we value and of how we should behave. Kierkegaard's insight led to the development of J.-P. Sartre's existentialism. Sartre not only repudiated any belief in divine law, but also seemed to regard human law as basically unjust, doubting whether individuals can formulate rules of conduct for themselves in advance of actual situations. Therefore, he asserted, in the face of the injustice and absurdity of life, we are entitled to respond to any situation in our own way, provided only that we accept full responsibility for our actions. Unfortunately, it seems to follow that, because there is no concept of decision making within community, existentialism cannot be made the basis of ethical medical practice. Nonetheless, reflection on the pain that is implicit in the assertions from which existentialism is derived can increase our empathy for the sick and disabled and our awareness of the loneliness in which disadvantaged people must too often make their decisions.

The final example of a twentieth-century development of moral philosophy with which we shall deal uses both cognitivist and emotivist ideas, and because it also takes a special cognizance of the inevitability of conflict, it seems well suited for use in the complex environment of health care delivery. It is called Good Reasons theory, and largely arises from the work of Stuart Hampshire (1983).

Hampshire begins by politely ridiculing such philosophers as Hegel, who wrote as though they stood within sight of the culmination of humanity's long journey, so that only a few modifications were necessary to assure a happy ending. Not so, says Hampshire; there are three serious limitations to our understanding, all of which will persist for the indefinite future: (1) Although neuroscience continues to make important strides in elucidating the organization of behavioral processes within the brain, we still have little understanding of the complex mental forces that drive us to our judgments and actions and of how they operate. (2) Whereas many philosophers of the last century looked forward to a unification of humankind, we now see that more and more ethnic and religious groups worldwide are ready to fight to the death to preserve their individuality. It is inconceivable that humankind will agree within the measurable future to assign primacy to one ethnic group or belief system over the others. (3) In any event, language differences offer an enormous barrier to complete understanding within the whole human family. Recent literary criticism has pointedly asked whether we ever fully understand what other people mean in their speech or writing, and the answer is not reassuring.

Next, Hampshire attacks utilitarianism, a system that—perhaps because it involves the same sort of risk-benefit calculation that we use in prescribing treatment—unfortunately tends to be widely held among health professionals. Its attractiveness is not diminished by the promise that, provided Reason is properly employed, one is assured of achieving an improvement in the human condition. Hampshire's first criticism is of the utilitarian tenet that when assessing the value of principles, rules, policies, or whatever, it is essential *and also sufficient* to consider their impact on the situations and feelings of those people who will be affected by them, for in that case, "the whole machinery of the natural order . . . is useful only in proportion as it promotes or prevents desired states of feeling." If the whole human species perished, the survival of the rest of creation would be of no consequence, and the survival of an endangered species is important only in so far as we humans take pleasure in its continued existence!

The preceding argument is relevant to medical practice. The utilitarian position implies, for example, that once a patient is in terminal coma, the way he or she is treated is of no account, except in so far as it causes satisfaction or pain to the family. Such a point of view represents a serious potential limitation of the principle of the sanctity of human life. Worse still, says Hampshire, utilitarianism makes morality "a kind of psychical engineering, which shows the way to induce desired or valued states of mind." What if people could be trained or bred to experience the kinds of feeling that would lend value to their surroundings? To subject human beings to such calculated behavioral modification would be a horror of Orwellian proportions. Lastly, the predictions of utilitarianism have not been fulfilled. At the turn of the nineteenth century, utilitarians felt confident that in an era of advancing technology, increasing education, and moral enlightenment, humankind would enjoy a new age of increasing happiness. In fact, this dream has been negated by, in Hampshire's words, "the hideous face of political events. Persecutions, wars and massacres have been coolly justified by considerations of (alleged) long-range benefit to mankind," and unique cultures and environments continue to be destroyed because of political calculations that have since proved to be wrong. In the field of medicine, too, there is little justification for the belief that we can devise a calculus to balance two qualitatively unrelated effects against the other. As long as such a calculation is believed to be valid, someone will devise a way to perform it, and the result will occasionally be disastrous. The alternative criterion need not be a strict deontology—many systems of ethics, both secular and religious, are teleologic without being utilitarian—but before we consider what it should be, we must return to the question of how we judge what is right and good.

Can we establish ethical truths by the same sort of deductive or inductive reasoning by which we establish logical or scientific truth? Hampshire says we cannot, because to establish the validity of an ethical position, we need to use arguments that are outside the stock in trade of logicians. For example, we can appeal to our cultural heritage. The argument that something has been believed to be true by people like ourselves at all times in all places, though not conclusive, is certainly not without force. Hence we may properly believe, without requiring logical proof, that there is such a thing as morally inexcusable conduct; the details of such conduct and the reasons for its rejection may differ between nations and generations, but the sense of outrage is constant. In any society, a full set of injunctions that would define absolutely prohibited types of conduct will also describe from the negative standpoint "the minimum general features of a respect-worthy way of life." The important thing is not that what we hold to be an ethically correct course will occasionally be different in apparently similar circumstances, but rather that it will much more often be the same. Good Reasons theory accepts the limitless complexity of the influences that act on us in all our moments of decision, and hence the inevitability of conflict.

## To Be or To Do Good?

T. S. Eliot reportedly once remarked that although for many of life's problems the proper question to ask is what we are going to do about them, there are also some problems in the face of which we instead need to ask how we should behave toward them. Is it sufficient for us to see our function as ethical professional people simply in terms of our reactions to specific problems, or should we not also pay attention to living in a virtuous way? Large organizations have power to hide the mistakes of their employees, and increasing specialization entails that few people really know what specialists are doing or to what end. May has commented that the solution is for those who wield institutional or professional power to cultivate virtue in themselves and their staff members. In many writings on medical ethics, however, when virtue is mentioned at all, it is defined in terms that subordinate virtues to principles, as when it is said that "virtues are settled habits and dispositions to do what we *ought* to do." May and also MacIntyre believe instead that virtues are those qualities that concentrate on "the goods internal to practices," rather than on those that flow from them. Of course, virtue cannot exist *in vacuo*, and Veatch has argued that virtue without well-conceived principles is ineffective and may be actually dangerous. Society must have a commitment to justice before personal virtue can flourish at all and it may be difficult for the individual to practice this within an organization that is dedicated to producing results. Nonetheless, it is a more respect-worthy reason to practice fidelity and truth-telling that these are virtues good in themselves, than that in so doing we make our patients feel happier, or protect ourselves from lawsuits.

## Turning Theory into Practice

Four models for decision making in the clinical situation will be mentioned. Two of these, The Four Principles and Siegler's model, are more particularly concerned with concepts and attitudes, while the other two, casuistry and Gert's model, offer a definite procedure.

## THE FOUR PRINCIPLES

At present, the Four Principles* model is probably the one most generally adopted in attempting to solve ethical dilemmas in clinical practice. It can be summarized as a system in which are balanced a respect for patient autonomy, non-maleficence, beneficence, and justice. It does not set up a procedure for resolving conflicting claims made within an overall principle or under different principles, although in a multiracial, multicultural, individualistic society, such dilemmas are to be expected.

Respect for the autonomy of others implies treating all people as ends, not means. In clinical practice, it involves four main elements: obtaining informed consent, confidentiality, promise keeping, and absence of deceit. Other issues of autonomy that need to be evaluated in individual cases are the amount of information that the physician should share with the patient and the degree of patient participation in the details of decision making. Consequently, every clinician is obligated to be a good communicator, and especially an active listener, for unless patients are able to communicate freely and without embarrassment with their physician, sick people and their families are in an abusive relationship, which the physician is morally obliged to correct or terminate. Several writers have suggested that respect for patient autonomy implies active patient empowerment.

The concepts of non-maleficence and beneficence are well defined, and the only points that need to be mentioned here are first, that avoidance of harm (*primum non nocere*) is ethically prior to attempting to do good, and second, that the principle of autonomy demands that the benefit the physician seeks to confer must be one that is apparent to, and desired by, the patient.

Three principles may be recognized as guides to achieving justice in health care: respect for people's rights and concerns, obedience to morally acceptable law, and a fair distribution of scarce resources. It must be noted, however, as Aristotle pointed out, that fair distribution means equal shares for those with equal needs, the rest receiving consideration in proportion to all morally relevant factors.

## SIEGLER'S MODEL

Siegler prefers an approach in which decision making is designed to take into account as many as possible of a fairly well-defined catalog of objectives. These comprise the medical indications, the patient's wishes, the resulting quality of life for the patient, and the socioeconomic factors that are operative in each

---

*It has been objected that the use of the word *principles* to denote the four headings under which Beauchamp and Childress (1983) teach that clinical practice should be conducted is misleading, and that they might better be described as *concepts*. The distinction is not trivial, but the phrase "The Four Principles" is well entrenched in the literature and will therefore be retained here.

individual situation. Most, if not all, of the items on Siegler's list, are explicit or implicit in the Four Principles scheme, but again no detailed insight is offered into the procedure by which conflicts are to be resolved.

## GERT'S APPROACH

Gert starts with five cardinal rules, all of which come under the general heading of *primum non nocere:* Do not kill, do not disable, do not cause (unnecessary) pain, do not deprive of freedom, do not deprive of pleasure. A second group of five rules, regarded as not quite so vital as the first, because breaches, although likely to cause harm, are not certain to do so, comprises the following: Do not deceive, do not cheat, do not break your promises, do not violate the law, do your professional duty.

Recognizing that there will be occasions when the good of the patient may seem to demand that one or more of these rules must be broken, Gert goes on to suggest that a "moral attitude" is required, and proposes a set of eleven questions that the members of the health care team should ask themselves to establish whether it is really justifiable to break the rule. The reader is referred to Gert's book for the questions and the ethical justification for them.

## CASUISTRY

Casuistry is a methodology that proceeds by induction. Having chosen a set of principles or rules according to one's convictions, one applies them to a particular dilemma, about which most people would be likely to agree on the solution. The method is then extended to similar but progressively more difficult cases, until (ideally) all possibilities are covered.

# Practical Decision Making

We have considered the way ethical judgments are thought to be made, and how normative rules can be formulated. It remains to be considered how the physician should function in the process, and only a few examples can be given here.

It is generally held that the physician's primary role is that of patient advocate. Other frequently mentioned duties include obedience to the law and the avoidance of self-interest, waste, and particularly, any tendency to refuse treatment because one disapproves of a patient's lifestyle.

In the context of practice within health care organizations, there is a good deal of unexplored territory, which is only now being broached (e.g., in *The Hastings Center Reports* [Wolf].) For example, physicians sitting on the boards of for-profit hospitals have a duty to their stockholders, but does this equal the duty they owe to the patients from whose payments the hospital will derive its profits? They may be the only board members able to interpret their colleagues' advocacy for their patients. How far should they go in reinforcing this advocacy? Can they ever (except in extreme circumstances of apparent waste, major expenditure on apparatus of unproved value, or proposed duplication of services covered elsewhere) ethically take a position that might diminish the benefit to patients?

The sort of problem that has been formulated here can be addressed by the doctrine of Scope, which attempts to delimit the field of a physician's moral obligations on the basis that "we clearly do not owe a duty of beneficence to everyone." In many religious traditions, however, the scope of beneficence includes one's neighbor, the definition of which has been widened without limit by Jesus of Nazareth. Even those who would deny the physician's duty to act with beneficence toward all, would surely agree that there is a duty to refrain from hurting anyone.

What, then, does "hurting" include? Most ethicists agree that there exists a right to life, and physicians certainly have no authority to kill anyone against his or her will, but are they always obliged to prolong the life of every patient? Perhaps the duty to prolong life ends when the patient has no chance of further fulfilling any part of life's purpose as he or she would be able to understand it, or when to continue attempting to do so would impose an intolerable burden of suffering.

Is there also a right to die? If there is, and if some of those wishing to exercise it are incapable of killing themselves, does someone have an obligation to kill them? It seems incredible that such a responsibility should be laid on any physician, any more than a physician can ever have a moral duty to participate in judicial execution. Tragic cases of failure of pain relief, of fear of slow suffocation, and so on, are often adduced in favor of allowing physicians to terminate life. Quite apart from the "slippery slope" argument that many physicians find compelling, however, some practitioners of clinical pastoral care hold that euthanasia represents the ultimate "quick fix," which prevents family and community from exercising their obligation of companionship-until-death with a suffering member. This does not imply that it is unethical to secure adequate relief of pain and distress at the risk of setting in motion a train of events that results in death. The acceptance of a risk of complications in this setting is only a special example of the normal process of decision making.

Finally, we must face the fact that there are occasions when the requirements of ethical clinical practice seem to conflict with those of the legal system. Some people believe that in an ideal world this would not happen, but this is only true (1) if one's concept of an ideal world includes complete uniformity of moral opinions and standards, which is certainly not everyone's view, nor, as has already been argued, is it attainable in the foreseeable future; and (2) if the legal code can be made a perfect reflection of these. One reason for the conflict may be that physicians still operate occasionally from an authoritarian and paternalistic position, against which the civil rights of individuals, as consumers and as beneficiaries of government-aided health programs, must be upheld. Unfortunately, the adversarial legal system and the unpleasantness of the process of litigation make it difficult for physicians to avoid demonizing their perceived tormentors. Most ethicists hold that there is no duty to obey a morally unacceptable law, but the definition of "morally unacceptable" must be considered carefully.

Hampshire's criterion is to ask whether one feels a sense of outrage, but it may be difficult to avoid confusing right thinking with prejudice. Gillon suggests that the test is whether the law was enacted through an acceptable (i.e., democratic) process, but this doctrine seems not to allow for bona fide minority viewpoints within a democracy, as well as denying the legitimacy of any other form of government. In practice, when a court interferes with a treatment decision that the physician believes to be in accordance with her or his patient's interests, the physician plainly has a moral right and duty to disobey in accordance with the dictates of conscience. To impede others from obeying, however, may be a breach of the patient's autonomy.

## Conclusion

Just as there are many different explanations of what is going on in our minds when we make ethical judgments, so also is there a diversity of possible approaches to the practical problems of decision making in medical ethics. Not only patients, but also health care workers come from a multitude of ethnic, linguistic, generational, and religious backgrounds. It is a sign of vitality and progress in medicine that the ethical issues it raises should continually be outstripping existing guidelines, and it is no less essential that these should generate debate and conflict, albeit always in an atmosphere of compassion and mutual respect. The important thing is that physicians as professional people should and must form a moral community, because health care is a moral enterprise.

## Acknowledgment

I am most grateful to the Very Rev. Dr. William Rankin and to the Rev. Fr. E. C. Vasek, SJ, for reading the manuscript and making numerous helpful criticisms. They are in no way responsible for the limitations of the ultimate product.

## References/Further Reading

Ashley BM, O'Rourke KD. Healthcare ethics: a theological analysis. 3rd Ed. St. Louis: Catholic Health Association of the United States, 1989.

Beauchamp TL, Childress JF. Principles of biomedical ethics. 2nd Ed. New York: Oxford University Press, 1983.

Bernat JL. Ethical issues in neurology. Boston: Butterworth-Heinemann, 1994.

Fulford KWM, Gillett G, Soskice JM. Medicine and moral reasoning. New York: Cambridge University Press, 1994.

Gert B. Morality: a new justification of moral rules. New York: Oxford University Press, 1988.

Gillon R. Medical ethics: four principles with attention to scope. Br Med J 1994;3:183–188.

Hampshire S. Morality and conflict. Cambridge, MA: Harvard University Press, 1983.

Hancock RN. Twentieth century ethics. New York: Columbia University Press, 1974.

MacIntyre A. A short history of ethics. London: Routledge, 1967.

May WF. The virtues in a professional setting. In Fulford KWM, Gillett G, Soskice JM, eds. Medicine and moral reasoning. New York: Cambridge University Press, 1994.

Pellegrino ED, Thomasma DC. The virtues in medical practice. New York: Oxford University Press, 1993.

Siegler M. Decision-making strategy for clinical-ethical problems in medicine. Arch Intern Med 1982;142:2178–2179.

Veatch RM. Medical ethics: an introduction. Boston: Jones and Bartlett, 1989.

Wolf SM. Health care reform and the future of physician ethics. Hastings Center Report 1994;24:28–41.

# V CEREBROVASCULAR PROBLEMS

# The Biology of Cerebral Arteries

14

Saadi Ghatan

Marc R. Mayberg

This chapter provides an overview of the basic physiology of the cerebral vessel wall, with emphasis on its response to pathologic conditions. Atherosclerosis, which damages coronary and peripheral vessels, affects the cerebral vasculature in the same way. Intimal hyperplasia resulting from injury to vessel wall is a common sequel to endarterectomy or other surgical intervention. Thrombosis related to atherosclerotic disease or cardioembolic events is a significant component of stroke and other cerebral vascular disorders. Cerebral arterial vasospasm is a common complication of aneurysmal subarachnoid hemorrhage, as well as of hemorrhage due to arteriovenous malformations, head trauma, or tumors.

## Anatomy of the Vessel Wall

The vessel wall has motor, sensory, and endocrine capabilities. Its four major constituents—endothelial cells, connective tissues, smooth muscle cells, and perivascular axons—function in concert to regulate the interaction between blood and the tissues with which it comes in contact and to modulate blood flow in response to a variety of physiologic and pathologic conditions (Fig. 14.1).

### ENDOTHELIAL CELLS

Endothelial cells form the inner lining of arteries and veins; their three primary functions are to provide a barrier between vessel lumen and extravascular tissue, to locally influence vascular tone, and to modulate pro- and anticoagulant properties. The macromolecules collagen, elastin, laminin, and the glycosaminoglycans synthesized by endothelial cells contribute to a relatively impermeable surface between vessel lumen and interstitium. Vasoconstriction and dilatation are modulated by endothelial cell products, depending on the functional state of the vessel wall. Vasomotor effectors of endothelial origin include prostacyclin, endothelial-derived relaxing factor (EDRF), renin, angiotensin-

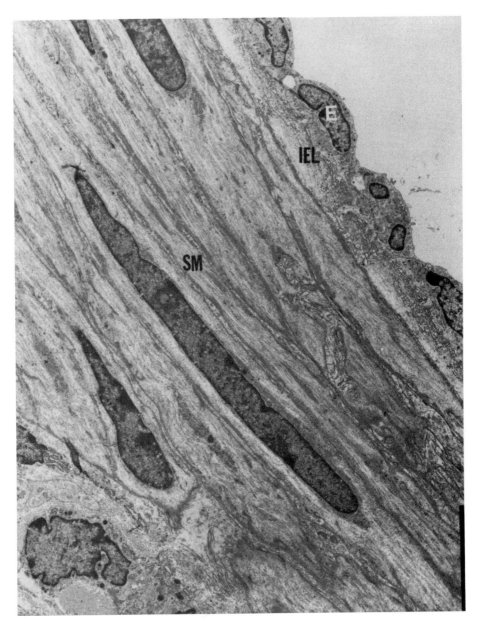

**Figure 14.1**
Electron micrograph of the normal cerebral vessel wall, demonstrating endothelial (E) and smooth muscle (SM) cells separated by an internal elastic lamina (IEL). Perivascular axons are not shown.

converting enzyme, endothelin, and adenosine nucleotides, which act in con-junction with alpha-adrenergic agents, thromboxane A2, and serotonin to modulate vessel tone.

Although procoagulant properties of endothelial cells compose a major portion of the vascular response to injury, other endothelial cell properties function to maintain normal and continuous blood flow under physiologic circumstances. By regulating platelet and procoagulant factor activity through such endothelial cell products as prostacyclin (PGI2), EDRF, thrombomodulin, and tissue plasminogen activator, blood fluidity is maintained and thrombotic events are inhibited.

## CONNECTIVE TISSUE LAYER

Composed primarily of collagen and elastin fibers in a ground substance matrix of glycoproteins and proteoglycans, the connective tissue layer plays structural and functional roles in the vessel wall. Strength and elasticity are provided through varying ratios of collagen to elastin to meet the flow and pressure needs of a particular vessel. Collagen types I, III, and IV are the primary collagens of the vascular system and are synthesized by endothelial and smooth muscle cells. The turnover of collagen is regulated by interstitial collagenases synthesized by smooth muscle cells and inflammatory cells in the vessel wall.

Beyond a structural role, macromolecular elements composing the ground substance of the vessel wall perform a diverse array of functional roles through cell-cell, molecule-cell, and intermolecular interactions. Fibronectin, a 45-kD protein with multiple binding domains, can bind collagen, heparin, fibrin, and cell surfaces simultaneously to facilitate cell-cell and macromolecular interactions. Laminin, another adhesive glycoprotein, is directed toward epithe-lial and endothelial cell binding to type IV collagen of the subendothelial basement membrane. Finally, the glycosaminoglycans (e.g., heparan sulfate) represent the major macromolecular constituents of the amorphous ground substance. They act structurally and functionally to modulate clotting and anticoagulation as well as cholesterol metabolism.

## SMOOTH MUSCLE CELL

The vascular smooth muscle cells function in several capacities, including contraction, proliferation, chemotaxis, secretion, adhesion, and other meta-bolic processes. Each function involves (1) recognition of a stimulus at the cell surface, (2) transduction of the stimulus through a second messenger system, and (3) production of the appropriate response. Surface receptor and signal transduction systems allow a common receptor-agonist interaction to influence a variety of functional responses.

The surface receptor systems of the smooth muscle cell include angiotensin, alpha- and beta-adrenergic, muscarinic, serotonergic, and histaminergic receptors. In the healthy state, the primary function of the smooth muscle cell is to maintain vascular tone via agonist-mediated signal transduction and second messenger systems such as g-proteins, phosphoinositide metabolism, cAMP, cGMP, and calcium.

The proliferative capacity of the vascular smooth muscle cell is well recognized in both healthy and pathologic conditions. Numerous cell-surface growth factor receptors are involved directly in the processes of smooth muscle cell hypertrophy and proliferation, but the mechanism by which common stimuli give rise to such diverse functional outcomes remains unknown.

## Perivascular Axons

Large cerebral arteries have a dense perivascular innervation and respond to inputs from four major neurotransmitters: norepinephrine, vasoactive intestinal peptide, acetylcholine, and substance P. The nerve fibers are situated in the adventitia and outer third of the media and contain varicosities apposed to smooth muscle cells, through which granular vesicles of neurotransmitters are released to influence vascular tone. Norepinephrine release classically stimulates alpha-adrenergic receptors to increase smooth muscle cell contraction and vasoconstriction. The sympathetic cerebral innervation has been hypothesized to provide protection in acute hypertension, as well as acting as a trophic influence for vessel wall growth and maintenance. Elsewhere in the body, norepinephrine release stimulates alpha1-adrenergic receptors to increase smooth muscle cell contraction and vasoconstriction; in the cerebral circulation, however, this response is limited. Vasoactive intestinal peptide is thought to be the major neurotransmitter in the cerebrovascular circulation responsible for vasodilation through smooth muscle cell relaxation. Cerebral arteries also contain cholinergic axons, but their role is uncertain because exogenous acetylcholine fails to elicit relaxation of cerebrovascular smooth muscle cells. Substance P, which is the neurotransmitter of primary sensory afferents involved in nociception, is present in large amounts in the cerebral circulation. Primary sensory afferents innervate the cerebral blood vessels, and may mediate the pain of vascular and stroke-associated headaches.

## Pathophysiology of the Vessel Wall

### Arteriosclerosis

Arteriosclerosis, or thickening of the arterial intima with subsequent narrowing of the arterial lumen, has been associated with a variety of vascular insults, including chemical, mechanical, immune, or infectious processes. Its most common forms are atherosclerosis and intimal hyperplasia secondary to vascular injury.

Endothelial injury is the common initiating event in both atherosclerosis and intimal hyperplasia. It is followed by release of a variety of mitogenic and chemotactic factors that act on smooth muscle cell to modulate proliferation in the media and migration of cells across the internal elastic membrane, thus stimulating further proliferation in the intima (intimal hyperplasia). Intimal smooth muscle cells subsequently produce extracellular matrix components and accumulate intra- and extracellular lipid. Finally, thrombus becomes attached to the area of injury and further influences these smooth muscle cell changes (discussion follows).

Atherosclerosis is commonly associated with hypercholesterolemia, diabetes, smoking, and hypertension, each of which may influence chemical, mechanical, and perhaps immunologic insults against the endothelium. The early atherosclerotic lesion is the *fatty streak*, composed of lipid, macrophage, and smooth muscle cells, which evolves into a *fibrous plaque*, characterized by proliferating smooth muscle cells and macrophages in a dense connective tissue matrix with variable amounts of intracellular and extracellular lipid (Fig. 14.2). Deep in this cellular mass may be an area of necrotic debris, cholesterol crystals, and calcification. The stenosis caused by these plaques and the thrombotic and embolic events generated through their disruption lead to the morbid events of stroke or myocardial infarction.

## Intimal Hyperplasia

Of the hundreds of thousands of vascular procedures performed each year in the United States to relieve arterial stenosis, 15% to 40% are complicated

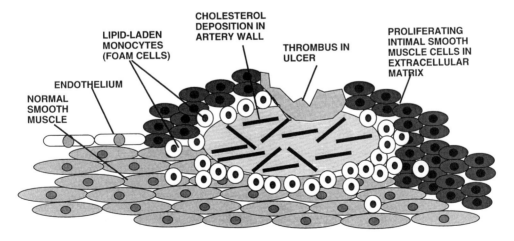

**Figure 14.2**
Schematic of a fully developed atheroma. See text for details.

by delayed arterial restenosis, the histopathologic correlate of which is intimal hyperplasia, consisting of proliferating smooth muscle cells in a connective tissue matrix (Fig. 14.3). Although smooth muscle cell proliferation is the basic mechanism causing intimal hyperplasia, this complex response also involves the interactions between smooth muscle cell and endothelial cells, extracellular matrix, platelets, thrombin, inflammatory cells, and various growth factors.

In the normal state, the smooth muscle cell maintains vascular tone through contraction and relaxation, and vessel wall integrity through the synthesis of extracellular matrix. Smooth muscle cells express differing phenotypes in response to proliferative stimuli, both in vivo and in vitro. They can be maintained in culture in a nonreplicating condition ($G_0$-$G_1$ phase), or they can be stimulated to initiate synchronous smooth muscle cell proliferation (S phase). Differentiated, quiescent smooth muscle cells (contractile phenotype) are relatively unresponsive to exogenous growth factors, but rapidly change in growth-promoting culture media to a synthetic phenotype characterized by proliferation and synthesis of extracellular matrix (notably collagen).

Phenotypic alterations associated with smooth muscle cell proliferation in vivo are comparable with those seen in vitro. Immediately after vessel injury, smooth muscle cells assume a synthetic phenotype. At the ultrastructural level, increased amounts of rough endoplasmic reticulum, free ribosomes, and mitochondria are present. This ultrastructural change is coincident with an increased proliferative rate for medial smooth muscle cells, which subsequently migrate through fenestrae in the internal elastic lamina to populate the intima (see Fig. 14.3). The neointimal smooth muscle cell population consists of cells that are proliferating within that area and those that have migrated there. After the initial phase of smooth muscle cell proliferation and migration,

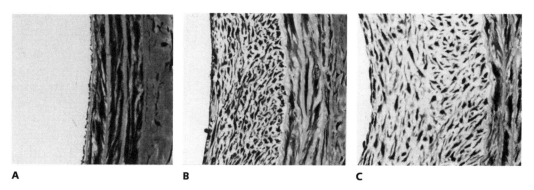

A    B    C

**Figure 14.3**
Intimal hyperplasia after arterial injury. These photomicrographs, taken (A) immediately after balloon angioplasty, (B) at 10 days after injury, and (C) at 20 days, demonstrate progressive intimal thickening with smooth muscle migration and cell proliferation to form a neointima.

which lasts 2 to 3 weeks, subsequent intimal thickening is secondary to connective tissue or extracellular matrix deposition.

## ENDOTHELIAL REGENERATION

Endothelium comprises a single layer of cells, so damage to it always involves subendothelial tissues as well. This results in the liberation of several growth factors and initiates thrombosis. Both sorts of injury modulate phenotypic changes, proliferation, and migration in medial smooth muscle cells. Minor injury to the endothelial cell layer is repaired through migration and local proliferation of endothelial cells in that area. When the injured surface is large, cell migration is insufficient and re-endothelialization does not occur. This may result in excessive smooth muscle cell proliferation.

## GROWTH FACTORS AND VESSEL WALL RESPONSE TO INJURY

Endothelial cells produce numerous peptide mitogens responsible for autologous growth regulation, and smooth muscle cell proliferation. Vessel wall injury causes release of basic fibroblast growth factor (bFGF), a heat-labile, heparin-binding, growth-promoting peptide produced by both endothelial cells and smooth muscle cells and sequestered in the subendothelial matrix. This growth factor stimulates smooth muscle cell proliferation in vivo in the acute phase of arterial wall injury.

Platelet-derived growth factor (PDGF) is a basic, 30-kD protein containing two distinct but homologous chains ($\alpha$ and $\beta$) that are secreted in various isoforms by activated platelets, endothelium, smooth muscle cells, and mononuclear phagocytes. The $\beta$ chain represents the transmembrane moiety of the protein that is similar in sequence and structure to the viral oncogene c-sis, a transforming protein with tyrosine kinase activity. The homodimeric form of PDGF-bb induces mitogenic, chemotactic, and vasoconstrictive effects. The influence of PDGF on intimal hyperplasia is related primarily to smooth muscle cell migration rather than proliferation.

Under certain conditions, endothelium and smooth muscle cells also synthesize transforming growth factor beta-1 (TGFβ1), which causes growth inhibition at high doses but growth promotion at low levels. It has only a weak chemotactic influence but may act to increase the extracellular matrix connective tissue components in the later stages of intimal proliferation. Transforming growth factor beta-1 mRNA levels are elevated after angioplasty in medial smooth muscle cell, and TGFβ1 likely stimulates smooth muscle cells by an autocrine mechanism.

Insulin-like growth factor 1 (IGF-1) plays a significant role in intimal proliferation, as a smooth muscle cell cofactor necessary for completion of the cell cycle after stimulation with other mitogens. Angiopeptin, a somatostatin analogue inhibitor of IGF-1, has been shown to limit intimal hyperplasia in several animal models. Heparin-like growth factor provides an inhibitory influence on smooth muscle cell proliferation, but its mechanism of action and the role it plays in intimal hyperplasia are not completely understood.

Angiotensin II has been demonstrated as a mitogen for smooth muscle cells in vitro and has been implicated in the pathogenesis of intimal hyperplasia. Angiotensin II is synthesized in the blood vessel wall by a locally active renin-angiotensin system, with angiotensin-converting enzyme (ACE) on the luminal surface of the endothelial cell; it has been proposed as a mediator of autocrine regulation of vascular wall tone and metabolic activity. Following vascular injury, the local renin-angiotensin system may be stimulated to produce greater mitogenic effect on smooth muscle cells and to activate platelets.

## THE ROLE OF PLATELETS AND THROMBOSIS IN INTIMAL HYPERPLASIA

The influence of thrombus formation on intimal hyperplasia is complex and involves interactions between platelets, subendothelial connective tissues, growth factors, and other platelet products (specifically PDGF) (Fig. 14.4). When the subendothelial connective tissue is exposed, platelets are activated to adhere to a previously nonthrombogenic surface. Platelet surface receptor GP1b binds to von Willebrand factor associated with type I collagen and fibronectin. This stimulates rapid expression of the GPIIb/GPIIIa platelet receptor complex, which binds fibrinogen to allow platelet-platelet bridging. The reaction is calcium dependent and may be modulated by A-granuloproteins, fibronectin, von Willebrand factor and thrombospondin.

In the next phase of thrombus generation, the activated platelet produces thrombin. Thrombin cleaves fibrinogen to form fibrin, which stabilizes the

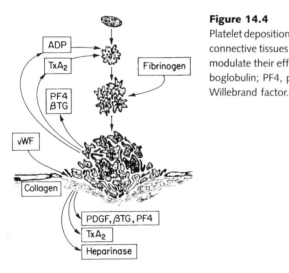

**Figure 14.4**
Platelet deposition and activation. Platelets adhere to the subendothelial connective tissues at the site of injury, and multiple interactive processes modulate their effects. ADP, adenosine diphosphate; βTG, beta thromboglobulin; PF4, platelet factor-4; TxA$_2$, thromboxane A$_2$; vWF, von Willebrand factor.

thrombus in a mechanism that is locally enhanced but controlled by systemic factors. Local activation of the clotting cascade by thrombin at the site of injury provides a positive feedback for thrombin production, whereas prevention of widespread thrombosis is accomplished through three primary systemic mechanisms of negative feedback: (1) Plasma protease inhibitors directly inactivate thrombin; (2) thrombin-antithrombin III–complex formation is enhanced; and (3) thrombin binds to endothelial surface thrombomodulin, which activates proteins C and S to destroy factors Va and VIIIa and therefore stop the clotting cascade (Fig. 14.5).

Although both fibrin and its precursor fibrinogen have been shown to stimulate migration of smooth muscle cells from the media into the intima, greater attention has been given to thrombin and its influence on restenosis. Hirudin, a potent thrombin inhibitor, limited intimal proliferation when given after injury in a rat model. Thrombin also has been shown to stimulate c-*sis* gene expression in endothelial cells, with the implications of PDGF expression described earlier.

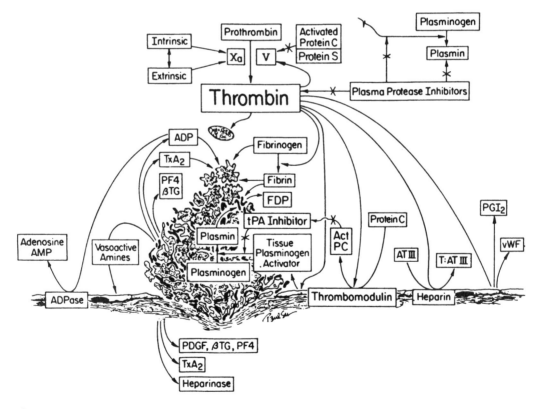

**Figure 14.5**

Thrombus formation. This complex process is regulated by multiple feedback mechanisms that limit its extension. (Reproduced by permission from Wilcox JN. Molecular biology. Am J Cardiol 1987;60(Suppl):20B–28B.)

## INFLAMMATORY CELLS IN THE PATHOGENESIS OF INTIMAL HYPERPLASIA

The association between inflammatory cells and atherosclerotic plaques has led to the hypothesis that inflammatory cells are involved in the development of intimal hyperplasia. Leukocyte adhesion molecules on the endothelial surface allow circulating polymorphonuclear leukocytes to bind to the vessel wall after injury and to become activated. Liberation of oxygen free radicals leads to further endothelial damage; expression of mitogens and chemotactic factors such as monocyte/macrophage-derived growth factor (MDGF) further stimulates smooth muscle cell proliferation. Finally, production of degradative enzymes and leukocyte infiltration into the vessel wall are thought to facilitate smooth muscle cell migration.

## ANIMAL MODELS FOR INTIMAL HYPERPLASIA

In animal models, balloon catheter injury produces both endothelial denudation and distention injury, the histologic response to which mimics the intimal hyperplasia that is seen after endarterectomy or angioplasty in humans. Platelet adherence commences without thrombus formation and is followed by smooth muscle cell DNA synthesis and smooth muscle cell proliferation within the media. The peak of this proliferative response occurs at 48 hours postinjury, and is followed by smooth muscle cell migration to the intima by day 4. Further smooth muscle cell proliferation occurs up to 2 weeks in the intima, thereby forming the neointimal lesion, which is expanded by connective tissue synthesis over many months. Although the balloon catheter injury lesion appears morphologically similar to the postendarterectomy or vein bypass graft lesion of humans, there exist important differences in response to therapeutic intervention.

## THERAPEUTIC STRATEGIES TO INHIBIT INTIMAL HYPERPLASIA: EXPERIMENTAL OBSERVATIONS

The multifactorial nature of the process of intimal proliferation reduces the likelihood that a single agent or intervention could limit intimal hyperplasia. Numerous pharmacologic strategies have been attempted in clinical trials after showing promise in vitro. Trials of antiplatelet agents such as aspirin and dipyridamole have yielded conflicting data with regard to the control of intimal hyperplasia. Aspirin decreases both platelet deposition on synthetic grafts in animal models and intimal hyperplasia after carotid endarterectomy in nonhuman primates. Nevertheless, studies in humans have not shown any beneficial effect for antiplatelet agents in reducing intimal hyperplasia. A large study of patients who had undergone carotid endarterectomy demonstrated that aspirin and dipyridimole did not reduce restenosis, and may have exacerbated it.

Anti-inflammatory agents have been used in an attempt to limit intimal hyperplasia through suppression of inflammatory mediators and consequent inhibition of leukocyte adherence, activation, and smooth muscle cell proliferation and migration. In rabbits, atherosclerosis was diminished by the administration of glucocorticoids, while dexamethasone injections given before and after injury also were effective in limiting neointimal hyperplasia after balloon catheter injury. However, the side effects of steroid therapy, namely systemic immunosuppression, wound breakdown, and electrolyte abnormalities induced by large doses of steroids in these experimental models preclude their use in humans. Calcium channel blockers have been shown in experimental models to inhibit smooth muscle cell proliferation, migration, and extracellular matrix production, while ACE inhibitors have blocked intimal hyperplasia and atherosclerosis experimentally, but neither drug has been shown to be effective in clinical trials.

Heparin, whether systemically administered or locally applied to the vessel, has been consistently successful in controlling smooth muscle cell proliferation in experimental models. Heparin directly inhibits smooth muscle cell proliferation by acting on the $G_1$ phase of the cell cycle, with a critical period of efficacy within the first 4 days after injury; its mechanism of action is probably multifactorial, because heparin inhibits several pertinent processes, including (1) smooth muscle cell binding of epidermal growth factor (EGF), PDGF, and bFGF; (2) oncogene expression; (3) deposition of extracellular matrix; and (4) expression of various proteases. Excessive bleeding, however, has precluded the routine clinical use of heparin following most vascular procedures, and a clinical study in which heparin was administered after coronary angioplasty showed no decrease in the incidence of restenosis.

A promising model for local and specific treatment without implications of systemic toxicity is local administration of oligonucleotides complementary to c-*myb*, the proto-oncogene involved in mitogen-induced proliferation of smooth muscle cells. It effectively controlled intimal thickening by binding the mRNA encoding this protein, which is synthesized in large amounts at the site of vessel injury.

Ionizing radiation has received little attention as a method to inhibit smooth muscle cell proliferation after vascular injury. Although high-dose (>50 Gy) radiation may *potentiate* long-term arterial narrowing, this sequel is relatively uncommon, while lower doses (600–800 cGy) can be employed as a means to *inhibit* nonneoplastic hyperplasia of mesenchymal tissue after injury (e.g., keloid and heterotopic bone formation). Gamma radiation at appropriate dose and timing effectively limited experimental intimal hyperplasia after arterial injury by killing proliferating smooth muscle cells in a time- and dose-responsive manner. The low doses used are not believed to have any adverse long-term effects on the vessel wall, and multifraction doses or brachytherapy may provide a safe means of administration, minimizing radiation damage to the surrounding tissues.

## Thrombosis

Arterial thrombosis associated with focal cerebral ischemia is classified according to the underlying pathogenetic mechanism responsible for creating the arterial occlusion. Atherothrombotic events are thromboses due to the sequelae of atherosclerotic disease within the intracranial cerebral circulation or the extracranial feeding arteries. These thromboses may follow arterio-arterial embolism from a proximal ruptured plaque, or local occlusion at a site of stenosis. Thromboembolic disease results from thrombus formation within the heart in the setting of atrial fibrillation, ventricular ischemia, or valvular vegetative disease. These thromboemboli are most commonly transmitted to the middle or anterior cerebral arteries. Small vessel thrombosis is usually associated with hyaline necrosis of penetrating arterioles, typically in patients with hypertension, and causes lacunar infarcts.

Cerebral venous thrombosis is a rare event affecting women predominantly. It is associated with those processes that promote thrombosis in any context (e.g., damage to the vessel wall, stasis, and hypercoagulability).

### THE CLOTTING CASCADE

Some understanding of the normal hemostatic mechanisms is necessary to understand the pathophysiology of thrombosis. Alteration of the delicate balance between procoagulant and anticoagulant mechanisms frequently results in the pathologic effects of intravascular thrombosis.

First, injury to the subendothelial tissues is followed by platelet adhesion and activation, so that a platelet plug (a rapidly progressive fibrin clot) is formed. Platelet-platelet adhesion is dependent on fibrinogen, fibronectin, thrombospondin, and von Willebrand factor, and the conversion of fibrinogen to fibrin stabilizes the otherwise fragile platelet plug, which is further stabilized by covalent cross-linkage between fibrin polymers. These culminating events of clot formation are mediated by thrombin.

In the second phase of clotting, the coagulation cascade, inactive plasma proteins (zymogens) are converted to active enzymes (serine proteases) in a series of step-wise catalytic events that receive their initiating stimulus from an either intrinsic or extrinsic coagulation pathway (Fig. 14.6).

The interaction between blood and the subendothelial structures at sites of endothelial injury initiates the intrinsic clotting pathway, which terminates in the activation of factor X to factor Xa. The extrinsic pathway is initiated by interaction of factor VII with tissue factor (thromboplastin), a group of

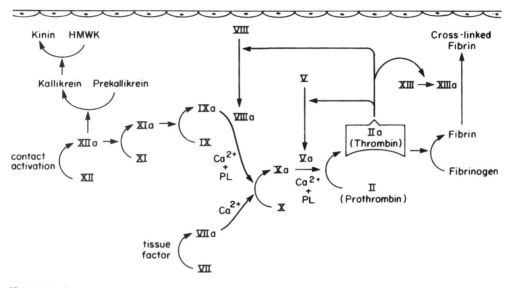

**Figure 14.6**
The clotting cascade. See text for details. (Reproduced by permission from Loscalzo J, Creager MA, Dzau VJ, eds. Vascular medicine. Boston: Little, Brown, 1992:235.)

membrane-associated glycoproteins expressed on the cell surface in response to inflammation and present in the subendothelial connective tissue. Again, catalytic conversions lead to activation of factor X, so that both pathways converge to cause thrombin formation and the conversion of fibrinogen to fibrin.

## FIBRINOLYSIS

Several anticoagulant mechanisms influence local thrombosis. Laminar blood flow depletes clotting factors around the site of injury and removes activated clotting factors. Endogenous local protease inhibitors such as antithrombin III and protein C inhibit one or more coagulation factors in the clotting cascade. Finally, the fibrinolytic system is primarily responsible for the dissolution of established fibrin clots. Plasminogen, tissue-type plasminogen activator (tPA), and single-chain urokinase-type plasminogen activator (scuPA) are the three zymogens involved in the breakdown of the fibrin clot. Each binds to fibrin and catalyzes the enzymatic formation of plasmin, which degrades fibrin and facilitates clot lysis.

Thrombin plays a central role in fibrinolysis. In addition to catalyzing fibrin formation and amplifying the clotting cascade, it stimulates the release of tPA from endothelial cells and activates protein C, which inhibits various steps in the coagulation cascade. Thus, by modulating the clotting response, thrombin prevents the unchecked spread of thrombosis.

## MECHANISMS OF THROMBOSIS

Endothelial injury, which is perhaps the most frequent mechanism whereby intravascular thrombi originate, has been discussed. Alterations in blood flow can potentiate the formation of arterial and venous thrombi at the site of vascular damage. Arterial turbulence and venous stasis predispose to thrombosis in their respective systems by interrupting normal laminar flow, wherein the central core of RBCs and WBCs is surrounded by an outer layer of platelets. Serum, which has the lowest thrombogenic potential, is in contact with the vessel wall and has the slowest laminar flow. Stasis and turbulence bring platelets into direct contact with the vessel wall, prevent clearance of activated coagulation factors, and retard the arrival of anticoagulant factors. In addition, turbulent shear stress on endothelium, as in hypertensive disease, causes endothelial cell injury and destroys the antithrombotic influences of tPA and prostacyclin. Turbulence also occurs over atherosclerotic plaques, where the smoothness of the endothelial surface has been altered. Ulceration of the plaque can create further turbulent flow, and this potentiates the interaction of platelets with the subendothelial connective tissue elements that is already in progress.

Hypercoagulable states can act alone or in concert with other mechanisms in the pathogenesis of arterial and venous thrombosis. Most primary hypercoagulable conditions are familial; they include antithrombin II deficiency, proteins C and S deficiencies, dysfibrinogenemias, and disorders of fibrinolysis. Primary hypercoagulable thromboses are often precipitated by an additional thrombogenic influence such as pregnancy, surgery, trauma, or infection.

Secondary hypercoagulable states (i.e., clinical conditions associated with an increased risk for the development of thrombosis) are usually acquired. These can be divided into categories based on abnormalities of coagulation and fibrinolysis, platelet abnormalities, or abnormalities of blood vessels and rheology. Although venous thrombosis is frequently associated with coagulation and fibrinolytic abnormalities, arterial thrombosis more often develops in the setting of platelet, vascular, and rheologic changes. Other secondary hypercoagulable states include myeloproliferative disorders, diabetes mellitus, and hyperlipidemia (which affect platelet function), and vascular and rheologic alterations such as vasculitides, chronic obliterative arterial diseases, homocystinuria, and blood hyperviscosity. There is a clear association between venous thrombosis and malignancy, pregnancy and the puerperium, nephrotic syndrome, and systemic lupus erythematosus. By contrast, the use of oral contraceptives is associated with a risk of arterial thrombotic stroke.

# Cerebral Vasospasm

## PATHOPHYSIOLOGY AND EXPERIMENTAL ASPECTS OF VASOSPASM

The precise mechanism by which subarachnoid hemorrhage elicits delayed arterial narrowing remains uncertain. In fact, controversy persists as to (1) whether arterial narrowing is a consequence of active vasoconstriction or passive structural changes in the vessel wall; (2) whether large vessel narrowing is integral to the pathologic process; and (3) what is the specific component of blood that elicits delayed arterial narrowing and by what process this occurs.

### Morphologic Changes in Vessel Wall

Both human and animal arteries exposed to blood over several days show characteristic ultrastructural changes in the vessel wall (Fig. 14.7). Alterations

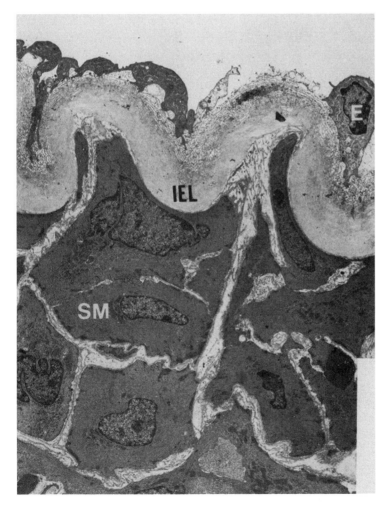

**Figure 14.7**
Cerebral vasospasm after experimental subarachnoid hemorrhage. This electron micrograph demonstrates a buckled internal elastic membrane (IEM) with endothelial cell (E) distortion, smooth muscle cells (SMC) in cross-section, and increased deposition of extracellular matrix in the media.

in endothelial cell morphology comprise thickening and discontinuities of the elastica, the appearance of smooth muscle vacuoles with occasional myo-necrosis, and peri-adventitial inflammation with loss of perivascular axons. In experimental vasospasm, ultrastructural changes in cerebral arteries did not give rise to increases in vessel wall mass and were associated with loss of contractile protein and apparent increases in vessel wall collagen. Degradation of cytoskeletal proteins (actin, myosin, desmin, a-actinin) is enhanced in chronic experimental vasospasm, suggesting that functional changes in cy-toskeletal structure may be important mediators of vasospasm. These structural changes may determine in part the unique physiologic abnormalities seen in cases of cerebral vasospasm.

Cerebral arteries exposed to subarachnoid blood for several days were less distensible than controls, and were relatively insensitive to both vasocon-strictor and vasodilator agents. Possibly prolonged vessel wall narrowing after exposure to blood is due to compaction of extracellular collagen by phenotypi-cally altered smooth muscle cells (myofibroblasts). Such a condition should respond to mechanical dilatation of vasospastic arteries by angioplasty, which in fact may yield dramatic improvement in vessel caliber, distal blood flow, and clinical status.

These observations are consonant with the idea that vasospasm after subarachnoid hemorrhage may involve boznth active and passive components. It is suggested that initial vasoconstriction is related to smooth muscle cell contraction stimulated by hemoglobin, endothelin, or other undefined factors; that it likely involves calcium-mediated processes; and that it may be reversible. Persistent exposure to hemoglobin, however, may facilitate a passive mainte-nance of arterial narrowing through calcium-independent means (possibly cytotoxic effects of free radicals or inflammatory cytokines on smooth muscle cells and endothelium), associated with structural changes in vessel wall, and reversible by mechanical dilatation (angioplasty). Spontaneous resolution of vasospasm occurs by indeterminate reversal of passive mechanisms, possibly related to reversion of phenotypic changes in the vessel wall.

## Conclusion

Cerebral arteries and veins are complex organs that regulate cerebral blood flow, selectively limit the access of intravascular substances to the brain, and determine in large part the volume of the intracranial contents. They respond to both normal and pathologic stimuli in a variety of distinct and interrelated ways, involving neurogenic, paracrine, autocrine, inflammatory, and reparative processes. It is these common elements of the cerebral response to injury that determine both the beneficial and the detrimental effects involved in thrombosis, intimal hyperplasia, and vasospasm. A better understanding of these basic responses may lead to effective therapies to prevent or ameliorate the pathologic consequences of diseases of the cerebrovascular system.

## References/Further Reading

Chervu A, Moore WS. An overview of intimal hyperplasia. Surg Gynecol Obstet 1990;171: 433–447.

Loscalzo J, Schafer AI, eds. Thrombosis and hemorrhage. Boston: Blackwell Scientific, 1994.

Loscalzo J, Creager MA, Dzau VJ, eds. Vascular medicine. Boston: Little, Brown, 1992.

Quinones-Baldrich WJ. Pharmacologic suppression of intimal hyperplasia. Austin, TX: Landes, 1993.

Ross R. The pathogenesis of atherosclerosis: an update. N Engl J Med 1986;314:488–500.

Wilcox JN. Molecular biology: insight into the causes and prevention of restenosis after arterial intervention. Am J Cardiol 1993;72:88E–95E.

# 15 Ischemic Stroke

Oscar M. Reinmuth

Stroke is the third leading cause of death in the United States, after cancer and heart disease. Despite the fact that over the past 40 to 50 years stroke incidence has fallen 50%, it is still the leading cause of chronic disability, with 3,000,000 patients surviving but affected. Approximately 500,000 new episodes of stroke occur yearly, with 150,000 deaths.

This chapter addresses the major categories of ischemic stroke that have features influencing the kind and intensity of application of medical care. Emphasis will be directed to the treatment choices, with recognition of important issues, about which there is still controversy.

## Principal Mechanisms of Ischemic Infarction

### ATHEROTHROMBOTIC STROKE

Atherothrombotic stroke most commonly results from the localized deposition of cholesterol-lipid substance subintimally. The deposition occurs more readily at arterial surfaces when flow direction changes at bifurcations or sites of arterial curvature (e.g., the carotid siphon). In the cerebral circulation, the large vessels are most involved; it is less common to see occluding atheroma in middle cerebral artery (MCA) beyond its bifurcation, in the anterior cerebral arteries (ACAs) past the genu of the corpus callosum, or in the posterior cerebral arteries (PCAs) dorsal to the brainstem.

In the presence of hypertension or diabetes, a significant involvement of small, short arterial branches occurs frequently. These vessels include the short midline and circumferential arteries of the basilar artery and the lenticulostriate vessels of the first part of the MCA. They have a very short passage from the high arterial pressure trunk of origin to the arteriolar termination into the capillary circulation. It is considered that these end arterioles are under higher pressure than those that originate downstream, where the large artery pressure is lower. In these vessels, excessive intimal thickening occurs, and often at the end-arteriolar level, the vessel develops an outpouching, the Charcot-Bouchard aneurysm, which, it is believed, may either rupture to produce hypertensive intracerebral hemorrhage or provide a locus of low blood flow predisposing to clot formation, arteriolar occlusion, and lacunar infarction.

In the intracerebral large vessels, slowly enlarging atheroma produces occlusion sufficient to compromise arterial pressure and flow distal to the stenotic area. The patient may have only vague, mild symptoms that do not alert anyone to the need for medical evaluation. In a few patients, this mild varying ischemic stress distal to the stenosing lesion will stimulate enlargement of existing but functionally inadequate collateral circulatory connections so that the eventual total occlusion of the stenosed vessel is accompanied by no neurologic deficit. If the stenotic lesion occurs in vessels proximal to the circle of Willis, there is a better chance of extracranial-intracranial collateral circulation developing. In one autopsy series, approximately 10% of patients over age 50 showed asymptomatic total occlusion of one of the two vertebrals or carotids.

The process by which atheromatous change in a vessel eventually leads to brain infarction is probably not, in most instances, simply that it reaches a size that occludes the vessels. Many of the larger atheromas have areas near their intimal surface that contain some semi-liquid, cholesterol-rich lipid deposits. As the atheroma narrows the vessel, the blood flow becomes turbulent distal to the narrow point, tearing the intimal surface and dislodging these deposits so that they block the distal arterial branches. Cholesterol mixtures are usually pushed through the arteriolar capillary bed within minutes and probably account for some first transient ischemic attacks (TIAs).

The further progress of atherothrombosis is fully described in chapter 14. Once the intima has been damaged, there is an outpouring of vessel tissue factors that attract platelets and activate them to become sticky, resulting in rapidly growing platelet fibrin clumps. These may be dislodged by the arterial flow, and like cholesterol deposits, embolize to the distal branches where they may also be capable of fragmentation and transit through the capillary network. The original site of damage may continue to evolve platelet fibrin clumps, which may grow in minutes or hours to occlude the vessel. If this happens, the clot may propagate both distally and proximally, and while doing this, the fibrin structure organizes; occlusion is then irreversible.

Very often the brain tissue surrounding the totally occluded area receives some supply from adjacent vessels, a supply that may allow preservation of cell viability but is inadequate for cell function. This area is called the penumbral zone, and it is the territory at which are aimed many of the new and still experimental therapies for stroke.

Many patients show progression of deficit for hours, or even 3 to 4 days after onset of first symptom. One mechanism that may operate is the growth of the clot proximally, which usually stops when the clot reaches the origin of the first main arterial branch proximal to the occlusion. It may extend, however, into a branch that is anastomotic to the occluded vessel, in which case not only the territory of infarction of the new vessel, but also brain tissue that had depended on the bordering zone of the first infarct for collateral supply, may now be irreversibly lost.

A further factor is that in the infarcted territory, and in some of the penumbral territory, cell membranes are damaged, metabolic maelstroms occur intracellularly, and brain edema evolves. The glial astrocytes begin to swell with water influx. At first, the neurons shrink and the overall mass of the infarcted area is only moderately enlarged, but the swelling contributes to further compromise of blood flow in the borderline penumbral area. Calcium released from endoplasmic reticulum increases in the extracellular space and, with the stress of other hypoxic liberated substances that open the membrane calcium pores, moves massively intracellularly. If low flow is maintained, fatty acids accumulate, including arachidonic acid, allowing production of leukotrienes and endoperoxides. These substances directly damage cells as free radicals. As a result, maintained low flow may at times produce more damage than the same duration of complete arrest of flow.

Cytotoxic edema is due to metabolic events within the brain cells. The capillary endothelium is more resistant to hypoxic injury, so edema due to leakage from this source is delayed. The water molecules in the injured cells tend to band loosely to cystosolic proteins and behave more like solids than liquids: If the brain surface is cut within the first few hours of this injury, it does not glisten and appears dry. The cytotoxic effect is progressively overshadowed by eventual capillary membrane destruction and further injury to glial and neuronal membranes. The resulting vasogenic edema produces gross swelling of the injured area, frequently causing the brain to herniate. This in turn significantly extends territories of brain injury and may be fatal.

## EMBOLIC INFARCTION

Formation of a proximal clot creates the possibility that a portion may fragment and be carried to a cerebral vessel. This event is at least as common as atherothrombotic stroke. There are some clinical distinctions between emboli from different sources.

### Artery to Artery Emboli

Most commonly, the source of the clot is a ruptured atheromatous plaque, ulceration of its surface being the immediate cause of platelet clumping, fibrin agglutination, and clot formation. Given the high flow rates of the major arteries, it is unusual for early platelet clumps to build before being washed away. Only if an arterial wall ulcer is present is it likely that an adequate-size embolus will be produced. Emboli from the aorta or heart, however, usually end up in organs other than the brain.

In cerebral vessels, clot formation on an ulcerated plaque lends itself to production of friable clots that embolize to the brain, producing TIAs. One of the most common sources of artery-to-artery brain embolization is from the origin of the internal carotid artery (ICA). At the bifurcation of the common carotid artery, the carotid bulb is a unique anatomic variant, the shape of which allows transient turbulent flow when BP changes, and probably with head and neck movement as well. These episodes of turbulence lead to early build-up of atheroma, which eventually results in stenosis.

The atheromatous process at the bulb can follow several courses. It may simply progressively narrow the lumen and eventually occlude the ICA. In this case, there is usually time for collateral circulation to be established, especially through the ocular vessels and external carotid artery. When the ICA finally occludes, it usually clots distal to its first significant branch. If the clot does not organize, further embolization may occur, but once the surface endothelializes, the patient is past danger until another atherosclerotic vessel is affected. Alternatively, the stenotic area may erode, creating an ulcer in the vessel wall, which becomes a source of repeated emboli, causing TIAs or distal vessel occlusion. The identification of an ICA source of embolization is an emergency because there is a high threat of major stroke, which can be removed or at least reduced by anticoagulation or, when appropriate, surgical repair.

## Cardioembolic Stroke

The most common contributory factors to cardiac clot and embolization are myocardial infarction (MI), atrial fibrillation, and paradoxical embolus. Acute and subacute bacterial endocarditis also may present with embolic stroke. Occasionally, drug users inject into the carotid directly, with foreign body embolism or damage to the artery wall by abscess, hematoma, and so on. Mitral stenosis with normal heart rhythm allows atrial thrombus formation. Prolapsed mitral valve is rarely the cause of embolus. Other occasional causes include nonbacterial thrombotic endocarditis and atrial myxoma. Prosthetic heart valves also have a high potential for embolus production. Diagnostic cardiac catheterization occasionally produces embolization from the catheter tip.

An embolus from a cardiac source is extremely unlikely to produce TIAs, being characteristically large and well organized. It commonly produces its full neurologic syndrome within seconds, so effective collateral development is unlikely. Possibly because of the acuteness of the ischemia, a seizure often occurs at the onset, which is rarely the case with atherothrombotic stroke. Occasionally, the embolic stroke will evolve in a pattern more like the atherothrombotic process—stuttering alternation and increase in deficit over time. It is considered likely that further extension of thrombosis accounts for this.

In perhaps 30% of cerebral emboli, the territory of the infarct converts to patchy or total hemorrhagic change. It is probable that this is caused by lysis of the clot by the body's own fibrinolytic mechanisms; blood returning to the damaged capillaries leaks into the tissue. Hemorrhage into an infarct poses a therapeutic dilemma, in that anticoagulation is often desired to reduce the risk of further emboli, and yet the fear of encouraging hemorrhage argues against it.

*Myocardial Infarction*    Severe MI may produce a region of akinetic cardiac wall, which predisposes to mural thrombus formation. Although the major risk period is in the first few weeks after MI, there are instances in which the embolus occurs months later. Tiny organized fragments of thrombus may be trapped in the endocardial surface structures and much later break free to embolize. Dysrhythmic events such as multiple premature ventricular contractions may contribute to the possibility of ventricular embolus.

*Atrial Fibrillation*    Atrial fibrillation contributes dramatically to the occurrence of embolization. Atrial fibrillation patients with rheumatic heart disease, and usually accompanying mitral stenosis, have nearly 20 times the stroke incidence of age-matched controls. The atrium has many irregular recesses, particularly in the atrial appendage, where the potential for stasis exists. Atrial fibrillation ordinarily prevents the atrium from making a successful emptying contraction, favoring stasis. In the rheumatic heart, mitral stenosius prevents efficient emptying, and endothelial inflammatory patches of rheumatic carditis aid in clot formation.

*Paradoxical Embolus*    Paradoxical embolus is so called because the clot arises in the venous circulation, usually of the legs or pelvis. Although an interatrial defect with a large patent foramen ovale is quite uncommon, a *potential* foramen ovale exists in 20% to 25% of normal individuals. The foramen is normally closed by a flap in the interatrial septum. It may open transiently, with momentary increases in pulmonary pressure, for example, during Valsalva's maneuver. In patients producing pulmonary emboli from femoral or pelvic thrombophlebitis, the repeated emboli produce pulmonary hypertension with resultant backup pressure in the right atrium. This may open the valvelike closure of the foramen ovale and allow shunting of the venous blood and clots into the left atrium, producing arterial embolization.

## Other Ischemic Mechanisms

### ARTERIAL DISSECTION

Arterial dissection is increasingly recognized because of improved physician awareness and new imaging techniques, including magnetic resonance angiography and digital subtraction angiography. The lesion results from a tear in the intimal arterial lining, sometimes at the edge of an atheromatous deposit, but frequently in structurally normal endothelium. There is often a minor trauma that may be overlooked because the neurologic symptoms may appear only days or even weeks later. The most common precipitating injuries involve vigorous turning or extending of head and neck. A variety of such maneuvers have been associated with the disorder, including vigorous swimming, backing up an automobile, painting the ceiling, yoga exercises, and chiropractic manipulation.

The arterial stream may dissect beneath the intima, in which case no damage is likely to result. (This often happens when an arterial catheter is passed.) Of more danger is extension of the dissection into the muscular coat of the artery. The hematoma produced may be large enough to occlude the vessel lumen and produce a stroke. Alternatively, the dissection may only narrow the vessel lumen, with varying effect on flow reduction, and then more distally rupture back into the lumen. It then may become the site of formation of a clot, which may embolize or totally occlude the artery or heal with no residual damage. The dissection occasionally extends into the adventitia, producing an aneurysmal out-pouching termed a *pseudo-aneurysm*. This ruptures occasionally, but more often it heals spontaneously. There is no generally agreed-upon therapy. The majority vote would probably favor anticoagulation for several weeks during presumed healing. Surgical repair of a pseudo-aneurysm is undertaken occasionally, but the risk of this lesion is less than that of a berry aneurysm and surgery is indicated only if the lesion is expanding or causing symptoms.

Carotid dissection often begins at the origin of the ICA or at its point of entry into the bony canal. Vertebral dissection most commonly begins at sites where neck movements stretch the vessel over an adjacent bony structure, namely, the point of entry into the vertebral foramen at C-6, osteophytes at vertebral foramina, the point of exit from the vertebral canal, the point where it crosses the arch of C-1, and especially the occipital condyles.

## ATRIAL MYXOMA

Although this atrial myxoma is a rare cause of stroke, it is often totally curable if recognized reasonably early by appropriate surgery. In 75% of cases, the tumor occurs in the left atrium, so that transient ischemic cerebral events are a likely early warning sign of its presence.

## ARTERITIS

The infectious arteritides are not a common cause of stroke, in part because there are usually systemic symptoms and signs that lead to diagnosis and treatment. Especially with syphilis, the large cerebral arteries develop granulomatous arteritis, and occasionally a warning TIA or minor stroke precedes a major event. The possibility of this disease is another reason for making an immediate diagnostic determination; the arteritis may respond dramatically to urgent antibiotic treatment.

Two major granulomatous infections, tuberculous and fungal, are prone to invade arteries and may produce stroke early. It is rare, however, not to have prominent preceding symptoms: daily fever, night sweats, and increasing headache. Temporal arteritis affects intracranial as well as extracranial blood vessels and is an occasional source of stroke.

## COLLAGEN VASCULAR DISEASE

Polyarteritis nodosa occasionally causes cerebral vessel occlusion, as do Wegener's granulomatosis and isolated CNS angiitis. Systemic lupus erythematosus may produce an encephalopathic state but rarely a cerebral infarct.

# Risk Factors and Their Control

Because definitive therapy of stroke when it occurs is unsatisfactory, the major drive from the health management standpoint is to eliminate as many contributory factors as possible. Age, however, is the factor that has the highest correlation with incidence, the stroke rate rising steeply with each decade past age 50, partly because most of the other risk factors (e.g., atrial fibrillation) are chronic, and passing time allows them to exert increasing effect.

## HYPERTENSION

Of specific conditions, hypertension has the most detrimental effect in accelerating atherosclerosis, as well as being specifically causative of primary intracerebral hemorrhage. It is also the factor that can be treated most effectively. It is now clear that systolic and diastolic pressures are independent risk factors for stroke. Reduction in diastolic pressure of only 5.8 mmHg has been shown to reduce stroke incidence. Reduction in systolic BP, however, is also important. Depending on the patient's age and degree of BP elevation, the goal is to achieve 140/90 or less. It should always be possible to maintain the systolic pressure under 160. The risk of stroke begins to lessen soon after pressure modification is achieved.

## HEART DISEASE

Any cardiac dysfunction increases the risk of stroke. Prior coronary artery disease and cardiac failure are two major warning elements, and ECG evidence of left ventricular hypertrophy has a very high risk ratio of stroke, probably because it reflects both degree and duration of hypertension. Patients with stroke complicating atrial fibrillation, or who have a prosthetic heart valve, should receive long-term anticoagulation.

## BLOOD LIPIDS

Low-density lipoprotein (LDL) levels that are high predispose to increased atherosclerosis, but correlate more closely with coronary artery occlusion than with stroke. High-density lipoprotein (HDL) effects on strokes are inversely related to those of LDL; that is, a high level of HDL reduces the risk of coronary artery disease and stroke. Curiously, levels of total serum cholesterol under 160 mg/dL are weakly associated with an *increased* risk of intracerebral hemorrhage. Triglyceride levels appear to have less importance for stroke.

Elevated cholesterol should be reduced and controlled as a preventive of heart disease, though the critical levels for stroke prevention are less certain. The lower atherothrombotic stroke risk in premenopausal women, compared with an age-matched male cohort, probably results from their generally higher HDL levels. Dietary alterations should be the first step, using pharmacotherapy if diet does not succeed. Although no absolute dictum exists, a total cholesterol of 200 mg/dL with the HDL faction over 50 mg/dL is desirable. Increasing regular exercise aids in lowering LDL and raising HDL.

## DIABETES MELLITUS

Diabetes mellitus is a risk factor because diabetics develop atherosclerosis at an accelerated rate in the small and large arteries of every vascular tree.

## OBESITY

Obesity associates commonly with hypertension and increased blood lipids, with a moderate risk relationship to stroke. The pattern of fat distribution affects the relationship, abdominal deposition being unfavorable.

## HEMATOCRIT

Hematocrit levels in the high normal range are associated significantly in men aged 35 to 64, but not in other age groups or women. The reason for this is not clearly understood.

## FIBRINOGEN

Elevation of blood fibrinogen has a significant effect in increasing stroke risk. In addition to increasing the risk of intravascular clot formation, it may have a role in facilitating atherogenesis.

## RACE

Large differences in stroke incidence exist between different races. Stroke due to both vascular occlusive disease and cerebral hemorrhage has a very high incidence in Chinese and Japanese as well as African-Americans. Since World War II, major changes have taken place in the Japanese diet, with considerable reduction in salt intake and increases in animal fat and protein. These changes have been associated with some decrease in stroke incidence. In Japanese and Black South Africans, extracranial atherosclerosis is strikingly less than in American whites, and atherothrombotic occlusive disease is almost entirely intracranial.

## CIGARETTE SMOKING

Smoking produces an approximately 50% increase in stroke incidence in both sexes at all ages. Cessation of smoking appears to reduce this increase significantly in a short time, and 5 years after cessation of smoking, the stroke incidence is the same as in age-matched individuals who never smoked. Smoking also increases the risk of cerebral hemorrhage, both intracerebral and subarachnoid.

## ORAL CONTRACEPTIVES

Regular use of oral contraceptives increases the risk of stroke, especially in smokers who are young women, a group with normally a very low incidence of stroke. When a stroke episode occurs in a woman taking oral contraceptives, most physicians advise discontinuing their usage, because this immediately reduces the risk of future stroke to its normally low level. If oral contracetives are continued, the preparation with lowest estrogen content should be used. Most neurologists also advise a woman who has an increase of migraine headaches or new onset of migraine-like headaches to discontinue oral contraceptives. The risk of stroke is greatest in those women on oral contraceptives who have increasing headaches and in those who smoke cigarettes.

## ALCOHOL CONSUMPTION

Light to moderate alcohol consumption reduces the incidence of coronary artery disease. The effect on stroke is less clear, although probably favorable. Heavy alcohol consumption encourages hypertension, weight gain, and serum lipid increase, and is associated with increased risk of cerebral hemorrhage. The level of alcohol consumption at which favorable effects cease and pathologic ones begin is not clear and probably varies between individuals. In any event, it is unlikely that a heavy drinker can succeed in reducing intake to the level of one to two drinks that produces a favorable effect on stroke incidence, so for such a patient, total abstinence should be advised.

### REDUCED PHYSICAL ACTIVITY

The sedentary individual is at risk of obesity with hypertension. In addition, LDL cholesterol may rise and HDL decrease. Regular vigorous exercise undoubtedly reduces the risk of coronary artery disease; its effect on stroke incidence is less convincing, but it is likely to be favorable, given the direct effect on other major risk factors.

## Definitive Management of Ischemic Stroke

No proved effective therapy has yet been identified for specifically treating an acute ischemic stroke. Evaluation of acute fibrobrinolysis of a thrombosed cerebral vessel has been hampered by the difficulty in instituting treatment early enough to allow ischemically damaged cells to survive, and before ischemic damage to the vessel increases the risk of hemorrhage. It is likely that some areas of a fresh infarct become irreversibly dead during the first 5 to 10 minutes of vessel occlusion. The surrounding penumbral area, however, may be salvaged if adequate reperfusion can be instituted.

Before thrombolytic therapy is begun, a computed tomography of the head must be obtained to exclude hemorrhage, and evaluation for contraindications to thrombolytics must be performed. There have been instances of cerebral hemorrhage following intracranial clot lysis, and whether risk overcomes benefit has not yet been determined. To produce thrombolysis, streptokinase and urokinase have been effective. Despite being much more costly, a tissue-type plasminogen activator has been considered preferable because of high fresh-clot specificity and because its 12-minute blood stream half-life allows fairly rapid reduction in fibrinolytic activity should hemorrhage occur.

Acute stroke characteristically results in BP higher than the patient's baseline level. If the patient has been chronically hypertensive, cerebral autoregulation may be at a high threshold, and even modest reduction may extend ischemic injury to areas of borderline supply. Although there is no absolute rule, many stroke experts believe that no reduction should be made in the first few days unless the systolic pressure exceeds 240 mm Hg and/or diastolic pressure is over 120 mm Hg. An exception is if the ischemia is due to aortic dissection.

With large hemisphere infarcts, cerebral edema may be severe enough to raise intracranial pressure and cause tentorial herniation, with brainstem ischemia and death. In the first few hours after onset of stroke signs, yawning and drowsiness may warn of the early progression of cerebral edema, which reaches a maximum in 24 to 96 hours. Osmotic diuresis should be instituted utilizing a 20% solution of mannitol; furosemide also may be given. Endotracheal intubation with mechanical hyperventilation to produce hypocarbia and reduced cerebral blood flow may be indicated. Corticosteroids are ineffective, and probably contraindicated.

In severe cases of cerebral edema with incipient brain herniation, neurosurgical resection of portions of the infarcted brain has sometimes been successful in allowing survival without increased neurologic deficit. A large infarct in the cerebellum leads to compression of the brainstem and/or acute hydrocephalus due to blockage of CSF outflow. Immediate surgical resection of the infarct gives excellent results.

Antithrombotic agents in the acute stroke patient were once considered important in preventing propagation of the occluding clot, and thus worsening the stroke. More commonly, however, worsening is due to edema and spreading toxic-metabolic effects. Even without the use of anticoagulant agents, the area of infarction may convert to a petechial pattern of hemorrhage throughout the necrotic area during the first several days. Concern over hemorrhage into an infarct has tempered enthusiasm for the use of antithrombotic agents.

Studies of the use of heparin in acute stroke have been contradictory. A randomized study of its intravenous (IV) use in embolic stroke patients was discontinued, probably prematurely, when after 2 weeks the 14 treated patients had no new events, but the 21 controls suffered two deaths and two strokes. Combining seven studies of the use of heparin produced the opposite result, with severe bleeding and fatal strokes more common in the anticoagulated patients. If heparin is used, attention must be given to the potential for severe thrombocytopenia and major bleeding. An initial bolus of 5000 IU may predispose to hemorrhagic conversion of the infarct, so many clinicians advise instituting a 1000 IU/hr IV drip, adjusting the dose to maintain the activated partial thromboplastin time (aPTT) at about twice its normal level.

Recent small trials of low-molecular-weight heparinoids have been promising in regard to safety and lack of complications. Their principal components are heparan sulfate, dermatan sulfate, and chondroitin sulfate. The main action is inhibition of thrombin. The aPTT is not significantly increased, so the activity of plasma anti-factor Xa is monitored instead.

When long-term anticoagulation is necessary, warfarin should be introduced. The full anticoagulation effect of warfarin depends on suppression of vitamin K–dependent factors, but measures of these do not all come to the desired levels within the same time span. Prothrombin time (PT) responds the earliest, and it may be an additional 3 to 7 days before the key additional factors have been suppressed. Heparin should be continued beyond the time that the PT has reached the projected chronic level. The International Normalized Ratio adds a correction to the prothrombin ratio value, depending on the different thromboplastins present in the different preparations used by laboratories.

There has been increasing recognition that lower levels of anticoagulant intensity continue to offer significant protection in many clinical situations. This reduces the danger of serious bleeding complications. Although strong evidence exists that aspirin is effective in reducing the incidence of stroke after TIA or minor stroke, there is little information on its effect in acute stroke. There is theoretical reason to question its importance in acute stroke because the platelets are principally affected by aspirin use, and aspirin is less effective in preventing the thrombin-fibrinogen-fibrin progression of an established clot. The most persuasive prevention studies used 1500 mg of aspirin daily, but at least one study showed similar benefit with 325 mg daily, which is probably the most popular choice.

Ticlopidine is a new platelet anti-aggregating drug that, unlike aspirin, does not block cyclo-oxygenase in the platelet membrane, but appears to bind an adenosine component that prevents platelet activation. It does not inactivate prostacyclin in the vessel wall, this being an undesirable effect of aspirin. Ticlopidine does not increase gastric acidity nor produce the risk of peptic ulcer or ulcer bleeding, but it does often produce diarrhea, skin rash, and, in 1% to 2% of patients, a reversible leukopenic reaction. Because this occurs during the first few months of treatment, careful monitoring of the WBC count is required. A close pharmacologic cousin of ticlopidine, clopidogrel, appears to have fewer side effects, is effective in a smaller dose, and is presently under assessment in a large multicenter trial.

During the past few years, many theoretically promising drugs have been tried in the therapy of stroke but have usually failed to produce adequate benefit. Some may have failed to show an effect for lack of very early administration, due to difficulties in rapid evaluation, randomization, obtaining of informed consent, and institution of treatment. Disappointing results have been obtained with barbiturates, prostacyclin, low-molecular-weight dextrans, and cerebral vasodilators.

Following introduction of carotid endarterectomy in 1954, its performance skyrocketed to a peak of 107,000 cases in 1985. Unfortunately, although the best surgeons had a major complication rate of less than 3%, the overall rate was 10%. Discussion in the medical and surgical community increased dramatically when a large randomized trial of bypass of symptomatic, obstructive, but inoperable carotid stenosis unexpectedly reported no benefit. The expectation of a benefit was so strong that surgeons and neurologists asked the question, "Could it be that carotid endarterectomy is also of slight or no benefit?" In 1988 and 1989 the number of carotid endarterectomies dropped to 70,000 yearly.

Had it not been for the unsuccessful bypass study, it is doubtful that one could have justified the North American Carotid Endarterectomy Trial (NACET). In 1991, the monitoring committee stopped randomizing patients with over 70% carotid stenosis to the group receiving medical treatment only, because the surgically treated group had a 65% relative risk reduction for ipsilateral stroke or stroke death. One week earlier, the European Carotid Surgery Trialists (ECST) group had reported similar results in the over-70% stenotic patients. Neither NACET nor ECST had reached a conclusion as to whether benefit or detriment existed in the 30% to 70% stenotic group.

Of great interest is the preliminary report of the results of the Asymptomatic Carotid Artery Surgery study released in late 1994. A modest but statistically significant reduction—from 10% to 5%—in later stroke risk was demonstrated in the patients randomized to endarterectomy. This reduction seems unlikely to produce a great rush to operate on asymptomatic patients.

## FUTURE MODIFICATION OF ISCHEMIC MECHANISMS

In the past it was presumed that ischemia was usually all-or-none in its fatal effects, and that there was little hope that there could be any therapy that worked rapidly enough to help. Increasingly, it has become known that unexpected complexity affects the process of cell damage. For example, complete circulatory arrest for 1 hour in normothermic animals allowed return of energy metabolism and electrical activity in the brain, provided care was taken to prevent the occlusion of the circulatory tree. Ion homeostasis and energy production returned to normal in approximately 1 hour, but protein synthesis did not recover for as long as 24 hours. Moreover, while most neurons recovered protein synthesis, some, in locations known to have cells very sensitive to ischemic death, were identified as subject to selective ischemic neuronal necrosis (SINN). These include the CA1 sector hippocampal cells; cortical neurons of layers 3, 5, and 6; pyramidal cells; cerebellar Purkinje cells; and small- and medium-sized striatal neurons.

Ischemic injury probably falls into a spectrum varying in intensity (completeness), duration, and distribution. With intense but brief global ischemia (e.g., cardiac arrest), varying amounts of neuronal survival occur, but with reperfusion a progressive SINN develops over a period of days. In focal cerebral ischemia there is likely to be an area of irreversible death, but with either re-established flow or with effective collateral flow, there may be areas of reversible function. Factors that are likely to increase total death of cells are duration of ischemia, hyperthermia, or increased tissue acidosis.

It appears that a major factor in ischemic cell death is excessive entry of $Ca^{2+}$ into cells, which is initiated by ischemia. Normally, intracytosolic $Ca^{2+}$ is $10^{-7}$ mol/liter (very low) and extracellular $Ca^{2+}$ is $10^{-3}$ mol/liter. Ischemia leads to the release of neurotransmitter glutamate, which functions at many CNS neuronal receptors as an excitatory stimulus. The neuron has three glutamate-activated ion channels: One is N-methyl-D-aspartate (NMDA), which normally allows $Ca^{2+}$ influx when glutamate-activated; the second is the kainate a-amino-3-hydroxy-5-methyl-4-isoxazole propionate

(AMPA), which normally functions to control $Na^+/K^+$ fluxes, but can also pass $Ca^{2+}$; the third is metabotropic and excites the release of the intracellular second messengers, inositol-1,4,5-triphosphate ($IP_3$) and diacylglycerol (DAG).

One scheme proposes that glutamate stimulates all three receptors. The NMDA channel allows $CA^{2+}$ entry, the kainate AMPA channel allows sodium $Na^+$ entry, and the metabotropic receptor stimulates $IP_3$, which releases more intracellular $Ca^{2+}$ from the endoplasmic reticulum and mitochondrial stores. Diacylglycerol also is released, which increases cell sensitivity to $Ca^{2+}$ and to other excitatory stimuli and increases the contribution of voltage-gated channels.

Increased $Ca^{2+}$ causes more glutamate release with spread to other cells and provokes the activation of enzymes that degrade DNA proteins and phospholipids. Breakdown of phospholipids leads to formation of arachidonic acid, which increases oxygen free radicals, in turn further damaging cell membranes. Eicosanoid molecules are produced, as well as platelet-activating factor, the latter from phospholipid breakdown. This encourages occlusion of previously healthy vessels and leads to a vicious cycle of spreading hypoxic destruction.

There has been much focus on the role of blockage of the NMDA ionophore with MK801 (now called dizocilpine). Although early animal experiments were encouraging, recent data suggest that the protective effect was due to hypothermia. A host of similarly acting agents have been investigated, but none has even reached the stage of limited clinical trial.

Excitement attends the introduction of the kainate AMPA receptor antagonist 2,3-dihydroxy-6-nitro-7-sulfamoyl-benzoquinoxaline, which has decreased cell death by as much as 50% when given after the ischemic injury. Even when administered 90 minutes after focal ischemic injury in a spontaneously hypertensive rat model, it reduced cortical infarction by 30% to 40%.

## Conclusion

Many new pharmacologic interventions are undergoing assessment. In addition, important knowledge is developing rapidly concerning numerous other mechanisms by which the hypoxic stress magnifies its harmful effects. Increasingly, it appears reasonable to anticipate the discovery of a safe medical agent, able to be given urgently, that can significantly reduce the damaging effect of ischemic stroke.

## References/Further Reading

Adams RD, Victor M. Cerebrovascular diseases. In: Cerebrovascular Diseases. Adams RD, Victor M, eds. Principles of neurology. New York: McGraw-Hill, 1993:669–748.

Amarenco P, Duyckaerts C, Tzourio C, et al. The prevalence of viscerated plaques in the aortic arch in patients with stroke. N Engl J Med 1992;326:221–225.

American Heart Association Medical/Scientific Statement. Guidelines for the management of transient ischemic attacks. Stroke 1994;25:1320–1335.

Barnett JJM, Hachinski V. Neurologic clinics, cerebral ischemia: treatment and prevention. Philadelphia: Saunders, 1992.

Caplan LR. Brain embolism, revisited. Neurology 1992;43:1281–1287.

Hachinski V, Norris JW. The acute stroke. Philadelphia: F.A. Davis, 1985.

Hossman KA. Disturbance of cerebral protein synthesis and ischemic cell death. Progr Brain Res 1993;96:161–177.

Kistler JP, Ropper AH, Martin JB. Cerebrovascular diseases. In: Isselbacher KJ, Braunweld E, Wilson JD, et al., eds. Harrison's principles of internal medicine. New York: McGraw-Hill, 1994:2233–2256.

Lipton SA, Rosenberg PA. Excitatory amino acids as a final common pathway for neurologic disorders. N Engl J Med 1994;330:613–622.

Nakaniski S. Molecular diversity of glutamate receptors and implications for brain function. Science 1992;258:597–603.

Pulsinelli W, Levy DE. Cerebrovascular diseases. In: Wyngarden JB, Smith LH Jr, Bennett JC, eds. Cecil textbook of medicine. Philadelphia: Saunders, 1992:2145–2170.

Pulsinelli W, Sarokin A, Buchan A. Antagonisms of the NMDA and non-NMDA receptors in global versus focal brain ischemia. Progr Brain Res 1993;96:125–135.

Rothrock JF, Hart RG. Antithrombotic therapy in cerebrovascular disease. Ann Intern Med 1991;115:885–895.

Siesjö BK. Pathophysiology and treatment of focal cerebral ischemia. Part I: pathophysiology. J Neurosurg 1992;77:169–184.

Siesjö BK. Pathophysiology and treatment of focal cerebral ischemia. Part II: mechanisms of damage and treatment. J Neurosurg 1992;77:337–354.

Warlow C. Disorders of cerebral circulation. In: Walton J, ed. Brain's diseases of the nervous system. New York: Oxford University Press, 1993:197–268.

# Complications of Subarachnoid Hemorrhage

Christopher S. Ogilvy

Robert M. Crowell

J. Philip Kistler

Nicholas T. Zervas

The diagnosis of subarachnoid hemorrhage (SAH) is made by computed tomographic (CT) scanning (Fig. 16.1, 16.2A) and confirmed, if necessary, by lumbar puncture. The cause is then established by angiography (Fig. 16.2B,C). The most common cause is ruptured aneurysm, management of which depends on the patient's neurologic grade (Table 16.1) and the length of time elapsed since the hemorrhage (Table 16.2). Patients in grades 1–3 who are seen within 24 hours are operated on as emergencies, because that is when rebleeding is most likely. Those seen after the first day are added to the next ordinary

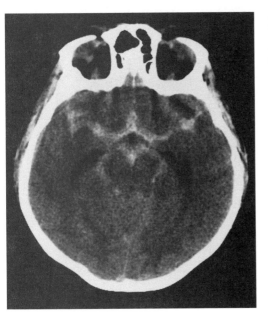

**Figure 16.1**
Typical CT scan in a patient with dense SAH. The subarachnoid blood is hyperdense on the scan and can be seen outlining the basal cisterns and the circle of Willis.

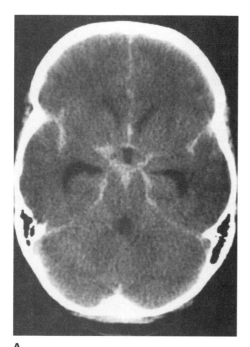

A

**Figure 16.2**

(A) Computed tomographic scan of a 35-year-old patient after SAH, demonstrating a dense distribution of subarachnoid blood in the basal cisterns as well as early hydrocephalus (increased size of the temporal horns of the lateral ventricles). (B) Lateral angiogram of the same patient shown in (A) with a ventriculostomy in place and a right posterior communicating artery aneurysm present. (C) The patient also harbored a basilar tip aneurysm, as seen on this AP projection of the vertebral angiogram. The patient initially was in Hunt-Hess grade 1 condition and initially did well after surgery, during which both aneurysms were obliterated with aneurysm clips. (D) On the seventh day after hemorrhage, however, she developed an aphasia and mild right hemiparesis, and angiography demonstrated vasospasm of the left internal carotid and middle cerebral artery. (E) The lateral angiogram demonstrates the intense spasm of the internal carotid artery as well. During the angiogram, intraarterial papaverine was administered with resultant improvement in the patient's aphasia and hemiparesis. (F) The postpapaverine angiogram demonstrates dilatation of the internal carotid artery, with some spasm still present on the middle cerebral artery segment. (G) The lateral angiogram also demonstrates dilatation of the internal carotid artery. The patient required several papaverine treatments over the course of 4 days and went on to make an excellent recovery with no neurologic deficit.

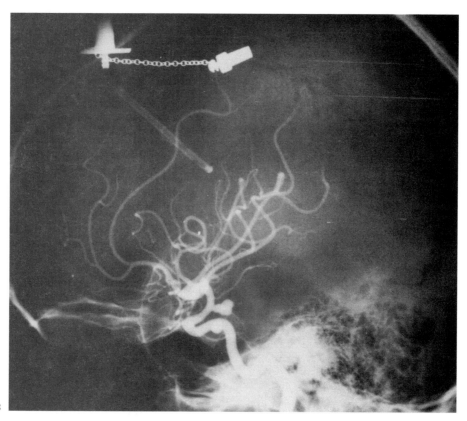

B

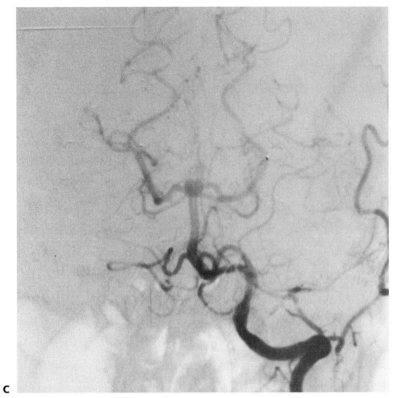

C

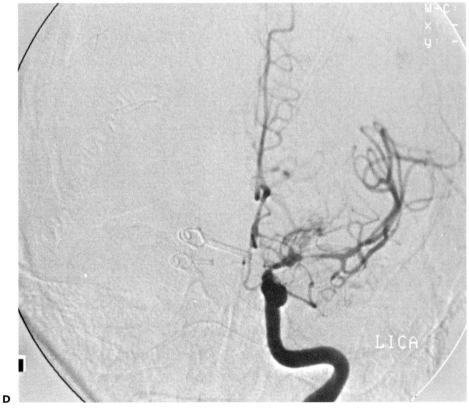

D

269

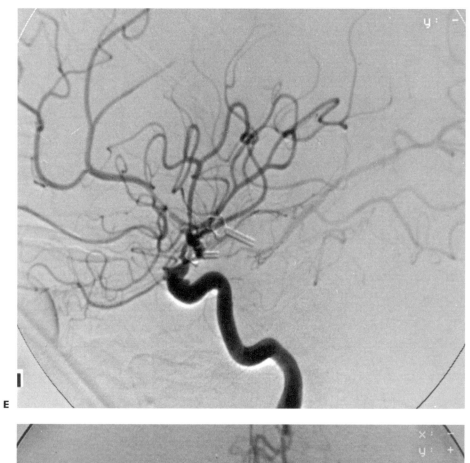

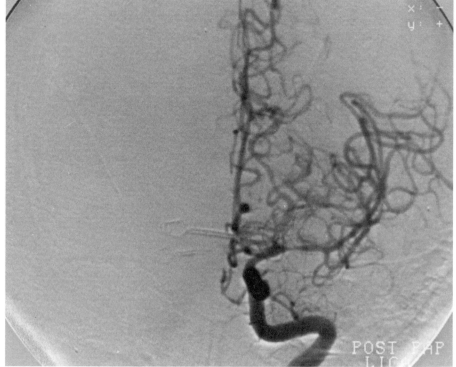

**Figure 16.2** *Continued*

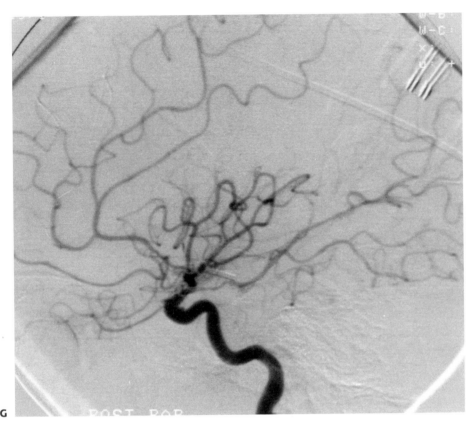

**Figure 16.2** *Continued*

**Table 16.1**  Hunt-Hess Classification of Subarachnoid Hemorrhage

| Grade | Description |
|---|---|
| 1 | Asymptomatic or mild headache with slight nuchal rigidity |
| 2 | Cranial nerve palsy with moderate to severe headache and nuchal rigidity |
| 3 | Mild focal deficit, lethargy, or confusion |
| 4 | Stupor, moderate to severe hemiparesis, decerebrate rigidity |
| 5 | Deep coma, decerebrate rigidity, moribund appearance |

Some authors add one grade for serious systemic disease (hypertension, diabetes mellitus, severe atherosclerosis, chronic obstructive pulmonary disease, or severe vasospasm on arteriography). This grading system is used to classify patients on presentation to the hospital, and subsequently prior to treatment.

**Table 16.2**   Goals for the Management of Subarachnoid Hemorrhage

Early (days 0–2)
 1. Patient stabilization (neurologic, cardiac, and pulmonary)
 2. Prevention of rebleeding
   a. Early surgery (or endovascular treatment)
   b. Control of hypertension
   c. Antifibrinolytic agents*
Later (day 3 and after)
 3. Vasospasm
   a. Prevention
     1) Hypervolemia and hemodilution
     2) Vasodilators
     3) Calcium channel blockers
     4) Phlebotomy*
   b. Therapy
     1) Hypertension
     2) Angioplasty*
 4. Treatment of symptomatic hydrocephalus
   a. Ventriculostomy
   b. Ventriculoperitoneal shunt*
 5. Prevention of seizures and systemic complications

*In selected cases.

operating list (Table 16.3). Management of of patients in grades 4 and 5 depends on the reason for their neurologic deficit. They are not subjected to definitive surgery unless this appears to be due to a treatable complication, and they respond favorably to treatment. In this chapter, we consider the complications of aneurysmal SAH and their management.

**Table 16.3**   Timing of Surgery

Emergency
 1. Hematoma evacuation and aneurysm obliteration, for deterioration from hematoma (even without angiography)
 2. Angiography and obliteration
   a. For SAH < 24 hr earlier
   b. For two or more SAHs
 3. Ventriculostomy (external ventricular drainage), for grade 4 or 5 with hydrocephalus
Urgent (within 24 hr)
 4. Angiography and obliteration
   a. For grades 1–3
   b. For grade 4 or 5 improved by external ventricular drainage

## Intracranial Hematoma

If a patient is deteriorating despite medical therapy and harbors a hematoma of sufficient size to produce symptoms of a mass lesion, craniotomy to evacuate the clot may be life-saving. Particularly in cases of carotid and middle cerebral artery aneurysms, emergency clot evacuation, even without angiography, may pave the way for clipping of the aneurysm in the same session.

## Vasospasm

### PATHOPHYSIOLOGY

After SAH, blood may be caked around basal arteries (see Fig. 16.1, 16.2A). Sometimes this clot lyses naturally, but in other cases it persists, and this has been correlated on the CT scan with the eventual development of delayed vasospasm and, in severe cases, downstream ischemia leading to infarction and permanent neurologic deficit. Much work has been done to identify the mechanism responsible for this phenomenon. There is evidence to suggest that hemoglobin breakdown products play an important role. In addition, the clot may prevent access of endothelium-derived relaxing factor to the muscularis, with resultant loss of normal vasodilatation.

### CLINICAL FEATURES AND DIAGNOSIS

Along with the damage caused by the primary SAH, brain damage due to vasospasm is an important cause of morbidity and mortality after SAH. Clinical signs of vasospasm are evident in about one-third of patients after SAH, while angiographic and transcranial Doppler studies reveal spasm in up to 70% of the cases; in 14% to 36% of patients, it leads to disability or even death.

When the condition of a patient deteriorates 3 to 14 days after SAH, vasospasm should be considered as a possible cause (see Fig. 16.2D, E). A CT scan should be performed promptly to rule out hydrocephalus, infarction, or rebleeding. Serum electrolytes and arterial blood gases should be checked to exclude abnormalities as a cause for deterioration. A decline in serum sodium level (sometimes precipitous) may presage symptomatic vasospasm; it is thought to be due to an atrial natriuretic factor that causes renal excretion of sodium. The net result is a hypovolemic hyponatremia (to be distinguished from the hypervolemic hyponatremia of inappropriate antidiuretic hormone [IADH] secretion). Serial transcranial Doppler examinations can detect progressive increase of cerebral blood flow velocity, an effect of vasospasm. In questionable cases, direct confirmation of vasospasm may be obtained by cerebral angiographic study.

## MANAGEMENT

Vasospasm may be prevented by maintaining cerebral perfusion through induction of moderate hypervolemia and hemodilution at normotension (Table 16.4). The previous common practice of dehydration for neurosurgical patients should be discouraged. Administration of hyposmolar fluids, such as Ringer's lactate (or 5% dextrose in 0.5% normal saline solution), beginning at admission, can gradually achieve hypervolemic hemodilution. In most cases, careful intake and output records and serum electrolyte values can guide therapy, with an aim of giving 2–3 liters of excess input over the initial 24 hours. It is possible that symptomatic vasospasm has become less common since hypervolemia has been used routinely in SAH cases. As described elsewhere, vasospasm may be prevented by injection of tissue plasminogen activator into the basal cisterns to dissolve the subarachnoid blood, but this method is still in the investigational phase.

For treatment of established symptomatic vasospasm, hypertensive hypervolemic hemodilution (triple-H therapy) is usually indicated. The central venous pressure is raised to 8–12 cm, or the pulmonary artery wedge pressure to 14–18 mm, in an effort to increase systemic blood flow. In patients with secured aneurysms, pressors are used to elevate the systolic BP to 200 mm Hg, using phenylephrine, dopamine, or dobutamine infusions, arterial pressure being monitored carefully. Extra caution is warranted in patients with cardiac, pulmonary, or renal dysfunction; central venous pressure monitoring, and occasionally insertion of pulmonary artery catheters, is indicated. At times, measurement of intracranial pressure (ICP) and cerebral perfusion pressure (CPP) has been useful to guide therapy, especially in the presence of hydrocephalus. The optimum hematocrit value is 30% to 33%, and occasionally phlebotomy is added to hydration to achieve this level. One must watch carefully for pulmonary edema, electrolyte abnormalities, and cerebral edema in such complex patients.

Mannitol also increases cerebral blood flow in the setting of vasospasm, with improvement in neurologic function. Recently, the calcium channel

**Table 16.4**   Protocol for Vasospasm Treatment

1. Maintain central venous pressure at 10–12 cm with albumin, blood, and fluids.
2. In selected patients with cardiopulmonary problems, monitor with pulmonary artery catheter. Maintain wedge pressure at 14–18 mm Hg, using albumin, blood, or fresh-frozen plasma.
3. Maintain blood glucose level at 100–150 mg/liter.
4. Elevate systolic BP to 160–180 mm Hg if aneurysm is secured, using phenylephrine or dopamine. If no improvement, increase BP to 200 mm Hg unless there is a cardiac complication or other contraindication.
5. If no improvement after maximal medical therapy, consider angioplasty.

blocker nimodipine has been shown to have a beneficial effect on stroke after SAH, but it is without documented effect on angiographic spasm.

Balloon angioplasty may have a beneficial effect on vasospasm in selected cases treated early. This technique involves endovascular placement of a balloon in the affected artery. Inflation of the balloon reverses the spasm. Numerous failures have been observed with angioplasty, however, and the indications for this form of treatment are not yet established. Presently, when severe deficit persists for 2 hours, despite maximal medical therapy, we consider angioplasty of clinically relevant proximal intracranial arteries.

Several reports indicated that intra-arterial papaverine may be of benefit in dilating distal intracranial arteries (see Fig. 16.2F, G), but controlled data are lacking. A host of other treatments have not been shown to provide a benefit, including trinitroglycerin, nitroprusside, aminophylline, isoproterenol, and reserpine/kanamycin.

## Hydrocephalus

Acute hydrocephalus after SAH (Fig. 16.3A) can lead to rapid decline and death. More often, gradually increasing hydrocephalus produces a diminished level of alertness and increased ICP. The condition can be diagnosed by careful clinical observation and correlation with a CT scan showing enlarged ventricles.

When an SAH patient arrives deeply comatose (Hunt-Hess grade 4 or 5) and the CT scan confirms hydrocephalus, it is worthwhile to insert a ventriculostomy (Fig. 16.3B). In a significant number of patients, ICP may be brought to normal levels and the clinical picture improved, with eventual useful recovery. Early angiography and surgery appear warranted in these cases. With an unsecured aneurysm, one should avoid overdrainage by gradually lowering the pressure to 15–20 cm. In most cases, the drain may be intermittently clamped over a few days and then removed. If the patient does not tolerate clamping of the ventricular drain, a ventriculoperitoneal shunt may be needed once the blood has cleared from the ventricular CSF. Only a few patients eventually require a shunt. At times, ventricular drainage helps maintain CPP at appropriate levels during the treatment of vasospasm. In such instances, it may be necessary to change the site of the ventricular drain to permit drainage for prolonged periods. Most patients with moderate ventriculomegaly do not develop significant symptoms, and ventricular size gradually returns to normal; ventriculostomy is not indicated for these patients.

## Increased Intracranial Pressure

Increased ICP, a common complication of SAH, results from rebleeding, intracranial hematoma, hydrocephalus, and cerebral edema or ischemia related to

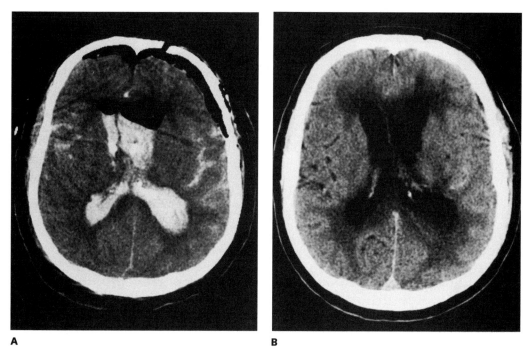

A                                    B

**Figure 16.3**
Hydrocephalus after subarachnoid hemorrhage. (A) The CT scan of a patient with a Hunt-Hess grade 4 subarachnoid hemorrhage. The CT was obtained 1 day after clipping of a left posterior communicating artery aneurysm with installation of intracisternal tissue plasminogen activator (tPA) at the time of surgery. The patient presented with dense intraventricular and subarachnoid hemorrhage. The day after surgery, tPA also was administered in the ventricle and a follow-up CT scan 2 and 1/2 weeks after surgery. (B) This CT scan demonstrates excellent resolution of subarachnoid and intraventricular hemorrhage with no cerebral infarction. This scan, however, demonstrates periventricular low density indicative of transependymal flow of CSF with continued ventriculomegaly. The patient responded to placement of a ventriculoperitoneal shunt, with resultant reduction in size of ventricles and clinical improvement.

vasospasm. Increased ICP can lead to impairment of cerebral perfusion, with exacerbation of ischemia and concomitant edema. Pressure gradients and intracranial compartmentalization may lead to transtentorial herniation and brainstem injury. Level of consciousness is the best clinical measure of ICP. Progressive stupor suggests critically elevated ICP. Increased ICP is less of a concern in alert patients with minimal deficit.

Measures to lower ICP include moderate restriction of free water, elevation of the head of the bed, avoidance of hypoventilation and hypercarbia, steroid therapy, diuretic therapy, and control of agitation and pain. Steroids (dexamethasone 4 mg every 6 hours) are used routinely. Hyperventilation is a temporarily effective method to lower ICP in a patient developing signs of herniation.

In patients without cerebral ischemia, $pCO_2$ can be safely lowered to 30 mm Hg, but further reduction can cause ischemia. The effects of hyperventilation, by reducing intracranial intravascular volume, are immediate but short-lived. Hyperventilation should be supplemented with mannitol (0.5 mg/kg in a 20% solution, given over 20 minutes). Mannitol is an osmotic agent that extracts water from normal brain. Effects begin in 20 minutes and persist for 4 to 6 hours. Additional doses of mannitol (0.25 mg/kg) can be given every 4 to 6 hours as needed. Monitoring of fluid status must go on serially, with serum osmolarity not to exceed 310 mosm/kg. Monitoring of ICP with a subdural transducer or ventricular catheter facilitates management. Drainage of CSF may help control ICP. Other diuretic agents include furosemide and acetazolamide (Diamox, Lederle, Pearl River, NY). Pentobarbital coma is rarely warranted for management of refractory elevated ICP.

## Seizures

Seizures occur in up to 13% of cases of SAH. They are considerably more common in middle cerebral artery aneurysms (up to 30% of middle cerebral artery aneurysms with intracerebral hematoma) and in 41% of patients with neurologic deficit. The first seizure usually occurs within 18 months. In a study of 53 patients, 83% had fewer than three seizures, indicating that these are usually not difficult to treat.

Following SAH, EEG changes are common, including slow waves and even spikes, and have been used as a guide to therapy. Richardson and Uttley recommended that even without seizures, patients with internal carotid artery or anterior communicating artery aneurysms should be treated for 6 to 12 months, but patients with middle cerebral artery aneurysms, a neurologic deficit, or intracerebral hemorrhage, should be treated for 2 to 3 years. Our practice is to treat all aneurysmal SAH patients with anticonvulsants for 3 months after obliteration and then taper the medication. If seizures occur, extended therapy and EEG monitoring are used. If rash or other reaction occurs, phenytoin is replaced with phenobarbital.

For the occasional case of status epilepticus, diazepam (10 mg, intravenously, over 5–10 min) is recommended, with repetition as required (up to 80 mg). Patients must be observed carefully for respiratory depression. Phenytoin loading at the same time is recommended, with adjustment of dosage according to serum concentration determinations.

## Psychiatric Alterations

Psychiatric alterations are common after SAH. In one study, although 45% of patients remained normal, in 24% the alterations were mild, in 18% moderate, in 10% severe, and in 3% very severe. Abnormal neurologic signs were associated with psychiatric alterations in 34% of the series.

Commonly reported changes include personality alteration, diminished intelligence quotient, anxiety, depression, loss of interest, and diminished initiative and energy. Occasionally observed are akinetic mutism and its lesser cousin, hypomania, which can be treated with lithium. After anterior communicating artery aneurysm surgery, Kodama and colleagues, reported that 9% of patients had difficulties with activities of daily living, but 60% had a full recovery. Personality changes, amnesia, and other such alterations seem to be more common in anterior communicating artery aneurysms. All of these alterations appear to diminish over time.

## Electrolyte Abnormalities

Hyponatremia after SAH is probably related to a salt-wasting syndrome, probably caused by centrally elaborated atrial natriuretic factor rather than IADH secretion, as originally supposed. This syndrome was initially identified after bleeding from an anterior communicating artery aneurysm, and fluid restriction was thought to be effective treatment. With declining sodium levels, there is reduction of central volume and falling BP with increases in hematocrit values. These changes tend to occur 3 to 15 days after SAH and may last 2 weeks. Free water restriction may be an effective treatment when combined with volume replacement. Replacement of lost volume with colloid (5% albumin solution) allows for replacement of salt along with replacement of volume. When the sodium level falls below 115 mEq/liter, 3% saline solution may be given, but only after careful consideration in patients with heart disease, advanced age, or hypertension. Hyponatremia and vasospasm have similar time courses. Therapy for the former could exacerbate the latter; therefore, to maintain intravascular volume despite fluid restriction, albumin should be administered.

Hypernatremia also may occur, especially from anterior communicating artery aneurysm surgery. A diagnosis of diabetes insipidus may be entertained when the hourly urine output is in excess of 250 mL/hr for 2 successive hours, but one should be aware of early postoperative mobilization of intraoperative intravenous (IV) fluids, which may pose as diabetes insipidus. Diabetes insipidus should be suspected with polyuria, polydipsia, and hypernatremia, in the presence of a serum osmolarity of 320–330 mosm/kg and a urine osmolarity of less than 1 mosm/kg. Treatment is with antidiuretic hormone.

## Cardiac Complications

A variety of cardiac complications can occur after SAH. In a series of 100 aneurysms, Weir noted cardiovascular complications in 23 (23%). In another review, congestive heart failure was noted in 1%, angina in 1%, myocardial infarction in 1%, (significant) arrhythmia in 2%, and hypertension in 16% of patients with SAH.

The frequency of significant cardiac injury has been debated. In some series, proven myocardial infarction has been documented in as many as 12% of patients. One study with careful ECG monitoring of 15 patients with SAH found that 3 (20%) had runs of ventricular tachycardia and 9 (60%) showed long QT intervals, while in another, with continuous ECG, all of the patients with SAH had arrhythmias in the first 48 hours, though their ECGs had all returned to normal after 10 days. Others have reported ECG abnormalities in only 50% of SAH patients.

Myocardial infarction documented by enzyme and serial ECG changes has been demonstrated in 12% of fatal intracranial hemorrhages. Evidence suggests that these ECG and cardiac abnormalities are related to subendocardial ischemia caused by increased levels of local and/or circulating norepinephrine, which in turn is due to alterations in the hypothalamus secondary to the SAH. Although subendocardial ischemia may be modest and reversible in most cases, pathologic studies of patients dying after SAH have shown myocardial hemorrhage, focal necrosis, and even overt myocardial infarction. Therapy for these conditions has logically focused on the use of beta-blockers, such as propranolol, which prevent ventricular arrhythmia and subendocardial lesions.

The question often arises as to whether patients with ECG abnormalities are stable enough to undergo general anesthesia and surgery. In consultation with our neuroanesthesia and cardiology colleagues, we have come to the conclusion that in the great majority of these cases, treatment can proceed, despite ECG changes, even when there has been a small enzyme leak, to procedures including angiography and surgery. Rarely, a patient with clear-cut, dramatic evidence of myocardial infarction may be better served by deferring anesthesia and surgery.

## Pulmonary Complications

Obtunded patients are vulnerable to respiratory obstruction and hypoxia, as well as atelectasis, aspiration, and pneumonia. The resulting hypoxia, hypercarbia, and acidosis may contribute significantly to intracranial complications, including cerebral ischemia, and increased ICP. Prompt intubation and ventilation are therefore indicated for all patients with SAH who are obtunded or seem likely to become so. Positive end-expiratory pressure (PEEP) is warranted for patients with a history of asthma, bronchitis, or emphysema; it has no adverse effect on ICP, provided peak pressure is maintained below 10 cm of water. In many patients with increased ICP, ventilation aimed to achieve hypocarbia in the range of 30–35 mm Hg is appropriate, but more extreme hypocarbia may bring unwanted cerebral ischemia by vasoconstriction. For comatose patients, frequent suctioning and chest physical therapy are helpful to keep the bronchial tree free of secretions.

Pulmonary care for SAH patients is a team effort involving nursing staff as well as the respiratory therapy service. Frequent monitoring of tidal volume and arterial blood gases permits the physician to tailor the best pulmonary program for each patient.

In unconscious patients, pneumonia is a common (~8%) complication after SAH. A high index of suspicion should be maintained and infectious disease consultation obtained at the earliest hint of trouble. Frequent evaluation of sputum by Gram's stain and culture, and of pulmonary status by chest x-ray, permits early diagnosis and treatment. With returning consciousness, the ability to fight off intercurrent pneumonitis is much improved.

Neurogenic pulmonary edema is a relatively common complication after SAH, particularly in poor-grade patients and those with increased ICP. It may be delayed in onset and appears to be related to increased levels of circulating catecholamines. It is usually a transient phenomenon in patients with SAH, but when it is severe, efforts to obliterate the aneurysm may have to be postponed. Pulmonary edema also can be iatrogenic in patients receiving hypervolemic hemodilution therapy. Management requires intubation, PEEP, and frequent suctioning. Helpful drugs include IV alpha-blockers, furosemide, and morphine.

## Hypertension

In the setting of increased ICP and vasospasm, moderate hypertension is a homeostatic response by the body to maintain cerebral perfusion and is probably mediated by an increase of catecholamines intracranially and systemically. With an unsecured aneurysm, elevation of the BP can lead to recurrent and potentially fatal SAH. Management must be calculated to avoid both recurrent bleeding and hypoperfusion. Often, sedation and bed rest are followed by gradual normalization of BP. Analgesia also should be offered to alert patients. When hypertension persists, administration of hydrochlorothiazide or furosemide and propranolol may be sufficient to normalize BP. In some cases, IV labetolol or oral nifedipine is appropriate. Hypotension must be avoided because it can lead to ischemia and neurologic deficit.

Occasionally, marked hypertension is encountered, with BPs of 200 mm Hg or above. This, in view of the threat of recurrent SAH, should be treated promptly. Often, IV nitroprusside is needed, along with an intra-arterial catheter for continuous monitoring of BP. Occasionally, ventricular drainage, by lowering increased ICP, can lead to reduction of hypertension, particularly in poor-grade patients.

It has already been noted that in the fluid management of patients with SAH, volume expansion is helpful in avoiding delayed ischemic deficits; this measure, however, tends to counteract efforts to lower BP. The impact of volume expansion on BP has been evaluated: In this study, one group of patients received volume expansion along with vasodilator and central antihypertensive drugs; the other received only the identical drugs. All patients' BPs rose above 150/95 mm Hg, so hydralazine and methydolpate were then given, and a few patients needed nitroprusside as well. A significantly greater proportion of the patients receiving volume expansion survived long-term, so volume expansion along with control of BP was recommended in the setting of aneurysmal SAH.

## Infection

The infectious complications of aneurysmal SAH are essentially those of post-operative and/or comatose patients and must be assiduously fought from the time of admission. Infection may occur in relation to a craniotomy wound or ventricular catheter, IV or intra-arterial catheters, bladder catheter, the bronchopulmonary tree, or pressure sores, and all of these sources must be monitored constantly. The emergence of rashes may indicate infection as well as medication reactions. The common use of steroids often complicates the situation by masking infection.

Infection of a ventriculostomy is relatively common (7% in one series). Therefore, an antibiotic regimen of IV ampicillin during maintenance of a ventriculostomy has been recommended.

When meningitis is diagnosed, the antibiotic should be selected on the basis of sensitivities. For *Staphylococcus aureus*, nafcillin, chloramphenicol, or vancomycin is generally effective. For *Streptococcus pneumoniae* or *pyogenes*, penicillin G or chloramphenicol is effective. Infection by gram-negative organisms such as *Escherichia coli*, requires ampicillin, chloramphenicol, or gentamicin given intravenously (and in some cases intraventricularly).

## Venous Thrombosis and Thromboembolism

About 2% of patients with aneurysmal SAH develop deep venous thrombosis and pulmonary embolus, and efforts to prevent these complications are mandatory. The basic regimen involves pneumatic-compression, thigh-high stockings and early mobilization, but although these measures diminish the frequency of deep venous thrombosis, they do not prevent pulmonary embolus from pelvic venous sources. Fortunately, it is now clear that full heparinization is safe 6 days after craniotomy. Moreover, subcutaneous heparin given in doses of 5000 units every 12 hours (mini-heparin), which significantly reduces the frequency of pulmonary embolus in general surgical patients and orthopedic surgical patients, can safely be instituted the day after neurosurgical operations. A controlled study is under way to determine the effectiveness of heparin in patients with SAH.

## Endocrine Abnormalities

Hypopituitarism is not uncommon in relation to giant aneurysms that compress the hypothalamic-pituitary axis, and, usually in a mild form, can also appear following SAH. It is usually more pronounced with ruptured anterior communicating artery aneurysms, poor neurologic status, vasospasm, and hydrocephalus.

Diabetes insipidus in patients with SAH usually occurs in relation to direct manipulation of the hypothalamic-pituitary axis, but also can occur in the setting of massive brain destruction, increased ICP, and a preterminal state. Management includes careful input and output measurements of fluids, monitoring of serum and urine electrolytes, and judicious administration of arginine-vasopressin in water subcutaneously. Sometimes, desmopressin acetate is useful. Rarely, when the serum sodium level falls below 115 mEq/liter, it is appropriate to give 3% hypertonic saline solution intravenously.

Iatrogenic Addison's syndrome may be induced by the abrupt withdrawal of steroid supplements in patients who have been receiving this type of medication for some weeks. It should be recalled that diabetes mellitus is exaggerated in the presence of corticosteroid therapy. Careful monitoring of blood sugar levels is required, with a sliding scale of insulin therapy as needed.

## Gastrointestinal Abnormalities

Hemorrhagic and ulcerative gastric mucosal abnormalities are common after SAH, especially from an anterior communicating artery aneurysm, and are exacerbated by administration of steroids and aspirin-containing compounds. The clinician should be alert to abdominal pain (though gastrointestinal [GI] bleeding may be painless) and to abdominal distension with ileus, anemia, and hypotension. Hematemesis and melena are more dramatic signs. Documented GI hemorrhage occurs in about 4% of patients with aneurysmal SAH. It is more common in comatose patients and may not occur until 7 days following SAH. To prevent significant GI bleeding, prophylaxis with antiacids or H2 blockers is recommended. Treatment of ongoing bleeding follows the usual lines.

## Genitourinary Complications

Genitourinary complications consist primarily of infections in obtunded patients. Initial treatment with antibiotics and continuous Foley catheter drainage is recommended. Once the patient reaches stable status, intermittent catheterization is probably a better approach. On rare occasions, there may be more complex problems. Andrew and colleagues reported six patients with anterior communicating artery aneurysms who developed incontinence, priapism, or impotence after SAH.

## Rehabilitation

A stroke or SAH is a catastrophe for the patient and the family. Rehabilitation is the key to optimizing the remainder of the patient's life and should be planned with the family, realistically and as soon as possible. A positive attitude by the principal physician is important in achieving the best outcome. Because depression is extremely common in patients following SAH, special attention should be given to this problem. The surgeon also should dispel undue fears of rebleeding for patients with satisfactory clipping of an aneurysm. Common problems, such as frozen shoulder or urinary tract infection, can be easily overlooked. A team of specialists, including physiatrist, occupational therapist, physical therapist, and psychologist, is required to assist in the rehabilitation of these patients on an inpatient or outpatient basis, according to need.

## Conclusion

Although great strides have been made in the treatment of SAH over the past decade, room for improvement exists. Vasospasm remains a problem despite the treatments outlined in this text. It has not been possible in this chapter to detail surgical procedures, but the endovascular techniques offer great promise in terms of direct aneurysmal treatment as well as treatment of vasospasm after hemorrhage. The goal of the next decade is to define the exact role of endovascular therapy and act accordingly. In addition, continued improved screening techniques with neuroimaging will lead, hopefully, to a larger number of aneurysms being discovered in the unruptured condition. Finally, genetic studies are underway to try to define the exact defect present in aneurysm patients that predisposes them to formation of aneurysms.

## References/Further Reading

Alekseeva VS, Karaseva A, Naidin VL, Shtamberg NA. Social and medical rehabilitation of patients subjected to surgery on account of aneurysms of the anterior connecting artery. Zh Nevropatol Psikhiatr 1977;77:1329–1333.

Andrew J, Nathan PW, Spanos NC. Disturbances of micturition and defecation due to aneurysms of the anterior communicating or anterior cerebral arteries. J Neurosurg 1966;24:1–10.

Bailes JE, Spetzler RF, Hadley MN, Baldwin HZ. Management and morbidity and mortality of poor-grade aneurysm patients. J Neurosurg 1990;72:559–566.

DiRicco G, Marini C, Rindi M, et al. Pulmonary embolism in neurosurgical patients: diagnosis and treatment. J Neurosurg 1984;60:972–975.

Findlay JM, Weir BK, Kassell NJ, et al. Intracisternal recombinant tissue plasminogen activator after aneurysmal SAH. J Neurosurg 1991;75:181–188.

Fisher CM, Kistler JP, Davis JM. Relation of cerebral vasospasm to SAH visualized by computerized tomographic scanning. Neurosurgery 1980;6:1–9.

Frim DM, Barker EG III, Poletti CE, Hamilton AJ. Postoperative low-dose heparin decreases thromboembolic complications in neurosurgical patients. Neurosurgery 1992;30:830–833.

Hunt WE, Hess RM. Surgical risk as related to time of intervention in the repair of intracranial aneurysms. J Neurosurg 1968;28:14–20.

Kassell NF, Boarini DJ. Perioperative care of the aneurysm patient. Contemp Neurosurg 1984; 6:1–6.

Kistler JP, Crowell RM, Davis KR, et al. The relation of cerebral vasospasm to the extent and location of subarachnoid blood visualized by CT scan: a prospective study. Neurology 1983;33:424–436.

Kodama T, Uemura S, Nonaka N, Sano Y, Nada S. The quantitative analysis of psychiatric sequelae after direct surgery of anterior communicating aneurysms: follow-up study. Neural Med Chir (Tokyo) 1977;17:327–333.

Neil-Dwyer G, Walter P, Cruickshank JM. Beta-blockade benefits patients following a SAH. Eur J Clin Pharmacol 1985;28 (Suppl 1):25–29.

Pickard JD, Murray GD, Illingsworth R, et al. Effect of oral nimodipine on cerebral infarction and outcome after SAH: British Aneurysm Nimodipine Trial. Br Med J [Clin Res] 1989; 298:636–642.

Richardson AE, Uttley D. Prevention of postoperative epilepsy. Lancet 1980;1:650.

Storey RB. Psychiatric sequelae of subarachnoid haemorrhage. Br Med J [Clin Res] 1967;3: 261–266.

Swann KW, Black PM. Deep vein thrombosis and pulmonary emboli in neurosurgical patients. J Neurosurg 1984;61:1055–1062.

Walton JN. The electroencephalographic sequelae of spontaneous SAH. Electrocencephalogr Clin Neurophysiol 1953;4:41–52.

Weir BK. Aneurysms affecting the nervous system. Baltimore: Williams & Wilkins, 1987.

Zubkov YN, Nikifovov BM, Shustin VA. Balloon catheter technique for dilation of constricted cerebral arteries after aneurysmal SAH. Acta Neurochir (Wien) 1984;70:65–79.

# Clinical Cerebral Hemometabolism in Severe Acute Brain Trauma

Julio Cruz

Thomas A. Gennarelli

Olle J. Hoffstad

A close linkage exists between cerebral hemodynamics and metabolism. Adjustment of hemodynamic variables, such as pressure and flow, represents a means of maintaining adequate metabolism, particularly in severe acute brain trauma, where the most comprehensive information has been obtained by uniting cerebral blood flow (CBF) data with the results of cerebral metabolic assessment; when this has not been done, ischemic thresholds for blood flow have been postulated without regard for cerebral metabolic needs. Because this approach might lead to biased conclusions, this chapter places emphasis on the value of combining data from measurements of both cerebral hemodynamics and metabolism, so as to comprehensively address clinically relevant issues in acute traumatic coma.

## Historical Background

In 1942, Gibbs and colleagues determined, from measurements obtained from a large group of healthy volunteers, the normal values for arteriovenous (i.e., arteriojugular) differences of oxygen ($AVDO_2$) content, glucose (AVDGL), and lactate (AVDL) concentrations, respectively. A few years later, Kety and Schmidt devised a way to measure CBF, and hence were able to quantify the cerebral metabolic rate of oxygen consumption ($CMRO_2$), calculated as the product of CBF and $AVDO_2$. It followed that the cerebral metabolic rate of any particular element also could be calculated as the product of CBF and the respective arteriojugular difference.

In 1951, Guillaume and Janny introduced monitoring of intracranial pressure (ICP) in humans, while in 1960 Lundberg described in detail the different patterns of variation of ICP in normal and abnormal conditions.

These major contributions paved the way for routine ICP monitoring in many different acute intracranial disorders. In the mid-1960s, Langfitt and his colleagues showed experimentally that cerebral hemodynamic changes could play a major role in intracranial hypertension. This work gave rise to the practice of monitoring cerebral perfusion pressure (CPP) (i.e., the difference between mean arterial pressure and mean ICP).

Other physiologic variables of cerebral hemometabolism that can be continuously monitored and managed have been proposed by the authors of this chapter. They include monitoring jugular bulb oxyhemoglobin saturation ($SjO_2$) as a means of physiologically and/or pharmacologically titrating cerebrovenous oxygenation in comatose patients, and calculating cerebral extraction of oxygen ($CEO_2$), cerebral consumption of oxygen ($CCO_2$), systemic-cerebral oxygenation index (SCOI), and systemic-cerebral ventilatory index (SCVI). To these we would add cerebral hemodynamic reserve (CHR), and cerebral hemometabolic regulation (CHMR).

## Main Hemometabolic Patterns in Severe Brain Trauma

Once the combined assessment of $CMRO_2$ and CBF was commonly adopted, it became clear that, in the presence of clinical signs of suppressed cerebral metabolic activity (i.e., coma), cerebral oxygen consumption was pathologically reduced, and that $CMRO_2$ could serve as an outcome predictor; these findings have been repeatedly confirmed. Care must be taken, however, in interpreting the $CMRO_2$ in relation to acute anemia, because $CMRO_2$ invariably decreases with decreasing total hemoglobin content, though this decrease is not always accompanied by parallel changes in level of consciousness as assessed by the Glasgow Coma Scale scores. It was on the basis of this discrepancy that we proposed the $CCO_2$, calculated as the product of CBF and $CEO_2$, where $CEO_2$ replaces $AVDO_2$ in the conventional $CMRO_2$ calculation, and we have shown that in acute traumatic coma, $CCO_2$ is not as dramatically reduced as $CMRO_2$ in relation to their respective normal mean values.

The relationship between cerebral hemodynamics and posttraumatic intracranial hypertension also has been established by jugular oxygen measurements. Fieschi and colleagues first demonstrated an association between elevated ICP and abnormally increased jugular oxygen tension ($PjO_2$). Later, Obrist and colleagues found a statistically significant association between intracranial hypertension and abnormally decreased $AVDO_2$. Both studies revealed a close relationship between posttraumatic intracranial hypertension and the phenomenon known as global cerebral luxury perfusion (relative cerebral hyperperfusion), in which the expected response of CBF to decreased cerebral oxygen demand does not occur.

## POSTTRAUMATIC CEREBRAL LUXURY PERFUSION

In relative cerebral hyperperfusion, CBF exceeds oxygen demand. This is reflected by increased cerebrovenous oxygenation and decreased cerebral oxygen extraction. In fact, most of the comatose patients classified as hyperemic in the study by Obrist et al. had normal CBF values, but because of decreased cerebral oxygen demand, these normal flow values (in absolute terms) were in excess of oxygen metabolic requirements. The explanation is that the cerebrovascular tree was not responding normally (by vasoconstriction) to a primary reduction in cerebral oxygen demand. This would indicate a derangement of the metabolic control of CBF.

These interrelationships are unequivocally demonstrated by independent measurements revealing decreased cerebral oxygen extraction (AVDO$_2$ or CEO$_2$), or increased cerebrovenous oxygenation (SjO$_2$ or PjO$_2$). Because AVDO$_2$ may be paradoxically decreased in acute anemia, however, caution should be exercised when assessing the significance of this variable during intensive care. In fact, as acute anemia develops in patients undergoing intensive care, AVDO$_2$ will most likely decrease, while CEO$_2$ may frequently increase. Under these apparently discrepant circumstances, CEO$_2$ is the more reliable variable, because CBF does not increase when the total hemoglobin content of the blood diminishes. Even though increases in CBF would be expected in normal brain with acute anemia, this response has not been observed in the severely traumatized human brain.

Cerebral oxygen extraction (CEO$_2$) monitoring has shown that global cerebral luxury perfusion, assessed as such without the confounding effects of acute anemia, contributes significantly to intracranial hypertension. Conversely, in the very early stage after severe acute brain trauma, elevated ICP also may be associated with relative cerebral hypoperfusion, or oligemic cerebral hypoxia (increased CEO$_2$). Nevertheless, in patients with predominantly diffuse traumatic brain swelling, the overall tendency after the first 6 to 12 postinjury hours will almost invariably be for the development of cerebral luxury perfusion. This condition may last for several days and may require prolonged, combined, and optimized management of ICP and CEO$_2$. So that the magnitude of global cerebral luxury perfusion can be evaluated, the authors of this chapter have published an objective grading of abnormally decreased CEO$_2$ levels.

## REDUCED CEREBRAL BLOOD FLOW VERSUS ISCHEMIA

It is well established that cerebral oxygen demand is pathologically decreased in acute traumatic coma, so the expected response of the cerebral circulation to such primary metabolic suppression would be a decrease in CBF due to homeostatic increases in cerebrovascular resistance (CVR). Decreasing tissue and pericapillary $PCO_2$ levels would lead to microcirculatory vasoconstriction, indicating *normal* (preserved) metabolic control of CBF rather than ischemia, and because $CMRO_2$ is usually 50% to 60% below normal in comatose patients, optimal hemodynamic-metabolic adjustments would imply CBF reductions of the same magnitude. Thus, for $CMRO_2$ levels reduced by 50% (1.65 mL/min/100 g), optimally adjusted CBF levels would be in the range of 25 mL/min/100 g (i.e., half the normal value of 50 mL/min/100 g). Because of statistical variability for CBF norms, however, a difference of $\pm 30\%$ from mean values would be within expected limits. Thus, ischemic thresholds for CBF in already comatose patients would be lower than the 18 mL/min/100 g reported from noninjured animal brain. In fact, in awake humans, intentional CBF decreases to so-called ischemic levels have not resulted in any neurologic deficit.

When acute traumatic coma is accompanied by acute anemia, however, a different interpretation of physiologic findings should be considered. This is because, while $CMRO_2$ may fall to as much as 5 SD below the normal mean in profound anemia, $CCO_2$ remains at only 2 SDs below normal. This would correspond to an approximate 30% decrease in cerebral oxygen demand, which is less pronounced than that assessed by $CMRO_2$.

It is plain from the foregoing that neither global nor regional cerebral ischemia can be established on the basis of CBF values alone. Metabolic information is mandatory in the assessment of cerebral ischemia in the traumatized brain.

## POSTTRAUMATIC CEREBRAL ISCHEMIA

In addition to primary global cerebral metabolic suppression in acute traumatic coma, even more pronounced regional metabolic changes may exist in the severely traumatized brain. This is particularly expected to occur in regions of hemorrhagic contusion (invariably showing as areas of encephalomalacia on follow-up computed tomographic [CT] scans of the head), where regional CBF levels may be even more dramatically reduced from normal than in noncontusional areas. Because the pathology of hemorrhagic contusions suggests primary traumatic regional neuronal loss, CBF in those areas might be absent or nearly so, without giving rise to the consequences of regional ischemia. Thus, a technique for ultra-early assessment of cerebral ischemia, region by region, would be a desirable future development. Perhaps PET, with combined measurements of regional CBF and oxygen extraction, will provide more definitive conclusions in this field. Such a technique would contribute more information than those in which only focal measurements (using near-infrared spectroscopy or microdialysis probes) are currently possible.

For bedside assessment, global arteriojugular oxygen and lactate differences constitute a practical means of evaluating ischemic changes. A ratio of AVDL/AVDO$_2$ (lactate-oxygen index, [LOI]) in the range of 0.08 or greater has been proposed as a reliable measure of global cerebral ischemia, but to exclude the effects of acute anemia, we proposed a modified cerebral LOI (MCLOI), in which $CEO_2$ replaces $AVDO_2$ in the denominator. Using this parameter, it has been found that while cerebral ischemia often appeared to be present according to the conventional LOI, true global cerebral ischemia was, in fact, a rare event. We do not deny the occurrence of infarction in rare cases, but when this has happened, CT evidence of ischemic infarction has been more convincing than $SjO_2$ or even AVDL.

## Cerebral Extraction and Consumption of Oxygen

From the foregoing considerations, $CEO_2$ would appear to be a most desirable parameter for bedside assessment, being a simple subtraction (arterial minus $SjO_2$ saturation), which strongly correlates with arteriojugular measurements of the cerebral oxygen extraction fraction ($O_2EF$). The latter is calculated as the ratio either of $AVDO_2$ to arterial oxygen content, or of $CEO_2$ to arterial oxyhemoglobin saturation. In a recent analysis of over 500 measurements, we found a high correlation coefficient ($r = 0.99$) between $CEO_2$ and $O_2EF$. As a simpler, more practical calculation than $O_2EF$, $CEO_2$ stands out as a formidably useful parameter for assessment of the global $CCO_2/CBF$ ratio (whether or not the patient is anemic).

In contrast to the strong correlation described, $CEO_2$ does not correlate well with $AVDO_2$ when the total hemoglobin content drops to ranges of 10–12 g/dL or less. In fact, only in the absence of anemia have we found these two variables to show good agreement. Because in this latter study CBF did not increase to compensate for the fall in total hemoglobin content, the increased $CEO_2$ values we observed were truly representative of the *expected* physiologic homeostatic response of the injured brain, whereas $AVDO_2$ showed a paradoxical large decrease under these circumstances. We therefore suggest that one should consider abandoning the use of $AVDO_2$ and its derived variables ($CMRO_2$ and LOI) during intensive care, because the patients so frequently develop acute anemia. In noninterventional research, $AVDO_2$ and $CMRO_2$ may continue to have a place, but both cerebral oxygen extraction and consumption appear to be proportionally much lower when estimated by $AVDO_2$ and $CMRO_2$, respectively, than when measured by $CEO_2$ and $CCO_2$. We believe that these two latter parameters are worthy of being introduced into clinical practice.

Most recently we have found a practical means of assessing global $CCO_2$ without the need to measure CBF. This novel parameter, the estimated cerebral metabolic rate of oxygen ($eCMRO_2$), is simply calculated as the product of $AVDO_2$ and arterial $PCO_2$ ($PaCO_2$). In our preliminary experience, the correlation between estimated and conventional $CMRO_2$ was strong, so that $eCMRO_2$ may be a practical replacement for $CMRO_2$ in daily bedside routine measurements. Future studies may evaluate an alternative for $eCMRO_2$, in which $CEO_2$ would replace $AVDO_2$ in the calculation of estimated cerebral oxygen demand, thus ruling out potential effects of anemia on the calculation.

## Systemic Cerebral Oxygenation

The systemic-cerebral oxygenation index and the systemic-cerebral ventilatory index represent the first-order parameters of combined systemic and cerebral oxygenation. Serial measurements of SCOI allow optimal adjustments of cardiac output in relation to $CEO_2$ in a variety of potential therapeutic applications. Measurements of SCVI allow optimal ventilatory adjustments for simultaneous maintenance of adequate cardiac output and $CEO_2$, thus permitting optimized hyperventilation for control of intracranial hypertension. The systemic-cerebral oxygenation index and SCVI do not rely on actual measurements of cardiac output, but rather on the ratio of systemic oxygen consumption to cardiac output. This ratio, the systemic extraction of oxygen ($SEO_2$), is calculated as a simple subtraction (arterial minus pulmonary artery–mixed venous oxyhemoglobin saturation).

We have recently modified treatment, using on-line optimization of these two novel physiologic parameters, in young comatose adults with predominantly diffuse posttraumatic brain swelling. The outcome figures at 6 months postinjury are extremely favorable. We are therefore pursuing further experience with SCOI and SCVI as indicators, even as we expand the spectrum of therapeutic options.

## Cerebral Hemodynamic Reserve

Cerebral hemodynamic reserve is a novel concept whereby it has become possible to estimate the global cerebral microcirculatory tolerance to increments of ICP. It has been quantified during spontaneous ICP rises accompanied by changes in $CEO_2$, using an equation based on the ratio of the differences between the baseline (initial) values and those obtained at maximum ICP (final) of simultaneous measurements of $CEO_2$ and CPP. As a general rule, if increasing ICP leads to decreasing $SjO_2$, CHR will be compromised, indicating that the global cerebral microcirculation will not tolerate further increases in ICP. Conversely, if $SjO_2$ does not change or even increases in response to the ICP increment, CHR may be considered to be preserved, indicating adequate global cerebral microcirculatory tolerance.

Cerebral hemodynamic reserve must be regarded as compromised when its value is less than zero, and preserved when its value is equal to or greater than zero. The CHR levels may serve as a guide to therapies aimed at optimizing the $CCO_2/CBF$ ratio (i.e., the $CEO_2$), as well as ICP and CPP, in relation to cerebral hemometabolism. Thus, the relevant data available when considering the effects of increasing or decreasing CBF can be supplemented by practical information on the status of the microcirculatory tolerance of such modulation.

When CHR is assessed by intermittent measurements, changes observed in $CEO_2$ may be associated, not just with CPP changes, but also with modifications in $PCO_2$. This is because relatively prolonged periods of time may elapse between initial (baseline) and final measurements (at maximum ICP) of $CEO_2$ and CPP. Therefore, a $PCO_2$ correction of the final $CEO_2$ is required. To make this correction, a practical equation has already been devised.

## Cerebral Hemometabolic Regulation

Cerebral hemometabolic regulation is the first clinicophysiologic concept to integrate cerebral pressure and metabolic autoregulation. Its practical advantages include the following: (1) Quantification is carried out during routine painful stimulation for assessment of level of consciousness; (2) CBF measurements are not required; (3) induced increases in blood pressure are short-lasting and strictly due to routine noxious stimulation; and (4) calculations can be performed at the bedside, without a computer.

The same equation previously devised for bedside assessment of CHR is used for calculations of CHMR. The final values, however, are those obtained at maximum BP (instead of at maximum ICP). In our preliminary report, a statistically significant association was found between pronounced ICP problems during the acute phase and at least one occurrence of compromised (less than zero) CHMR during the first 48 postinjury hours. This novel concept, therefore, allows practical, fast, ethical prediction of the onset of intracranial hypertension, thus allowing for future therapeutic planning.

Figures 17.1 through 17.4 illustrate traditional and novel cerebral hemometabolic parameters and their basic interrelationships: CPP is the basic pressure gradient, while CVR appears to be the main modulator of cerebral hemometabolism. Oxygen demand plays a direct role in modulating CVR, and therefore in the metabolic control of CBF.

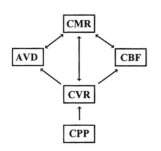

**Figure 17.1**
Diagram illustrating conventional variables of cerebral hemometabolism. AVD, cerebral arteriovenous difference; CMR, cerebral metabolic rate.

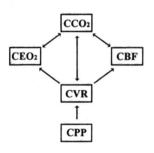

**Figure 17.2**
Diagram illustrating new variables of cerebral hemometabolism.

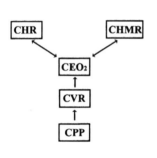

**Figure 17.3**
Diagram illustrating new concepts of cerebral hemometabolism.

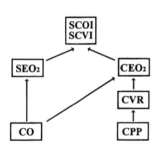

**Figure 17.4**
Diagram illustrating combined systemic and cerebral hemometabolism. CO, cardiac output.

## General Therapeutic Considerations

On the basis of our understanding of cerebral hemometabolism in comatose patients, we have developed a cumulative treatment protocol that differs from any others that have been proposed, in that it is not based on management of ICP and CPP alone. In our protocol, concomitant optimization of ICP, CPP, and $CEO_2$ represents the cornerstone of the treatment paradigm. Up to nine treatment steps may be sequentially required. A discussion of each follows.

### SEDATION AND MUSCLE PARALYSIS

Our routine use of sedatives and muscle relaxants is primarily by means of intermittent injections. If patients require aggressive pulmonary toilette, muscle relaxants are discontinued. Whether or not these measures directly contribute to sustained control of elevated ICP, they certainly allow optimization of mechanical ventilation. The latter, in turn, constitutes an underlying strategy for the next step of the cumulative treatment protocol, namely *optimized* hyperventilation.

A recent study based on the Traumatic Coma Data Bank has sought to prove that muscle relaxants may be deleterious in severely brain-injured patients. In that study, however, the mortality rate was significantly higher in patients who did not receive muscle relaxants, although the length of stay in the intensive care unit was significantly greater for those who did receive them. This effect could have arisen as a result of more frequent early deaths among those patients who were not subjected to muscle paralysis, and our overall impression is that the routine use of muscle relaxants by intermittent administration has yet to be definitively demonstrated as having significant adverse effects.

## OPTIMIZED VERSUS CONVENTIONAL HYPERVENTILATION

For decades, hyperventilation has been adopted as a standard procedure for management of acute intracranial insults. In the first study addressing outcome, major mortality reduction was reported for patients undergoing prolonged hypocapnia, in comparison with nonhyperventilated patients. These positive findings, however, lacked any attempt at cerebral hemometabolic assessment.

In the mid-1980s, while studying combined measurements of CBF and $AVDO_2$ in severe acute head injury patients, we found that patients with reduced baseline flow developed abnormally increased levels of $AVDO_2$ while hyperventilated. At the time, we described those findings as *suggestive* (but not necessarily indicative) of global cerebral ischemia. Two statistically significant associations were reported: (1) Patients with elevated ICP tended to present with cerebral luxury perfusion, that is, with CBF values that, although predominantly normal in absolute terms, were associated with decreased cerebral oxygen extraction, while (2) patients with normal ICP tended to have reduced flow, usually coupled with reduced $CMRO_2$, and therefore associated with normal $AVDO_2$ values.

These results were at least suggestive of the idea that there is a physiologic basis for hyperventilation in head-injured patients whose ICP is increased, but hardly when it is normal. Thus, a randomized trial, carried out by Muizelaar and colleagues on a group of patients in whom stratification was not done according to lesion type based on CT scans of the head, and of whom no more than 14% initially had ICP greater than 20 mm Hg though approximately 64% were immediately started on hyperventilation, seems methodologically unsound. It is irrelevant to try to identify so-called adverse effects of prolonged hyperventilation in a population for whom hyperventilation is not indicated.

In sharp contrast to the adverse effects of *indiscriminate* hyperventilation, *optimized* hyperventilation can lead to an extremely satisfactory outcome. This is because a main feature of optimized hyperventilation is that it is based on normalization of $CEO_2$ (as well as of ICP and CPP), and allows optimal physiologic-therapeutic $PCO_2$ adjustmemts to be made with a view not only to ICP control, but also to an adequate match between global cerebral oxygen demand and CBF.

Additional positive physiologic patterns associated with hyperventilation also may partly explain the aforementioned satisfactory outcome figures. They include (1) restoration of impaired cerebral pressure autoregulation, (2) restoration of pathologically reduced $CMRO_2$ levels, (3) resolution of global cerebral luxury perfusion and of the markedly increased ICP levels associated with this hemodynamic abnormality, (4) improvement or maintenance of adequate systemic hemometabolism, and (5) normalization of abnormally reduced cerebral glucose uptake, leading to improved cerebral aerobic metabolism. Thus we regard optimized hyperventilation as the primary and most powerful therapy among all currently available treatment modalities for management of posttraumatic brain swelling and elevated ICP, onto which other treatment strategies can be subsequently added as necessary.

## MANNITOL ADMINISTRATION

Hypertonic mannitol solution has been used routinely for decades in acute brain injuries. It used to be thought that mannitol exerted its effects primarily by means of acutely dehydrating the brain, but recently it has been shown that mannitol increases CBF, both in animal and clinical studies. In animal experiments, at least, it also increases (improves) $CMRO_2$.

Now that continuous monitoring of $SjO_2$ and $CEO_2$ is available, it has become clear that fast intravenous boluses of 25% mannitol, given while patients are being hyperventilated, almost invariably lead to significant decreases in ICP and increases in cerebrovenous oxygenation. Such changes would accord with the aforementioned effects of mannitol on CBF.

Despite its positive effects, however, two relevant side effects of mannitol administration still pose some limitations to its prolonged, frequent use. Specifically, serum hyperosmolality and systemic hypovolemia (occasionally leading to arterial hypotension) will frequently indicate discontinuation of mannitol therapy, at least until serum osmolality returns to normal levels and/or systemic hemodynamic stability is attained.

## CEREBROSPINAL FLUID DRAINAGE

Cerebrospinal fluid (CSF) drainage is still relied on by some neurotraumatologists as a means of controlling ICP. In one report, all consecutively admitted comatose patients were said to have undergone placement of lateral ventricular catheters. In our experience, however, CSF spaces are always moderately or markedly decreased in patients with diffuse brain swelling, which frequently makes placement of ventricular catheters impossible. Furthermore, in patients with hemispheral brain swelling and unilateral ventricular obliteration, CSF drainage from the contralateral side may potentially lead to further midline displacement and tentorial herniation. Therefore, our practice has tended to be *not* to place (or to repeatedly attempt to place) ventricular catheters in patients with severe diffuse or hemispheral brain swelling.

We concede that there are circumstances of normal or near-normal ventricular size in which ventricular CSF drainage may be useful, particularly when there is associated intraventricular hemorrhage. Such cases, however, represent a minority of consecutively admitted comatose patients with severe acute brain trauma. Computed tomography scans obtained throughout the first week after injury reveal that acute hydrocephalus is even less likely to be seen. Thus, for the most part, our current practice has been to place ventricular catheters primarily as a means of draining intraventricular hemorrhage. We are not satisfied, however, that this treatment modality has proved to be effective, or even logical, in patients whose CT scans show virtually absent CSF spaces.

## BARBITURATE THERAPY

The use of barbiturates in acute brain injuries was first proposed for the management of brain swelling by Ishii in 1966 and has continued to be advocated over the years. Whether or not barbiturate therapy contributes to controlling elevated ICP, its tendency to decrease systemic arterial pressure is undesirable, and while it may be well tolerated in patients with baseline arterial hypertension, there would seem to be a risk that patients with initially normotensive levels might become so hypotensive as to develop global cerebral ischemia.

Barbiturate therapy also may decrease cerebrovascular $CO_2$ reactivity, and so interfere with the hypocapnic effects of optimized hyperventilation. We consider that the potential benefits versus the side effects of barbiturate therapy should be weighed carefully on a case-by-case basis.

## HYPOTHERMIA

Induced hypothermia has been practiced ever since ICP monitoring was first evaluated in acute brain trauma. While previous experience has been less than encouraging, hypothermia has recently been revisited. In one recent randomized study, preliminary data have failed to demonstrate statistically significant outcome differences between treated and untreated patients. Evaluation of the outcome in larger numbers of patients is needed to produce a final statistical assessment of its value, and it would be helpful to include $CEO_2$ measurements, as well as those of ICP and CPP, for comparability.

## MANNITOL GIVEN IRRESPECTIVE OF OSMOLALITY

This infrequent therapeutic approach has been shown to be effective under life-threatening conditions. In two case reports, mannitol administration despite high serum osmolality was felt to be useful in reversing pupillary signs of tentorial herniation, and thereby "buy time" to have the patients taken for decompressive surgery.

## DECOMPRESSIVE SURGERY

Because of the availability of a broad spectrum of nonsurgical options for combined optimization of ICP, CPP, and $CEO_2$, the role of decompressive craniotomies or craniectomies has lessened somewhat over the years. It is still our impression, however, that decompressive surgery should be considered in, for example, patients with large temporal contusions and pericontusional edema, who appear to respond more satisfactorily to supplementary surgical decompression than to nonsurgical measures alone, and in patients with predominantly diffuse brain swelling, which becomes refractory to clinical management. Indeed, when, despite aggressive nonsurgical management, there are evolving signs of tentorial herniation, decompressive surgery may still be the single most powerful means of preventing definitive, irreversible herniation and death.

# Future Therapeutic Options

Because uncontrollable intracranial hypertension still represents a major challenge in the management of acute craniocerebral trauma, alternative therapeutic modalities deserve careful attention. Among these, dihydroergotamine has recently been proposed. In this study, however, the drug was claimed to be useful in maintaining cerebral hemodynamic adequacy in comatose patients when, in fact, CBF was normal-to-increased, indicating cerebral luxury perfusion.

Furthermore, while CBF values were consistent with relative cerebral hyperperfusion, jugular oxyhemoglobin saturation was reportedly normal, rather than increased (as expected). This discrepancy needs further clarification.

In any event, and irrespective of novel therapeutic options, we believe that the combined optimization of ICP, CPP, and $CEO_2$ should serve as the first step in the management of severe brain injury, on top of which a variety of therapies can be adopted in a most logical sequence.

## References/Further Reading

Bruce DA, Langfitt TW, Miller JD, et al. Regional cerebral blood flow, intracranial pressure, and brain metabolism in comatose patients. J Neurosurg 1973;38:131–144.

Clifton GL, Allen S, Barrodale P, et al. A phase II study of moderate hypothermia in severe brain injury. J Neurotrauma 1993;10:263–271.

Cruz J. An additional therapeutic effect of adequate hyperventilation in severe acute brain trauma: normalization of cerebral glucose uptake. J Neurosurg 1995;82:379–385.

Cruz J. Combined continuous monitoring of systemic and cerebral oxygenation in acute brain injury: preliminary observations. Crit Care Med 1993;21:1225–1232.

Cruz J. Continuous monitoring of cerebral oxygenation in acute brain injury: assessment of cerebral hemometabolic regulation. Minerva Anestesiol 1993;59:555–562.

Cruz J, Gennarelli TA, Alves WM. Continuous monitoring of cerebral hemodynamic reserve in acute brain injury: relationship to changes in brain swelling. J Trauma 1992;32:629–635.

Cruz J, Gennarelli TA, Alves WM. Continuous monitoring of cerebral oxygenation in acute brain injury: multivariate assessment of severe intracranial "plateau" wave. Case report. J Trauma 1992;32:401–403.

Cruz J, Hoffstad OJ, Jaggi JL. Cerebral lactate-oxygen index in acute brain injury with acute anemia: assessment of false versus true ischemia. Crit Care Med 1994;22:1465–1470.

Cruz J, Jaggi JL, Hoffstad OJ. Cerebral blood flow, vascular resistance, and oxygen metabolism in acute brain trauma: redefining the role of cerebral perfusion pressure? Crit Care Med 1995;23:1412–1417.

Cruz J, Jaggi JL, Hoffstad OJ. Cerebral blood flow and oxygen consumption in acute brain injury with acute anemia: an alternative for the cerebral metabolic rate of oxygen consumption? Crit Care Med 1993;21:1218–1224.

Cruz J, Miner ME, Allen SJ, et al. Continuous monitoring of cerebral oxygenation in acute brain injury: assessment of cerebral hemodynamic reserve. Neurosurgery 1991;29:743–749.

Cruz J, Zager EL, Schnee CL, et al. Failure of jugular lactate determinations to disclose cerebral ischemia in posttraumatic cerebral infarction: case report. J Trauma 1993;35:805–807.

Fieschi C, Battistini N, Beduschi A, Boselli L, Rossanda M. Regional cerebral blood flow and intraventricular pressure in acute head injuries. J Neurol Neurosurg Psychiatry 1974;37:1378–1388.

Gibbs EL, Lennox WG, Nims LF, Gibbs FA. Arterial and cerebral venous blood. Arterial venous differences in man. J Biol Chem 1942;144:325–332.

Gopinath SP, Robertson CS, Narayan RK. Cerebrovascular effects of dihydroergotamine in head injured patients. Crit Care Med 1994;22:204. Abstract.

Guillaume J, Janny P. Manometrie intracrânienne continue: intérêt de la méthode et premiers résultats. Rev Neurol 1951;85:131–142.

Ishii S. Brain-swelling. Studies of structural, physiologic, and biochemical alterations. In: Caveness WF, Walker AE, eds. Head injury. Conference proceedings. Philadelphia: JB Lippincott, 1966:276–299.

Jaggi JL, Cruz J, Gennarelli TA. Estimated cerebral metabolic rate of oxygen in severely brain-injured patients: a valuable tool for clinical monitoring. Crit Care Med 1995;23:66–70.

Kety SS, Schmidt CF. The nitrous oxide method for the quantitative determination of cerebral blood flow in man: theory, practice, and normal values. J Clin Invest 1948;27:476–483.

Langfitt TW, Weinstein JD, Kassell NF. Cerebral vasomotor paralysis produced by intracranial hypertension. Neurology 1965;15:622–641.

Lundberg N. Continuous recording and control of ventricular fluid pressure in neurosurgical practice. Acta Psychiatr Neurol Scand 1960 (Suppl 149);36:1–193.

Muizelaar JP, Marmarou A, Ward JD, et al. Adverse effects of prolonged hyperventilation in patients with severe head injury: a randomized clinical trial. J Neurosurg 1991;75:731–739.

Nordstrom C-H, Messeter K, Sundbarg G, et al. Cerebral blood flow, vasoreactivity, and oxygen consumption during barbiturate therapy in severe traumatic brain lesions. J Neurosurg 1988;68:424–431.

Obrist WD, Clifton GL, Robertson CS, Langfitt TW. Cerebral metabolic changes induced by hyperventilation in acute head injury. In: Meyer JS, Lechner H, Reivich M, et al., eds. Cerebral vascular disease. vol. 6. Amsterdam: Excerpta Medica, 1987:251–255.

Obrist WD, Langfitt TW, Jaggi JL, et al. Cerebral blood flow and metabolism in comatose patients with acute head injury. Relationship to intracranial hypertension. J Neurosurg 1984;61:241–253.

# Surgery of Arteriovenous Malformations

## 18

B. Gregory Thompson
Robert F. Spetzler

The management of patients with arteriovenous malformations of the brain must be guided by a thorough knowledge of the risks and efficacy of the treatment alternatives. Prior to implementation of any treatment regimen, the clinician must clearly recognize the specific type of vascular malformation and carefully consider (1) the likely clinical course based on the natural history of the lesion at hand, (2) the age and overall medical condition of the patient, (3) the specific vascular anatomy of the lesion and surrounding brain, and (4) the risk-benefit ratio of the proposed treatment for the patient.

## Classification of Cerebrovascular Malformations

Although a variety of classification schemes have been offered to categorize vascular malformations of the brain, the clinicopathologic classification offered by McCormick 30 years ago has proven the most accurate and is the most widely used (Table 18.1). The McCormick classification categorized cerebrovascular malformations into four groups based on the type of component vessels and the presence or lack of intervening neural tissue: (1) arteriovenous malformations (AVMs), (2) venous malformations, (3) cavernous malformations, and (4) capillary telangiectasia. The usefulness of McCormick's classification is reflected by the fact that each type of vascular malformation is distinguished by typical clinical characteristics, specific imaging features, and a distinct natural history.

Arteriovenous malformations and cavernous malformations, between them, account for only one-sixth of the total incidence of cerebral vascular malformations, but are responsible for more than 99% of clinically significant hemorrhages caused by these.

**Table 18.1** Histologic and Gross Pathology of Cerebrovascular Malformations

| | AVM | Cavernous Malformation | Venous Malformation | Telangiectasia |
|---|---|---|---|---|
| Vascular component | Arteries, thin-walled dysplastic vessels, arterialized veins | Thin- and thick-walled hyalinized vessels | Multiple small venous channels | Multiple enlarged capillaries |
| Parenchymal component | Gliosis and dysplastic brain adjacent to AVM | Gliosis surrounding tightly packed hyalinized vessels without intervening parenchyma | Normal | Normal |
| Hemosiderin | Frequently | Typical | Rare | Rare |
| Size | <1->10 cm | Commonly 1–2 cm, rarely up to 6 cm | Usually <6 cm | <1 cm |
| Shape | Typically wedge-shaped, with apex toward ventricle | Typically spherical or mulberry-like | Typically radial-patterned, or like "caput medusae" | Diffuse, vessels more sparsely dispersed |
| Location | All areas; always has a pial and/or ependymal margin | All areas; often completely intraparenchymal | Cerebellum and cerebral hemispheres | Typically pons |

# Arteriovenous Malformations

## EPIDEMIOLOGY AND CLINICOANATOMIC FEATURES

The prevalence of AVMs (Table 18.2) is approximately 0.5%. The incidence of subarachnoid hemorrhage (SAH) is approximately 10 per 100,000 per year, and 8.6% of all SAHs is caused by rupture of an AVM, hence, the estimated incidence of AVM hemorrhage is approximately 1 per 100,000 annually.

There are three distinct anatomic components of true AVMs: the feeding arteries; the nidus, or core, of arteriovenous fistula; and the draining "arterialized" veins. The nidus is composed of a compact tangle of abnormal vessels that shunt blood directly from feeding arteries to draining veins. The normal capillary "resistance" vessels are absent in this circuit, resulting in shunting of abnormally high flow, and transmission of proportionately greater pressure to the venous outflow system, with resulting vascular dilatation.

The compact nidus generally has little intervening neural tissue. A thin rim of adjacent brain tissue is typically dysplastic and gliotic, facilitating microsurgical dissection of the AVM. Although brain parenchyma just a few millimeters away may appear normal, adjacent areas often demonstrate ischemic injury, suggestive of vascular steal, or hemosiderin deposition, indicative of prior hemorrhage. Feeding arteries, which typically enlarge as they approach the nidus, harbor aneurysms in 7% to 10% of AVMs. Draining veins may be superficial or deep, and stenosis or occlusion of venous outflow has been implicated as a risk factor for hemorrhage.

## CLINICAL PRESENTATION

Hemorrhage (approximately 50%), seizure (25%–30%), and headache (10%–15%) are the three most common presenting symptoms of AVMs. The occurrence of these cardinal symptoms is influenced by the location and size of the lesion and the age of the patient. Less common presentations include progressive neurologic deficit, obstructive hydrocephalus, and high-output cardiac failure. Since the advent of computed tomography (CT) and magnetic resonance imaging (MRI), an increasing number of asymptomatic patients have been identified.

Although hemorrhage is the most common and dangerous presentation of AVMs, the risk of major morbidity and mortality is much lower with rupture of cerebral AVMs than with that of aneurysms. Each AVM hemorrhage carries a risk of 15% mortality and 30% major morbidity, and a markedly lower risk of delayed neurologic deterioration due to early rebleeding, vasospasm, and hydrocephalus. Arteriovenous malformations typically bleed into the subarachnoid, intracerebral, and intraventricular spaces. Although less than a third of them occur in the posterior fossa, AVMs cause the majority of cerebellar hematomas in patients under the age of 40.

**Table 18.2** Clinical and Imaging Characteristics of Cerebrovascular Malformations

| | AVM | Cavernous Malformation | Venous Malformation | Telangiectasia |
|---|---|---|---|---|
| Frequency | 0.5% | 0.3% | 3% | 0.9% |
| Annual risk of hemorrhage | 3%–4% | 0.6%–4.5% per lesion | None, unless associated with other vascular malformation | Rare |
| Clinical presentation | Hemorrhage, seizure, headache | Seizure more often than hemorrhage | Usually incidental | None |
| Flow rate | Medium-high | Low | Low | Very low |
| Familial | Rare | Often | No | No |
| Multiplicity | Rare | Frequent in familial cases | No | No |
| Imaging characteristics | | | | |
| CT | Isodense, contrast enhancing | Iso- or hyperdense with some contrast enhancement | Isodense, contrast-enhancing | Negative |
| MRI | Flow void signal | Heterogeneous signal intensity with low-signal (hemosiderin) ring producing "target" appearance | Flow void signal, often in radial pattern | Negative |
| Angiography | Enlarged feeding vessels, early draining vein, nidus | Usually not diagnostic; may demonstrate venous pooling | Radial caput medusae in venous phase | Negative |

The issue of timing of surgery following hemorrhage from cerebral AVMs is also less vexing than that following aneurysmal rupture. Because the relative incidence of early rebleeding and delayed neurologic deterioration due to vasospasm is much lower following AVM rupture, early surgery is indicated only when the hematoma itself warrants surgical evacuation. Otherwise, surgery can be delayed 2 to 4 weeks, until the patient has stabilized and improved neurologically.

In patients with AVMs presenting with *hemorrhage*, particularly SAH, one must always consider the possibility of aneurysmal rupture, because 7% to 10% of such patients will also harbor an aneurysm. When these lesions coexist, the aneurysm is more often the source of hemorrhage than the AVM and should usually be treated first.

The incidence of *seizures* in patients harboring AVMs is influenced greatly by the age of the patient and the location of the lesion. The proportion of patients with AVMs in whom seizures is the presenting symptom is inversely related to their age at symptom onset; and patients with lobar AVMs have a high incidence of epilepsy (approaching 70% for medial temporal AVMs), while those with deep or infratentorial AVMs virtually never present in this way. The location of the AVM also tends to influence the type of seizures. Frontal lobe malformations most often produce generalized seizures; AVMs near the primary motor and sensory gyri generally present with jacksonian fits; mesial temporal lobe lesions typically cause partial complex seizures.

*Headache* occurring as the presenting symptom in patients harboring true AVMs is often described as having a migrainous quality. Arteriovenous malformations that have recurrent headache as the sole or predominant symptom generally tend to have either a meningeal supply or an occipital location.

## NATURAL HISTORY

Appropriate treatment of a patient with an AVM is guided by an accurate assessment of the risk-benefit ratio of the intervention, judged from the future risk of hemorrhage, morbidity, and death if the lesion is not treated. The ideal treatment should remove the risk of future bleeding, with little or no morbidity due to the intervention.

Using the life-table method of analysis, the annual rate of hemorrhage from ruptured AVMs is approximately 3% annually, and virtually the same (2%–3%) for unruptured and asymptomatic AVMs. Ondra and associates found an annual hemorrhage rate of 4% by calculating the rate of hemorrhage per person-year of risk. The rate of hemorrhage generally doubles in the first year after a hemorrhage but returns to 3% to 4% thereafter. The rate of rebleeding also appears to be higher in patients who have had more than one bleed.

The annual rate of major morbidity and death due to AVM rupture was well described in Ondra's Finnish study, in which 96% of patients were followed for periods up to 24 years. The annual risk of death (for patients presenting with either ruptured or unruptured AVMs) was 1%, and the annual risk of major morbidity was 2.7%. The natural history of AVMs in children appears to be somewhat worse, regardless of the presenting symptom: Their annual risk of hemorrhage was 5%, and their annual mortality rate was greater than 2%.

Patients with AVMs associated with arterial aneurysms have a markedly higher annual risk of hemorrhage (7%) than patients with AVMs alone (3%–4%). Retrospective evidence also exists to suggest that draining vein stenosis and deep venous drainage are risk factors for cerebral hemorrhage. There is, however, conflicting evidence concerning the influence of AVM size, AVM location, pregnancy, and the presence of intranidal aneurysm on risk for rupture.

## DIAGNOSTIC AND IMAGING STUDIES

Cerebral angiography usually displays three typical components: (1) *Feeding arteries,* which enlarge as they approach the lesion, and (2) a compact *arteriovenous nidus,* which drains via (3) the angiographic hallmark of an AVM, *the early draining vein* (Fig. 18.1). Early venous drainage, which is visualized during the arterial and early capillary phases of cerebral angiography, signifies shunting of arterialized flow through a circuit lacking the normal capillary resistance vessels. The identification of early draining veins is particularly important when postoperative angiography is performed: Their presence or absence defines whether an AVM shunt has been totally excised.

Magnetic resonance imaging, though not the modality of choice, is a valuable imaging adjunct. With high-resolution multiplanar MRI, juxtaposition of flow void signal with brain provides exquisite anatomic detail of the relationship of the AVM to adjacent brain (Fig. 18.2). Compared with angiography, therefore, MRI allows for more accurate preoperative assessment of how closely eloquent brain is adjacent to an AVM, and enhances operative or radiosurgical planning. Magnetic resonance imaging also may help to substantiate whether there has been a remote hemorrhage: A thin circumferential rim of low signal indicates hemosiderin deposition and prior bleeding.

Computed tomography has become less useful since the advent of MRI. Arteriovenous malformations typically are isodense to brain, but their dense enhancement with contrast occasionally proves useful in patients presenting as moribund with a surgical hemorrhage.

## SURGICAL MANAGEMENT

### AVM Grading Scheme and Prediction of Surgical Risk

Because the natural history of AVMs is fairly well understood, an accurate preoperative assessment of the risks of treatment for a given AVM would allow

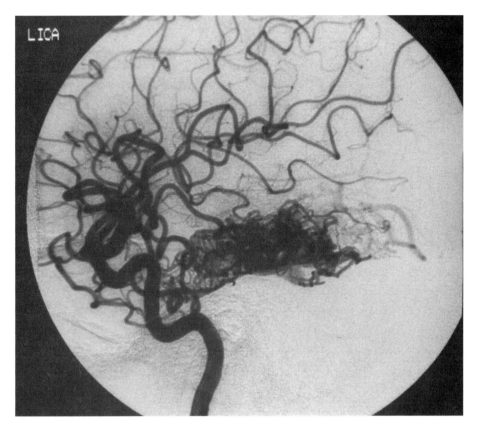

**Figure 18.1**
Anteroposterior cerebral angiography demonstrating feeding arteries, arteriovenous nidus, and an early draining vein.

one to weigh the risks of surgical intervention against the expected clinical behavior of the lesion. Hence, several AVM grading schemes have been proposed to predict the technical difficulty and risks associated with surgical intervention.

The Spetzler-Martin AVM grading system (Fig. 18.3) uses three broadly important characteristics of the AVM to predict surgical risk: the size of the AVM, its location, and the pattern of its venous drainage. The size, measured as the greatest diameter of the nidus on angiography, MRI, or CT, is scored as small (<3 cm), medium (3–6 cm), or large (>6 cm). Size is related to amount of flow, number of feeding arteries, and extent of hemodynamic effect on adjacent brain. The significance of AVM *location* is related to the functional importance (eloquence) of adjacent brain.

The pattern of *venous drainage* is related to the technical difficulty of surgical access to the AVM. Venous drainage is scored as superficial if *all* of

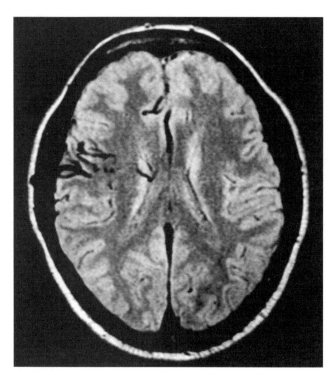

**Figure 18.2**
Magnetic resonance image demonstrating anatomic relationship of AVM to adjacent brain.

the veins drain to the cortical mantle. Drainage is counted as deep if *any* of the veins drain into the galenic system or internal cerebral veins.

The AVM grade is determined by the sum total of points from each of the three categories. The technically straightforward grade 1 AVM is small (1 point), located in non-eloquent brain (0 points), and has superficial venous drainage (0 points). However the grade 5 AVM, which is large (3 points), located in eloquent brain (1 point), and has deep venous drainage (1 point), is technically demanding and carries a higher risk of morbidity and mortality. Large AVMs which involve extensive areas of eloquent brain are classified as grade 4 (Fig. 18.4). Their attempted surgical removal results in severe neurologic disability.

### Results of Operative Treatment
The predictive value of the aforementioned grading schema has been demonstrated prospectively (Table 18.3). The operative results indicate that the Spetzler-Martin grading scale is a reliable predictor of surgical risk, and is therefore a useful adjunct in the preoperative evaluation of surgical risk-benefit ratio.

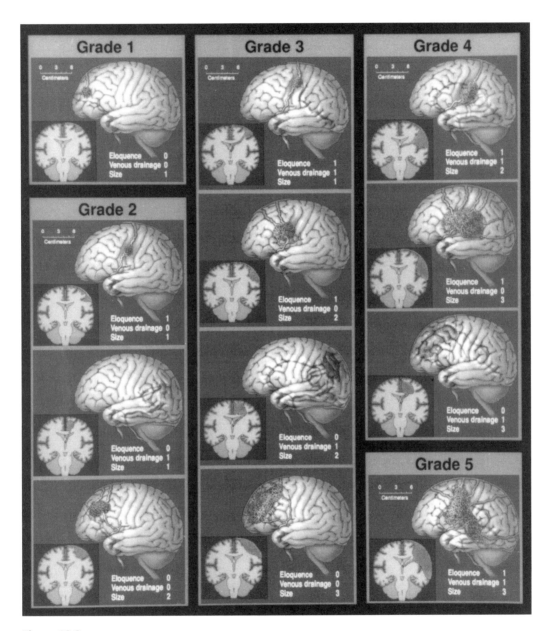

**Figure 18.3**
Spetzler-Martin AVM grading schema.

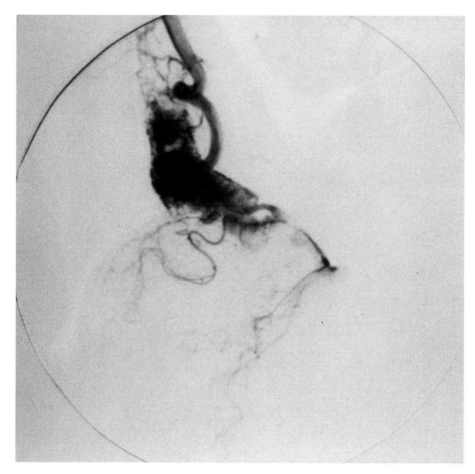

**Figure 18.4**
Spetzler-Martin grade 4 (brainstem) AVM.

**Table 18.3**  Arteriovenous Malformations: Surgical Risk in 100 Consecutive Patients

| Grade | Morbidity (%) |
|-------|---------------|
| 1 | 0 |
| 2 | 0 |
| 3 | 4 |
| 4 | 20 |
| 5 | 22 |
| 6 | Inoperable |

## Surgical Indications and Guidelines for Operative Management

By weighing the predicted risks of surgery against the known natural history of AVMs, it is possible to establish guidelines for surgical management (Table 18.4). In general, grades 1 and 2 AVMs carry a very low surgical risk, and patients in these categories, unless they are in poor medical condition, are suitable candidates for microsurgical excision. Similarly, patients harboring grade 3 AVMs, unless elderly (carrying a lower lifetime risk of hemorrhage) or medically infirm, also are generally good candidates for microsurgical AVM resection. Surgery on grades 4 and 5 AVMs should be reserved for those patients with a high lifetime risk of hemorrhage (children and young adults), a history of repetitive hemorrhage, or a fixed or progressive neurologic deficit.

The place of stereotactic radiosurgery (see chapter 26) in the treatment of AVMs also must be discussed. It offers the advantages of a shorter hospital stay at the time of surgery, decreased initial medical and neurologic morbidity, and lower initial cost. It is best suited to the treatment of small, deep-seated lesions, because the maximal radiosurgery target diameter is approximately 3 cm and in the most experienced hands results in an AVM obliteration rate of approximately 80% to 85% at 2 years following treatment.

Because the annual rate of hemorrhage from an AVM does not diminish until the lesion has become completely obliterated, however, the risk of hemorrhage persists during the interim 2 years and may actually be doubled in the first year after treatment. At 2 years after surgery, 15% to 20% of patients still harbor AVMs, which then require additional surgical or repeat radiosurgical treatment. For patients harboring grades 1 and 2, and perhaps even grade 3 AVMs, the extra 2 years' risk for hemorrhage and the 15% to 20% rate of failure of complete obliteration may render the initial advantages of stereotactic surgery insufficient to replace the greater ultimate certainty of microsurgical resection.

# Cavernous Malformations

Cavernous malformations (CMs) are multiple in more than a third of cases. In the pre-MRI era, CMs were frequently not well visualized on CT and

**Table 18.4**  Surgical Indications for Excision of Cerebral Arteriovenous Malformations

Grades 1 and 2: Low risk except in elderly and medically infirm (candidates for stereotactic radiosurgery)

Grade 3: Most are good candidates, but with careful consideration of surgical risks

Grade 4: Only in patients with fixed or progressive neurologic deficit, high risk of hemorrhage, or AVMs in the non-eloquent cortex

Grade 5: Only patients with repetitive hemorrhage or significant fixed or progressive neurologic deficit angiographic studies, and were therefore very likely the most common underlying lesions in that group known as "angiographically occult vascular malformations" (AOVMs). Between 10% and 30% of them are located within the brainstem and cerebellum. These posterior fossa CMs are more often symptomatic than their supratentorial counterparts and account for 13% of all infratentorial vascular malformations. There appears to be an autosomal-dominant inheritance pattern in 20% to 50% of patients with CMs, particularly in families of Hispanic descent.

Symptomatic CMs, particularly those located in the brainstem, have a distinctive clinical presentation and clinical course that may result in severe neurologic disability if they are left untreated.

## CLINICAL PRESENTATION

The symptomatic presentation of CMs is most closely tied to the location of the lesion, the anatomic distribution of which is roughly proportional to the relative sizes of the supra- and infratentorial spaces. Epilepsy is the most common presentation (38%–55% of cases). Focal neurologic deficit (15%–45%) and headache (12%–34%) are the next most frequently presenting symptoms. The occurrence of hemorrhage varied considerably in these studies (4%–32%), suggesting that perhaps small hemorrhages, which may result in focal neurologic deficit or seizures, often go unrecognized.

Cavernous malformations of the brainstem typically present in one of two ways: with acute deficits due to hemorrhage or with progressive neurologic deficits secondary to mass effect on adjacent brainstem pathways. Because of their location, even small brainstem CMs may be associated with significant neurologic deficits. The most common signs and symptoms associated with brainstem CMs are listed in Table 18.5.

Unlike true AVMs, CMs infrequently bleed into the ventricular or subarachnoid space. Most often, they bleed into the sinusoidal space of the malformation itself, resulting in acute enlargement with mass effect. This typically results in the acute onset of headache, vertigo, nausea, and vomiting, together with the onset or exacerbation of neurologic deficit.

**Table 18.5**  Brainstem Cavernous Malformations: Clinical Presentation

| Most Common Signs and Symptoms in 38 Patients | % |
| --- | --- |
| Facial pain or hypesthesia | 53 |
| Hemiparesis and spasticity | 47 |
| Disorders of ocular motility | 47 |
| Hemisensory deficits and dysesthesia | 45 |
| Headache | 42 |

## NATURAL HISTORY

Until recent years, the natural history of CMs was poorly understood. Since the advent of MRI, which provides superior sensitivity in diagnosis and monitoring of CMs, it has been possible to diagnose and follow patients prospectively to examine the issue of annual hemorrhage rate. Nonfamilial CMs appear to bleed at a rate of 0.7% to 2.6% per patient-year, while patients with familial CMs have a proportionately higher annualized rate of 6.5% per patient (1.1% per lesion) per year. In addition, the risk of hemorrhage in supratentorial CMs is more than doubled by a history of prior bleed. When infratentorial as well as supratentorial CMs are considered, the overall annual bleed rate rises from 2.6% to 4.5% with a prior history of hemorrhage.

Prospective tracking of patients with these lesions demonstrates that change is possible in both the number and size of CMs and in their MRI signal characteristics. The appearance of "new" lesions may actually represent the growth of previously small, asymptomatic lesions. "Growth" of these lesions may be secondary to single or repetitive hemorrhage into the fragile, thin-walled sinusoids of the CM.

Brainstem CMs, likely due to their anatomically critical location, tend to follow a more progressive clinical course when they become symptomatic. The stuttering, progressive course of neurologic deterioration in patients with brainstem CMs, interspersed with acute exacerbations and subsequent periods of neurologic recovery, may mimic multiple sclerosis.

## DIAGNOSTIC STUDIES

Magnetic resonance imaging is the best way to image CMs (Fig. 18.5): T2-weighted images reveal a reticulated core of mixed signal intensity surrounded by a ring of decreased signal intensity, producing the typical target appearance. The rim of decreased signal intensity is thought to be related to the deposition of iron from hemoglobin degradation following microhemorrhage.

The angiographic features of CMs are inconsistent and nonspecific. They include the presence of an avascular mass during the capillary phase, dense venous pooling in the venous phase, and the association of a venous angioma (malformation) in more than 30% of cases (Fig. 18.6).

Computed tomography has a sensitivity of less than 50% for CMs. These lesions may enhance on contrasted CT, but the majority of CMs will be missed or misdiagnosed with routine CT imaging.

## SURGICAL INDICATIONS

In general, operation on *lobar* CMs carries a markedly lower risk of morbidity than does surgery for deep-seated CMs. Surgical intervention may be offered

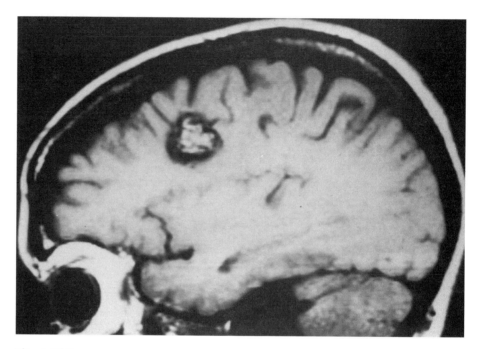

**Figure 18.5**
Magnetic resonance imaging characteristics of CM. T2-weighted axial MRI demonstrating the typical target appearance: rim of low-signal intensity (hemosiderin) surrounding reticulated core.

to medically suitable patients presenting with (1) intractable seizures, (2) symptomatic but not devastating hemorrhage, (3) recurrent hemorrhage, (4) progressive neurologic deficit, and (5) symptomatic mass or hydrocephalus.

Because symptomatic *brainstem* CMs tend to follow a progressive clinical course, the appropriate management of these lesions is excision, provided only that this is technically possible without undue risk. Radiosurgery offers no benefit for patients with CMs.

Candidates for operative microsurgical resection of brainstem CMs must (1) be of suitable medical status and neurologic function to warrant surgical intervention, and they must harbor lesions that (2) abut a pial or ependymal surface, and (3) are accessible through a surgical approach that spares eloquent brain tissue. Brainstem CMs that are entirely buried or surrounded by brain provide no window of entry into the lesion and, like asymptomatic brainstem CMs, should be treated expectantly (Fig. 18.7). Careful clinical follow-up often reveals that these lesions bleed or rebleed, and the resultant hemorrhage frequently opens up a pial or ependymal portal of entry to the lesion.

Correct timing of the operation after a hemorrhage is important. Patients with minimal posthemorrhagic neurologic dysfunction are good candidates for early surgery. Severely impaired patients, however, will often make a gratifying

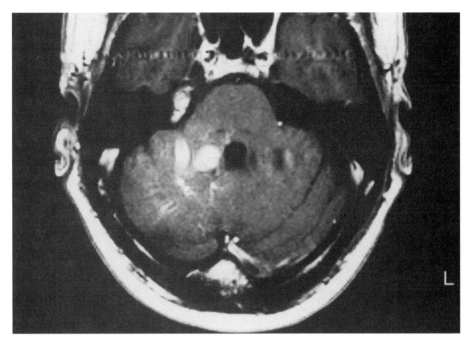

**Figure 18.6**
Magnetic resonance imaging of CM associated with a venous malformation.

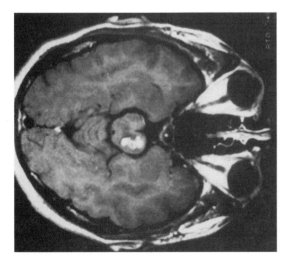

**Figure 18.7**
Cavernous malformation completely enveloped in brainstem, without pial or ependymal surface.

neurologic recovery, and it is beneficial to delay surgery until their optimal neurologic function is manifest.

## PRINCIPLES OF OPERATIVE MANAGEMENT

The technical basis for successful surgical management of brainstem CMs is the selection of an appropriate operative approach. The ideal operative approach optimizes exposure of the lesion by (1) taking advantage of the pial or ependymal "window" and (2) minimizing distortion of the normal neurovascular anatomic relationships. A simple two-point method may be used to help determine this optimal approach: After constructing a line drawn through (1) the central point on the pial or ependymal surface of the lesion and (2) the epicenter of the cavernous malformation, the surgeon selects the surgical approach that most closely approximates this line (Fig. 18.8). Often, this line mandates an approach through the cranial base, and the surgeon must therefore be competent in the performance of the full range of skull base approaches that are available for operation on each level of the brainstem. From these can be chosen the one that is most appropriate to the individual case (Fig. 18.9)

Midbrain CMs may be approached from several angles, depending on the criteria described earlier (Fig. 18.10). Lesions located in the anterior midbrain–interpeduncular fossa are best approached by way of a fronto-temporal/pterional craniotomy. Removal of the posterior clinoid process, performance

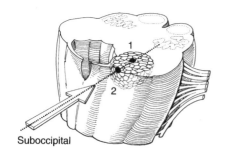

Suboccipital

**Figure 18.8**
The two-point method for determining surgical approach to brainstem CMs.

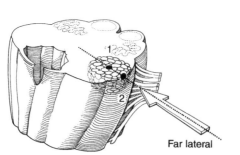

Far lateral

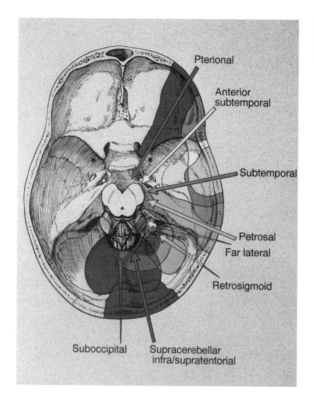

**Figure 18.9**
Radial array of skull base approaches to the brainstem.

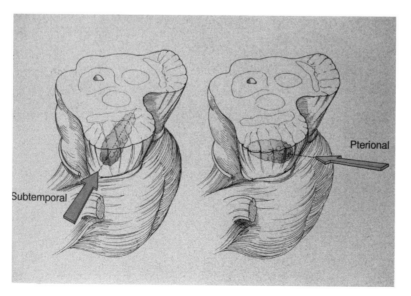

**Figure 18.10**
Pterional-transsylvian versus subtemporal-zygomatic approach to a midbrain CM.

of an orbitozygomatic osteotomy, or splitting of the fissure of Sylvius may enhance this exposure (Fig. 18.11). Lesions of the lateral midbrain (crus cerebri) may be accessed by the pterional or subtemporal approach, and these exposures may be enhanced by division of the tentorium and/or drilling of the petrous apex, as described by Kawase and colleagues. Lesions of the posterior mesencephalon and tectum may be accessed by either an infratentorial-supracerebellar or an occipital-transtentorial approach.

Pontine CMs most commonly occur in the floor of the fourth ventricle, and are typically approached with a standard suboccipital craniotomy. Lateral lesions are inaccessible via this route, but may be accessed via the retrosigmoid or presigmoid approach. Anterior pontine lesions may be approached either by a far lateral, transtemporal, or combined supra- and infratentorial approach.

Cavernous malformations of the dorsal medulla also are approached via the standard suboccipital route. Lesions situated anterolaterally in the medulla are readily accessed via the far lateral approach (Fig. 18.12). A detailed technical description of these operative skull base techniques is available elsewhere.

Accurate localization of the CM is imperative, and this involves a detailed knowledge of the normal anatomy of the brainstem. There are two important surgical adjuncts: (1) preoperative imaging with high-resolution, multiplanar, thin-cut MRI of the brain stem and, if available, (2) intraoperative frameless stereotaxy (see chapter 26).

"Safe entry zones" for excision of brainstem lesions have recently been described: a "suprafacial triangle" (bordered medially by the medial longitudinal fasciculus, laterally by the cerebellar peduncle, and caudally by the facial trigone

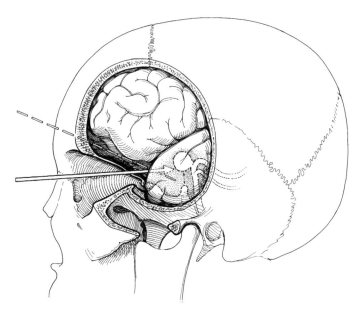

**Figure 18.11**

The orbitozygomatic osteotomy enhances the pterional-transsylvian exposure to the anterior upper brainstem.

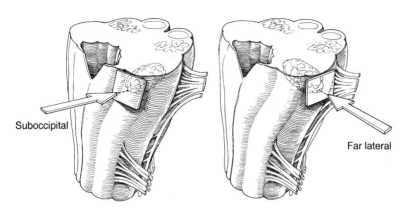

and nerve) and an "infrafacial triangle" (bordered medially by the medial longitudinal fasciculus, laterally by the facial nerve, and caudally by the striae medullares) for access to pontine lesions through the floor of the fourth ventricle.

Preoperative, high-resolution, thin-cut MRI of the brainstem not only allows one to discern whether a given lesion is truly accessible through a pial or ependymal surface, but also provides exact anatomic detail with respect to adjacent structures (Fig. 18.13). Having preoperatively determined the relationship of the lesion to a particular neural or vascular structure, one may prepare an appropriate surgical plan. A typical crucial preoperative finding is the anatomic relationship of the lesion to essential vascular structures, such as adjacent arterial branches or venous malformations (Fig. 18.6). Venous malformations (which are the anomalous but necessary venous drainage for that region of brain) will accompany brainstem CMs in more than 30% of cases and must be meticulously preserved, as must arterial perforating arteries, to prevent cerebral infarction.

Intraoperative stereotactic localization is an additional highly useful surgical adjunct, making possible real-time feedback to confirm accuracy of localization and extent of resection. Frameless stereotaxic techniques are the least burdensome and provide for maximal operative flexibility of approach. Several different frameless stereotaxis systems have been employed successfully. Some investigators advocate intraoperative ultrasonography to enhance real-time localization.

A brainstem CM is usually found to be a mulberry-like lesion surrounded by a thin rim of gliotic, hemosiderin-stained tissue. It is typically composed of sinusoidal vascular spaces and the degradation products of prior hemorrhages. Circumferential dissection will separate the CM from the surrounding zone of gliosis. This technique separates the lesion from its microvascular blood supply and avoids the frank hemorrhage often encountered if an attempt to core out the lesion is made first, but because of restricted access, it may not be feasible or

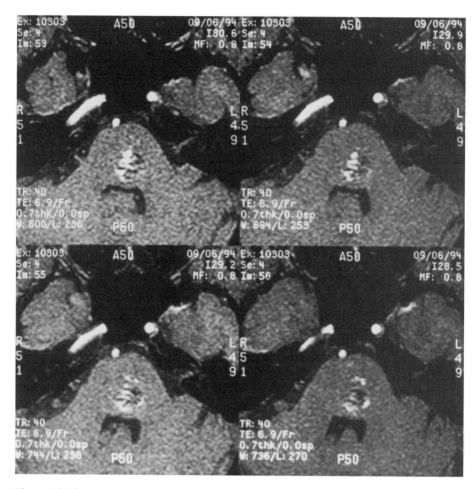

**Figure 18.13**
Axial, thin-cut, high-resolution MRI demonstrating pontomedullary CM without ependymal presentation.

advisable to remove the lesion as a whole. Hence, after maximal circumferential dissection, the lesion is often removed piecemeal, carefully preserving perforating arterial vessels and draining veins.

## OUTCOME OF SURGICAL MANAGEMENT

Between 1985 and 1994 there were 38 patients with CMs of the brainstem operated at the Barrow Neurological Institute. There were 14 males and 24 females, and the average age was 35.7 years (range 4–73 years). Thirty-two (84%) patients described an apoplectic onset of symptoms. Nineteen (37%) had multiple CMs, and nine (26%) had a known familial history of CMs.

Twelve patients (31.5%) had an associated venous malformation. Complete resection of the lesion (defined by both intraoperative impression and postoperative MRI) was achieved in 35 (92%) of 38 patients, and greater than 90% resection was achieved in the remaining three. Median follow-up was 46 months.

Table 18.6 lists the transient and permanent complications incurred from surgery. Overall, there was a 5% risk of major morbidity/mortality (including one patient without neurologic morbidity who died of a presumed pulmonary embolus 28 days after surgery), and a 20% risk of permanent minor morbidity; 31 (81%) of 38 patients were improved from their preoperative status at the time of their last follow-up. These results demonstrate that with careful patient selection, individualized planning of operative approach, and meticulous microsurgical technique, CMs may be removed from the brainstem with reasonably low morbidity.

## References/Further Reading

Celli P, Ferrante L, Palma L, et al. Cerebral arteriovenous malformations. Clinical features and outcome of treatment in children and in adults. Surg Neurol 1994;22:43.

Crawford P, West C, Chadwick D, et al. Arteriovenous malformations of the brain: natural history in unoperated patients. J Neurol Neurosurg Psychiatry 1986;49:1–6.

Forster D, Steiner L, Hakanson S. Arteriovenous malformations of the brain: a long-term clinical study. Neurosurgery 1972;37:562–566.

Friedman W, Bova F. Linear accelerator radiosurgery for arteriovenous malformations. J Neurosurg 1992;77:832–836.

**Table 18.6**  Brainstem Cavernous Malformations in 38 Patients

|  | % |
| --- | --- |
| Transient Complications | |
| Hemiparesis | 21 |
| Diplopia | 21 |
| VII, VIII neuropathy | 16 |
| Hydrocephalus | 13 |
| Meningitis | 8 |
| Deep venous thrombosis | 8 |
| IX, X, XI, XII | 8 |
| Gait ataxia | 5 |
| CSF leak | 2.5 |
| Permanent complications | |
| Death (cerebellar infarction and pulmonary embolism) | 5 |
| Internuclear ophthalmoplegia and dysmetria | 8 |
| Partial third nerve palsy | 5 |
| Incomplete INTO | 5 |
| Mild upper and lower limb dysmetria | 2.5 |

Fults D, Kelly DJ. Natural history of arteriovenous malformations of the brain: a clinical study. Neurosurgery 1984;15:658–663.

Kawase T, Toya S, Shiobara R. Anterior transpetrosal-transtentorial approach for sphenopetroclival meningiomas: surgical method and results in 10 cases. Neurosurgery 1991;28:869–876.

Kondziolka D, Lunsford D. The natural history of cavernous malformations. In: American Association of Neurological Surgeons, Orlando: 1995.

Kyoshima K, Kobayashi S, Gibo H, et al. A study of safe entry zones via the floor of the fourth ventricle for brain-stem lesions. J Neurosurg 1993;78:987–993.

Lunsford D, Kondziolka D, Bissonette D, et al. Stereotactic radiosurgery of brain vascular malformations. Neurosurg Clin North Am 1992;3:79–85.

McCormick P, Michelsen W. Management of intracranial cavernous and venous malformations. In: Barrow D, ed. Intracranial vascular malformations. Park Ridge, IL: American Association of Neurological Surgeons, 1990:197–217.

McCormick W. Pathology of vascular malformations of the brain. In: Wilson CB, ed. Intracranial vascular malformations. Baltimore: Williams & Wilkins, 1984:44–63.

McCormick W. The pathology of vascular ("arteriovenous") malformations. J Neurosurg 1966;24:807.

Ondra S, Troupp H, George E, et al. The natural history of symptomatic arteriovenous malformations of the brain: A 24 year follow up assessment. J Neurosurg 1990;73:387–394.

Rigamonti D, Hadley M, Drayer B, et al. Cerebral cavernous malformations. Incidence and familial ocurrence. N Engl J Med 1988;319:343–347.

Robinson J, Awad I, Little J. Natural history of the cavernous angioma. J Neurosurg 1991;75:709–714.

So S. Cerebral arteriovenous malformations in children. Childs Brain 1978;4:242.

Spetzler R, Daspit C, Pappas C. The combined supra- and infratentorial approach for lesions of the petrous and clival region: experience with 46 cases. J Neurosurg 1992;76:588–599.

Spetzler R, Grahm T. The far-lateral approach to the inferior clivus and upper cervical region: technical note. BNI Quarterly 1990;6:35–38.

Spetzler R, Martin N. A proposed grading system for arteriovenous malformations of the brain. J Neurosurg 1986;65:476–483.

Thompson B, Kraus G, Hamilton M, et al. Special techniques for managing giant intracranial aneurysms. In: Tindall G, Cooper P, Barrow D, eds. The practice of neurosurgery. Baltimore: Williams & Wilkins, 1995.

Wascher T, Spetzler R. Cavernous malformations of the brain stem. In: Carter L, Spetzler R, eds. Neurovascular surgery. New York: McGraw-Hill, 1995:541–555.

Zabramski J, Wascher T, Spetzler R, et al. The natural history of familial cavernous malformations: the results of an ongoing study. J Neurosurg 1994;80:422–432.

Zimmerman R, Spetzler R, Lee, et al. Cavernous malformations of the brain stem. J Neurosurg 1991;75:32–39.

# VI CEREBROSPINAL FLUID CIRCULATION

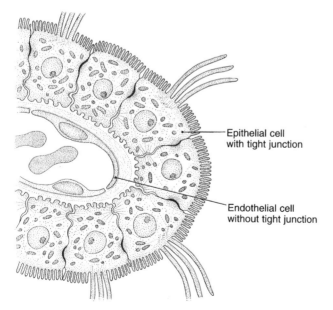

# Cerebrospinal Fluid Circulation

**19**

Harold L. Rekate

## Anatomy of Cerebrospinal Fluid Flow

Cerebrospinal fluid (CSF) is formed in two distinct ways. At least half, and possibly as much as 80% of CSF, is formed in an energy-requiring process by specialized organs within the lateral, third, and fourth ventricles, called choroid plexuses. Figure 19.1 is a schematic of the choroid plexus, showing frond-like out-pouchings of capillaries. The endothelial cells are unusual in the CNS in that they lack the tight junctions that are the substrate of the blood-brain barrier. Capping these endothelial cells is a specialized monolayer of ciliated columnar or epithelial cells. Tight junctions do exist between the epithelial cells.

An ultrafiltrate of plasma derived from the capillary lumen passes through the spaces between the endothelial cells where it is presented to the interior of the epithelial cells. Cerebrospinal fluid differs from plasma in its concentra-

**Figure 19.1**
Schematic diagram of a villus of the choroid plexus showing apical tight junctions between the epithelial cells rather than between capillary endothelial cells, as found in the rest of the CNS.

Epithelial cell with tight junction

Endothelial cell without tight junction

tion of various electrolytes, particularly sodium. In his standing gradient hypothesis, Pollay stated that the plasma ultrafiltrate entered the basal unfoldings of the choroidal epithelial cells by an active transport process linking sodium and water metabolism by a sodium-potassium adenosine triphosphatase (ATPase) pump. This pump can be blocked by the cardiac glycoside, ouabain, and CSF production will then decrease. The carbonic anhydrose inhibitor, acetazolamide, also blocks choroidal CSF production, but the mechanism is poorly understood.

Cerebrospinal fluid also is produced as a by-product of metabolism in the brain and spinal cord. This mode of CSF production does not depend on active transport and is unaffected by carbonic anhydrase inhibitors. The fluid produced thus is identical to brain extracellular fluid. The extracellular fluid passes by bulk flow into the ventricular fluid where it mixes with choroidally produced CSF. Studies of CSF production using tracer infusion techniques have failed to show CSF production within the spinal canal. Recent experiments by Milhorat and associates, however, in which syringomyelia is produced by occluding the central canal in the upper cervical spinal cord, imply that spinal cord extracellular fluid is a by-product of metabolism that passes into the central canal and exits through the fourth ventricle.

Figure 19.2 is a schematic diagram of the CSF pathways passing among the cerebral ventricles through named orifices (e.g., foramina of Monro, aqueduct of Sylvius) to exit the ventricular system through the paired foramina of Luschka laterally and through the inconsistently open foramen of Magendie medially. It then percolates throughout the spinal and cortical subarachnoid spaces. Finally, it is absorbed, primarily through specialized organelles termed *arachnoid villi* (microscopic). When confluent, the arachnoid villi become arachnoid granulations, which are visible macroscopically either grossly or radiographically.

Absorption of CSF occurs primarily by bulk flow. Although some elements may be absorbed by energy-requiring active transport, the contribution of this mechanism must be minimal, because CSF absorption is found to be no less rapidly absorbed in postmortem infusion studies than it is in vivo (McComb JG, personal communication, 1994).

At normal intracranial pressures (ICPs), most of the CSF is absorbed (by bulk flow) into the dural venous sinuses. At high ICPs, alternative pathways for CSF absorption may be recruited. These include the lymphatic system, the root sleeves of exiting cranial and spinal nerves, and possibly the paranasal sinuses. It also has been speculated that the CSF is forced to move in a retrograde fashion into the cerebral parenchymal capillaries and veins, a process termed *transependymal absorption;* the evidence for this, however, is tenuous. Furthermore, because the capillaries and veins have no muscularis layer, they are easily collapsible tubes and flow through them is impeded as intraparenchymal tissue pressure rises above cerebral venous pressure.

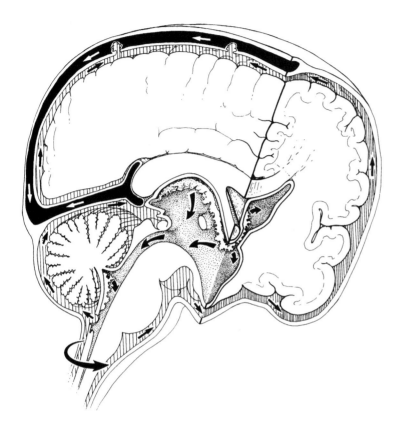

**Figure 19.2**
The anatomy of CSF fluid pathways. (Reproduced with permission from Rekate HL, Olivero W: Current concepts of CSF production and absorption. In: Wirth FP, Ratcheson RA, Grubb RL Jr, et al., Hydrocephalus. Baltimore: Williams & Wilkins, 1990:11–22. Copyright © 1990, Williams & Wilkins Co.)

## Flow of Cerebrospinal Fluid as a Circuit

From the perspective of biophysics, CSF flow can be considered a circuit, with the heart analogous to a pump in a fluid circuit or to a battery in an electrical circuit (Fig. 19.3). The brain receives energy in the form of a pulsatile fluid wave passing through the carotid and vertebral arteries at systemic arterial pressure. This energy is received by both the choroid plexuses and the parenchymal precapillary arterioles. These elements serve as the most important resistors in the system whereby arterial pressure is converted to ICP, which in turn is identical to intraventricular pressure (IVP). For the purpose of discussion, mean pressures will be described as an analog to a direct-current (DC) circuit. The importance of the pulsatile nature of these pressures is discussed next.

Cerebrospinal fluid flows through resistors, including the foramina of Monro, aqueduct of Sylvius, and foramina of Luschka and Magendie, to the spinal subarachnoid space (SSAS), thence through the basal cisterns to the cortical subarachnoid space (CSAS), and finally through the arachnoid villi into the sagittal sinus. In a parallel circuit, the arterial blood passes through the parenchymal capillaries to the intraparenchymal and then to the cortical

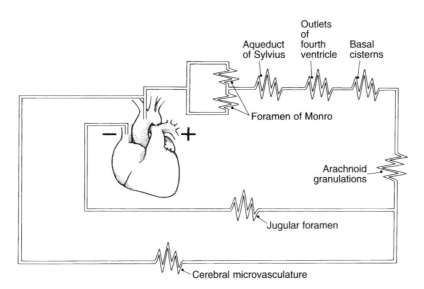

**Figure 19.3**
Theoretical circuit diagram of the CSF pathways as an electrical analog. (Reprinted with permission of the Barrow Neurological Institute.)

veins. Finally, venous blood and CSF mix in the dural venous sinuses. The combined circuit then travels through the jugular foramina and foramen magnum (azygous vein) to join the systemic circulation flowing into the right atrium of the heart at central venous pressure.

Cerebrospinal fluid is produced at approximately 0.3 mL/min in humans who are older than 2 years. Compared with the rate at which blood traverses the brain (about 1 liter/min), the flow rate of CSF is trivial, which makes accurate measurement of resistances within these pathways difficult. This problem is compounded when one considers that most commercially available pressure transducers are only accurate to ±1 mm Hg. Measurements of CSF pressures within the CSF pathways in normal laboratory animals have failed to disclose measurable resistances (pressure drops) at any point within the pathways, even when artificial CSF infusions are employed.

Under normal conditions, then, resistances can be measured only between the arterial pressure and ICP and between ICP and the pressure in the superior sagittal sinus. Absorption of CSF seems to require a pressure gradient of 7 mm Hg in our greyhound dog model and 5 mm Hg in humans. In addition, pressure differentials have been recorded between the cortical veins and the sagittal sinus. This differential maintains an equilibrium between CSF pressure and the pressure in these thin-walled venous structures.

Figure 19.4 is a graph of CSF flow, both absorption and production, as described by Cutler and colleagues. The rate of CSF production is unaffected by ICP, except in the case of extremely elevated pressures. No absorption of CSF occurs below a pressure of 70 mm $H_2O$ (~5 mm Hg). The more that ICP rises above this opening pressure, the greater the rate of CSF absorption. The lines of production and absorption rates intersect at approximately 140 mm $H_2O$ (10 mm Hg), which is normal ICP.

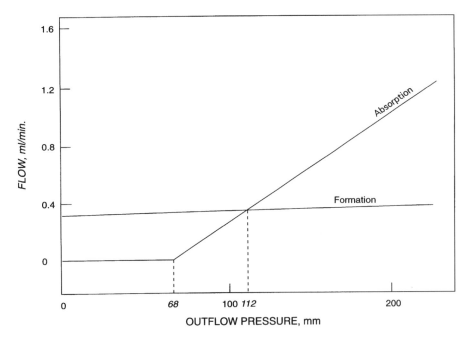

**Figure 19.4**
Diagram of CSF formation over a range of outflow pressures. (Reproduced with permission from Cutler RW, Page L, Galicich J, Watters GV. Formation and absorption of cerebrospinal fluid in man. Brain 1968;91:707–720.)

Under normal conditions, changes in ICP are not reflected in sagittal sinus pressure changes until extremely high pressures are reached. With increased ICP, a steeper gradient is therefore maintained across the arachnoid villi, resulting in increased CSF absorption until either normal conditions are resumed or no further CSF can be accessed at the sagittal sinus. As stated, when ICP rises above 15 mm Hg, alternative routes of CSF absorption are recruited.

In normal mammals, the arachnoid villi function as a shunt, from the CSAS (the pressure within which it is undistinguishable from IVP) to the sagittal sinus, equipped with a medium pressure valve having a closing pressure of 70 mm $H_2O$ (5 mm Hg). Because the proximal and distal ends of the shunt are always close together despite changes in posture, for the most part hydrostatic pressure does not play a role in normal CSF absorption. As a human assumes an erect position, sagittal sinus pressure becomes negative with respect to atmospheric pressure and right atrial pressure; but because the jugular veins collapse when the pressure within them becomes negative relative to atmospheric pressure, the degree of negativity of ICP in the erect position is limited to approximately −1 to −3 torr.

## Pathophysiology of Abnormalities of Cerebrospinal Fluid Flow

Most, if not all, abnormalities of CSF flow can be understood qualitatively as obstruction or high-level resistance to the flow of CSF through one of the resistance elements proposed in Figure 19.4. In this section, several pathologic states are discussed relative to their effect on the CSF-flow circuit diagram.

### POSTSHUNT VENTRICULAR ASYMMETRY

In his classic experiments defining the choroid plexus as the generator of CSF, Dandy created hydrocephalus by the intracisternal injection of lampblack. He then performed a unilateral choroid plexectomy and occluded the foramen of Monro. The ventricle without the choroid plexus became much smaller than the ventricle containing the choroid plexus. Bering later repeated Dandy's experiments, but without occluding the foramen of Monro, and obtained the same results. Bering measured IVP bilaterally and determined that while mean IVP was the same in each ventricle, pulse pressure was greater in the ventricle containing the choroid plexus. He then postulated that the CSF pulse pressure was the cause of ventriculomegaly in hydrocephalus, and that the pulse pressure generated by the expansion of the choroid plexus was damped by its passage through the foramen of Monro.

In describing the long-term radiologic findings in chronically shunted children, Kaufman et al. documented that postshunt ventricular asymmetries could be expected in all children whose ventricles returned to normal size and whose septa pellucida were intact. In a series of experiments on hydrocephalic dogs, they measured IVP in both lateral ventricles. Shunted ventricles had the same mean IVP as the unshunted contralateral ventricle but a substantially diminished pulse pressure. This finding supports Bering's hypothesis.

In our laboratory at the Barrow Neurological Institute, we performed infusion studies on normal and hydrocephalic anesthetized greyhound dogs. Despite rapid rates of infusion of artificial CSF into one lateral ventricle, we were unable to record pressure differentials across the foramina of Monro–third ventricle complex. When nonhydrocephalic dogs (in which there was no distension of the foramina of Monro) underwent external drainage procedures, however, 2- to 3-mm Hg pressure differentials could be recorded from one lateral ventricle to the other. Anatomic studies (Fig. 19.5) showed that as the pressure in one lateral ventricle is decreased by drainage or shunt, the septum pellucidum is distorted and comes to lie against the head of the caudate nucleus.

This configuration creates a high level of obstruction at the foramen of Monro and thus postshunt ventricular asymmetry. If the ventricles remain large, the foramina of Monro remain distended and postshunt ventricular asymmetry rarely, if ever, occurs. Figure 19.6 shows the electrical circuit analog for postshunt ventricular asymmetry in which the shunt acts as an electrical ground creating a large resistance at the foramen of Monro.

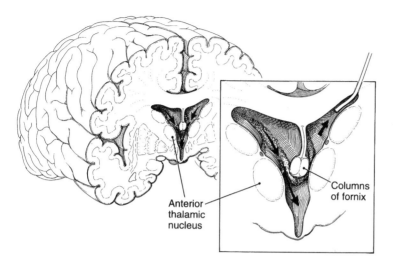

**Figure 19.5**
Anatomy of the foramen of Monro, suggesting its function as a ball valve. (Reproduced with permission from Rekate HL, Williams FC Jr, Brodkey JA, et al. Resistance of the foramen of Monro. Pediatr Neurosci 1988;14:85–89.)

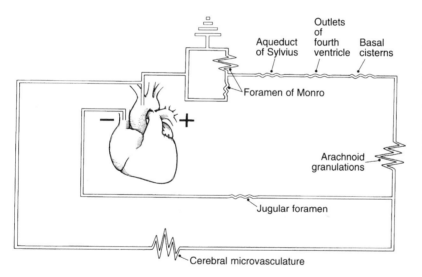

**Figure 19.6**
Circuit diagram of CSF flow with shunting of one lateral ventricle, leading to postshunt ventricular asymmetries. (Reprinted with permission of the Barrow Neurological Institute.)

## COMMUNICATING HYDROCEPHALUS AND PSEUDOTUMOR CEREBRI

Communicating hydrocephalus, also called extraventricular obstructive hydrocephalus, results from the failure of CSF to be absorbed, even when ventricular CSF can flow from the ventricular system to the lumbar theca. This condition has many causes and is associated with many clinical presentations. The most common cause of this condition is subarachnoid hemorrhage from a variety of etiologies, including neonatal intraventricular hemorrhage, trauma, and aneurysmal subarachnoid hemorrhage. Other causes of communicating hydrocephalus include infection, congenital absence of the arachnoid villi, and intracranial venous hypertension.

Communicating hydrocephalus, therefore, results from an increase in the resistance between the SSAS and CSAS, between the subarachnoid space and the sagittal sinus, or within the venous system. Of these points of high resistance, the venous system creates the most complex results, because in the mature organism (i.e., human or laboratory animal) venous obstruction without mass lesion does not lead to hydrocephalus but to high ICP. This condition is called *pseudotumor cerebri* or *benign intracranial hypertension*. With all other causes of hydrocephalus, the increased resistance affects only CSF flow and only indirectly affects the brain parenchyma by causing it to be compressed. When the pressure within the sagittal sinus is high, as in venous sinus thrombosis, anatomic abnormalities (i.e., stenosis of the jugular foramen as seen in Crouzon's syndrome and achondroplasia), or severe right heart failure, not

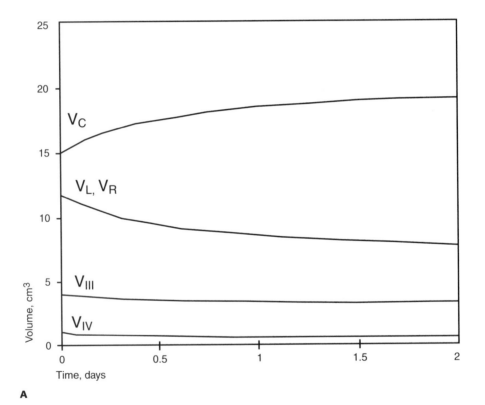

**A**

**Figure 19.7**
Computer-generated graphic representation of the changes in (A) the volume of CSF within the cerebral subarachnoid space ($V_C$) and (B) the pressures within all the various CSF compartments produced by increasing the stiffness of the brain from its normal value of 0.34 to 0.99 while simultaneously blocking outflow from the CSAS to the sagittal sinus (thus simulating pseudotumor cerebri). The volume of CSF within the lateral ventricles ($V_L$, $V_R$) decreases, while changes in the size of the third and fourth ventricles ($V_{III}$, $V_{IV}$) are minimal. The pressures within the various compartments increase with time at the same rate. (Adapted with permission from Rekate HL, Brodkey JA, Chizeck HJ, et al. Ventricular volume regulation: a mathematical model and computer simulation. Pediatr Neurosci 1988;14:77–84.)

only is the absorption of CSF affected but also the pressure within the cortical and parenchymal veins.

As has already been noted, for CSF to be absorbed, CSF pressure must be about 5 mm Hg greater than sagittal sinus pressure. This is the valve mechanism at the arachnoid villi and is one of the features that controls ICP. Simultaneously, because the cortical veins are collapsible tubes, CSF pressure and cortical vein pressure must be the same, and a similar valve mechanism must exist between the cortical veins and the sagittal sinus, as demonstrated by Portnoy. As venous sinus pressure increases, both CSF absorption pressure and cortical vein pressures must increase to maintain flow. Both the cortical and, secondarily, the intraparenchymal venous structures become engorged. Cerebral venous blood volume therefore increases, and because water and therefore blood are incompressible, brain turgor increases.

In most of its forms, pseudotumor cerebri results from increased venous sinus pressure. Using a mathematical model of ICP and ventricular volume interaction, pseudotumor cerebri should result either from a blockage of CSF flow at the arachnoid villi associated with increased brain turgor or simply from increasing intracranial venous sinus pressure (Fig. 19.7). Active measurement of

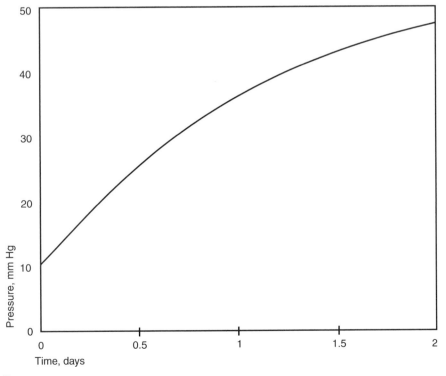

**B**

**Figure 19.7** *Continued*

venous pressures in obese young women with idiopathic pseudotumor cerebri has shown not only high sagittal sinus pressures but also very high right atrial pressures. This finding suggests that pseudotumor cerebri is actually a systemic disease in these women.

If venous obstructive disease causes pseudotumor cerebri, how can it also cause hydrocephalus? The answer lies in the role of the container (skull). Epstein and colleagues have demonstrated the essential role of the skull and, to some extent, of the dura in the pathophysiology of hydrocephalus. In their experiments, intracisternal injection of kaolin, which creates an intense basilar meningeal thickening, blocking the flow of CSF from the SSAS to the CSAS, created only a mild degree of hydrocephalus in intact cats. Massive hydrocephalus developed in craniectomized cats under the same conditions. By analogy, increased brain turgor due to increased venous pressure leads to pseudotumor cerebri when the skull sutures are closed, allowing ICP to rise sufficiently to permit CSF absorption to continue at a normal rate. If, however, the skull is distensible, as in babies, the intracranial compartment is linked to atmospheric pressure and can expand, so that hydrocephalus results. The final piece of the puzzle is provided by the work of Olivero and Asner, who produced hydrocephalus in craniectomized rabbits by occluding the sagittal sinus.

When high resistance to venous outflow occurs before the cranial sutures close, as in Crouzon's syndrome or achondroplasia, a condition that is analogous to craniectomy in the experimental animal exists. In such situations, the open fontanel, which reflects atmospheric pressure, can enlarge as CSF accumulates and hydrocephalus develops.

## NORMAL-PRESSURE HYDROCEPHALUS

Normal-pressure hydrocephalus is a subcategory of communicating hydrocephalus that occurs in the elderly. First described by Hakim and Adams, this syndrome consists of the clinical triad of gait disturbance, urinary incontinence, and dementia in elderly patients who have large ventricles on imaging studies. Originally, the imaging studies were pneumoencephalograms, but now computerized tomography (CT) or magnetic resonance imaging (MRI) is employed. These patients show no papilledema or overt signs of increased ICP. Intracranial pressure as reflected by lumbar puncture is normal.

These patients improve after either ventriculoperitoneal or lumboperitoneal shunting. A variety of tests have been advocated for selecting patients who may respond to shunting: chronic ICP monitoring, CSF infusion studies, CSF flow studies using radioactive tracers, and lumbar puncture in which a large volume of CSF is withdrawn and the patient is observed for signs of clinical improvement. No test is a perfect predictor of response to shunting, but patients who objectively respond to fluid removal are very likely to improve.

There are many theories about the pathophysiology of normal-pressure hydrocephalus. Postmortem studies of patients who were documented to have responded positively to shunting have shown periventricular spongiosis and gliosis as well as marked thickening of the arachnoid membrane at the skull base. Using a mathematical model to simulate the behavior of the lateral ventricles, we made the following assumptions based on the pathology described earlier.

A marked increase in resistance to CSF flow is assumed to exist between the SSAS and CSAS. This finding is common with CSF tracer studies in clinical situations. Periventricular spongiosis also can be seen on MRI in elderly patients without normal-pressure hydrocephalus and can be assumed to be a normal process of aging described in classical neuropathologic terms as *état criblé*. As a result of this "spongiosis," the brain may exhibit less turgor. Figure 19.8 is a computer simulation of the pressure and volume changes seen over time if CSF flow is restricted from the SSAS to CSAS and the turgor of the brain decreases with age. These findings help clarify the dynamics of normal-pressure hydrocephalus.

## Nature and Importance of Cerebrospinal Fluid Pulsations

To this point, the discussion has focused on CSF flow as if it were a constant, thereby yielding a DC model. Considerable energy, however, is derived from the pulsatile nature of CSF—a situation more analogous to an alternating current (AC) electrical system. The potential importance of the intraventricular pulse pressure was shown in a series of experiments by DiRocco and colleagues in Rome. They implanted a balloon into the lateral ventricle of lambs and timed the expansion of the balloon to the cardiac cycle through an ECG. When pulse pressure was augmented chronically, the lambs developed ventriculomegaly without an increase in outflow resistance.

The energy derived from the pulsatile CSF flow (AC component), as well as changes in the erectile nature of the brain, exceeds the energy derived from the flow of CSF (DC component) by at least an order of magnitude.

What is the source of the intraventricular fluid pulsation? Using a specialized pump designed at the Electronics Design Center at the School of Engineering, Case Western Reserve University, in Cleveland, we created intracranial perturbations with complex waveforms. The pump was used in anesthetized dogs to create a waveform that was 180° out of phase with the normal intraventricular pulse wave. To damp the pulse wave to almost a straight line, the to-and-fro volume of fluid required was more than an order of magnitude greater than the rate of CSF production per stroke (personal observation). The implication is that the entire brain expands and contracts with every heartbeat and that the role played by the choroid plexus in this phenomenon is likely to be trivial.

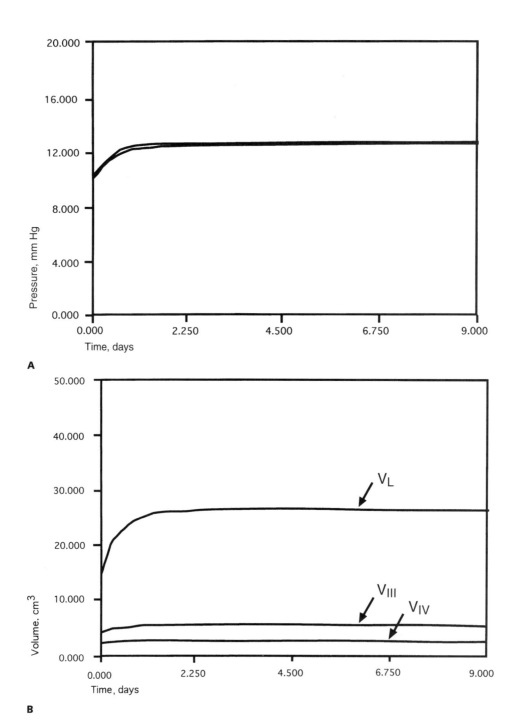

**Figure 19.8**
Computer-generated graphic simulates normal-pressure hydrocephalus. The changes in pressure and volume with respect to time are produced by increasing the resistance to flow between the SSAS and the CSAS while at the same time decreasing brain turgor from its normal value of 0.34 to 0.01. (A) The ICP rises slightly at first but, as in Figure 19.7A, there are no differences between the pressures within the various CSF compartments. (B) The volume of each lateral ventricle ($V_L$) increases markedly, but the volumes of the third ($V_{III}$) and fourth ($V_{IV}$) ventricles change very little. (Reproduced with permission from Rekate HL. Circuit diagram of the circulation of cerebrospinal fluid. Concepts Pediatr Neurosurg 1989;9:46–56.)

The pulsatile flow of CSF is important for two reasons. The energy generated by the wide to-and-fro swings of CSF as seen on cine MRI radiographs is likely to be more effective than DC flow in creating the parenchymal distortions that occur in hydrocephalus and syringomyelia. Figure 19.9 is a cine MRI in a normal patient showing downward displacement of CSF (white) and recovery of CSF in an upward direction (black).

The second clinical implication of the AC model of CSF is that it reflects the state of brain compliance or elastance as described by Avezaat and van Eijndhoven. As ICP increases and the intracranial buffering mechanism is exhausted, the pulse pressure increases and is therefore a relative measure of cerebral compliance. The intraventricular pulse wave also can be analyzed mathematically to analyze the vascular reactivity of the brain or CSF autoregulation.

As the ability to visualize and quantify the pulsatile nature of CSF flow improves, the importance of this element of CSF dynamics will be better understood. For now, however, abnormal CSF flow and ventricular volume regulation lend themselves to a DC-analog model in subjective explanations of cause and effect.

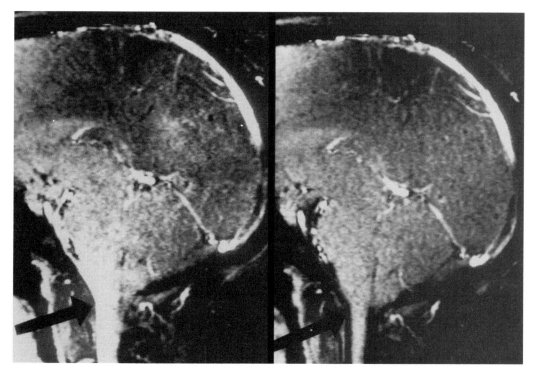

**Figure 19.9**

Images generated by CSF flow study using cine MRI show downward (white) displacement of CSF during systole and upward (black) movement during diastole, documenting the pulsatile nature of CSF flow. (Reprinted with permission of the Barrow Neurological Institute.)

## References/Further Reading

Avezaat CJ, van Eijndhoven JH. Clinical observations on the relationship between cerebrospinal fluid pulse pressure and intracranial pressure. Acta Neurochir (Wien) 1986;79:13–29.

Awad IA, Johnson PC, Spetzler RF, Hodak JA. Incidental subcortical lesions identified on magnetic resonance imaging in the elderly. II. Postmortem pathological correlations. Stroke 1986;17:1090–1097.

Black PM, Ojemann RG, Tzouras A. CSF shunts for dementia, incontinence, and gait disturbance. Clin Neurosurg 1985;32:632–651.

Castro ME, Portnoy HD, Maesaka J. Elevated cortical venous pressure in hydrocephalus. Neurosurgery 1991;29:232–238.

Di Rocco C, DiTrapani G, Petterorossi VE, et al. On the pathology of experimental hydrocephalus induced by artificial increase in endovascular CSF pulse pressure. *Child's Brain* 1979;5:81–95.

Epstein FJ, Hochwald GM, Wald A, Ransohoff J. Avoidance of shunt dependency in hydrocephalus. *Dev Med Child Neurol Suppl* 1975;35:71–77.

Kaufman B, Weiss MH, Nulsen FE. Effects of prolonged cerebrospinal fluid shunting on the skull and brain. *J Neurosurg* 1973;38:288–298.

McQuarrie IG, Saint-Louis L, Scherer PB. Treatment of normal pressure hydrocephalus with low versus medium pressure cerebrospinal fluid shunts. Neurosurgery 1984;15:484–488.

Milhorat TH, Nobandegani F, Miller JI, et al. Non-communicating syringomyelia following occlusion of central canal in rats: experimental model and histological findings. *J Neurosurg* 1993;78:274–279.

Olivero WC, Asner N. Occlusion of the sagittal sinus in craniectomized rabbits. Childs Nerv Syst 1992;8:307–309.

Olivero WC, Rekate HL, Chizeck HJ, et al. Relationship between intracranial and sagittal sinus pressure in normal and hydrocephalic dogs. Pediatr Neurosci 1988;14:196–201.

Pollay, M. Formation of cerebrospinal fluid: relation of studies of isolated choroid plexus to the standing gradient hypothesis. *J Neurosurg* 1975;42:665–673.

Portnoy HD, Branch C, Castro ME. The relationship of intracranial venous pressure to hydrocephalus. Childs Nerv Syst 1994;10:29–35.

Portnoy HD, Chopp M, Branch C, Shannon MB. Cerebrospinal fluid pulse waveform as an indicator of cerebral autoregulation. J Neurosurg 1982;56:666–678.

Rekate HL. The usefulness of mathematical modeling in hydrocephalus research. Childs Nerv Syst 1994;10:13–18.

Rekate HL. Brain turgor (Kb): intrinsic property of the brain to resist distortion. Concepts Pediatr Neurosurg 1991;12:1–11.

Rekate HL. Circuit diagram of the circulation of cerebrospinal fluid. Concepts Pediatr Neurosurg 1989;9:46–56.

Rekate HL, Brodkey JA, Chizeck HJ, et al. Ventricular volume regulation: a mathematical model and computer simulation. Pediatr Neurosci 1988;14:77–84.

Rekate HL, Olivero WM, McCormick J, et al. Resistance elements within the cerebrospinal fluid circulation. In: Gjerris F, Borgesen SE, Sorensen PS, eds. Copenhagen: Outflow of cerebrospinal fluid. Alfred Benzon Symposium 27, 1989:45–57.

Rekate HL, Williams FC Jr, Brodkey JA, et al. Resistance of the foramen of Monro. Pediatr Neurosci 1988;14:85–89.

Rekate HL, Williams F, Chizeck HJ, et al. The application of mathematical modeling to hydrocephalus research. Concepts Pediatr Neurosurg 1988;8:1–14.

# 20 Syringomyelia and Syringobulbia

Bernard Williams[†]

*Syringomyelia* is a condition in which longitudinal cavities extending over several of its segments develop within the spinal cord. Its most common cause is a structural lesion at the foramen magnum (Figs. 20.1, 20.2), in which case it is termed *hindbrain-related syringomyelia* (HRS).

Extension of the cerebellar tonsils through the foramen magnum, sometimes called the Arnold Chiari deformity, is a *herniation of the hindbrain* (HBH). Intermediate forms exist between the adult form and the severe deformities

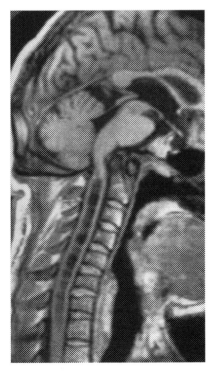

**Figure 20.1**

Hindbrain-related syringomyelia without hindbrain herniation. Note the narrowing of the foramen magnum produced by the gross up-turning of the lower edge of the occipital bone and the backward displacement of the odontoid peg. The cerebellum is confined within a small space. Observe septations in the syringomyelia. The patient suffers from fragilitas ossium.

†Deceased.

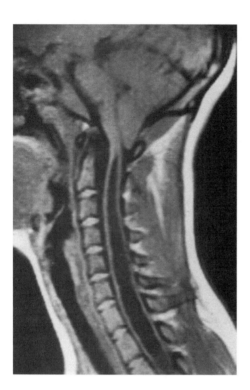

**Figure 20.2**
Hindbrain-related syringomyelia without any bony deformities. Observe that the hindbrain is down to the arch of C1. There is a hint of septation in the syrinx.

of intrauterine life found in infants with spina bifida aperta. Herniation of the hindbrain both causes and is exacerbated by pressure differences between the head and the spinal compartment.

*Syringobulbia* is the name given to bulbar symptoms—cranial nerve dysfunction, giddiness, syncope, nystagmus, and so on—arising in conjunction with clefts running from the floor of the fourth ventricle (Fig. 20.3) into the lower pons and medulla. Its *ascending form* occurs when a syringomyelia cavity tracks upward into the grey matter of the medulla, the pons, and occasionally even higher.

*Subarachnoid meningeal fibrosis* (arachnoiditis) is a condition in which the arachnoid becomes thickened and adhesions develop between it and the dura and sometimes the pia. These adhesions may result in almost complete blockage of the subarachnoid space. Among its causes are meningitis, chemical insults, and the late effects of injury, especially birth-related head injury.

Syringomyelia due to occlusion of the CSF pathways in the spine may be called *primary*, or *non–hindbrain-related, spinal syringomyelia*. It is usually secondary to spinal fracture.

The term *tidal flow* is used to describe the mass movement of fluid within a partly filled syrinx, occurring when the intrathoracic and/or intra-abdominal pressure is raised. As the intrathecal pressure rises, the fluid within the syrinx

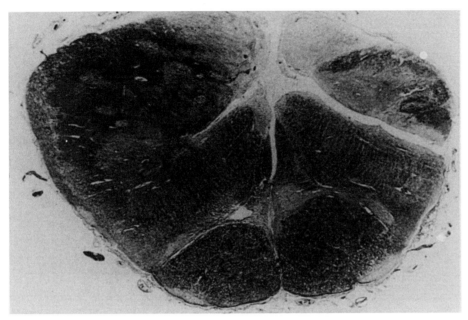

**Figure 20.3**
Fourth ventricular clefts are present in this section of the medulla opposite the trigeminal nucleus. The region of the obex shows a particularly deep groove. On the right side of the illustration, a cleft runs out through the trigeminal nerve nucleus. There is a similar but less well formed cleft on the left and a prominent midline cleft. On the dorsum of the left pyramidal tract there is another cavity, which may communicate with a syrinx below; this might therefore be an ascending form of syringomyelia.

surges in either direction, potentially tearing its walls. The occurrence of septations within a syrinx cavity (see Fig. 20.1) suggests that tidal flow may be limited.

Abnormal collections of CSF near the foramen magnum are commonly called *arachnoid cysts,* but because most of them have a single entrance and form a diverticulum rather than a cyst, the term *pouch* may be preferable.

## "Congenital" Syringomyelia

Syringomyelia is commonly associated with congenital problems such as segmentation abnormalities, spina bifida, widespread mesodermal abnormalities, and so on. Spina bifida with hydrocephalus is often accompanied by abnormal CSF dynamics and by a communication between the fourth ventricle and the syrinx, sometimes termed a *patent central canal.*

There seems to be a subset of cases that might be called *primarily dysraphic syrinxes.* They are seen in association with spina bifida occulta or spina bifida manqué. They may become symptomatic, the most common symptom being pain. Some are found to be tense and respond to treatment (Fig. 20.4). Others are not tense, extend only through two spinal segments, and taper at each end, suggesting that no tidal flow is occurring; in these cases, surgical treatment is often ineffective.

Small cysts in the cord opposite the site of trauma occur in approximately half of cord injuries, causing paraplegia; they are not syringomyelia and may be termed *primary cysts* (Fig. 20.5) They do not usually communicate with an adjacent syrinx, if one is present, remaining full even if the syrinx is emptied.

## Gardner's Concept

Cord cavitation develops at sites of weakness, and fluid from within the cavity may track through the cord. Gardner suggested that most cases had a communi-

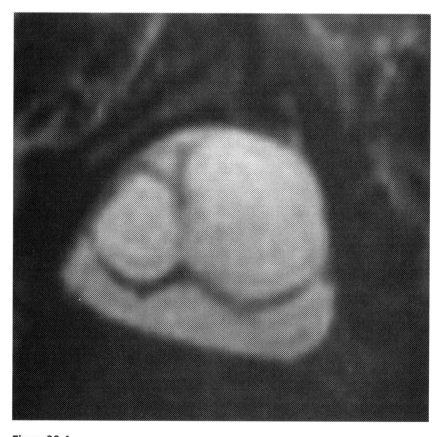

**Figure 20.4**
T2-weighted axial magnetic resonance imaging of the fifth lumbar level. This patient had a massive syringomyelia running from the top of the cervical spine to a diplomyelia in the lumbosacral region. Each leg of the bifid cord is substantially filled with syrinx having only a very thin surrounding shell of nervous tissue. The hindbrain was normal. The patient responded well to myelotomy combined with the coperitoneal shunting.

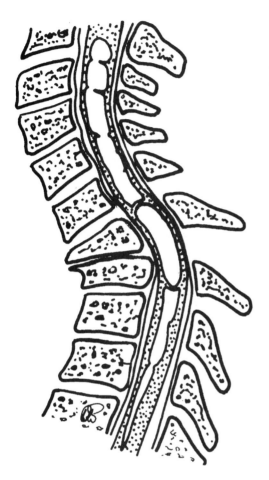

**Figure 20.5**
Diagram of a primary cyst opposite a site of injury. Syringomyelia may be communicating with such a cyst, separated from it by a complete septum, as shown here, or by a short (up to 2-cm long) zone of almost normal cord.

cation from the fourth ventricle to the syrinx, and he noted that abnormalities of the base of the brain might cause pressure differences between the intracranial and intrathecal cavities. Gardner showed that craniovertebral decompression often produced clinical improvement.

## Pathogenesis

### THE HYDRODYNAMIC FORCES

Cardiac pulsation in the capillary bed of the brain constantly impacts the nervous system, but of greater significance are pressure changes within the *venous plexuses*. An increase in pressure in the abdominal and thoracic cavities is transmitted into the spine when venous distension of the epidural plexus causes compression of the dura. This in turn produces rapid movements of CSF. The pressures produced by a cough massively exceed those produced by cardiac pulsation; imposed on the tissues of the neuraxis, they affect the formation of a syringomyelic cavity and perpetuate it.

## CRANIOSPINAL PRESSURE DISSOCIATION

Pressure differences between CSF compartments may be developed by venous pulsation. When the pressures are measured at rest in a normal adult subject and a hydrostatic correction is made, these pressures are equal, but when the subject coughs or sneezes, the pressure of the spinal CSF rises sooner, more steeply, and higher than it does in the head. Cerebrospinal fluid therefore travels upward into the cranial cavity, returning as the intraspinal pressure falls.

In patients with HBH, the CSF is delayed in its return downward through the foramen magnum. These pressure differences have been studied in erect adults by simultaneous intraventricular and intraspinal pressure recordings. Analysis of heartbeat activity and response to coughing and Valsalva maneuvers have confirmed the mechanisms of this phenomenon, which may be termed *craniospinal pressure dissociation* (CPD) (Figs. 20.6, 20.7).

Normally the half-life of the return of pressure differences between the top and bottom of the spine to normal is less than one tenth of a second; with HBH it may exceed 30 seconds (e.g., during a post-Valsalva rebound after straining). This molds the HBH tightly within the foramen magnum and may sometimes be concerned with the pathogenesis of syringomyelia. Correction of the HBH commonly produces radiologic and clinical improvement in the features of syringomyelia and syringobulbia.

## THE COMMUNICATING HYPOTHESIS

Gardner believed that the embryologic communication between the fourth ventricle and the central canal of the spinal cord was the path along which the syrinx was filled, but in fact only about 10% of cases of HRS have a communication. In some other such cases, perhaps, there was a communication in the past that had been closed, perhaps by compression of tissues at the level of the foramen magnum, but often it cannot ever have been present (e.g., in hindbrain tumor cases and those where the top of the syrinx is well below the hindbrain).

## TRANSMURAL PRESSURE GRADIENTS

It may be simplistic to think of "filling mechanisms." Syrinxes are not necessarily always in a state of active filling. It may be more reasonable to seek an analogy, with the Starling equations describing the behavior of tissue fluid as a state of equilibrium between several interacting forces. The cord is a porous

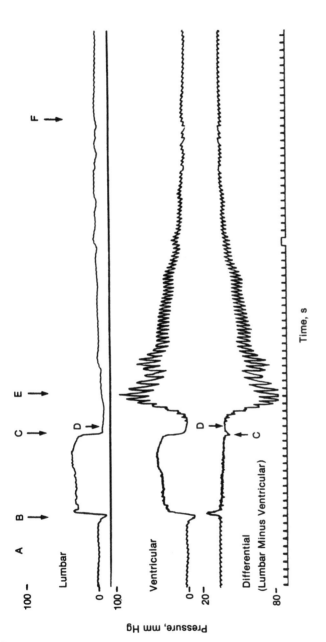

**Figure 20.6**

Craniospinal pressure dissociation associated with a marked post-Valsalva rebound. The patient shows normal lumbar and ventricular pressures at A. Notice that the ventricular arterial pulse is slightly higher. At B, the patient breathed in (causing a downward deflection in both traces) and then performed a Valsalva maneuver (caused a rise in pressure in both). Observe that the rise of pressure in the lumbar sac occurred earlier than in the ventricle. This produced an upward blip in the differential trace, indicating an obstruction to upward flow. The differential trace then moved slightly downward. At the conclusion of the Valsalva's maneuver, which lasted about 13 seconds, there was a downward blip at C, indicating impediment of flow downward, followed by a massive post-Valsalva rebound at E, signaling sudden return of blood to the heart. There was slowing of the pulse rate and increase in CSF pressure, confined to the intracranial compartment. The pressures did not return to normal until F.

343

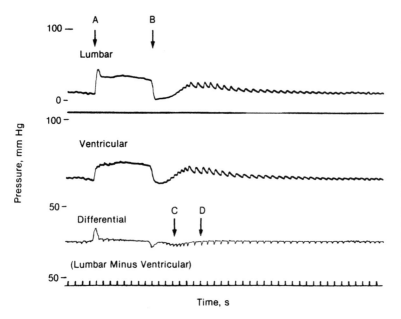

**Figure 20.7**
Same case as in Figure 20.4 after craniovertebral decompression (CVD). In this case, the tonsils were not removed and the arachnoid was left intact. The upward blip at A is much the same as before, but the post-Valsalva rebound was shared between the intracranial and intraspinal compartments, so that only a slight pressure difference developed at C, and the ventricular and lumbar pressures are again equal at D. This patient has been clinically cured.

structure. There are no tight junctions in the gliotic lining of the syrinx cavity and fluid can enter readily, as in the case of water-soluble contrast material in postmyelography computed tomography (CT) scans. Ball and Dayan have emphasized the porous state of the cord, particularly the sizable Virchow Robin spaces alongside the vessels. They suggested that the fluid is driven into the syrinx due to blockage at the foramen magnum combined with upward movement of CSF in the spinal subarachnoid spaces. Imagining the cord as being porous does not imply that it would inflate if the surrounding CSF developed raised pressure.

## Clinical Presentation

### HINDBRAIN-RELATED SYRINGOMYELIA

Patients may present with features of the HBH rather than the syrinx. These include vertigo, oscillopsia, and lower cranial nerve involvement, and are often thought of as syringobulbia, but the majority of cases with brainstem signs have no clefts, only dysfunction due to compression of tissues in the foramen magnum or traction on lower cranial nerves aggravated by CPD.

In other cases, any manifestation of spinal cord disorder may be the first symptom, including scoliosis, trophic, sensory, or motor features. Most patients develop sensory changes in the upper limbs, but it is not always *dissociated*. Stiffness in the legs is a frequent symptom, and presentation with leg symptoms alone is not uncommon. Most patients experience pain in the trunk or limbs at some time; a sudden pain in the trunk or limbs after sneezing may be associated with tidal flow within the syrinx.

Hemiatrophy, asymmetries of the face and upper limbs, wasting, Charcot's joints, and trophic changes, often with severe finger involvement, may all occur. Scoliosis is present in almost half of the patients with HRS. It is correlated with early onset and usually precedes other neurologic features. The side and level of the cord cavity are not related to the curve.

A birth problem is reported in almost half of all cases of adult HRS. Of those who have no other cause detectable, it is probable that the majority are birth-related.

## HINDBRAIN HERNIA HEADACHE

Strain-related headache is characteristic of HBH. The pain is usually nuchal and bilateral, sometimes one sided, radiating into the occiput or vertex. It is commonly brought on by rising suddenly from the lying position, coughing, straining, lifting, or shouting, and is characteristically pounding in quality. Its features are similar to those that occur in HBH due to tumor.

## NON–HINDBRAIN-RELATED SYRINGOMYELIA

In patients with complete traumatic paraplegia, symptoms tend to ascend above the level of injury. Pain, dissociated loss, numbness, loss of tendon reflexes, anisocoria, weakness, and sweating disturbances are common. Below the level of injury, there may be sweating changes, improvement in spasms, and worsening of automatic bladder and bowel control. In incomplete paraplegics, too, there may be ascending symptoms, but the significance of neurologic features below the injury is more difficult to assess because deterioration may occur—due to collagenosis, gliosis, or fracture instability at the site of injury—without syringomyelia. In patients with meningeal fibrosis without bony damage to the spine, neurologic deterioration is often due to the meningeal fibrosis itself or to gliosis rather than to syringomyelia.

# Radiologic Assessment

## PLAIN RADIOGRAPHS

In HRS, plain radiographs of the skull and spine may show effects of hydrocephalus, platybasia, or basilar invagination. There may be segmentation abnormalities with fusion of the occiput and atlas or fusions of cervical vertebrae; encephalocele and spina bifida are uncommonly seen. Enlargement of the vertebral canal is frequent, but also may be due to intraspinal tumors without syringomyelia or to neurofibromatosis.

## COMPUTED TOMOGRAPHY

Cerebral ventricle size should be checked, even in patients with a spinal presentation. Patients with walking difficulty, for example, may have hydrocephalus. In approximately one third of cases of syringomyelia there is some ventricular enlargement, and in 10% or so there may be symptomatic hydrocephalus. Intracranial subarachnoid pouches show well on CT. Midbrain and posterior fossa tumors, and especially foramen magnum tumors, may cause syringomyelia.

## WATER-SOLUBLE MYELOGRAPHY

Meningeal fibrosis may be better revealed by water-soluble myelography than by magnetic resonance imaging (MRI) (Fig. 20.8). The radiologist should look for a temporary hold-up either at the hindbrain or in the spinal canal. The

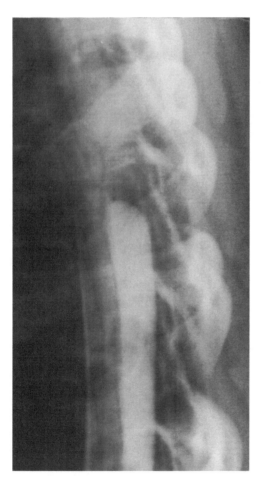

**Figure 20.8**
Water-soluble myelography in the same case as shown in Figure 20.4. An arachnoid condensation (meningeal fibrosis) is shown by a hold-up of contrast opposite T7. This web was not visible on MRI.

flexibility of the cord and its susceptibility to the effects of tidal flow are readily demonstrated.

Postmyelography CT scanning shows hindbrain herniation well, and also deformities associated with syringobulbia. It is best performed at 1 hour, at 4 to 6 hours, and then again at about 24 hours after myelography. The cavity within the spinal cord will often opacify clearly, at 6 hours, but occasionally only after 24 hours.

These investigations have largely been supplanted by MRI, but patients with severe claustrophobia, electronic implants, and severe kyphosis who cannot tolerate MRI, may require myelography and follow-on CT.

## MAGNETIC RESONANCE IMAGING

Magnetic resonance imaging shows the ventricles, the hindbrain descent, and the configuration of the spinal cord, and is invaluable for use in outpatients. It seems likely that improvements in MRI technology will show movement of CSF increasingly clearly and so increase understanding of the pathogenesis of syringomyelia as well as clarify indications for operation.

# Operation

## INDICATIONS

### Hindbrain-Related Syringomyelia

The younger the patients, the better they do; early operation is best even for asymptomatic patients. Because the progression of syringomyelia is often slow, patients may become accustomed to their disabilities and not want an operation. It is necessary not to minimize difficulties. Some cases are self-cured: MRI showing tonsils that are no longer impacted together with collapsed syrinx cavities. If the syrinx is small and occluded in parts along its length, there is no hydrocephalus, and the HBH seems disimpacted, then not even recent neurologic deterioration is an indication for operation. This outcome, however, is rare and not to be anticipated.

Sizable hydrocephalus with neurologic symptoms is an indication for ventriculo-extrathecal shunting. Objective improvement of syrinx size often may be achieved. Ventricular shunting reduces the overall volume of CSF and pre-distends the cerebral veins. Thus, when the patient coughs or sneezes, the veins are already almost full; they have a lessened capacity for further distension, and CSF movement is therefore much reduced (Fig. 20.9).

Hydrocephalus or a communication from the fourth ventricle to the syrinx favors shunting, and measurement of CSF conductance to outflow may predict cases that will benefit from this, being a more thorough test than even a protracted recording of the patient's intracranial pressure (ICP) or intraspinal

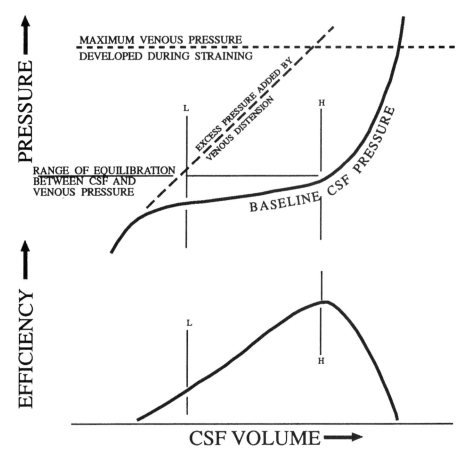

**Figure 20.9**
A diagram to show the efficiency of mechanisms inflicting pulsation on the CSF. The top curve is the familiar CSF volume/pressure curve. The zone between L and H marks the region of normal pressure equilibration between the CSF and the veins. L marks the point at which all of the veins are full, and H marks the level at which they are flat. The diminution in efficiency to the right of H is not so much limited by the maximum pressure developed during straining as by the patients' becoming ill.

pressure. Neither hydrocephalus nor a communication from the fourth ventricle to the syrinx is necessary for benefits to result from ventriculo-extrathecal shunting, but the likelihood of hindbrain impaction is a contraindication to shunting CSF from the lumbar sac.

Indications for operation on hindbrain abnormalities include syringomyelia and HBH features such as headache. Operation should be simple. When it includes HBH decompression, plugging of the central canal at the obex, a drainage tube from the fourth ventricle, a dural graft, and a syringostomy, it is difficult to know to what the surgical result—good, bad, or mixed—is attributable, nor, if there is a postoperative problem, what should be done.

## Non–Hindbrain-Related Syringomyelia

In posttraumatic cases of non-HRS, the indications may be difficult. If the syrinxes are less than two segments in length, if they extend predominantly downward below the level of the fracture, or if the patients are partial paraplegics with no progressive features, then it may be best simply to follow their progress with frequent scanning.

## Transpharyngeal Removal of the Odontoid Peg

When there is compression of the pons anteriorly, then transpharyngeal removal of the odontoid peg has been recommended. However, interference with the odontoid articulation requires a later posterior approach to attain fusion, and an initial posterior approach gives a better decompression, affords the opportunity to deal with the tonsils, and so corrects the CSF pathway abnormalities more effectively.

## When Should a Hindbrain Hernia be Left Alone?

The minimum degree of descent of the hindbrain that can cause symptoms is difficult to determine; at times, one can look at the shape of the herniation and decide whether it seems likely to be responsible for CPD. Barkovitch et al. suggested that 5 mm of descent below the foramen magnum may be normal, but minor degrees of tonsillar herniation may be symptomatic if they are shaped so as to cause CPD and if there is no cisterna magna (see Fig. 20.1), especially if there is adhesive meningeal fibrosis present.

## Which Operation Is Best for Patients with Hindbrain Hernia?

The best operation is craniovertebral decompression (CVD). Hydrocephalus or raised ICP invite a valved ventriculoatrial or ventriculoperitoneal shunt before the hindbrain is decompressed. This may be done at the same time as a CVD, but the effects of the individual procedures are then more difficult to disentangle. There are often problems in deciding whether ventricular enlargement is relevant. A suggested decision tree is given as Figure 20.10. There are risks involved in lowering the CSF pathway pressure by, for example, thecoperitoneal shunting. If this is done, even in patients with no preexisting herniation, then HBH can develop.

If the syrinx is big and there is a problem with the hindbrain, especially dense fibrosis of the cisterna magna, it may be tempting to shunt the syrinx first, but this is dangerous in the presence of CPD and is not recommended for HRS until the HBH has been corrected.

## TECHNICAL CONSIDERATIONS

### Ventricular Shunting

The techniques for ventricular shunting are the same as for hydrocephalus. The author presently favors adjustable Medos valves (Medos, SA, Le Locle,

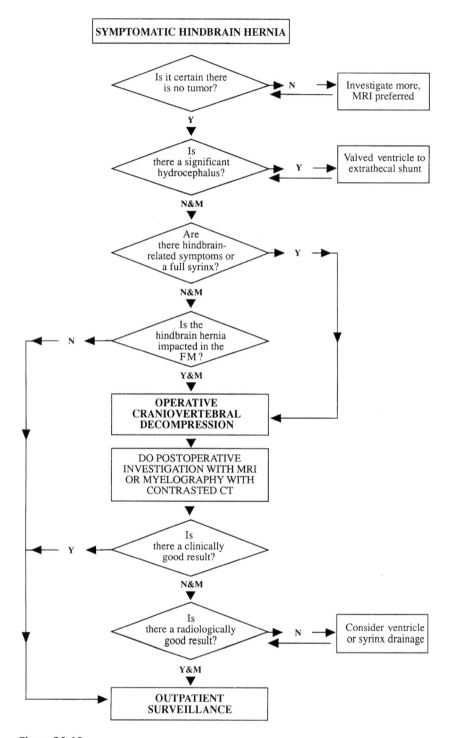

**Figure 20.10**

A decision tree for hindbrain hernia surgery. N & M, no and maybe; FM, foramen magnum; Y & M, yes and maybe.

Switzerland), which do not degrade MRI of the hindbrain and the syrinx unacceptably, but which require checking after MRI. The objective is to create the lowest tolerable pressure without producing slit ventricles or intracranial hemorrhage.

## Hindbrain Decompression

The objective of hindbrain decompression is to provide a communication between all the normal CSF-containing spaces; the cisterna magna should be in communication with the subarachnoid spaces around the cord, with the fourth ventricle and also with the spaces alongside the upper medulla leading up to the cerebello-pontine angles (Fig. 20.11). The damping effect of the decompression prevents a sudden wave of increased ICP; the energy is dispersed in moving the occipital musculature or lifting the cerebellum, neither of which does any harm, instead of being imposed on CNS structures.

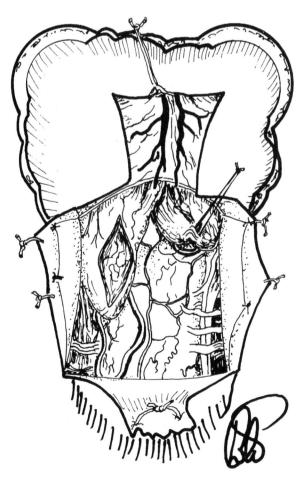

**Figure 20.11**

Craniovertebral decompression. Observe that the dura may be preserved at the top of the exposure. This will help to hold the cerebellum in position. Nevertheless, there must be good opening up of the CSF spaces below, and this is obtained by incising the tonsil, as shown on the left, and removing parenchymal tissue by suction and bipolar cautery. The partly empty pial shell may then be sutured upward and laterally, as shown on the right, opening up the spaces alongside the vertebral artery and the eleventh nerve. The foramen of Magendie should be opened if possible.

Blockage of a presumed communication between the fourth ventricle and the central canal of the cord is not recommended. The plug is likely to become dislodged by CSF movement, and there is a complication rate attached to fixing of tissue into the region of the obex. More importantly, such a putative communication is irrelevant.

Too small a decompression may cause impaction of tonsils and postoperative respiratory arrest. The use of too large a decompression in the occipital bone may lead to the lower parts of the cerebellum prolapsing into the bony decompression, where they reimpact, so that CPD persists. A very large decompression, however, has been recommended on the basis that CPD may thereby be more surely abolished, and good results have been claimed.

*Exposure*    A vertical midline incision exposes the occiput and posterior lamina of the atlas. Fat and some deep muscle close to the midline may be removed to help provide an artificial cisterna magna. Removal of about 2–3 cm of occipital bone from side to side, and a little more from the foramen magnum upwards, gives a good decompression without provoking slump. The posterior arch of the atlas need not be removed if the tonsils can be adequately exposed with the atlas in place. It is usually possible to leave the spinous process of the axis in position.

The incision in the dura may terminate a little way inside the craniectomy so that it continues to support the cerebellum, but at the lower edge of the wound, a tongue of dura should be hinged downward and stitched to the spinous process of the axis. The lateral is sutured to the suboccipital muscles.

*Dealing with the Arachnoid*    Shreds of arachnoid left to float about in the CSF tend to adhere to whatever they can, and then to proliferate. Arachnoid should be stitched to dura at the sides of the decompression to ensure permanent communication between the spinal spaces and the artificial cisterna magna (see Fig. 20.11). Leaving the arachnoid intact is occasionally advocated, but correction of CPD is probably less effective if this is done. It is important not to leave arachnoid with a small hole in it; CSF will collect between the arachnoid and the inside of the artificial cisterna magna, creating a pouch that presses against the back of the tonsils and forms dense adhesions.

*Dealing with the Tonsils*    The cerebellar tonsils form the valve that maintains CPD, causing downward traction, compressing the brainstem, and sometimes obstructing the outlet of the fourth ventricle. The more radical the clearance of the tonsils, the more certainly are these factors dealt with.

In the majority of cases, the tonsils are not bound down and they may be picked up, sometimes after division of a few strands of arachnoid, to expose the obex. It is tempting to remove either the largest tonsil or both; their removal does not result in any morbidity. A vertical incision into the tonsil allows it to be removed by suction. A suture may then be passed once or twice through the pia of the medial wall of the tonsil and attached to the pericranium or dura to pull the tonsillar remnants upwards, backwards, and laterally. This results in clearance of the lateral CSF pathways.

The tonsils may, through necessity, be left alone. If they are bound together and to the back of the medulla with arachnoid adhesions, then dissection may imperil, for example, branches of the posterior inferior cerebellar arteries. If dense adhesions are present, the cerebellum is not likely to prolapse, and the benefits of the operation are not thereby negated.

*Closure* The dura should be left open. In the author's experience, grafts have many complications: All that is required is that the suboccipital muscles are closed in a water-tight manner away from the brain. If a CSF leak develops, or the decompression bulges, then the patient may need a valved shunt, but by paying attention to hydrocephalus before CVD in the manner discussed earlier, the author has had no CSF leaks in 387 posterior fossa decompressions.

Aseptic meningitis may occur, but there is no evidence that it is less common when a graft is used. A few lumbar punctures and a brief course of corticoids are better than a recurrence of fibrosis at the foramen magnum.

## Operation for Primarily Spinal Syringomyelia
If there is a small zone of meningeal fibrosis, such as occurs following a fracture, then decompressive laminectomy and opening up of the subarachnoid space above and below is recommended. The spaces may be made to communicate with each other via a surgical meningocele, analogous to the artificial cisterna magna (Fig. 20.12). Preserving the pathways from the spinal subarachnoid space to the meningocele may be aided by the use of stents, which are removed a few days later (Fig. 20.13). Details of surgical technique are given elsewhere. If there is hydrocephalus or extensive and undissectable meningeal fibrosis, then CSF shunting from the ventricle, the basal cisterns, or the lumbar sac is recommended.

## Syrinx Drainage
The natural history of drains is that they become blocked, because if a drain works, it flattens the walls of the cord around it. No variety of syrinx drainage will address the problems of hindbrain compression; this is justified only if CVD leaves the syrinx largely full and the patient continues to have syrinx symptoms.

*Shunting from Syrinx to Subarachnoid Space* Myelotomy is occasionally successful. Syringosubarachnoid shunts have been praised, and a wide variety of technical variants have been proposed, including endoscopy of the syrinx. However, the only reason why a syrinx should empty rather than fill through

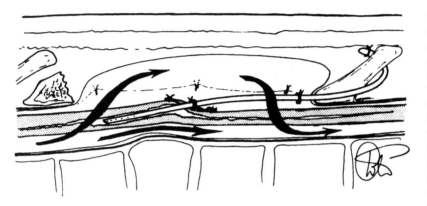

**Figure 20.12**
Diagram to show the idealized situation after formation of a surgical meningocele. The drain is not necessary. Important features are the opening from the subarachnoid space into both ends of the artificial meningocele. Under these circumstances, the equalization of pressure, analogous to that seen after decompression of the hindbrain, appears to afford protection against the perpetuation of syringomyelia.

a syringostomy is that there is some residual elasticity in its wall. The majority of syrinxes that come to operation are too big to empty in this way.

Excision of the tip of the spinal cord, called by Gardner "terminal ventriculostomy," should be considered only when MRI shows a well-filled syrinx extending into the conus.

*Syringopleural Shunting*    Drainage to low-pressure areas outside the theca has been used, to both peritoneum and pleura. The pleura is convenient; there is no need to change the operating position; the pressure is low; and the absence of omentum lessens the risk of susequent occlusion. Use of a valve is an unnecessary complication, but care should be taken to close the pathways through which lumbar CSF from outside the syrinx might enter the pleura. If this is not done, the hindbrain may become impacted. In any event, it seems likely that such a drain will work for only a few days before the tissues obliterate the end of the catheter.

## Outcome

### COMPLICATIONS

#### Nonneurologic Complications
*Respiratory Problems*    Respiratory problems include sudden death due to sleep apnea, so an apnea alarm or anoxia monitor is recommended. Additional

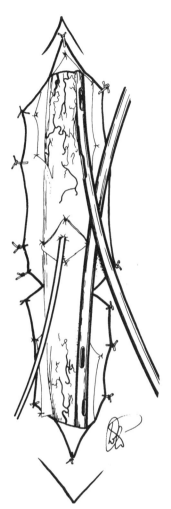

**Figure 20.13**
Non-HRS with a localized zone of meningeal fibrosis, such as is produced by a fracture, may be treated by suturing the dura back, as shown. The arachnoid also should be carefully sutured back to open up the subarachnoid spaces. If a temporary drain is used, as shown on the left side, it must not allow fluid from the subarachnoid spaces to drain to either pleura or peritoneum. For this reason, a patch is shown. The author's practice, however, is to dispense with such drainage and to rely on stents.

problems arise from impaired lower cranial nerve function in many cases, sometimes with the risk of inhalation. Careful postoperative observation is required, with physiotherapy and vigilant trials of drinking clear water before thick fluids or solids are tried. Tracheostomy for bronchial lavage may be necessary.

*Aseptic Meningitis*    A delayed meningitic response may occur occasionally, presenting with pain in the lower back and legs. Lumbar puncture will confirm that no organisms are present, and may be repeated if the pressure is high and to help to clear proteinous CSF; corticosteroids also speed resolution.

## Neurologic Deterioration

*Inadequate Decompression*    Inadequate decompression may be followed by reimpaction of the hindbrain, leading to progressive neurologic deficit and respiratory arrest. Infarcts of the lower parts of the cerebellum may occur due to vascular damage. Shunting and tracheostomy may be helpful. Reoperation rarely should be required.

*Hydrocephalus*    Postoperative hydrocephalus due to reimpaction at the foramen magnum, and so re-establishing CPD, can lead to the development of fourth ventricular clefts and clinical features of syringobulbia. Early recourse should be made to valved ventriculoperitoneal shunting.

*Prolapse of the Cerebellum*    Over-decompression of the hindbrain, combined with lack of support of the cerebellum, may lead to the cerebellum slumping into the decompression, especially if CPD is not eliminated. The most common symptom is that of headache. Hitching up what remains of the tonsils may help to prevent this, but the size of decompression is probably primary.

*Persistent Tension in the Syrinx*    A syrinx that has not collapsed poses problems. If the hindbrain is still causing CPD, it is unlikely that success will follow a second attempt at dissection. A ventriculo-extrathecal shunt is the safest next step, or a syringopleural drain may be used. In the majority of cases, the syrinx is still visible, but relatively slack; any attempt to insert a drainage tube may encounter haustra and wall-to-wall adhesions and be technically difficult.

## THERAPEUTIC RESULTS

There are few reports on outcome over satisfactory periods of time. Chronic neurologic deficits are difficult to record objectively, placebo effects of operation are powerful, and the natural history of even untreated syringomyelia may be long and uneventful.

The results of HBH decompression are often good. The symptoms that resolve well are those of the long tracts, pain (especially neck pain and headache), and bulbar features. Signs related to the cavitation of the cord are less favorably influenced; loss of tendon reflexes, muscle wasting, and sensory loss seldom improve. Morgan and Williams, reviewing 54 patients with bulbar or hindbrain compression features with follow-up of over 3 years, found that 31 (57%) of the patients graded themselves as having "great improvement," 14 (26%) claimed "some improvement," 5 (9%) claimed to be unchanged, and 4 (7%) thought that they were worse.

Changes in the MRI scans may provide the best objective measure of the lower border of the tonsils, the opening up of the midline fourth ventricular drainage canal, the avoidance of slump, and improvement in the appearance of the syrinx. Duddy and Williams analyzed 17 patients with hindbrain decompression, 11 of whom had syringomyelia. They found that downward movement of the hindbrain was usual. The cerebellum tended to descend more than the brainstem: Five of their patients had a descent of more than 6 mm, as measured from the fastigium, but of only 4 mm as measured from the pons. All 17 patients showed reduction in either the width or the length of the syrinx, usually both.

Prolonged outpatient surveillance commonly discloses a progression of symptoms over many years. Re-investigation to establish the state of the ventricles is often rewarding, and MRI is helpful, although the appearances may be inconclusive. It is sometimes tempting, on reviewing a patient with a persisting cord cavity and tonsils still jammed in an inadequately decompressed foramen magnum, to recommence hindbrain surgery, even years after an earlier attempt. The clinical progression, however, may be no greater than in patients with a surgically perfect result, and patients may restabilize for long periods, so enthusiasm needs to be checked.

## COUNSELLING AND SUPPORT

Patients with syringomyelia of all kinds have to anticipate the problems of chronic and slowly progressive neurologic disease, even if operation produced improvement. They may need psychological and social support. A self-help group may be helpful for some patients.*

# Future Developments

The problems of syringomyelia would be better prevented than cured; early diagnosis is best. Children with idiopathic scoliosis and all forms of dysraphism constitute a high-risk group, as do patients with complete or incomplete post-traumatic paraplegia. Magnetic resonance imaging is the investigation of choice and may show unexpected HRS.

Improvement in the quality of MRIs, leading to the ability to assess the movement of CSF and thus the energy that it imposes on vulnerable structures, may be of value in the future and may help to clarify indications for operation as well as elucidate the mystery of the filling mechanism.

The discovery that in HRS there is an association with difficult birth, suggests that screening of infants at risk from birth trauma and early surgery could almost eliminate HRS.

*Syringomyelia Patient Support Groups: Ann's Neurological Trust Society (ANTS), c/o Anne Lane, Midland Centre for Neurosurgery and Neurology, Holly Lane, Warley, West Midlands B67 7JX; American Syringomyelia Alliance Project, PO Box 1586, Texas 75606-1586, USA; and Irish ANTS, c/o Gena Scott, Castleblayney, Ireland.

## Acknowledgment

This work has been supported financially by ANTS.

## References/Further Reading

Abe T, Nakamura N, Tashibu K, et al. Role of birth injury in syringomyelia. In: Nakamura N, Hashimoto T, Yasue M, eds. Recent advances in neurotraumatology. Tokyo: Springer Verlag, 1993:436–493.

Backe HA, Betz RR, Mezgarzadeh M, et al. Post traumatic spinal cord cysts evaluated by magnetic resonance imaging. Paraplegia 1991;29:607–612.

Ball MJ, Dayan AD. Pathogenesis of syringomyelia. Lancet 1972;ii:799–801.

Barkovitch AJ, Sherman JL, Citrin CM, Wippold FJ. MR of postoperative syringomyelia. Am J Neuroradiol 1987;8:319–327.

Batzdorf U. Syringomyelia related to abnormalities at the level of the cranio-vertebral junction. In: Batzdorf U, ed. Syringomyelia: current concepts in diagnosis and treatment. Baltimore: Williams & Wilkins, 1991:163–182.

Duddy M, Williams B. Hindbrain migration after decompression for hindbrain hernia: a quantitative assessment using MRI. Br J Neurosurg 1991;5:129–140.

Dyste GN, Menezes AH. Presentation and management of paediatric Chiari malformations without myelodyspasia. Neurosurgery 1988;23:589–597.

Gardner WJ. The dysraphic states: from syringomyelia to anencephaly. Amsterdam: Excerpta Medica, 1973.

Kruse A, Rasmussen G, Borgesen SE. CSF dynamics in syringomyelia: intra-cranial pressure and resistance to outflow. Br J Neurosurg 1987;1:477–484.

Morgan DW, Williams B. Syringobulbia: a surgical reappraisal. J Neurol Neurosurg Psychiat 1992;55:1132–1141.

Sahuquillo J, Rubio E, Poca MA, et al. Posterior fossa reconstruction: a surgical technique for the treatment of Chiari 1 malformation and Chiari 1 syringomyelia complex—preliminary results and quantitative MRI of hindbrain migration. Neurosurgery 1994;35:874–885.

Sgouros S, Williams B. A critical appraisal of drainage for syringomyelia. J Neurosurg 1995 (in press).

Williams B. Surgical management of non-hindbrain-related and post-traumatic syringomyelia. In: Schmidek HH, Sweet WH, eds. Operative neurosurgical techniques: indications, methods, and results. 3rd ed. Philadelphia: Saunders, 1995:2119–2138.

Williams B. Surgery for hindbrain related syringomyelia. Ad Tech Stan Neurosurgery 1993;20:107–164.

Williams B, Fahy G. A critical appraisal of "terminal ventriculostomy" for the treatment of syringomyelia. J Neurosurg 1983;58:188–197.

Williams B. Simultaneous cerebral and spinal fluid pressure recordings: I. Technique, physiology and normal results. Acta Neurochir (Wien) 1981;58:167–185.

Williams B. Simultaneous cerebral and spinal fluid pressure recordings: II. Cerebrospinal dissociation with lesions at the foramen magnum. 1981;59:123–142.

Williams B. Subarachnoid pouches of the posterior fossa with syringomyelia. Acta Neurochir (Wien) 1979;47:187–217.

Williams B. Difficult labour as a cause of communicating syringomyelia. Lancet 1977;ii:51–53.

# VII NEURO-ONCOLOGY

# Malignant Tumors of the Nervous System

## Biologic Considerations and Treatment Strategies

William R. Shapiro

Joan Rankin Shapiro

Primary tumors of the CNS are the second most common cancer in children, and in adults are more common than systemic Hodgkin's disease. Traditional treatment has included biopsy and surgical resection, when possible, followed by radiation therapy. Chemotherapy is increasingly offered to patients with malignant tumors. Recent developments in cellular and genetic molecular biology have fostered a new scientific interest in brain tumors; several new discoveries about the biology of brain tumor growth have resulted. Further research has been aided by new techniques of diagnosis and therapy, including magnetic resonance imaging (MRI), positron emission tomography (PET), stereotactic laser surgery, interstitial radiotherapy, and stereotactic radiosurgery.

## Etiology of Neuroectodermal Tumors

The specific etiology of neuroectodermal tumors remains unknown, but like that of cancer in general, is thought to be related to altered genetic information, caused either directly by mutation or through epigenetic events such as defects in mitosis that produce segregational errors. There has been concern about the etiologic role of electromagnetic waves in the production of brain tumors; cranial radiation for tinea capitis has been associated with the later development of brain tumors. Both RNA and DNA viruses can induce animal brain tumors, but few viruses have been found to account for specific human tumors, HIV being an exception. A study of children with medulloblastoma suggested a possible relationship between the incidence of this disease and prior exposure to polio vaccine contaminated with $SV_{40}$ virus. Human brain tumors may be inherited as part of several familial diseases, notably von Hippel-Lindau syndrome, tuberous sclerosis, and von Recklinghausen's neurofibromatosis.

## Epidemiology of Brain Tumors

The median age-adjusted incidence rate for primary brain tumors is between 4 and 5 cases per 100,000 per year. Brain tumors are more common in men than in women, the notable exception being meningiomas. Gliomas account for about half of primary intracranial tumors. Glioblastoma multiforme (GBM) is the most common tumor, followed equally by meningioma and astrocytoma. A general age-specific incidence is found, with a small peak occurring in childhood and a higher peak occurring later, reaching a maximum between 60 and 80 years of age. The incidence of gliomas may be increasing in the elderly, especially GBM, but some of this apparent increase may be due to newer imaging techniques.

## Classification and Pathology of Neuroectodermal Tumors

Historically, Kernohan and Sayre graded astrocytomas according to their cellular morphology, from grade I, the slowest growing tumor, to grade IV, the highly malignant GBM: A three-tiered system defined by Burger and Vogel placed well-differentiated astrocytomas at one end of the spectrum of malignancy, glioblastomas at the other end, and anaplastic astrocytomas in the middle; this system is favored by neuropathologists working with cooperative brain tumor groups (Radiation Therapy Oncology Group, Eastern Cooperative Oncology Group, Brain Tumor Study/Cooperative Group [BTSG], Southwest Oncology Group), because it more accurately correlates with prognosis.

One problem inherent in all pathologic grading systems is the fact that astrocytomas are quite heterogeneous. Generally, tumor grades are underestimated by needle biopsy, but this error rate can be minimized by multiple samples. A classification proposed by Daumas-Duport and her co-workers lends itself well to distinguishing between especially astrocytoma and anaplastic astrocytoma, and correlates well with prognosis: Histopathologic criteria include the presence or absence of nuclear atypia, mitoses, endothelial proliferation, and necrosis. Grading is based on the occurrence of these features in the tumor: grade 1: none present; grade 2: one feature present; grade 3: two features present; grade 4: three or four features present. The system was reproducible in predicting survival outcome in a large series of astrocytomas.

A grading system for oligodendrogliomas has been developed, which attempts to distinguish the relatively rare malignant tumor from the slower-growing forms. Recent criticisms of these systems are based on the lack of importance of necrosis in prognosis and the frequency of mixed astrocytic/oligodendrogliotic tumors.

## Biology of Gliomas

Brain tumors differ from systemic neoplasms, in that much smaller tumor mass may have devastating or lethal effects on the patient. In addition to size, other important factors are location, rate of growth, and histologic grade. Histologically benign intracranial tumors (e.g., meningiomas) are slow-growing, have few mitoses, no necrosis, and no vascular proliferation. They may achieve considerable size before producing symptoms, in part because there is often no associated cerebral edema. Malignant tumors are characterized by more rapid growth, invasiveness, frequent mitotic figures, vascular proliferation, endothelial hyperplasia, and necrosis. However, any brain tumor that cannot be entirely excised because of size or location usually proves lethal. "Malignant" brain tumors rarely metastasize out of the CNS, but instead cause death by inexorable local growth. Thus, the distinction between benign and malignant tumors is less important for intracranial tumors than for systemic cancer.

## Cellular Characteristics

### CELLULAR HETEROGENEITY

Gliomas are heterogeneous in their cellular content. Karyotypically, the chromosomal complement of each cell type ranges from near diploid (2n) to hypo- or hypertetraploid (4n) in chromosome number, and the distribution of cell types varies with each tumor. Phenotypic heterogeneity includes variation in cell kinetics, content of glial fibrillary acidic protein, and chemosensitivity.

New molecular biology studies are now defining how the aberrancy of oncogenes, tumor suppressor genes, growth factors, growth factor receptors, DNA repair and drug resistance genes, extracellular matrices, and adhesion molecules contribute to the heterogeneous phenotypes observed in human gliomas.

### MUTATIONS CONTRIBUTING TO MALIGNANT PROGRESSION OF HUMAN GLIOMAS

While most tumors are clonal in origin, malignant transformation is considered to be a multistep process, beginning with somatic mutations. Some of these changes cause the cells to disregard the controls that normally limit proliferation and metastasis. Genes that act in a dominant manner are generally termed *oncogenes*, while recessive genes are designated *tumor suppressor genes* (growth suppressor genes, recessive oncogenes, anti-oncogenes). Both oncogenes and tumor suppressor genes can contribute to the disparate phenotypes observed in neoplastic disease.

Central nervous system cancer most typically affects the glial cell, which is programmed to undergo cell division throughout its life cycle. Once transformation and immortalization of a glial cell is established, additional changes (genotypic and epigenetic) usually result in the evolution of this cell(s) to a more malignant state.

## Oncogenes

Oncogenes that participate in the transformation of a cell are in fact derived from normal cellular genes (proto-oncogenes). Some mutable event occurs that releases the cell's oncogenic potential either by aberrantly expressing the normal gene product (amplification) or by producing an abnormal gene product. In either circumstance, the normal expression or activity of the proto-oncogene is disrupted, and the cell is released from normal cellular control. Non-random chromosomal changes have been linked cytogenetically with specific malignancies. In addition to rearranged chromosomes, the loss and gain of a whole chromosome has provided the first clues for the map position of cellular oncogenes and tumor suppressor genes in human malignant disease, including human gliomas.

The carcinogenic potential of the dominantly acting oncogenes suggests that such genes function as key components in cellular pathways. To date, most proto-oncogenes have been found to encode genes related to some aspect of signal transduction pathways. Proto-oncogenes found to be aberrant in gliomas include growth factors (c-*sis*, *hst*/KFGF, *int-2*), tyrosine kinase receptors (*erbB*, *fms*, *neu*, *kit*, *ros*, *trk*, *ros*, *met*), nonreceptor tyrosine kinases (*src*, *yes*, *abl*), serine/threonine protein kinases (*raf*-1, *mos*), guanine nucleotide-binding proteins (Ha-*ras*, Ki-*ras*, N-*ras*), and nuclear proteins (c-*myc*, N-*myc*, *myb*, *fos*, *gli*, *p53*, *jun*, *erbA*). Each group of genes mediates signal transduction as a ligand, receptor or transducer molecule (which propagates or amplifies a signal from the cell's surface to its interior), or as a regulator of transcription.

Despite the large number of oncogenes that have been identified as aberrant in gliomas, only the *erbB* oncogene (epidermal growth factor receptor, EGFR) is amplified in an appreciable number of gliomas (approximately 80%), usually of advanced grade. If the aberrant expression of one or more of these genes is a significant event in gliomas, the genes could be used as markers of diagnosis and/or prognosis of these tumors.

## Tumor Suppressor Genes

Tumor suppressor genes most frequently involve recessive changes. These genes code for proteins thought to be involved in normal negative control mechanisms (inhibition) that regulate cellular proliferation. A mutation or deletion of a tumor suppressor gene(s) is associated with the loss of a function, and this loss in turn contributes to the breakdown of the normal constraints that function to control cell behavior.

The technique of restriction fragment-length polymorphism analysis allows one to determine if a gene is lost or mutated. The loss of the genetic material (loss of heterozygosity, LOH) is confined to tumor cells, and the genomic alterations are frequently complex. Loss of five specific tumor suppressor genes located on different autosomes is associated consistently with the progression of malignant gliomas. The most frequent abnormality is loss of genetic information from chromosome 17p. The loss or mutation of chromosome 17p occurs in astrocytomas, anaplastic astrocytomas, and glioblastomas, suggesting that this loss is an early event in the evolution of the glial malignancy. The LOH on chromosome 10 is most frequently associated with GBM, a finding that led to the postulation that the loss of this gene permits the tumor cells to evolve to a more malignant state.

## Regulatory Genes

Products of a cellular regulatory gene also can induce an aberrant phenotype. The p53 tumor suppressor gene is frequently associated with the loss of one allele in malignant gliomas, although a large number of malignant gliomas have no p53 mutations. The explanation for this appears to be a cellular protein, murine double minute 2, that can form a complex with the p53 and suppress its activity.

## Tumor Induction and Progression

The current hypothesis for glial tumor progression is that the controlling genetic events occur in a step-wise fashion similar to that described for other neoplasms. Amplification of the EGFR is observed in only intermediate and high-grade primary gliomas. Loss of heterozygosity in 9p is similarly restricted, and appears to occur most frequently in the highest grade gliomas.

## Growth Factors and Growth Factor Receptors

Transformation and tumor progression also can occur when proto-oncogenes, coding for proteins that are part of the growth stimulatory pathway or that stimulate growth regulatory pathways, are mutated. These genetic lesions appear to unmask or activate the oncogenic potential of that gene.

Five modes of growth factor action have been postulated as mechanisms involved in stimulating the diverse cellular behaviors that follow ligand receptor/binding. Sporn and Todaro suggested that all growth factor activities could be classified into three categories: *endocrine* (factor is secreted into the bloodstream to reach its target cell), *paracrine* (factor from one cell can stimulate adjacent cells), and *autocrine* (the cell produces factor to which it is itself responsive). A fourth category has been described, *juxtacrine*, in which a membrane-bound growth factor of one cell stimulates the cognate receptor on an adjacent cell. Lastly, the *intracrine* mode involves an internal loop in which the ligand activates the receptor within intracellular compartments. Irrespective of the mode of interaction of ligand/receptor, it has been recognized that high-grade malignant glioma cells grown in vitro require few growth factors in order to proliferate, despite their often rapid growth rates, indicating the presence of intrinsic mechanisms of growth promotion.

*Epidermal Growth Factor Receptor*    Epidermal growth factor is observed in small amounts in the adult human brain. It acts as a mitogen in glial cells and has the ability to make cells migrate. Epidermal growth factor receptor is a 170-kd membrane receptor with three domains: an extracellular binding domain that binds EGF and transforms growth factor-α (TGF-α), a transmembrane domain, and an intracellular tyrosine kinase domain.

Amplification of EGFR is well documented in malignant gliomas. Alterations in specific regions of its gene have been described. These alterations are clustered within three regions of the EGFR, one being on the amino terminal of the extracellular domain and two within the intracellular domain. The actual role of the EGFR amplification in malignant progression is unknown, although this truncated receptor presents a unique tumor antigen at the cell surface. Such tumor antigens can provide targets for antibody-related therapy.

*Transforming Growth Factors*    The transforming growth factors (TGF-α and TGF-β) are part of a family of EGF-like molecules. When processed, TGF-α binds to EGFR. In normal cells, TGF-α is observed only during development. Numerous tumors secrete this growth factor, however, including high-grade malignant gliomas. Because amplification of the EGFR occurs in approximately 50% of human gliomas, it is thought that the TGF-α produced by these tumors acts to stimulate an autocrine growth pathway but does not participate directly in tumorigenicity.

While TGF-α has been associated with proliferation, another member of this gene family, TGF-β, has been found to inhibit certain immune reactions, and such inhibition may influence tumor growth. There are two forms of TGF-β (TGF-β1 and TGF-β2), and both have been identified in human gliomas. The type 2 molecule is closely related to the polypeptide described as glioblastoma cell-derived T-cell suppressor factor (G-TsF) because of its immunosuppressive activity. Human gliomas secrete this factor; the factor can significantly reduce the cytotoxic properties of tumor-infiltrating lymphocytes. In addition, TGF-β depresses the activity of natural killer cells. Both cytotoxic T-cell and natural killer cell activities are thought to be important in destroying tumor cells. Therefore, it is postulated that the secretion of TGF-β in gliomas could allow this tumor to grow out of control by inhibiting cells that would kill it.

*Platelet-Derived Growth Factor*    Platelet-derived growth factor (PDGF) is a stimulant to a number of connective tissue cells, glia, and glioma cells. Three bioactive forms of PDGF have been described: AA, AB, BB. Each of these isoforms is thought to have a different biologic activity. The receptor for PDGF has a cytoplasmic tyrosine kinase domain that is responsible for the phosphorylation of cellular proteins. The PDGF receptor has an extracellular ligand-binding portion containing five immunoglobulin-like binding domains. The binding of any one of the three isoforms of PDGF produces a ligand-receptor complex that is internalized (down-regulation). This signals the phos-phorylation process, which involves a series of second messenger molecules, that ultimately stimulates proliferation. Several studies demonstrate that there are different populations of PDGF receptors. There are two receptors—α-PDGF and β-PDGF receptors—that bind differentially the three isoforms of PDGF. The PDGF receptors appear to have different functions. The α-receptor mediates the actions of PDGF on cellular metabolism, while both α-PDGF and β-PDGF receptors initiate mitogenic action.

The PDGF receptors have been mapped to a multigene complex that includes the *c-kit* proto-oncogene on chromosome 4. The viral homologue, *v-kit*, is the feline transforming virus in the cat. Thus, mutations occurring in this multigene complex could be involved with the pathogenesis of CNS malignancies despite normal function of the PDGF isoforms.

Platelet-derived growth factor (ligand and receptor) function is an im-portant component in the repair of vascular injury, cellular differentiation, and chemotaxis. The inappropriate expression of this growth factor and/or its receptors has been associated with human diseases such as atherosclerosis and cancer. The working hypothesis states that the aberrant expression of PDGF (ligand and/or receptor) provides cells with a selective growth advantage. In gliomas, aberrant expression of the PDGF ligand and receptors has been found both in vitro and in vivo. Several reports indicate that there are differences in the expression of the PDGF AA, BB, and AB chains and the α-, β-receptors in glioma tissue; the most malignant tissues generally express the highest concentration of PDGF BB form. Each of the PDGF isoforms possesses the ability to stimulate glioma cells, reactive astrocytes, and endothelial cells, in an autocrine and paracrine manner both in vitro and in vivo. Genetic aberra-tions in any or all of these ligands and receptors, along with the possibility of an intracrine loop, may be key factors in the pathogenesis of gliomas.

*Fibroblast Growth Factor*    Fibroblast growth factor (FGF) is another mitogen produced by primary astrocyte cultures and malignant gliomas. This mitogen has an acidic (aFGF) and a basic (bFGF) isoform. The genes for these isoforms share extensive homology in sequence and function, but bFGF is 30 to 100 times more active than aFGF. Unlike EGFR and PDGF receptors, FGF receptors do not act through protein kinases. Although a secretory autocrine loop has not been described in gliomas, Libermann and co-workers proposed that an autocrine loop exists in glioma cells and is a factor in the transformation process. Further, the tumor angiogenesis factor isolated from gliomas has been found to be aFGF and bFGF. The release of these growth factors by glioma cells implicates FGF as a factor contributing to neovascularization in these tumors, as FGF is a potent mitogen for endothelial cells.

The interrelationship of FGF and other growth factors may be important in tumor progression. For example, FGF mRNA can be stimulated by PDGF.

Thus, the interaction between growth factors like FGF and PDGF may synergistically contribute to the proliferative potential of several cell types. Fibroblast growth factors also may be oncoproteins; aFGF and bFGF have extensive sequence homology with c-*hst* and c-*int-2* proto-oncogenes, respectively. The viral counterparts of these proto-oncogenes all exhibit transforming potential.

*Insulin-Like Growth Factor*     There are two forms of IGF, IGF-I and IGF-II. The IGFs are polypeptides that resemble proinsulin in structure and are capable of stimulating cellular proliferation in many tissues. They have been identified in a variety of tumors, including gliomas, and function as autocrine simulators of cell growth. Both normal and glioma tissues transcribe mRNA for both IGF-I and IGF-II, suggesting that the ligands and/or receptors have a function in normal growth and maturation of the brain, which if aberrant, could be involved in the pathogenesis of gliomas.

The IGF polypeptides differ from most other growth factors in that they bind to specific binding proteins, which can modulate the effect of these growth factors on target tissue. In tumors, a variety of binding proteins have been identified, including those observed in normal cells. The proteins are thought to modulate the effects of these growth factors, although the exact role in development and tumor progression has not been defined. Inappropriate induction of an autocrine mechanism could contribute to the neoplastic process; glioma cells show elevated production of IGFs and IGF receptors.

In summary, gliomas like other neoplasms, are characterized by loss of cellular growth control, occurring as a result of aberrancy of specific growth factors, oncogenes, and tumor suppressor genes. Inactivation or over-expression of such gene(s) could be sufficient to disrupt the function of that pathway, removing it as a regulator in normal cell growth.

## The Blood-Brain Barrier in Intracranial Tumors

The term *blood-brain barrier* (BBB) defines the relatively restricted transport between blood and the CNS of water-soluble, ionized molecules larger than about 200 daltons. The anatomic substrate of the BBB resides in the endothelial cells of brain capillaries, which differ from capillaries in the rest of the body in being joined together by tight intercellular junctions and being devoid of fenestrae. The BBB presents a problem in pharmacologic therapy of CNS disease, because many drugs do not cross the barrier in sufficient quantity to be effective. The BBB almost entirely excludes entry of large molecules like proteins, and limits entry of smaller molecules because the ease with which these cross the lipid-bilayer of the plasma membranes of the cells varies according to their physical–chemical properties. Lipid-soluble molecules readily enter the brain; examples are nicotine, ethanol, heroin, and the chemotherapeutic agent, BCNU. Water-soluble molecules cross the barriers very poorly. Molecules like glucose enter the brain by facilitated transport through the endothelial cells using a glucose transporter; amino acids must utilize active transport systems.

In brain tumors, the BBB is substantially altered; tight endothelial cell junctions are disrupted, fenestrations appear within the vessels, and there is an increase in pinocytotic vesicles which permit transport of macromolecules across the barrier. The BBB is not completely broken in a brain tumor, and the question remains quantitative as to the physiologic nature of the blood-tumor barrier.

One immediate consequence of the alteration in the BBB induced by brain tumors is the production of cerebral edema. The edema is confined to the cerebral white matter and comes from an increase in permeability of the tumor's capillary endothelial cells so that plasma enters the extracellular space. Recent observations suggest that brain tumor cells make a material that can open the BBB, permitting fluid to escape as "edema." This vascular *permeability factor* has precedent in animal models as well as other human cancers.

## Therapy of Malignant Glioma

Multimodality therapy for malignant glioma consists of cytoreductive surgery, radiation therapy, and chemotherapy. Each of the modalities are reviewed here, with concentration on recent results and controversies.

### PROGNOSTIC VARIABLES

Before reviewing therapy, note should be taken of certain clinical characteristics that relate to a patient's prognosis—the so-called prognostic variables—which often influence the outcome more than the treatment. Young patients live significantly longer than elderly patients. Patients with GBM survive half as long as those harboring anaplastic astrocytomas. Patients with postoperative Karnofsky ratings of 90 to 100 live much longer than those with ratings of 40 or below. Patients whose symptoms have been present longer than 6 months when they are first seen have a death rate half that of patients whose symptoms have lasted less than 6 months. A long duration of symptoms correlates with the occurrence of seizures; hence the presence of seizures is a good prognostic sign, and their absence a bad prognostic sign. An abnormal level of consciousness after surgery is associated with twice the death rate observed when the postoperative level of consciousness is normal. It is thus clear that many prognostic variables in combination influence survival at least as much as does therapy. In addition, as noted, the amount of residual postoperative tumor inversely correlates with survival, as does growth of the tumor during radiotherapy.

## SURGERY

The role of surgical resection in the treatment of malignant gliomas remains controversial. No one disagrees that only surgery permits a pathologic diagnosis to be established during life. Many physicians, however, consider that current methods of radiologic diagnosis, including the CT scan and MRI, permit a diagnosis of malignant brain tumor without the necessity for attempted tumor resection, thus avoiding the risks of surgery. While stereotactic neurosurgery usually provides enough tissue to make a diagnosis of primary glioma, the amount of tissue may be inadequate to grade the tumors.

Stereotactic biopsy alone denies a role for surgery as "cancer" therapy. There is evidence that surgical reduction of tumor to very small residual amounts can prolong survival and permit the patient to return to an active life for a year or longer. Survival correlates best not with preoperative tumor size, but with postoperative tumor size, so the goal of surgery is to leave as little residual tumor as possible. These points are more fully discussed in chapter 15. All retrospective studies are subject to the criticism that the extent of attempted resection depends on the condition of the patient at the time of surgery (age, tumor location, clinical state), and that favorable conditions usually lead the surgeon to attempt a greater resection. Therefore, in such studies it is not clear that the *extent* of surgery was important to survival, but rather the more *favorable prognostic variables*. Nevertheless, most studies support surgical removal of the largest possible volume of tumor that safe operation allows.

## RADIATION THERAPY

The proper portals and doses of radiation therapy in the treatment of brain tumor have changed with the advent of better imaging techniques. The BTSG first reported, in controlled studies, that whole-brain radiation therapy increases the survival of patients over that which follows surgery alone. Other data showed that patients receiving 5500–6000 cGy live significantly longer than those receiving 5000 cGy or less. Patients with larger tumors that shrank by more than 50% survived longer than those whose tumors shrank less than 50% or those whose tumors actually increased in size. Those patients whose tumors enlarged during radiation therapy had a substantially worse outcome. Partial whole-brain radiotherapy plus coned-down boost was as effective as total whole-brain irradiation. No benefits have been seen with radiation sensitizers or increased fractionation of radiotherapy.

Among newer techniques in radiotherapy is that of interstitial implantation of radioactive seeds. Prolonged survival has been reported in patients with recurrent malignant gliomas treated with temporarily implanted Iodine 125 sources. The beneficial effect of implants may be greater for GBM than anaplastic astrocytoma.

Radiosurgery, either by gamma knife or by linear accelerator, has been shown to be effective in the treatment of arteriovenous malformations and small primary and metastatic brain tumors; its use in the treatment of gliomas has only recently been addressed. While data are conflicting, radiosurgery is probably indicated for a small group of good-prognosis patients with small tumors. It is hoped that a study currently underway will determine whether radiosurgery has an added beneficial effect over conventional radiotherapy and BCNU chemotherapy.

## CHEMOTHERAPY

In 1983 the BTSG reported that surgery plus radiation therapy and chemotherapy with BCNU added significantly to the survival of patients harboring malignant glioma, in comparison with surgery plus radiation therapy without chemotherapy. Patients receiving all three modalities had a median survival of a year, whereas those receiving only surgery and irradiation survived a median of 10 months. The 18-month survivorship was 2.5-fold greater among patients receiving BCNU than among those not treated with chemotherapy. Procarbazine and streptozocin have each demonstrated effectiveness similar to that of BCNU. Combination chemotherapy studies showed no advantage of adding procarbazine, hydroxyurea, or VM-26 to BCNU. Methyl prednisolone had no effect on survival.

Because BCNU is not curative, intra-arterial (IA) administration has been studied. A large phase III study of IA versus intravenous (IV) BCNU plus 5-FU and radiotherapy showed serious toxicity for the IA group, along with worse survival than the IV group. Intra-arterial BCNU produced white-matter necrosis. Intra-arterial BCNU, as used in this protocol, was neither safe nor effective. The BTSG also has tested IA cisplatin, IV BCNU, 10-ethyl-10-deaza-aminopterin and piroxantrone. Results from these studies will be presented in the near future.

Over the past several years, there has been increasing interest in the use of targeted interstitial drug delivery using biodegradable microspheres and wafers. A controlled trial (Brem et al., 1994) of such wafers in patients with recurrent malignant gliomas showed increased median survival in patients who had BCNU wafers rather than placebo wafers implanted at the time of reoperation.

Although many chemotherapeutic agents have been examined, none has been found to be more effective than nitrosoureas. In general, response to chemotherapy is best for oligodendrogliomas, intermediate for anaplastic astrocytomas, and worst for GBMs.

## FUTURE THERAPEUTIC STRATEGIES: GENE THERAPY

Gene therapy introduces genetic information into the cell(s) of interest. Central to the development of this technology is how the gene will be delivered and how it will be expressed. To some extent, the renewed interest in the development of vaccines against cancer cells stems from the vision that such cells can be altered to make them immunogenic. Another approach is the introduction of a gene that will make the tumor cells more susceptible to a treatment agent.

The vehicle that has generated the greatest interest in being able to get foreign genes into tumor cells has been the retrovirus. For example, insertion of a gene for herpes thymidine kinase makes the cells susceptible to ganciclovir, which blocks the kinase. It is thought that modifying the surface of some tumor cells may render others susceptible to either the drugs or the host immune system, thus enhancing the cytotoxic effect for cells not directly infected with the vectors, the so-called bystander effect.

# Therapy of Low-Grade Gliomas

The advent of CT and MRI has had substantial impact on our ability to diagnose and follow brain tumors. While malignant gliomas are often of such size as to be readily apparent on angiography, low-grade gliomas may present only with a seizure and rarely possess sufficient mass to be diagnosable by angiography. It is for diagnosing the latter tumors that modern imaging techniques, especially MRI, are so valuable.

## CEREBELLAR ASTROCYTOMAS

Cerebellar astrocytomas, especially in children, may be cystic and are usually pilocytic in histopathology. They carry a good prognosis. They are treated primarily by surgical removal, and even when all of the tumor is not entirely removed, they recur either not at all or only after many years. The most common of these tumors has a 10-year survival rate of greater than 90%. Cystic pilocytic astrocytomas also occur in the cerebral hemispheres, primarily in children and young adults. The tumor is usually cystic with a mural nodule that contrast-enhances on CT scan; it is often located in the temporal lobe and with its medial wall in contiguity with the lateral ventricle. Excision of the mural nodule is associated with very good outcome. Radiation does not improve survival.

## OTHER ASTROCYTOMAS

Low-grade astrocytomas of the cerebral hemisphere occurring in adults generally have a good prognosis, with expected survival of 3 to 7 years. Patients with low-grade astrocytomas commonly present with a single seizure, and the tumor is found on MRI. Although the diagnosis can be suspected from MRI appearance, biopsy should be performed, because diagnosis of low-grade glioma by MRI has an approximate 50% error rate.

There has been considerable discussion as to when to treat and with which modalities. Removal may be curative, but when residual tumor is present after attempted resection, the patient may benefit from radiation therapy. Positive prognostic factors include young age, location not in the temporal or frontal region, absence of postoperative neurologic deficit, and extent of resection. Radiotherapy is especially valuable for patients over 40 years of age with more extensive tumors.

The aim of surgery should be to totally resect the tumor, if possible, and if not, to leave minimal residual tumor. Preoperative volumes of less than 10 $cm^3$ were not associated with recurrence. Postoperative tumor volumes of less than 10 $cm^3$ were not associated with tumor recurrence. The role of radiation therapy in the treatment of low-grade astrocytoma remains controversial. Data from the Mayo Clinic suggests that radiotherapy should be performed for patients with astrocytomas that were incompletely resected.

One consideration in treating low-grade gliomas is the rate of malignant transformation. There is no evidence that the rate of malignant transformation depends on whether the low-grade glioma is treated immediately at diagnosis or at times of progression. In general, patients who have malignant transformation of low-grade gliomas have longer survival than patients who have anaplastic glioma arising de novo. Thus, radiating low-grade gliomas early might extend overall survival. Concern remains over long-term effects of radiation on normal cells, but cognitive deficits have not been observed.

## OLIGODENDROGLIOMAS

Oligodendrogliomas occur mostly in middle-aged adults, although there also is a small peak of incidence in children. The most common clinical manifestation is seizures, and many of the tumors demonstrate calcification on routine skull films or CT scans. This tumor has a less favorable prognosis than generally thought: In surgically treated cases, median survival time from onset of symptoms was 74 months in one study. Several reports have related survival to grading; median survival is 94 months for the lowest grade and only 17 months for the highest grade.

The role for radiation therapy in treating oligodendroglioma is not established. Most studies have found that survival correlates best with histologic grade, and that radiation had little effect on outcome. Chemotherapy may be effective for oligodendrogliomas, as mentioned previously.

## Conclusion

Neurologists and neural scientists are taking an increasing interest in the problems of cancer of the nervous system. This has come about because of the recognition that cancer biology involves fundamental molecular events that control the tumor both in its inception and in its evolution. In this circumstance, it is unlikely that the standard therapeutic techniques of surgical resection, radiation therapy, and chemotherapy will cure such disease, and, just as in the therapy of systemic cancer, newer techniques, founded in molecular principles, will be needed. As these techniques are being developed, the current methods of therapy must be applied. Fortunately, these have proved beneficial in extending useful survival. In this chapter, we have reviewed both the new molecular methodology and the available therapeutic modalities. This represents a beginning; our students need to continue the task.

## References/Further Reading

Brem H, Piantadosi S, Burger PC, et al. Intraoperative chemotherapy using biodegradable polymers: safety and effectiveness for recurrent glioma evaluated by a prospective, multi-institutional placebo-controlled clinical trial. Proc Am Assoc Clin Oncol 1994;13:174. Abstract.

Burger PC, Vogel FS. Surgical pathology of the nervous system and its coverings. 2nd ed. New York: John Wiley & Sons, 1982.

Daumas-Duport C. Histological grading of gliomas. Curr Opin Neurol Neurosurg 1992;5: 924–931.

Daumas-Duport C, Scheithauer B, Aphelian J, et al. Grading of astrocytomas: a simple and reproducible method. Cancer 1988;62:2152–2165.

Ekstrand AJ, James CD, Cavenee WK, et al. Genes for epidermal growth factor receptor transforming growth factor alpha and epidermal growth factor and their expression in human gliomas in vivo. Cancer Res 1991;51:2164–2172.

Kernohan JW, Sayre GP. Tumors of the central nervous system. Fascicle 35. Atlas of Tumor Pathology. Washington: Armed Forces Institute of Pathology, 1952.

Kondziolka D, Lunsford LD, Martinez AJ. Unreliability of contemporary neurodiagnostic imaging in evaluating suspected adult supratentorial (low-grade) astrocytoma. J Neurosurg 1993; 79:533–536.

Libermann TA, Razon N, Jaye M, et al. An angiogenic growth factor is expressed in human glioma cells. EMBO J 1987;6:1627–1632.

Libermann TA, Razon N, Bartal AD, et al. Expression of epidermal growth factor receptors in human brain tumors. Cancer Res 1984;44:753–760.

Shapiro WR, Green SB, Burger P, et al. A randomized trial of interstitial radiotherapy (IRT) boost for the treatment of newly diagnosed malignant glioma (glioblastoma multiforme, anaplastic astrocytoma, anaplastic oligodendroglioma, malignant mixed glioma): BTCG study 8701. Neurology 1994;44 (Suppl 2):263. Abstract.

Shapiro WR, Green SB, Burger PC, et al. A randomized comparison of intra-arterial versus intravenous BCNU, with or without intravenous 5-fluorouracil, for newly diagnosed patients with malignant glioma. J Neurosurg 1992;76:772–781.

Shapiro JR, Mehta BM, Ebrahim SAD, et al. Tumor heterogeneity and intrinsically chemoresistant subpopulations in freshly resected human malignant gliomas. In: Sudilovsky O, Pitot

HC, Liotta LA, eds. Boundaries between promotion and progression during carcinogenesis. New York: Plenum, 1991:243–262.

Shapiro WR, Green SB, Burger PC, et al. Randomized trial of three chemotherapy regimens and two radiotherapy regimens in postoperative treatment of malignant glioma: Brain Tumor Cooperative Group Trial 8001. J Neurosurg 1989;71:1–9.

Shapiro WR, Voorhies RM, Hiesiger EM, et al. Pharmacokinetics of tumor cell exposure to [¹⁴C] methotrexate after intracarotid administration without and with hyperosmotic opening of the blood-brain and blood-tumor barriers in rat brain tumors: a quantitative autoradiographic study. Cancer Res 1988;48:694–701.

Shapiro WR. Therapy of adult malignant brain tumors: what have the clinical trials taught us? Semin Oncol 1986;13:38–45.

Shapiro JR, Yung WA, Shapiro WR. Isolation, karyotype and clonal growth of heterogeneous subpopulations of human malignant gliomas. Cancer Res 1981;41:2349–2359.

Simpson JR, Horton J, Scott C, et al. Influence of location and extent of surgical resection on survival of patients with glioblastoma multiforme: results of three consecutive Radiation Therapy Oncology Group (RTOG) clinical trials. Int J Radiat Oncol Biol Phys 1993; 26:239–244.

Sporn MB, Todaro GJ. Autocrine secretion and malignant transformation of cells. N Engl J Med 1980;303:878–880.

# Surgical Management of Malignant Brain Tumors

**22**

Ronald E. Warnick

The *goals* of surgery, when offered to patients with malignant brain tumors, are establishing a tissue diagnosis, improving neurologic symptoms, and prolonging survival; the *expectations* depend on what is done. Stereotactic biopsy can accomplish the goal of providing a histologic diagnosis, and in the case of cystic tumors, can decrease the overall mass effect and improve the neurologic condition of the patient. In most cases, however, stereotactic biopsy has little effect on the patient's neurologic symptoms nor does it change the natural course of the tumor.

In contrast, craniotomy with cytoreductive surgery (Table 22.1) allows a representative histologic sampling of the tumor, which is essential to accurate diagnosis because of the known heterogeneity of these tumors, and it assures that there is sufficient tumor tissue for tissue culture assays, molecular genetic

**Table 22.1**  The Advantages of Cytoreductive Surgery

Diagnostic
  Extensive tissue sampling for histologic study
  Routine and special studies
    Labeling index
    Tissue culture
    Molecular studies
Neurologic
  Improved neurologic status
  Reduced intracranial pressure
  Steroid dose reduction
  Improved seizures
Oncologic
  Rapid decrease in tumor burden
  Removal of resistant cells
  Recruitment of nonproliferating cells
  Increased survival

SOURCE: Adapted by permission from Salcman M. Malignant glioma management. In: Rosenblum ML, ed. The role of surgery in brain tumor management. Philadelphia: Saunders, 1990:49–63.

studies, and measurement of proliferative potential to predict outcome. Most importantly, it can immediately improve the neurologic condition of symptomatic patients and relieve increased intracranial pressure. This allows reduction of the corticosteroid dose, minimizing the adverse effects associated with long-term steroid use. Surgical resection also reduces tumor burden immediately and removes neoplastic cells that may be resistant to adjuvant therapy because of their hypoxic and ischemic microenvironment. Debulking also may stimulate the recruitment of quiescent cells into the proliferative pool, where they are more susceptible to cytotoxic therapy.

The theoretical advantages of open resection, however, are significant only if they lead to increased survival for the patient. The scientific evidence linking extent of tumor resection with length of survival is examined later.

## Selection of a Surgical Approach

Determining the appropriate surgical approach for a particular patient with a malignant brain tumor is dependent on factors related not only to the tumor, but also to the patient and the surgeon (Table 22.2). Factors related to the tumor are readily apparent on, or can be inferred from, imaging studies (e.g., magnetic resonance imaging [MRI]), exact tumor location being critical to the decision-making process: (1) Superficial tumors are generally amenable to craniotomy and resection, whereas deep tumors may require a stereotactic approach. (2) If located in silent brain areas (e.g., a frontal pole), they can be removed without neurologic deficits, whereas resection of tumors located in eloquent brain areas carries a significant risk of neurologic complications, unless cortical mapping procedures are used. (3) Large tumors with significant mass effect generally require craniotomy to relieve symptoms of increased intracranial pressure and improve the patient's neurologic condition. Smaller tumors with no or minimal mass effect can be approached by either stereotactic or open procedures. (4) A single, discrete tumor should usually be resected, whereas for a highly invasive or multifocal tumor, resection may be no more

**Table 22.2**  Factors Influencing the Surgical Approach

| Tumor | Patient | Surgeon |
|---|---|---|
| Location | Neurologic status | Ability |
| Size | Age | Experience |
| Mass effect | Medical condition | Resources |
| Multiplicity | Patient choice | Beliefs |
| Invasiveness | | |
| Presumed histology | | |

SOURCE: Adapted by permission from Salcman M. Malignant glioma management. In: Rosenblum ML, ed. The role of surgery in brain tumor management. Philadelphia: Saunders, 1990:49–63.

beneficial than stereotactic biopsy. (5) When preoperative radiographic studies favor a particular histology (e.g., lymphoma), which is known not to benefit from cytoreductive surgery, stereotactic biopsy is clearly preferable.

Clinical information also influences the choice of surgical procedure. Patients with significant neurologic deficits may benefit from radical tumor removal. Age is an important prognostic factor for most malignant brain tumors and may determine the aggressiveness of the surgical approach. Both age and general medical condition significantly affect the operative and anesthetic risks of any proposed surgery. In the end, the wishes of the patient or family finally determine the surgical approach.

The choice may or may not be easy. At one extreme is an elderly, medically ill patient with a small tumor having minimal mass effect and located in a deep and eloquent area (Fig. 22.1A). There would be universal agreement to perform only a stereotactic biopsy for tissue diagnosis rather than attempt an open tumor resection. At the opposite end of the spectrum would be a young, healthy patient with symptoms of intracranial hypertension who has a large tumor exerting mass effect and located in a superficial, non-eloquent cortical area (Fig. 22.1B). Nearly all neurosurgeons would advocate craniotomy and tumor resection.

The majority of patients with malignant brain tumors, however, fall between these two extremes. Significant controversy exists regarding the appropriate management of patients who do not require radical tumor resection for relief of mass effect or neurologic symptoms: Their problem can be approached either by stereotactic biopsy, to obtain a tissue diagnosis as a preliminary to considering all the possible treatment modalities, or by immediate craniotomy and tumor resection (Fig. 22.1C). For these patients, the experience and beliefs of the individual neurosurgeon may determine the ultimate surgical approach, but there is now considerable scientific evidence linking the extent of tumor resection with length of survival in patients with malignant brain tumors.

## The Role of Cytoreductive Surgery

### SCIENTIFIC VALIDITY OF CLINICAL STUDIES

Green has analyzed the various types of observational and randomized studies and has developed a hierarchy that describes the strength of evidence of these studies (Table 22.3). Observational studies of malignant brain tumor patients allow us to understand the course of the disease, identify important prognostic factors, and help select groups of patients that are appropriate for randomized trials; for the assessment of surgical results, however, they have limitations. Unequal distribution of known (or unknown) prognostic factors between surgical groups may mask a true survival difference, or it may demonstrate an apparent difference when none really exists. Sophisticated multivariate analyses

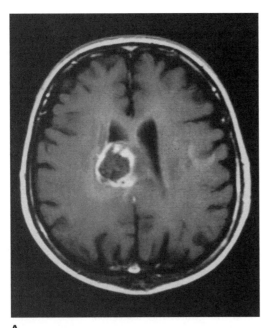

**Figure 22.1**
Choice of surgical procedure. (A) An elderly patient with a progressive decline in cognitive ability. Axial MRI with gadolinium demonstrates a deep hemispheric tumor with minimal mass effect. Stereotactic biopsy to establish tissue diagnosis is indicated. (B) A young patient presents with fulminant headache, nausea/vomiting, and a right hemiparesis. Axial MRI with gadolinium shows a large tumor with significant mass effect, which is located superficially in the frontal lobe. Craniotomy and tumor resection will reduce intracranial pressure and improve neurologic function. (C) A middle-aged patient presents with a single seizure and is neurologically intact. Axial MRI with gadolinium reveals a small, superficial tumor in the dominant hemisphere. Treatment options include stereotactic biopsy and tumor resection with cortical mapping.

A

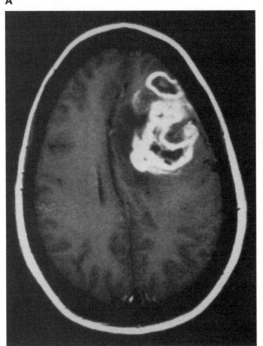

B

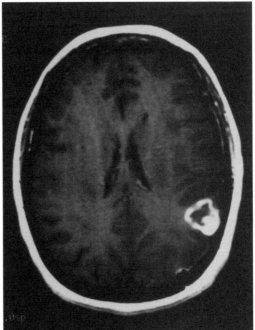

C

**Table 22.3**   Hierarchy of Strength of Evidence Concerning Efficacy of Treatment

1. Anecdotal case reports
2. Series without controls
3. Series with literature controls
4. Series with historical controls
7. Single randomized clinical trials
8. Confirmed randomized clinical trials

SOURCE: Adapted by permission of the publisher from Green SB. Patient heterogeneity and the need for randomized clinical trials. Controlled Clinical Trials, vol. 3, pp. 189–198. Copyright 1982 by Elsevier Science, Inc.

can control for *known* prognostic factors (e.g., age, neurologic status, histology, adjuvant treatment) but not for factors unknown to the investigator. There may be insufficient information regarding the reasons why a specific surgical approach was selected (see Table 22.2). If the selection of the operative approach is related to a conscious (or subconscious) decision regarding the prognosis of the patient, then the treatment groups will already include an inherent bias. In addition, observational studies often span long periods or utilize historical control groups; they cannot take into account changes in diagnostic studies, staging criteria, or adjuvant treatment. Lastly, their results may be significantly affected by missing data or by loss of patients to follow-up.

In randomized studies (see Table 22.3), known and unknown prognostic factors are balanced between the treatment groups. Time trends in diagnosis and adjuvant therapy affect all groups equally, and the formalized nature of most randomized trials usually results in data that are more than 90% complete, and follow-up of all patients. The randomized trial is the ideal way to assess efficacy of treatment, in this case, cytoreductive surgery. Unfortunately, it has so far been accomplished only for single brain metastasis (discussion follows).

## LOW-GRADE ASTROCYTOMA

The term *low-grade astrocytoma* is generally intended to refer to astrocytoma, as defined in three-tier grading systems, and grades I and II in the Kernohan classification. Since the advent of MRI, this tumor can be detected earlier. The patient often presents with a single seizure and normal neurologic examination, so immediate relief of neurologic symptoms and increased intracranial pressure are not significant considerations. Some patients with low-grade astrocytoma, however, present with intractable seizures, in which case craniotomy and resection of the tumor may sufficiently relieve the seizures to improve the patient's quality of life. Such astrocytomas may contain focal areas of anaplasia, which can be missed even by multiple stereotactic biopsies. No one doubts that extensive histologic sampling is important in formulating a treatment strategy for this sort of tumor. There also are two important theoretical issues:

(1) Astrocytomas may develop histologic evidence of anaplasia, presumably reflecting the evolution of clonal subpopulations. Assuming that the risk of malignant transformation relates to the number of clonogenic cells and to time, radical tumor resection should reduce it. A recent study demonstrates that the risk of higher grade recurrence is related directly to the volume of residual low-grade astrocytoma after surgery. (2) Typical low-grade astrocytomas contain a large pool of nonproliferating neoplastic cells and have an intact blood-brain barrier, giving them a relatively high resistance to cytotoxic therapy. Surgery may be the most effective treatment for this quiescent portion of the tumor. Neither of these theoretical advantages to cytoreductive surgery, however, is significant to the patient unless it contributes to a longer survival.

Several retrospective studies have attempted to determine the usefulness of cytoreductive surgery in the treatment of low-grade astrocytoma. Gol, in a series of 194 patients with low-grade astrocytomas who underwent surgery and radiation therapy, found that patients treated by tumor resection had better outcomes than those undergoing simple biopsy. Gol's two groups, however, were not comparable in terms of major known risk factors such as age, neurologic status, and tumor location and extent. Laws et al., analyzing six decades of experience at the Mayo Clinic, used multivariate analysis to show that the extent of resection was an independent prognostic factor. The extremely long period encompassed by this retrospective review, however, may have introduced unknown factors of preselection and time trends not accounted for by the statistical analysis. Soffietti et al. found a 51% 5-year survival rate for patients treated by macroscopically complete resection, compared with 24% for those undergoing subtotal removal.

Other studies have denied any association between extent of surgery and outcome. Scanlon and Taylor, reviewing 134 patients with astrocytoma referred to the Mayo Clinic for radiation therapy, found no difference in survival between patients undergoing subtotal resection and those undergoing simple biopsy, but patients who were thought to have undergone complete tumor removal were excluded from the study. Shaw et al., using multivariate analysis, did not find extent of tumor resection to be an independent prognostic factor, but patients with "mixed" gliomas were included in the study. Piepmeyer, reporting a series of patients treated since the introduction of the CT scan, found no correlation between the extent of resection (as determined by the operative report) and survival time.

Recently, Lunsford et al. reported a series of 35 patients with low-grade astrocytomas with minimal mass effect (i.e., not requiring craniotomy for improvement of neurologic status). They were treated by stereotactic biopsy followed by focal external beam radiotherapy and had a median follow-up of 62 months. Median survival was 10 years, with 5- and 10-year survivals of 88% and 47%, respectively. These results compare favorably with a modern series of low-grade astrocytoma utilizing cytoreductive surgery in which the median survival was 5.5 years, with 5- and 10-year survivals of 64% and 50%, respectively.

None of these studies, in which surgical resection was performed, used postoperative imaging by either computed tomography (CT) or MRI to assess objectively the degree of resection, yet this is especially important in patients with low-grade astrocytomas because these may be diffuse and often blend imperceptibly with the surrounding white matter, thus limiting the accuracy of subjective assessments made by the surgeon. The only study in which this was done was reported by Berger et al., who analyzed pre- and postoperative imaging studies (CT and/or MRI) in 53 patients with low-grade gliomas, with a mean follow-up of 42 months. Their data suggested that the *volume of residual tumor* imaging studies is an independent predictor of recurrence and time to tumor progression. Patients with a residual tumor volume of greater than 10 cc had a higher rate of recurrence (46.2%) and shorter time to tumor progression (30 months) than patients with less than 10 cc of residual tumor (14.8% and 50 months, respectively). No patient with complete resection of the radiographic abnormality had evidence of tumor recurrence within the timeframe of the study. The results must be interpreted cautiously, however, because the follow-up was not long enough to determine whether postoperative residual tumor volume predicts patient survival, only that it correlates with recurrence rate and time to progression.

In summary, significant disagreement still exists regarding the influence of cytoreductive surgery on outcome for patients with low-grade astrocytoma. Although the most recent studies have adequately accounted for the known prognostic factors, they cannot adjust for imbalance of prognostic factors that are still unknown, nor for all possible sources of bias in treatment assignment, nor for time trends over the course of the study. Nevertheless, it seems that the bulk of the evidence supports the role of cytoreductive surgery in the initial management of low-grade astrocytoma. A randomized trial comparing stereotactic biopsy with maximal tumor resection is needed to definitively answer this important question, but because many physicians (and some patients) already hold strong opinions on the matter, it may not be feasible.

## HIGH-GRADE ASTROCYTOMA

Malignant gliomas include anaplastic astrocytoma and glioblastoma, as defined by the three-tier classification. Open tumor resection has many of the same advantages as it does for low-grade astrocytoma (see Table 22.1). In addition, glioblastomas contain microscopic areas of necrosis, which typically coalesce to form the nonenhancing necrotic core seen on imaging studies and encountered during surgery. Resection of this reduces mass effect, thereby allowing more rapid tapering of corticosteroid dosage. This is an important advantage: In a recent study, 76% of patients with tumors of the CNS, who received dexamethasone for over 3 weeks, experienced corticosteroid toxicity, compared with 5% of those treated for less than 3 weeks.

Does cytoreductive surgery affect survival? In 1967, Jelsma and Bucy reported the outcome of surgery for 162 patients with glioblastoma. Gross total resection consistently resulted in improved neurologic status and longer independent and absolute survival than biopsy alone; the patients undergoing biopsy, however, were highly preselected and included mainly poor-risk patients who were not candidates for more extensive surgery. In addition, most of the biopsy procedures were performed before routine use of corticosteroids perioperatively, which doubtless contributed to the high morbidity and mortality associated with tumor biopsy. Salcman summarized studies involving a total of 603 patients treated by surgery without adjuvant treatment. The extent of resection was correlated with survival time, but this study also suffered from an inherent preselection bias—those receiving aggressive surgery being more likely to be good-risk patients.

Numerous clinical trials have used modern statistical methods to analyze the relationship between extent of tumor resection and other prognostic factors. In some of these studies, gross total resection brought about better results than subtotal resection, whereas others showed differences in survival only between patients undergoing total or subtotal resection and those undergoing biopsy. Interestingly, a retrospective review of prognostic factors by the Radiation Therapy Oncology Group demonstrated that extent of resection was an independent predictor of survival in patients with glioblastoma, but not in patients with anaplastic astrocytoma. Because anaplastic astrocytomas respond better to radiation and to chemotherapy than do glioblastomas, the amount of residual tumor may be a more important prognostic factor in the latter.

Other studies, however, have failed to identify extent of resection as an independent prognostic factor. In all of these, the operative report was used to assess the extent of tumor resection, thus introducing a significant degree of subjectivity.

More recent studies have sought to measure the extent of tumor resection by postoperative imaging. In a series reported by Levin et al. of 61 patients who underwent surgery, radiation therapy, and nitrosourea chemotherapy, the time to tumor progression was inversely correlated with the volume of enhancing tumor seen on the postoperative CT scan. Similarly, in 115 patients entered in the Brain Tumor Study Group (BTSG) trials, the volume of enhancing tumor on postoperative and postirradiation CT scans was inversely correlated with patient outcome. These are important conclusions, but neither study took into account such important prognostic factors as age, neurologic status, histologic grade, and differences in adjuvant treatment.

Ammirati et al. analyzed a series of 31 consecutive patients with malignant glioma who preoperatively were considered candidates for complete resection. Using CT scans to assess the degree of resection, the authors compared patients who had no residual enhancing tumor postoperatively with those with up to 20% of the presurgical enhancing volume; the two groups were well balanced in terms of important prognostic factors and adjuvant treatment. Survival was significantly better in the patients who underwent gross total resection than in the patients who were left with residual tumor (Fig. 22.2), and of those who underwent gross total resection, 97% had improved or stable neurologic status after surgery; in contrast, 40% of patients undergoing subtotal resection experienced neurologic morbidity as a result of surgery. The differences are noteworthy in light of the relatively small differences in the degree of resection (0% vs. 20% residual tumor). However, (1) the small number of patients precluded a multivariate analysis of prognostic factors, and (2) the patients in whom only subtotal resection was achieved may have been different from the others in terms of tumor location and/or size.

The most convincing study supporting the use of cytoreductive surgery in the treatment of malignant glioma was Wood et al.'s report of 510 patients from a previous BTSG cooperative trial. This randomized study, which divided patients into two radiation therapy arms and three chemotherapy arms, had shown no difference in survival between patients in the different treatment arms. When preoperative, postoperative, and postirradiation CT scans were retrospectively reviewed, however, the volume of residual tumor estimated from postoperative CT scans was found to be inversely correlated with survival time (Fig. 22.3). Similarly, on postirradiation CT scans, a smaller residual tumor volume was associated with a longer survival time. Most importantly,

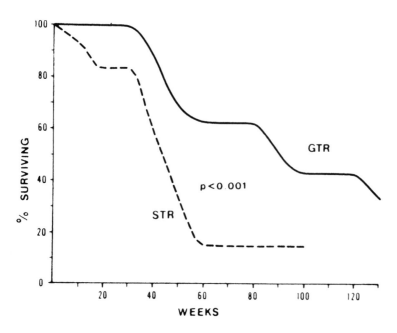

**Figure 22.2**
Survival related to extent of resection. The graph illustrates the cumulative probability of survival after operation in patients with malignant glioma. GTR, gross total resection (no residual tumor); STR, subtotal resection (less than 20% residual tumor). (Reproduced by permission from Ammirati M, Vick N, Liao Y, et al. Effect of the extent of surgical resection on survival and quality of life in patients with supratentorial glioblastomas and anaplastic astrocytomas. Neurosurgery 1987;21:201–206.)

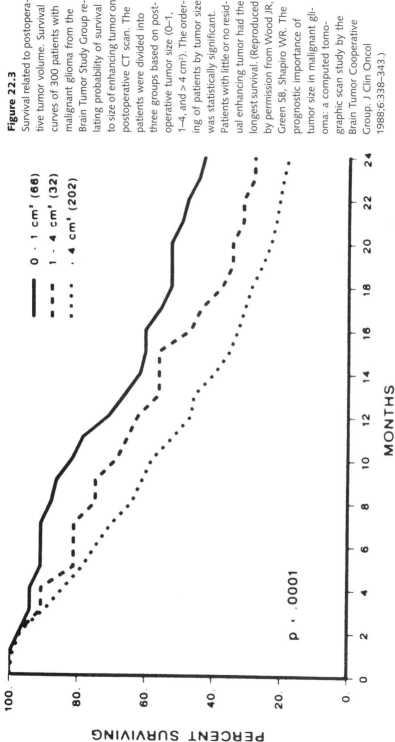

**Figure 22.3**

Survival related to postoperative tumor volume. Survival curves of 300 patients with malignant glioma from the Brain Tumor Study Group relating probability of survival to size of enhancing tumor on postoperative CT scan. The patients were divided into three groups based on postoperative tumor size (0–1, 1–4, and > 4 cm²). The ordering of patients by tumor size was statistically significant. Patients with little or no residual enhancing tumor had the longest survival. (Reproduced by permission from Wood JR, Green SB, Shapiro WR. The prognostic importance of tumor size in malignant glioma: a computed tomographic scan study by the Brain Tumor Cooperative Group. J Clin Oncol 1988;6:338–343.)

tumor size after surgery and after radiation therapy was a significant prognostic factor, even after adjustment for age, neurologic condition, and histologic grade. Although the study did not control for steroid use and differences in treatment at tumor recurrence, it was the first to use the objective information obtained from CT scans as well as multivariate statistical techniques to analyze the extent of resection as an independent prognostic factor.

Wood et al.'s work also established the importance of distinguishing *extent of resection* from *absolute volume of residual tumor* on the postoperative scan; despite the clear predictive value of residual tumor volume, neither the preoperative tumor size nor the percentage reduction in tumor size postoperatively was a significant prognostic factor. These findings can be interpreted oncologically as showing that the goal of surgery in malignant astrocytoma should be to *minimize* the volume of residual tumor. This is what must be controlled by adjuvant treatment to prolong survival. The inconsistent results of previous studies may be partially explained by their reliance on the surgeon's estimate of the percentage of tumor removed rather than on actual residual tumor volume.

In conclusion, the more reliable observational studies discussed here support the view that cytoreductive surgery has a positive influence on outcome in the treatment of patients with malignant astrocytoma. Only a randomized trial, however, can adjust for imbalance of prognostic factors that are unknown to its authors, for bias in treatment assignment and for time trends occurring over the course of the study. As with treatment of low-grade astrocytomas, the strong, preconceived opinions of most physicians (and some patients) makes the performance of such a trial, at least in the near future, unlikely on ethical as well as practical grounds.

## SINGLE BRAIN METASTASIS

Approximately 50% of patients who present with metastatic brain tumors will be found to have a single metastasis. Most of these have a known history of systemic cancer, or a systemic work-up reveals the primary site and allows determination of the tissue diagnosis. In approximately 15% of patients with a single brain metastasis, however, no primary site of cancer is found and surgery is therefore indicated to establish a tissue diagnosis. In addition, most neurosurgeons believe that patients with single brain metastasis have a longer and better quality survival when treated by craniotomy and tumor resection (with radiation therapy if indicated) than if they undergo stereotactic biopsy or no surgery. This consensus is based on both retrospective studies, and the results of two well-designed randomized trials.

The landmark randomized trial was reported by Patchell et al. and was the logical extension of his previous observational study. A total of 48 patients with single brain metastases was randomized to either open tumor resection or stereotactic biopsy, followed by whole brain radiotherapy. Patients were further stratified by tumor type, tumor location, and extent of systemic disease. The study excluded patients with radiosensitive tumors such as small cell lung cancer, lymphoma, and germ cell tumor. Local tumor recurrence was significantly lower in the group undergoing open tumor resection (20% vs. 52%), and this translated into a longer median survival (40 vs. 15 weeks) and a superior quality of life. A subsequent randomized trial by Noordijk et al. confirmed that resection of a single brain metastasis followed by whole brain radiotherapy was superior to radiotherapy alone.

In conclusion, two randomized trials have confirmed the efficacy of cytoreductive surgery for patients with single brain metastasis who have controlled systemic disease. The role of surgery for radiosensitive single brain metastases (e.g., small cell lung cancer, lymphoma, and germ cell tumor) has not been firmly established but is generally held to be limited to stereotactic biopsy when a tissue diagnosis is necessary. Whether stereotactic radiosurgery can replace surgery in the initial management of patients with single brain metastases is being studied in a multicenter, randomized trial.

## MULTIPLE BRAIN METASTASES

In contrast to its value in the management of single brain metastasis, the role of surgery in the management of multiple brain metastases is less conclusive. Surgery is used occasionally to establish a histologic diagnosis when the systemic evaluation is negative, to prevent neurologic deterioration from a dominant mass lesion, and as primary treatment for radioresistant tumors (e.g., melanoma, renal cell carcinoma).

It has been generally believed that for patients with multiple brain metastases, no survival advantage results from an aggressive surgical approach. A retrospective study by Bindal et al., however, has provided some support for cytoreductive surgery. They analyzed 56 patients with multiple brain metastases who underwent craniotomy and resection of one or more metastatic lesions, followed by whole brain radiation in approximately half of the patients. Group A (30 patients) underwent craniotomy and removal of at least one metastatic tumor, leaving one or more lesions unresected. Group B (26 patients) underwent resection of all intracranial metastases, requiring between one and three craniotomies. Known prognostic indicators were statistically equivalent between groups A and B. Both of these groups were compared with 26 patients with single brain metastases (group C) who underwent complete resection and were carefully matched for the same prognostic factors. For patients in groups B and C, median and 5-year survivals were about the same; survival of patients in group A was significantly poorer and similar to that of patients with multiple brain metastases undergoing only whole brain radiation (i.e., 6 months). The authors concluded that patients with multiple brain metastases, in whom all lesions are surgically accessible, should undergo removal of all lesions through multiple craniotomies.

This recommendation, however, must be balanced by three considerations. First, most such patients will require two or more craniotomies, and the cumulative morbidity of these must be weighed against potential improved survival; second, the study was retrospective and therefore subject to the limitations already discussed; and third, the patients in group A may have been different from the patients in groups B and C *because* they harbored unresectable tumors. In any event, because stereotactic radiosurgery has been shown to have a local control rate similar to complete tumor resection, it is unclear whether radiosurgery or an aggressive surgical approach is the best option for patients with multiple brain metastases.

A randomized trial comparing the results of craniotomy with radiation therapy to radiation therapy alone is being organized and, it is hoped, will provide definitive information regarding the role of cytoreductive surgery in these patients. Until its results are available, patients who are candidates for multiple craniotomies should be screened carefully to be certain that their systemic disease is controlled, thus maximizing the likelihood that they will benefit from an aggressive surgical approach.

## Conclusion

To date, most studies evaluating the role of cytoreductive surgery in the management of malignant brain tumors have been observational and have certain limitations. Several conclusions, however, can be reached on the basis of the available retrospective studies.

1. There is significant disagreement regarding the influence of cytoreductive surgery on outcome for patients with low-grade astrocytoma, but the bulk of the evidence supports its performance as soon as the diagnosis is made.

2. Cytoreductive surgery is essential in the treatment of patients with malignant astrocytoma to relieve neurologic symptoms, reduce steroid dependency, and minimize the tumor burden remaining after surgery, thus enhancing the benefit of adjuvant therapy. The available observational studies support the belief that it has a favorable influence on patient survival.

3. Observational studies and two randomized trials have confirmed the efficacy of cytoreductive surgery for patients with single brain metastases whose systemic disease has been controlled.

4. Surgery is indicated for selected patients with multiple brain metastases when a tissue diagnosis is needed, for removal of a large, symptomatic tumor, and for radioresistant tumors. A single observational study supports an aggressive surgical approach for patients with multiple brain metastases, in whom all demonstrable lesions are surgically accessible.

5. Finally, the surgical approach to a patient with a malignant brain tumor must be individualized on the basis of the patient's age and neurologic condition, the location and extent of the tumor, the neurosurgeon's experience and resources, and the wishes of the patient.

## References/Further Reading

Ammirati M, Vick N, Liao Y, et al. Effect of the extent of surgical resection on survival and quality of life in patients with supratentorial glioblastomas and anaplastic astrocytomas. Neurosurgery 1987;21:201–206.

Andreou J, George AE, Wise A, et al. CT prognostic criteria of survival after malignant glioma surgery. AJNR 1983;4:488–490.

Berger MS, Deliganis AV, Dobbins J, et al. The effect of extent of resection on recurrence in patients with low grade cerebral hemisphere gliomas. Cancer 1994;74:1784–1791.

Bindal RK, Sawaya R, Leavens ME, et al. Surgical treatment of multiple brain metastases. J Neurosurg 1993;79:210–216.

Ciric I, Ammirati M, Vick N, et al. Supratentorial gliomas: surgical considerations and immediate postoperative results. Neurosurgery 1987;21:21–26.

Curran WJ, Scott CB, Horton J, et al. Does extent of resection influence outcome for astrocytoma with atypical or anaplastic foci (AAF)? A report from three Radiation Therapy Oncology Group (RTOG) trials. J Neurooncol 1992;12:219–227.

Gol A. The relatively benign astrocytomas of the cerebrum: a clinical study of 194 verified cases. J Neurosurg 1961;18:501–506.

Green SB. Patient heterogeneity and the need for randomized clinical trials. Controlled Clin Trials 1982;3:189–198.

Jelsma R, Bucy PC. The treatment of glioblastoma multiforme of the brain. J Neurosurg 1967;27:388–400.

Laws ER, Taylor WF, Clifton MB, et al. Neurosurgical management of low-grade astrocytoma of the cerebral hemispheres. J Neurosurg 1984;61:665–673.

Levin VA, Hoffman WF, Heilbron DC, et al. Prognostic significance of the pretreatment CT scan on time to progression for patients with malignant glioma. J Neurosurg 1980;52:642–647.

Loeffler JS, Alexander E. Radiosurgery for the treatment of intracranial metastases. In: Alexander E, Loeffler JS, Lunsford LD, eds. Stereotactic radiosurgery. New York: McGraw-Hill, 1993:197–206.

Lunsford LD, Somaza S, Kondziolka D, et al. Survival after stereotactic biopsy and irradiation of cerebral nonanaplastic, nonpilocytic astrocytoma. J Neurosurg 1995;82:523–529.

McCormack BM, Miller DC, Budzilovich GN, et al. Treatment and survival of low-grade astrocytoma in adults: 1977–1988. Neurosurgery 1992;31:636–642.

Noordijk EM, Vecht CJ, Haaxma-Reiche J, et al. The choice of treatment of single brain metastasis should be based on extracranial tumor activity and age. Int J Radiat Oncol Biol Phys 1994;29:711–717.

Patchell RA, Tibbs PA, Walsh JW, et al. A randomized trial of surgery in the treatment of single metastases to the brain. N Engl J Med 1990;322:494–500.

Patchell RA, Cirrincione C, Thaler HT, et al. Single brain metastases: surgery plus radiation or radiation alone. Neurology 1986;36:447–453.

Piepmeyer JM. Observations on the current treatment of low-grade astrocytic tumors of the cerebral hemispheres. J Neurosurg 1987;67:177–181.

Ransohoff J, Kelly PJ, Laws ER. The role of intracranial surgery for the treatment of malignant gliomas. Semin Oncol 1986;13:27–37.

Salcman M. Malignant glioma management. In: Rosenblum ML, ed. The role of surgery in brain tumor management. Philadelphia: Saunders, 1990:49–63.

Scanlon PW, Taylor WF. Radiotherapy of intracranial astrocytomas: analysis of 417 cases treated from 1960 through 1969. Neurosurgery 1979;5:301–307.

Shaw EG, Daumas-Duport C, Scheithauer BW, et al. Radiation therapy in the management of low-grade supratentorial astrocytomas. J Neurosurg 1989;70:853–861.

Smalley SR, Law ER, O'Fallon JR, et al. Resection for solitary brain metastasis: role of adjuvant radiation and prognostic variables in 229 patients. J Neurosurg 1992;77:531–540.

Soffietti R, Chio A, Giordana MT, et al. Prognostic factors in well-differentiated astrocytomas in the adult. Neurosurgery 1989;24:686–692.

Sposto R, Ertel IJ, Jenkin RDT, et al. The effectiveness of chemotherapy for treatment of high grade astrocytoma in children: results of a randomized trial. J Neurooncol 1989;7:165–177.

Walker MD, Alexander E, Hunt WE, et al. Evaluation of BCNU and/or radiotherapy in the treatment of anaplastic gliomas: a cooperative clinical trial. J Neurosurg 1978;49:333–343.

Warnick RE. The role of cytoreductive surgery in the treatment of intracranial gliomas. Semin Radiat Oncol 1991;1:10–16

Wood JR, Green SB, Shapiro WR. The prognostic importance of tumor size in malignant glioma: a computed tomographic scan study by the Brain Tumor Cooperative Group. J Clin Oncol 1988;6:338–343.

# Ionizing Radiation in the Central Nervous System

J. Robert Cassady

Baldassarre Stea

When presented with a patient suffering from a tumor originating from or affecting the CNS, the oncologist ideally desires to eradicate all clonogenically viable tumor cells with trivial adverse normal tissue consequences or none at all. This ideal is never entirely attained: It can be approached only by thorough awareness of the physical and biologic principles of radiation.

## Physical Principles

Clinical treatment of tumors with radiation uses beams of high-energy photons or subatomic particles (e.g., electrons, protons, or neutrons) directed at the tumor site. Alternatively, beams of gamma rays emanating from a radioactive source such as $^{60}$Co may be used, or one or more such radioactive sources may be implanted directly into the tumor-bearing tissue for variable periods of time. Photons differ from gamma rays only in their source of origin, gamma rays originating intranuclearly, photons extranuclearly.

Radiation beams produce physical changes in tissues by excitation and ionization. Orbital electrons ejected from their source atoms may act directly on critical molecules and damage them, or they may produce highly reactive species that interact with water, producing free radicals. These in turn damage critical molecules such as DNA, probably by breaking a single or a double strand in the DNA molecule.

Treatment of tumors by utilizing machines that produce beams of ionizing radiation is termed *teletherapy*. Linear accelerators represent the most common teletherapy equipment used today, although $^{60}$Co sources and betatrons are still frequently employed. Direct implantation of sources of radiation into a tumor or a body cavity is termed *brachytherapy*. Occasionally, radioisotopic solutions may be instilled directly into tumor tissue, or the patient may be given a radioisotopically labeled antibody targeted against tumor tissue.

Radiation is absorbed by any material in one of three ways, primarily dependent on the energy of the photon or gamma ray and on the composition of the absorbing material:

1. At low energy levels (< 80 keV), photoelectric absorption occurs. Usually, all or nearly all of the incident photon energy is lost in this process, and a tightly bound orbital electron is ejected. A less tightly bound electron fills the vacancy so created, with release of characteristic radiation. Photoelectric energy absorption varies with the third power of the atomic number of the absorbing material. This explains the differential absorption of diagnostic x-rays and accounts for the ease with which bony or metallic structures are imaged at these energies.

2. With megavoltage beams, the predominant mechanism of radiation absorption is Compton scatter. In this process, the incident energetic photon interacts with a lightly bound outer shell electron imparting energy and producing a secondary photon. The probability of interaction depends on electron density and not on atomic number.

3. The third process, pair production, in which an incident photon with an energy greater than 1.02 MeV is converted to an electron and a positron (or positively charged electron), is not prominent with the photon energies used in conventional treatment approaches.

Almost all current treatment approaches for deep-seated tumors use megavoltage ($^{3}1$MeV) beams. These spare the skin, because ionized electrons arising from superficial interactions travel some distance in tissue, only producing peak doses of radiation at an appreciable distance from the surface (Fig. 23.1). The percent of photons traveling and depositing their energy at a given distance within a material (tissue) is dependent on beam energy and the nature of the absorbing tissue.

Optimization of treatment requires *treatment planning*. First, precise localization of the gross tumor and of any surrounding tissue likely to contain microscopic tumor is established utilizing clinical observations, imaging techniques, laboratory studies, and knowledge of the natural history of the type of tumor in question. The patient is then taken to a *treatment simulator*, a radiographic unit wherein physical parameters of the treatment unit can be reproduced (or simulated), diagnostic quality radiographs obtained, and/or fluoroscopy utilized. Many modern simulators are coupled to a computerized tomographic (CT) unit to more precisely identify gross tumor margins in three dimensions. Images of the relevant body part (i.e., brain and/or spinal cord) are made and diagrams created in which the tumor and critical normal tissues are accurately placed. The planning team (radiation oncologist, radiologic physicist, and dosimetrist) then decides which arrangement of treatment beams is optimal.

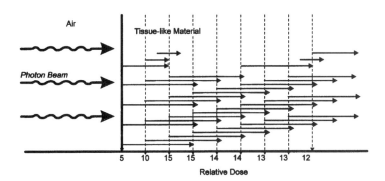

**Figure 23.1**
A build-up of radiation dose occurs with depth when an incident *megavoltage* photon (x-ray) beam strikes the skin of a patient. Due to forward scatter of freed energetic electrons, the maximum ion density occurs at a depth from the surface (i.e., skin). This density gradually decreases due to absorption in the deeper tissues.

**Schematic Plot of Absorbed Dose as Function of Depth**

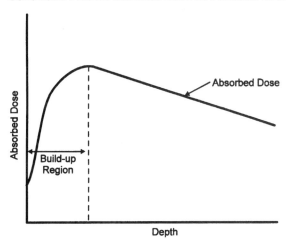

In the ideal situation, radiation would be homogeneously delivered to the tumor-bearing volume of tissue, while normal, non-tumor-bearing tissue peripheral to this volume would receive no irradiation at all. It is the task of the planning team to approximate this ideal as nearly as possible in an easily reproducible fashion. Two types of normal tissue can be identified (Fig. 23.2): *Transit* normal tissue represents normal tissue peripheral to the tumor-bearing volume through which the treatment beam must pass to reach the tumor; *matrix* is normal tissue surviving inside the tumor.

It is generally possible to reduce the dose of radiation to transit normal tissue significantly below the desired tumor dose by the use of multiple radiation beams and by selecting beams of appropriate energies, but matrix represents the dose-limiting structure for the radiation oncologist. Because of the exponential nature of radiation effects on tissue, optimal treatment planning attempts to

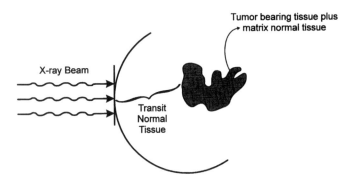

X-ray Beam

Transit Normal Tissue

Tumor bearing tissue plus matrix normal tissue

**Figure 23.2**

Two classes of normal tissue. Transit normal tissue represents normal tissue that an external radiation beam must traverse to interact with the tumor. In general, radiation dose to transit tissue can be minimized and kept below the tumor dose by appropriate treatment planning. In contrast, *matrix* normal tissue is normal tissue admixed with the tumor. Treatment planning minimizes radiation dose to this tissue by maintaining homogeneity of dose throughout the tumor volume. Radiation dose to this tissue, however, will necessarily equal or surpass the minimum tumor dose. Many radiation oncology departments require tumor-volume heterogeneity of less than 15%. Late radiation toxicity to this matrix normal tissue often represents the dose-limiting problem in clinical treatment.

minimize transit normal tissue radiation dose and make normal tissue dose inside the target as homogeneous as possible. It usually requires use of two or more radiation beams and representative isodose contours for a single field; various multiple-field arrangements are shown in Figure 23.3. Beam modifiers (wedges and compensators) may provide significant benefit for the planning process (Fig. 23.4).

Electron beam therapy also may be an option in certain settings. Figure 23.5 demonstrates a typical depth-dose distribution for two electron energies and, for comparison, for a frequently used photon beam. Using electrons, a relatively sharp breakpoint between high- and low-dose radiation areas can be achieved with virtually no radiation dose beyond a certain point. Two potential clinical settings for which (in view of the sharply defined depth of tissue that can be irradiated with sparing of more distal normal tissue) electron beam treatment has been considered for CNS tumor treatment are (1) intraoperative therapy and (2) spinal irradiation in craniospinal treatment approaches for malignant "seeding" tumors such as medulloblastoma. The second of these, however, may pose significant treatment planning problems: Figure 23.6 shows craniospinal irradiation fields used in the treatment of those malignant tumors of the CNS that may spread by CSF seeding.

Craniospinal treatment approaches are considerably facilitated by the precise and predictable angulations that are possible with current treatment-machine gantries and collimators and by the ability to tilt the treatment table

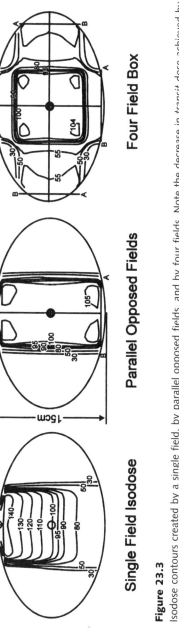

**Single Field Isodose**    **Parallel Opposed Fields**    **Four Field Box**

**Figure 23.3**

Isodose contours created by a single field, by parallel opposed fields, and by four fields. Note the decrease in *transit* dose achieved by the four-field approach, minimizing radiation dose (for a given tumor dose) to transit normal tissue. Note also the greater homogeneity throughout a deep-seated potential tumor volume, which is achieved by the two-field plan, compared with a single field. The use of a single field maximizes both maximum transit tissue dose and tumor-dose heterogeneity, hence this technique is seldom used in clinical treatment. All beam isodoses represent 10 × 10-cm fields using a 10-MeV photon beam (100-cm target to skin distance [TSD]) from a linear accelerator.

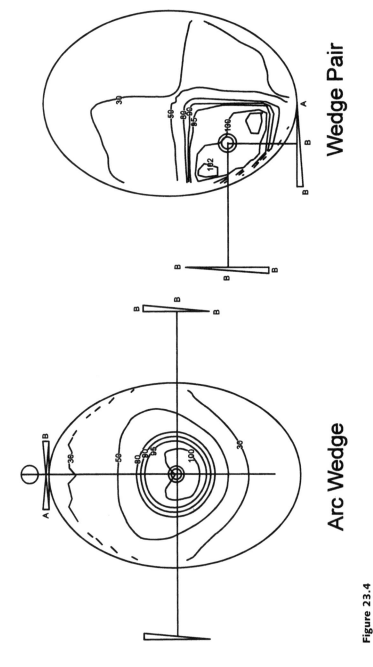

**Arc Wedge**

**Wedge Pair**

**Figure 23.4**

Graphic depiction of arc wedge and wedge pair field arrangements. The arc wedge permits excellent dose homogeneity throughout a deep-seated central (e.g., pituitary or pineal) tumor volume in the head and minimizes transit dose. The wedge pair arrangement is ideal for an eccentrically placed tumor volume (i.e., frontal lobe, occipital lobe). Without the interposed wedges, the dose at the medial intersection point of the two beams would be excessive.

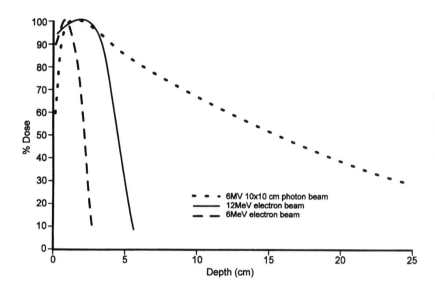

**23.5**
Depth-dose distribution of two electron beams (6 MeV and 12 MeV at 100 cm) and one photon beam (10 × 10-cm field of 6-mV photons from a 100-cm TSD linear accelerator).

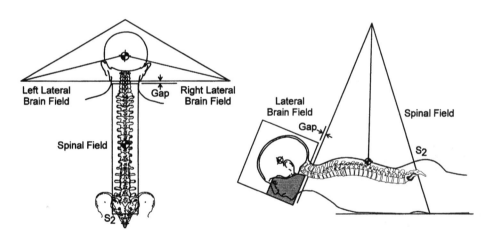

**Figure 23.6**
Fields utilized in craniospinal treatment. Note that the beam and couch must be angulated to avoid overlap or underlap areas, due to the necessity of multiple junctioned beams coming from two directions.

or couch. Using these machine capabilities, areas of increased (overlap) or decreased (underlap) dose can generally be avoided.

## Clinical Principles

As noted in the introduction, the ideal clinical outcome in patients with CNS tumors would constitute uniform tumor control with no normal tissue toxicity, a goal that currently cannot be met.

It is important at this point to define cell *death* as "loss of clonogenic viability." As long as a tumor no longer contains any cells capable of indefinite replication, it rarely poses a threat to the host. It also is important to separate the *radioresponsiveness* (rapidity of disappearance) of a tumor from its *radiocurability*. It was formerly commonly assumed that these two were the same: tumors that decreased rapidly in size following treatment being deemed *radiosensitive* in contrast to those that showed little immediate response.

We now know that many factors contribute to the rate of tumor regression. These include the rate of cell division within a tumor, the percent of tumor cells actively undergoing cell division, the percent of these that die following division or due to nontreatment causes (cell loss factor), the percent of cells that are not clonogenically viable, the likelihood of intermitotic death (apoptosis) in a tumor, and a variety of other factors. Thus, there is no direct relationship between the rate of regression of a tumor following irradiation and that tumor's radiocurability.

## MECHANISM OF CELL DEATH

Except in certain highly sensitive classes of cells (e.g., lymphocytes, spermatozoa), cell death occasioned by irradiation is caused by lethal damage to the reproductive apparatus of a cell that is expressed following one or more cell divisions (mitotic death). Recently, it has become apparent that intermitotic or "programmed" cell death (apoptosis), which is the primary mode by which radiation kills lymphocytes, also occurs in many tumors (perhaps all) and in normal tissues following irradiation. This phenomenon currently is an active area of biologic research.

## EXPONENTIAL NATURE OF EFFECTS

When the rate of normal tissue damage or injury (the probability of a complication) and tumor control probability are plotted as a function of radiation dose, a predictable pattern is seen (Fig. 23.7). Following an initial dose region where no measurable effect is seen (threshold region), a steeply rising exponential region is apparent, where small increments of additional irradiation are accompanied by a marked increase in effect (i.e., tumor control). This region is followed by a so-called plateau region where significant increases in radiation yield a smaller or no increase in the effect being measured. In most tumor settings, critical normal tissue damage (in contrast to transient, reparable acute effects such as skin erythema, hair loss, nausea, emesis, etc.) lies beyond (i.e., requires more dose than) tumor control on such a probability curve. Unfortunately, in many CNS tumors, such as glioblastoma, normal tissue damage (brain

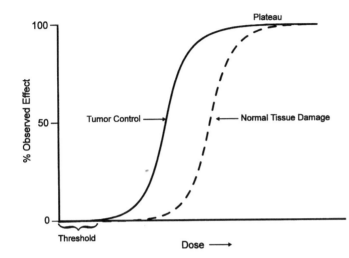

**Figure 23.7**
The exponential nature of tumor response (or normal tissue damage) with increasing radiation dose. Note the importance of even very small, incremental increases of dose in outcome; omission of even a single fraction of radiation can have a significant bearing on tumor control outcome. In this depiction, serious normal tissue injury occurs at higher doses than those necessary to achieve a reasonable probability of tumor control. With some CNS tumors (e.g., glioblastoma), these curves (normal tissue damage and tumor control) may be superimposed or reversed.

necrosis) occurs at a dose similar to or less than the dose necessary for tumor control.

## NORMAL TISSUE EFFECTS

The timing and clinical consequences of normal tissue damage depend on the tissue being considered, its rate of cell division, its ability to repair sublethal radiation damage (vide infra), the pattern in which functional units are organized within that tissue (i.e., in parallel or in series), and the extent to which functionally redundant units are present within it.

1. Acute effects (skin erythema, hair loss, mucositis, nausea, vomiting, diarrhea, and lymphopenia) are caused by irradiating tissues such as skin, mucous membranes, and gastrointestinal mucosa, which are composed largely of cells that undergo rapid cell division. These tissues rapidly (normally during treatment) manifest radiation damage and often, equally rapidly, repair that damage.

2. Subacute normal tissue effects such as radiation pneumonitis, hepatitis, and postirradiation somnolence syndrome commonly occur 4 to 8 weeks following a course of fractionated treatment, and are thought to be caused by damage to the endothelial cells of small blood vessels (e.g., arterioles, capillaries), in which the percent of cells in division at any point in time is small and overall cell turnover rates are slow.

3. The cause of late tissue damage (e.g., bone or brain necrosis) is controversial. It appears likely that these serious and permanently damaging results of irradiation are caused by both intrinsic cellular damage within the target

organ and by vascular damage and subsequent ischemia. Second tumor production caused by irradiation has a latent period of 5 to 10 years and also is commonly considered to be a late effect.

## THERAPEUTIC RATIO

The relationship between the effect on a tumor (i.e., the desired result) of a given treatment compared with the effects of that treatment on normal tissue (undesired effects) has been termed the *therapeutic ratio* or *therapeutic index*. Many approaches, including sequential or concurrent chemotherapy, inhalation of oxygen or of carbon dioxide at high concentrations, alterations in surrounding atmospheric pressure, hypoxic sensitizers, and hyperthermic treatment, have been utilized in an attempt to improve the therapeutic ratio in the treatment of a given tumor. Importantly, unless an endpoint of equal toxicity with improved tumor control can be demonstrated, an improvement in therapeutic ratio cannot be claimed.

## FRACTIONATION

Division of a radiation treatment course into many daily increments (as opposed to delivery of one or a few large treatments) improves the therapeutic ratio in many clinical settings. Development of megavoltage radiation equipment that generally eliminated the effect on the skin as the critical dose-limiting factor heightened interest in development of different fractionation approaches.

Parameters important to the efficacy/toxicity of a course of radiation include the total radiation dose delivered, the total elapsed time over which the treatment course has been delivered, the size of the individual dose fractions, the dose rate at which the radiation was delivered (in cGy/minute), and, finally, the volume of tissue irradiated.

In the past 1 to 2 decades, important differences between so-called early-reacting (skin, mucosa, bone marrow, etc.) and late-reacting tissues (brain, fibrous supporting tissue) have been recognized, and the disproportionate impact of fraction size on the amount of late-reacting normal tissue damage has been acknowledged. When fewer and larger fractions of radiation are given, the effects on late-reacting normal tissues are more severe than when these are of greater number and smaller size, even when the two courses are matched for early-reacting tissue changes.

Effects on early-reacting tissues have been termed *alpha effects* and those on late-reacting tissues have been termed *beta effects*. A formula has been proposed that would equate them:

$$\log_e S = -n(aD + bD^2)$$

where n is the number of equal fractions, S is survival of clonogenic cells, and D is dose. In most instances, tumors have been considered to respond like alpha-reacting tissues.

## IMPORTANCE OF FRACTIONATION PATTERNS

With the recognition that the larger the fractional doses in which radiation therapy is delivered, the greater the late damage to normal tissues, combined with the realization that so-called sublethal cellular damage repair is largely completed by 2 hours after irradiation (although significant repair continues for at least 4 more hours), considerable clinical experimentation with fractionation patterns has occurred. The goal of hyperfractionated regimens is to decrease late-reacting tissue reactions, while achieving the same tumor control with the same (or only slightly increased) early-reacting tissue reaction.

Rather than utilizing a so-called standard fractionation pattern (e.g., 6000 cGy delivered in 30 daily fractions of 200 cGy with 5 daily fractions/wk), hyperfractionation (increasing the total number of fractions and total dose, while duration of treatment remains unchanged) and accelerated fractionation techniques (delivering the same total dose in a larger number of fractions of equal or slightly decreased size and in a shorter overall period) have been introduced. The results of these regimens, compared with those of hypofractionated (> 200 cGy fractions) treatment regimes, have confirmed the importance of fraction size on the likelihood and extent of neurotoxicity and other late radiation damage.

## ORGANIZATION OF NORMAL TISSUE

It was noted earlier that the organization of normal tissue can vary from organ to organ or, for that matter, within a given organ. Three examples will illustrate this point and demonstrate that a given rate of normal cell damage will have graver functional consequences in a tissue in which the elements are organized in series than in one that is organized in parallel.

Skin is a parallel-organized tissue (Fig. 23.8). If one assumes that a certain dose of irradiation damages 1% of cells and more than 100 cells are present, then one or a small number of these normal cells will probably be damaged. This damage, however, will have little functional consequence because an overwhelming number of the remaining cells are unaffected and can initiate cell division to replace damaged cell(s) and repair the damage. In contrast, the spinal cord represents a tissue organized in series. If the same treatment is given as with skin, and similar cell sensitivity and cell numbers are assumed, then the number of damaged normal cells will be the same, but the death of each will result in a functional (nerve conduction) loss that is irreparable. An organ such as the kidney represents an intermediate case: It comprises many

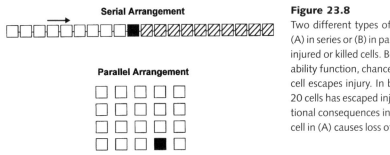

**Serial Arrangement**

**Parallel Arrangement**

**Figure 23.8**

Two different types of normal tissue organization: (A) in series or (B) in parallel. Black squares represent injured or killed cells. Because cell damage is a probability function, chance determines whether a given cell escapes injury. In both of the examples, one in 20 cells has escaped injury. This has no serious functional consequences in (B), but damage to only one cell in (A) causes loss of function of the entire series.

functional units (nephrons) organized in parallel, but each unit is itself partly composed of cells organized in series (as in the proximal tubule).

## STEREOTACTIC RADIATION THERAPY

During the past few years, considerable interest has developed in techniques that deliver radiation through a large number of radiation fields of very limited size (often < 2 cm) directed in an arc at a single location within the brain or skull base. Generally, a single dose or limited number of large doses is given to this extremely restricted volume with highly precise localization and fixation techniques. This approach (stereotaxic radiosurgery; see chapter 26) has shown beneficial results in a number of clinical settings (e.g., arteriovenous malformations, single metastases, acoustic neuromas, and meningiomas). It is currently being investigated as boost therapy for gliomas after completion of conventional fractionated external beam radiotherapy.

## TREATMENT VOLUME

Because the total radiation dose necessary to ensure a high probability of tumor control is a function of the total number of tumor cells, smaller tumors can be controlled by using lower total doses of irradiation. This conclusion has led to the general concept of "shrinking field" approaches in most tumor settings. The small number of tumor cells present at the periphery of a tumor is treated by a relatively small total dose of radiation to an initial large field. A "cone-down" technique is then used to treat the smaller volume of tissue, which contains a larger number of tumor cells. This approach allows the radiation oncologist to reduce the treatment volume (and thus quantity of normal tissue) that receives a high dose of radiation, thereby reducing the incidence of normal tissue damage.

## RADIATION BIOLOGY

Following the development of cell culture techniques that permitted mammalian cells to be grown in vitro, considerable insight was rapidly gained into the effect of radiation on proliferating cells. Early experiments produced a characteristic complex survival curve (Fig. 23.9) that can be separated into two principal regions; when the data are plotted semi-logarithmically, an initial "shoulder" region is followed by a linear exponential region. Various terms and abbreviations used to describe portions of this characteristic curve are shown in the illustration.

The amount of irradiation necessary to reduce the fraction of surviving cells by 63% in the exponential or straight-line portion of the curve has been termed the $D_o$. This parameter represents the survival that results each time one additional lethal radiation event per cell has been delivered; because the probability of interaction is random, on average, 37% of cells will *not* suffer a lethal event in such circumstances. When a shoulder region is present, $D_o$ will always be smaller than $D_{37}$ (the dose required to reduce an initial cell population to 0.37 of its original number). When no shoulder region is present, $D_o$ will equal $D_{37}$.

The radiation sensitivity of a given cell population is generally expressed by the magnitude of $D_o$. Sensitive cells have smaller $D_o$'s than resistant populations. In a few cases, striking differences in $D_o$'s have been ascertained between populations of both normal cells and different tumor cells in the tissues that have been studied. Representative $D_o$ values for each are usually in the 130–250-cGy region and range from approximately 90–200 cGy.

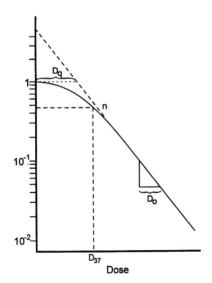

**Figure 23.9**

Typical semi-logarithmic cell survival curve following radiation treatment. An initial shoulder region is followed by an exponential, straight-line portion. The extrapolation number, n, found by extrapolating the straight-line portion of the curve to the ordinate, reflects the width of the shoulder region and is a measure of capability to repair sublethally damaged cells. $D_{37}$ represents the radiation dose necessary to kill 63% of an *initial* number of tumor cells. $D_o$ represents the dose necessary to kill 63% of a population of cells on the straight-line portion of the curve. This figure (or the corresponding slope) is indicative of true cellular radiosensitivity. Whenever repair of sublethal damage can occur in a cell population, $D_{37}$ will always be greater than $D_o$. $D_q$ represents the quasi-threshold dose.

## Radiation Repair

Two principal types of radiation damage repair have been described:

1. The fraction of cells surviving a single dose of irradiation is smaller than that which results when the same total dose is given in two equal fractions separated by a variable time interval (Fig. 23.10). This type of cellular recovery, termed *sublethal radiation damage repair*, is rapid, almost all possible repair occurring within 6 hours following irradiation. The shoulder region of the single-dose radiation survival curve in Figure 23.9 is related to the sublethal repair potential of a cell population, and is absent in cells lacking that capacity, which varies widely between different tumor and normal cell populations and can be quantified by the parameters $D_q$ (the quasi-threshold dose) and n (extrapolation number) (see Fig. 23.9).

2. A second form of cellular radiation repair has been termed *potentially lethal damage repair.* If irradiated cells are placed in a nutritionally deprived medium for a period of some hours after treatment with irradiation before they are stimulated to divide and grow, a greater number survive than when they are placed immediately after irradiation in a nutritionally rich medium that encourages them to divide and express imparted damage.

Both types of repair exist in tumors and normal tissues, and variations in the relative extent of each may account in part for clinically observed differences in radiation effects on different cell populations, despite similar $D_o$'s or other survival curve characteristics.

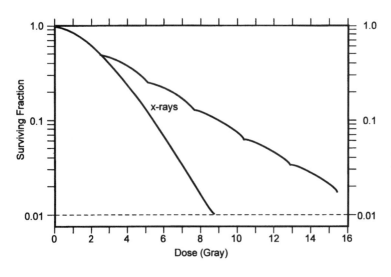

**Figure 23.10**
Using a semi-log scale, the relative cell survival occurring after a single large fraction of radiation, compared with the same dose delivered in multiple fractions separated by significant (> 6 hours) time intervals. Because sublethal damage repair occurs during this interval, fewer cells are killed in the multiple-fraction model, despite delivery of the same total dose.

## OXYGEN EFFECT

The presence of oxygen enhances cellular damage caused by irradiation. Figure 23.11 demonstrates representative survival curves for a cell population irradiated under oxic or hypoxic conditions. An approximate 2.5- to 3-fold decrease in $D_o$ (or the slope of the survival curve) is regularly observed in cells irradiated under hypoxic conditions, compared with what is found in well-oxygenated cells. Powers and Tolmach first observed a change in the slope of the exponential portion of the survival curve following irradiation of a tumor (Fig. 23.12).

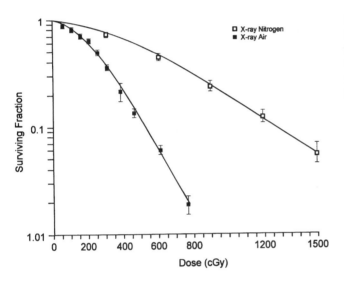

**Figure 23.11**

Cell survival as a function of dose in cells treated in air and in nitrogen (or under anoxic conditions). The slope of the survival curve is significantly steeper for cells treated in air.

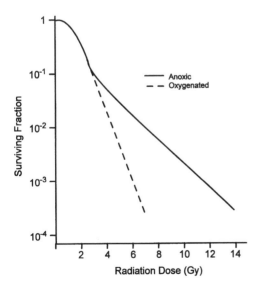

**Figure 23.12**

Biphasic exponential cell survival curve. The shallower portion of the slope represents survival of the hypoxic cell fraction in the irradiated tumor. By extrapolating the curve of this more resistant portion to the ordinate, one can estimate the percent of hypoxic cells present in the initial tumor. This figure varies but is commonly in the 5% to 15% range.

This second portion of the survival curve, indicating greater resistance to irradiation, is presumed to be due to the presence of hypoxic tumor cells. Extrapolation of this slope to the ordinate suggests that from 10% to 20% of cells in most tumors are hypoxic under experimental conditions.

Many experiments that demonstrated an oxygen enhancement ratio (OER) of 2.5:3.0 have utilized a single or a relatively small number of radiation fractions. In contrast, most curative human clinical treatment approaches comprise 25 to 30 radiation fractions over 5.5 to 7+ weeks. Although some positive results have been seen in clinical trials that attempted to improve tumor oxygenation by various means, many have been negative, and in the rest, the magnitude of the benefit has been modest. Trials using hypoxic cell sensitizers also have had mostly negative results. It is possible that reoxygenation occurs in most tumors following treatment, leaving an ever-decreasing number of hypoxic cells to be controlled. Nonetheless, attempts to exploit the oxygen effect for clinical benefit continue.

## CELL CYCLE EFFECTS

The cell cycle of dividing mammalian cells can be divided into four phases: $G_1$ (for gap), S (for synthesis of DNA), $G_2$, and M (for mitosis). Unlike certain chemotherapeutic agents, such as methotrexate and vincristine, which are only active in the synthetic phase of cycling cells, radiation can damage or kill cells in all phases of the cell cycle. This sensitivity, however, varies over the cell cycle, cells in mitosis and $G_2$ being most sensitive, whereas as cells proceed from $G_1$ to S, increasing resistance is usually noted.

There is an approximate two- to threefold difference in radiation sensitivity between the most sensitive and resistant phases of the cell cycle. Some degree of cell synchrony can be induced by fractionated irradiation, but it is usually rapidly lost, and no successful attempts to clinically exploit this phenomenon have been reported.

## CONTROLLING REPOPULATION

Even though radiation induces a short period of delay in cell division following each treatment, tumor cell proliferation occurs during a fractionated course of treatment. This cell proliferation adds to the total number of clonogens that must be killed to sterilize a tumor. Clinical techniques such as treatment breaks given to permit recovery of normal tissues, weekend breaks, and split-course treatment approaches all permit proliferation of a tumor during treatment, thereby making local control more difficult. In certain tumors, such as Burkitt's lymphoma, this intertreatment tumor-cell proliferation is so notable that actual tumor growth can occur during daily fractionated irradiation, despite relative tumor-cell sensitivity to treatment. To combat this adverse effect, two or even three fractions of irradiation have been administered daily. The partial success of this approach has led to other attempts to alter fractionation patterns in an effort to minimize the clinical impact of intertreatment tumor-cell proliferation.

When hyperfractionation is used, the same or a slightly higher total dose of irradiation is delivered in the same overall treatment time as in conventional treatment (180–200 cGy/fraction/day and 900–1000 cGy/5 fx/wk) but in a larger number (usually 2 or 3 times) of radiation fractions per day (110–120 cGy/fraction) separated by intervals of only 4 to 6 hours. With accelerated hyperfractionation, the same or a larger total radiation dose is delivered in a shorter period using a larger number of radiation fractions somewhat smaller than is conventional (i.e., 140–160 cGy/fraction).

In the treatment of malignancies of the head and neck region, treatment approaches using 110–120 cGy/fraction delivered twice daily with an interfraction interval of at least 6 hours, up to total doses of 7000–7500 cGy have demonstrated clinical benefit over conventional radiation techniques, and are believed to be beneficial by reducing intertreatment tumor-cell proliferation as well as by reducing late normal tissue damage, such as subcutaneous fibrosis. Effects on normal tissue, such as subcutaneous fibrosis, are minimized by allowing sufficient time between treatments for the repair of sublethal tissue damage, while reducing late damage by utilizing smaller fractions of radiation.

## LINEAR ENERGY TRANSFER AND RELATIVE BIOLOGIC EFFECTIVENESS

The rate of energy deposition or loss along the path of photons or particles depends, in general, on the atomic number of the particle or photon and/or its velocity, and is described as linear energy transfer (LET). Photons and high-energy electrons are sparsely ionizing, whereas alpha particles are densely ionizing. Protons or neutrons are intermediate in their ionizations.

The rate of energy deposition has biologic implications. Relative biologic effectiveness (RBE) is a comparison of the relative doses of two different radiation beams required to produce a given biologic effect or endpoint. High LET or densely ionizing radiation (e.g., alpha particles) is not oxygen-dependent for its effect. High LET also alters both the shoulder and the exponential portion of the cell survival curve, presumably because the density of ionization is sufficient to greatly reduce the effectiveness of cellular repair processes, as well as eliminating the need for free radicals containing molecular oxygen to be present to produce a biologic effect.

As both the slope and shoulder portions of the survival curve are affected, a complex relationship exists between the biologic endpoint measured and the RBE. At very small dose increments, the relative absence of a shoulder for high-LET radiation will make the RBE for this beam very large, compared with the RBE of a sparsely ionizing beam of photons. As the dose increases,

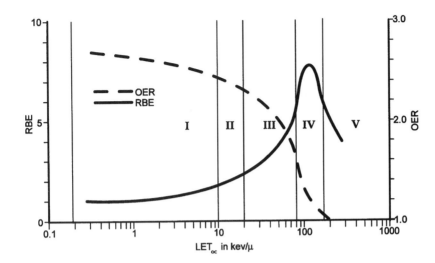

**Figure 23.13**
Relationship between RBE, OER, and LET in keV/μm of path. Note that as LET increases, OER decreases to become insignificant at LET greater than ~110 keV/μm. In addition, as LET increases, RBE increases. The late fall in RBE with further increases in LET results from overkill due to the extensive destruction of normal tissue at high LETs.

the RBE of the densely ionizing beam will decrease. The relationship between LET, RBE, and OER is shown in Figure 23.13.

## Conclusion

As knowledge of the ways in which cells are damaged by radiation increases, and as equipment and techniques to deliver radiation in ways that minimize the dose to normal tissues improve, we may hope to be able to further increase the therapeutic ratio. The fractionated stereotaxic approaches currently utilized at certain specialty clinics appear to offer this benefit to many patients with CNS tumors.

## References/Further Reading

Barendsen GW. Response of cultured cells, tumors and normal tissues to radiations of different linear energy transfer. Curr Top Radiat Res 1968;4:293–356.

Buck B, Siddon R, Svensson G. A beam alignment device for matching fields. Int J Radiat Oncol Biol Phys 1984;11:1039–1043.

Collins MK, Perkins GR, Rodriguez-Tarduchy G, et al. Growth factors as survival factors: regulation of apoptosis. Bioassays 1994;16:133–138.

Hornsey S. Iso effect relationships for normal tissue damage high and low linear energy transfer radiations. In: The biological basis of radiotherapy. Steel GG, Adams GE, Eckham MJ, eds. Amsterdam: Elsevier, 1983:167–179.

Martin SJ, Green DR, Cotter TG. Dicing with death: dissecting the components of the apoptosis machinery. Trends Biochem Sci 1994;19:26–30.

Powers WE, Tolmach LV. A multicomponent x-ray survival curve for mouse lymphosarcoma cells irradiation in vitro. Nature 1963;197:710–711.

Stewart BW. Mechanisms of apoptosis: integration of genetic, biochemical, and cellular indicators. Natl Cancer Inst 1994;86:1286–1296.

# VIII BIOMECHANICS AND THE SPINE

# Biomechanics of the Spine

Dan M. Spengler
John M. Dawson

Knowledge of the principles of spinal mechanics is essential to understand and manage patients with spinal injuries. Of course, the surgeon must appreciate the implications of neurologic injury when formulating a rational treatment approach for the spine-injured patient, but a thorough understanding of the biology of fracture repair, coupled with the natural history of nonoperative management of patients with spine trauma, is also necessary.

## Disorders of the Spine

### FRACTURE CLASSIFICATION

Thoracolumbar spine fractures have been classified recently by Gertzbein and the AO group* into a system relevant to spinal mechanics (Table 24.1), which relates the severity of the injury to the fracture type or to the prognosis or both. Type A fractures are compression injuries primarily involving the vertebral body. Type B fractures are distraction injuries involving anterior and

**Table 24.1** Classification of Thoracolumbar Fractures

A. Compression injuries
  1. Impaction (wedge)
  2. Splitting (coronal)
  3. Burst (complete fragmentation)
B. Distraction injuries
  1. Through posterior soft tissue (subluxation)
  2. Through posterior arch (chance fracture)
  3. Through anterior disc (extension spondylolisthesis)
C. Multidirectional injuries with translation
  1. Anteroposterior (dislocation)
  2. Lateral (lateral shear)
  3. Rotational (rotational burst)

*The Association for the Study of Problems of Internal Fixation (AO/ASIF) posterior vertebral elements. Type C fractures are multidirectional injuries with translation, also involving anterior and posterior vertebral elements. Injury types include compression fractures, burst fractures, seat-belt type injuries, and fracture dislocations. Denis proposes the biomechanical concept of three columns: anterior, middle, and posterior.

## Compression Injuries

Middle-column and burst fractures of the spine are associated with axial loading of the spine with or without flexion, lateral bending, and rotation (Fig. 24.1). These injuries are common (73% of all fractures) and are frequently associated with significant canal compromise, which may or may not be accompanied by neurologic injury. Controversy surrounds the ideal management for a patient with a thoracolumbar burst fracture when canal compromise is greater than 50% but there is no neurologic deficit. If surgical management is selected, instrumentation and fusion are essential in addition to decompression.

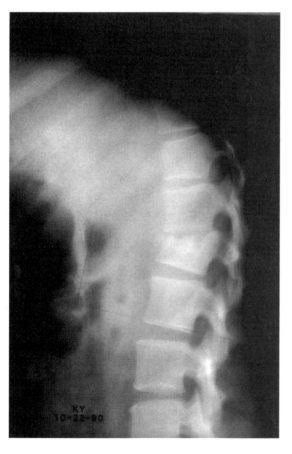

**Figure 24.1**
Lateral radiograph of a 25-year-old woman with an L-1 burst fracture. Note mild retropulsion of superior portion of L-1 end plate into spinal canal.

## Distraction Injuries

The seat-belt type injury is a common example of a distraction injury to the spine. Flexion-distraction injuries are frequently missed because the only abnormality may be a widening of the distance between the spinous processes (Figs. 24.2, 24.3). Evidence of widening may be determined in an anterior-posterior (AP) radiograph of the spine. Spinal injuries that include the disruption of the posterior ligamentous complex are highly likely to deform further because ligamentous healing will not reestablish spinal stability.

## Rotational Injuries

Rotational injuries are characterized by involvement of the lateral masses, ribs, and/or transverse processes, depending on the portion of the spine in which the injury occurred. Severity of injury is high in patients with rotational injury. The incidence of neurologic involvement is 64%, compared with 13% for patients with compression injuries. Rotational injuries are normally associated with failure of all three columns: anterior, posterior, and middle.

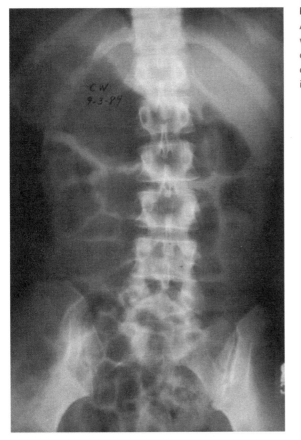

**Figure 24.2**
Anterior-posterior radiograph of a 22-year-old woman who was involved in a motor vehicle accident. Note widening between the spinous process of L-2 and L-3 consistent with a flexion-distraction injury.

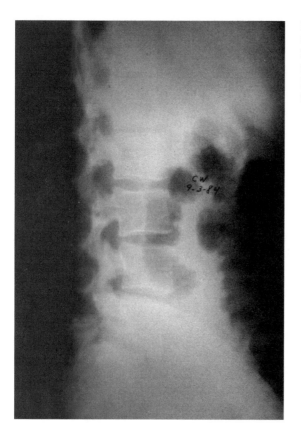

**Figure 24.3**
Lateral radiograph of patient of Figure 24.2. There is no evidence of compression deformity of vertebral bodies. This type of injury is often unrecognized because of the subtle nature of the widening of the interspinous ligament, seen only in the AP view.

## SPINAL DEFORMITIES

### Kyphotic Deformities

Kyphotic deformities of the spine often require surgical intervention to stabilize the spine and to prevent progression of deformity. Early kyphotic deformities of the spine may be managed by posterior instrumentation and fusion alone, but to halt the progression of severe deformities, a deficient anterior column must also be reinforced by anterior grafting techniques using ribs, fibulae, and/or allografts (Figs. 24.4, 24.5). These patients are at risk for neurologic injury if excessive corrective forces are applied. Intraoperative cord monitoring techniques and/or wake-up tests are essential to reduce the risk of neurologic injury.

### Lateral Deformities

A patient with thoracic scoliosis usually has an associated thoracic lordosis. Both curves must be addressed in any stabilization and/or correction procedure, using a distraction construct on the concavity of the curve and a compression construct on the convexity. A modular device for which multiple anchoring devices are available is often recommended. For example, a construct may

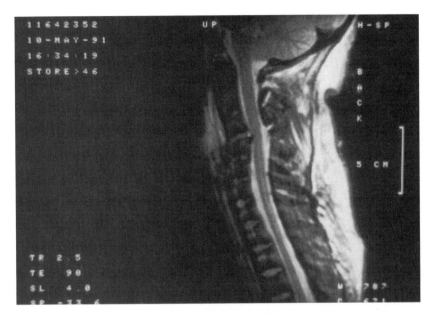

**Figure 24.4**
Sagittal magnetic resonance imaging of a patient with complete disruption of the C3-4 motion segment. Note mild subarachnoid compromise anteriorly and evidence of posterior disruption.

comprise rods in combination with laminar and pedicle hooks, pedicle screws, iliac screws, and/or transverse connectors (Figs. 24.6 through 24.8). The age of the patient and the compliance (flexibility) of the spinal deformity determine the amount of correction obtainable.

Anterior instrumentation with fusion also can be considered for selected scoliotic deformities, especially the lumbar curves where few motion segments (functional spinal units [FSUs]) need to be fused. Most anterior devices are placed on the curve convexity as compression constructs. In extensive multiplanar deformities, curves are best managed by both anterior and posterior instrumentation and fusion.

## DISC DISORDERS

Low back pain is a common complaint and usually associated with intervertebral disc disorder. Few patients require surgical intervention because most can be improved with mild pain-controlling medications and exercise to facilitate dynamic stability of the intervertebral unit.

## SEGMENTAL INSTABILITIES

Degenerative disorders of the spine can produce segmental hypermobility resulting in nerve root compression and/or irritability. Such patients are generally

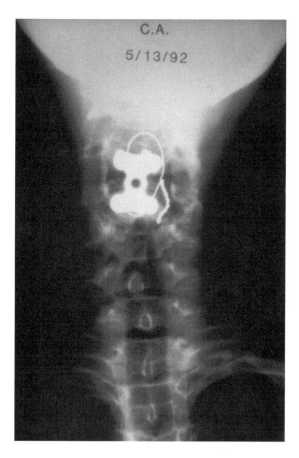

**Figure 24.5**
Postoperative AP radiograph view of the patient of
Figure 24.4. Stability has been restored by anterior
and posterior fusion and instrumentation.

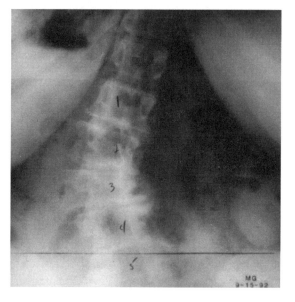

**Figure 24.6**
Anterior-posterior radiograph of a 65-year-old
woman with degenerative scoliosis and lateral lis-
thesis of the third and the fourth lumbar vertebra.

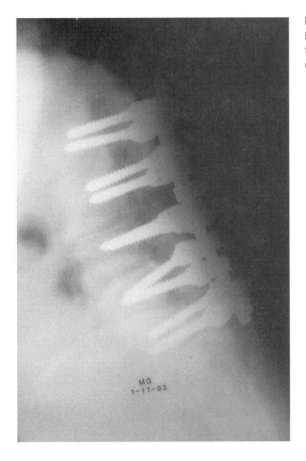

**Figure 24.7**
Lateral radiograph of the patient of Figure 24.6 following instrumentation of five vertebral bodies (four FSUs).

improved by active exercises that focus on trunk strengthening, both in flexion and extension; if there is documented mechanical instability of the spine, persistent pain, and a diminished quality of life, spinal fusion can be considered. Those with degenerative spondylolisthesis do best when decompression is combined with fusion. Spinal instrumentation enhances fusion rates.

## Mechanics of the Spine

### FORCES AND DEFORMATIONS

An FSU is a segment of the spine that represents inherent biomechanical characteristics of the spine at one spinal level. It consists of the two adjacent vertebrae and the interconnecting soft tissue without musculature.

An FSU is subjected to three types of forces in vivo: ligament forces, disc pressures, and facet contact forces. These forces cause internal loads, in resisting which internal stresses are developed within the vertebra (Fig. 24.9).

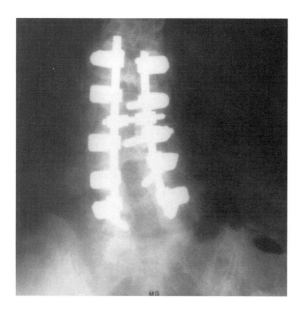

**Figure 24.8**
Anterior-posterior radiograph of the patient of Figure 24.6 following stabilization with a rod and pedicle screw construct.

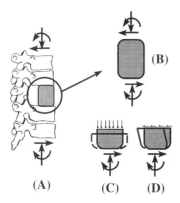

(A)     (C)     (D)

**Figure 24.9**
(A) A structure such as the spine may be subjected to external load, axial and shear loads, flexion, lateral bending, and torsion moments. (B) The external loads have an external effect (e.g., producing flexion) and an internal effect. Components that compose the structure must resist the external loads. (C) The internal force per unit area that one component or part of a component exerts on another is the stress. A distributed internal force acting perpendicular to a given surface is the normal stress. Normal stresses produce changes in length and volume (normal strains). (D) A distributed internal force acting tangential to a given surface is the shear stress. Shear stresses produce changes in shape, termed *shear strains.*

Stresses may be compressive, tensile, shear, or a combination of these. Vertebral deformations induced by these stresses are termed *strains*. The magnitude of strains is proportional to the stresses in the material and to the material's physical properties. Under identical stresses, cartilage will undergo larger strains than cortical bone.

## MECHANICAL BEHAVIOR OF TISSUES

Biologic materials (bone, cartilage, ligament, and muscle) are different than most implant materials (stainless steel and titanium) in that their mechanical responses to loading are a function not only of load, but also of the rate of loading and unloading. This behavior is termed *viscoelastic*. For example, when axially loaded, an intervertebral disc begins to decrease in height, continuing to do so until, after a long period, its height reaches a constant value (Fig. 24.10). This test, which is termed a *creep test*, characterizes the viscoelastic behavior of a given material. A *stress relaxation test* is also used to characterize this behavior. Here, a constant deformation (change of height) is imposed and the load developed in the disc is recorded: It is greatest when the disc is first compressed and then gradually diminishes over time.

The use of surgical instrumentation to distract the spine is itself a stress relaxation test. As the instrumentation is inserted, the height of the spine is instantaneously changed and thereafter held constant; over the next few days or weeks, however, the distraction forces on the spine diminish, because the ligaments, intervertebral discs, and bone exhibit stress relaxation.

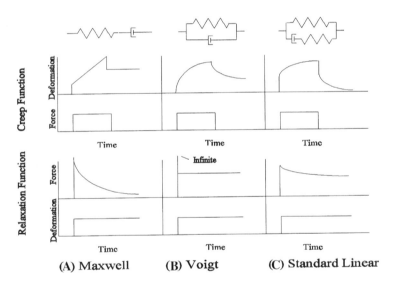

**(A) Maxwell**    **(B) Voigt**    **(C) Standard Linear**

**Figure 24.10**
Materials exhibiting viscoelastic behavior can be modeled as combinations of elastic springs and viscous dashpot. The different models produce different force-time and deformation-time responses during creep and relaxation tests. (A) In a creep test, under application of a constant force, the Maxwell model produces instantaneous deformation (elastic response) followed by linear lengthening (viscous response). When the load is removed, the material will be permanently deformed in proportion to the viscous deformation. In a relaxation test, under application of a constant deformation, the Maxwell model produces an instantaneous force (elastic response) followed by a gradual decay in the force to zero (viscous response). (B) In a creep test, the Voigt model responds with gradual lengthening. When the load is removed, the material gradually returns to its original shape. In a relaxation test, the Voigt model produces an instantaneous infinite force (because the viscous dashpot cannot respond to elongation instantaneously), after which the material maintains a constant resistive force. (C) In a creep test, the standard linear model produces an instantaneous deformation followed by gradual lengthening. When the load is removed, the material will gradually return to its undeformed shape. In a relaxation test, the standard linear model produces an instantaneous force followed by a gradual decay in the force to a constant nonzero value.

Viscoelastic materials also exhibit hysteresis. Hysteresis is the tendency of a material to store energy during loading and unloading. The energy stored represents the amount of work done.

### Bone

*Dependence on Strain Rate*   One test used to characterize a material such as bone is a uniaxial tension test: A cylindrically shaped specimen is subjected to axial tension and its elongation is recorded. When the results are graphed, the slope of the force-elongation curve defines the stiffness of the material. If the tensile force is normalized by area and the elongation by original length, then the curve describes the stress-strain behavior of the material. The slope of the stress-strain curve is the material's *modulus*. With increasing rates of strain, bone becomes (1) *stiffer* and *stronger* (Fig. 24.11), its strength being characterized by the load at which it breaks, but also (2) more *brittle*. A brittle material is one that breaks at a low amount of strain (e.g., glass). A creep test may be conducted on a sample of bone to show its viscoelastic nature. Under a constant applied load, the specimen will initially elongate (elastic part of behavior); then, over time, the bone will gradually stretch (viscous part of behavior). After a period ranging from hours to days, creep begins to increase rapidly because significant damage has occurred within the material. The sample then breaks.

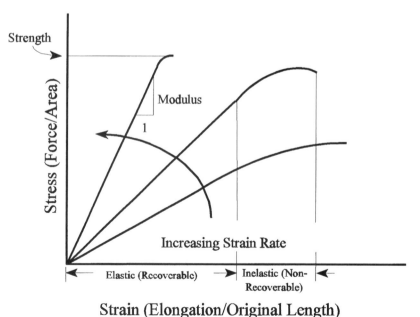

**Figure 24.11**
Results for uniaxial tension tests of dense bone. As the strain rate increases, the strength of the bone material increases, as suggested by the increasing peak stresses of the curves. Its modulus also increases, as suggested by the increasingly steeper curves, as does its brittleness, suggested by the decreasing proportion of plastic strain to elastic strain.

*Dependence on Density*   Another factor affecting the mechanical response of bone is its *density*. The stiffness of bone is roughly a function of the cube of bone density, and the strength of bone is generally a function of the square of bone density. That the mechanical properties of bone depend on its density has significant clinical implications. In a patient suffering from osteoporosis, a decrease in the density of bone by a factor of 2 will be accompanied by an eightfold decrease in its stiffness and a fourfold decrease in strength.

## Soft Tissues

*Influence of Collagen Content*   The mechanical properties of soft tissues depend on their structure. Tissues with an ordered (or an oriented) collagen fiber network (e.g., ligaments) exhibit greater initial tensile stiffnesses and greater tensile strengths than tissues with less-ordered collagen fibers (e.g., skin). At large strains, when tissues are extremely stretched, the stiffnesses of the materials may be quite similar because the less-ordered collagen fibers become aligned under these forces.

*Articular Cartilage*   The spinal zygopophyseal joints are diarthrodial, being surfaced with articular cartilage and lubricated by synovial fluid. Articular cartilage contains substantial amounts of collagen throughout the different regions of its structure. The structure of articular cartilage is arranged in zones (Fig. 24.12). In the *superficial zone*, the collagen fibers are arranged tangential to the surface. Collagen content is greatest in this zone; proteoglycan content is least. In the *middle zone*, which comprises the bulk of the tissue, collagen fibers are arranged in a three-dimensional network. Large collagen fibers are perpendicular to the surface; small collagen fibers are arranged between them like branches on a tree. Cells in this area are spherical and evenly distributed.

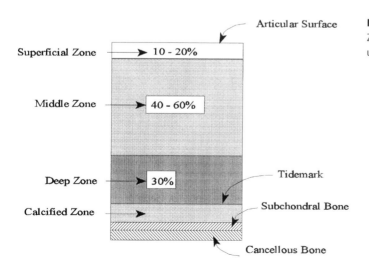

**Figure 24.12**
Zone system of architecture for articular cartilage.

Proteoglycan content is greatest in the middle zone. In the *deep zone*, collagen fibers are arranged in a tight mesh and are generally perpendicular to the surface. Cells are clustered in columns of four to eight cells. Finally, in the *calcified zone*, the calcium content is high, with few cells present.

The structure of articular cartilage relates to its function of transmitting load and facilitating motion. The superficial layer of the articular cartilage is strong in tension and resists wear. The middle and deep layers are more subject to compressive loading (hydrostatic pressure). Within the middle zone, a greater proteoglycan concentration helps to resist compressive stresses.

Articular cartilage is *non-homogeneous* and *anisotropic*. A nonhomogeneous material (bone is another example) exhibits different material properties at different points. Anisotropic materials possess different material properties in different directions at a particular point. On the articular surface of cartilage, properties are different between the direction tangential to the surface and the direction perpendicular to the surface.

**Intervertebral Disc**    The superior and inferior vertebral end plates of adjacent vertebrae and the interposed discs compose the *intervertebral joints*. The intervertebral discs comprise three distinct parts: the *nucleus pulposus*, the *annulus fibrosus*, and the *cartilaginous end plates*.

The nucleus pulposus is a hydrated gel, surrounded by the laminated bands of the annulus, and occupying 40% to 60% of the disc volume. It plays an important role in the transmission of load: Pressure developed within it pushes the annular fibers and the cartilaginous end plates outward. Tensile stresses (hoop stresses) developed within the annular fibers stiffen the disc.

The annulus fibrosus comprises fibrous tissue in concentric laminated bands. Within a band, the fibers are oriented at $\pm 30°$ from the plane of the disc, the fibers of adjacent bands being oriented in opposite directions. The mechanical properties of the disc annulus are direction-dependent, its tensile stiffness at $\pm 15°$ from the plane of the disc being about 3 times as great as in the axial direction.

The annulus fibrosus and nucleus pulposus are separated from the vertebral bodies by the cartilaginous end plates, which are composed of hyaline cartilage.

The collagen that comprises the intervertebral disc is type I and type II. Type I collagen is typically found in tissues that resist tension (e.g., skin, tendon), and type II in tissues that must resist compression (e.g., articular cartilage). Significantly, the collagen of the nucleus pulposus is primarily type II, while that of the annulus fibrosus is primarily type I. These are the types of stresses borne by each.

## LIGAMENTOUS CONSTRAINTS OF THE SPINE

There are six main ligaments associated with the articulations of the intervertebral and facet joints (Figure 24.13). These ligaments establish *syndesmotic*

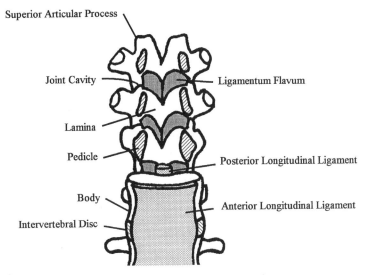

**A**

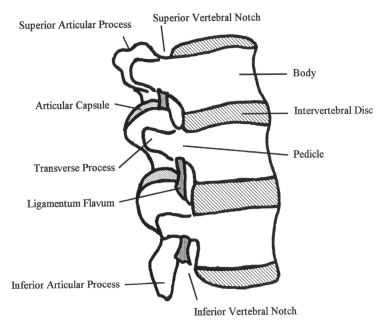

**B**

**Figure 24.13**
The ligamentous anatomy of the spine can be divided into intrasegmental ligaments (those that connect two adjacent vertebrae) and intersegmental ligaments (those that connect two or more vertebrae). Intrasegmental ligaments include the ligamentum flavum, the interspinous ligaments, the intertransverse ligaments, and the supraspinous ligament. Intersegmental ligaments include the anterior longitudinal ligament and the posterior longitudinal ligament.

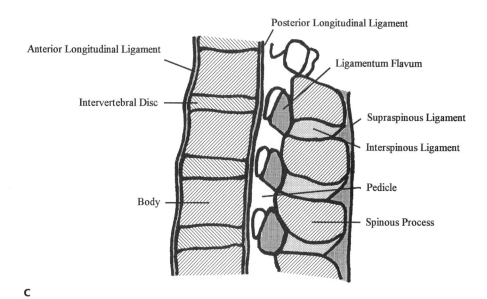

**C**

**Figure 24.13** *continued*

*amphiarthroses* between the processes and the bodies of the vertebrae: The articulations of the relatively immobile joints between the bony surfaces of the vertebrae are governed by the ligaments. The processes of the vertebrae are connected by the *ligamentum flavum*, the *interspinous ligaments*, the *intertransverse ligaments*, and the *supraspinous ligament*. Vertebral bodies are connected by the *anterior longitudinal ligament* and the *posterior longitudinal ligament*.

## Ligamentum Flavum
The ligamentum flavum is the most purely elastic tissue in the body. When the spine is in neutral flexion, the ligamentum flavum is under tension, thus maintaining compression on the discs, stiffening them and so allowing the spine to be more self-supporting.

## Interspinous Ligaments
The interspinous ligaments are best developed in the lumbar region. They resist separation of the spinous processes during flexion.

## Intertransverse Ligaments
The mechanical function of the intertransverse ligaments is uncertain because of their small cross-sectional areas, but their anatomic location in relation to the axis of the vertebral column dictates that lateral flexion and torsion activities will tense them.

## Supraspinous Ligament
The supraspinous ligament runs along the tips of the spinous processes. It is indistinct from muscle insertions in the lumbar region and is thicker and broader in the lumbar region than in the thoracic region. In flexion, this ligament stretches and, if flexion is excessive, is the first structure to fail.

## Anterior Longitudinal Ligament
The superficial fibers of the anterior longitudinal ligament span the vertebral column; its deep fibers bridge single pairs of vertebrae. Both are strongly attached to the vertebral bodies but not to the intervertebral discs.

## Posterior Longitudinal Ligament
The deep fibers of the posterior longitudinal ligament join adjacent vertebrae; its superficial fibers connect several vertebrae. In contrast to the anterior longitudinal ligament, the posterior longitudinal ligament is weakly attached to the vertebral bodies but strongly attached to the intervertebral discs. Its tensile strength is only one-sixth that of the anterior longitudinal ligament.

## MECHANICAL FUNCTIONS OF THE SPINE

### Definitions
The *stiffness* of an FSU is its ability to resist displacement under load, expressed as the ratio of the applied load to the resultant displacement. The neutral zone is that range of motion through which a motion segment may be displaced from the neutral position to one at which elastic deformation of the ligaments begins. *Stability* is the inverse of the amount of motion that an FSU exhibits. An unstable FSU exhibits greater than normal range of motion (or has a greater than normal neutral zone) under physiologic loads. The *strength* of an FSU is the load that the motion segment can sustain before failing, due, for example, to fracture of the bone, rupture of the disc, and/or rupture of the restraining ligaments.

### Resistance to Loads
*Axial Loading*   Axial loading is produced by a force acting through the long axis of the spine at right angles to the discs (Fig. 24.14). Compressive axial loading occurs due to gravity and to forces produced by ligaments and muscular contractions. Axial loads are carried mainly by discs and vertebral bodies. In the body of the vertebra, trabecular bone of the centrum carries 25% to 55% of the load, while the cortical bone of the shell carries the remainder. In some body postures, the facet joints may carry load in compression (0% to 33%).

Under compression, the nucleus pulposus loses height and tries to expand outward toward the annulus, causing compression of the vertebral end plates and tension in the annulus. The discs and trabecular bone of the centrum,

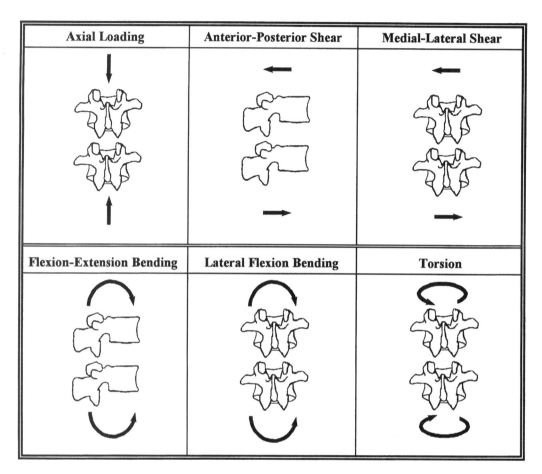

**Figure 24.14**
Loading modes to which a functional spinal unit may be subjected.

when compressed, can sustain larger deformation before failure than can the cartilaginous end plates or the cortical shell. The end plates are the first to fail (fracture), and the discs are the last to fail (rupture).

*Shear Loading*   Shear loads acting on the spine cause each vertebra to move anterior to posterior or side to side (medial-lateral) in relation to each other (see Fig. 24.14). Shear is resisted by the ligaments and, especially in the lumbar spine, by the discs and the facet joints.

*Flexion and Extension Bending Movements*   Flexion and extension bending movements cause both compression and tension in the spine (see Fig. 24.14). In forward flexion, the anterior structures (anterior aspect of the disc and anterior ligaments) are in compression. The posterior structures are in tension. In backward extension, the posterior structures are in compression (or unloaded), while the anterior structures are in tension. Resistance to flexion is provided by tension in the posterior outer annulus, across the facet joints, and in the posterior ligaments. Resistance to extension is provided by tension in the anterior outer annulus, by compression across the facet joints, by tension in the anterior longitudinal ligament, and (in extreme extension) by contact between the spinous processes.

***Lateral Flexion Bending Movements***    During lateral flexion, the ipsilateral side is in compression, while the contralateral side is in tension (see Fig. 24.14). Lateral flexion is resisted by the contralateral outer fibers of the annulus and by the contralateral intertransverse ligament.

***Torsional Movements***    Torsion is resisted by the outer layers of the intervertebral discs, and the facets (see Fig. 24.14). In the annulus, one half of the fibers resist clockwise rotations while the other half resists counterclockwise rotations. Torsional rigidity is greatest in the lumbar spine (especially at the lumbosacral junction), less in the thoracic spine, and least in the cervical spine.

## Motion

A *main motion* is the primary motion produced by the application of an external load. *Coupled motions* accompany the main motion and result from kinematic constraints such as are produced by the facets and ligaments; they occur in all regions of the spine:

- In the cervical spine, axial rotation produces the coupled motion of axial translation of C1 relative to C2. In addition, right or left lateral bending produces the coupled motion of right or left axial rotation, and rotation produces the coupled motion of lateral bending.
- In the thoracic spine, coupling patterns of motion are similar to those in the cervical spine.
- In the lumbar spine, left axial rotation produces the coupled motion of right lateral bending at L1-2 and L2-3 and left lateral bending at L4-5 and L5-S1. Flexion is coupled with rotation at all levels. Right lateral bending produces the coupled motion of left axial rotation from L2-3 to L5-S1.
- Lateral bending also produces flexion.
- At the lumbosacral joint, right axial rotation produces right lateral bending.

# Treatment of Spinal Disorders

## QUANTITATIVE DIAGNOSIS

A knowledge of biomechanics can be helpful in assessing the large number of patients who have clinical symptoms of low back pain, with or without sciatica, but who have few, if any, objective findings on examination. In particular, the use of a multi-axis dynamometer can be worthwhile. Most patients with low back pain exhibit a marked decrease in lumbar spine motion on physical examination, but assessment of this requires patient compliance and effort and the examiner may be uncertain whether pain is the cause or there is a lack of effort. Isometric and dynamic activity are assessed during dynamometer testing (Fig. 24.15), primary and secondary axes of motion being described in terms of the range of motion and the torque produced.

The concept of coupled motion helps interpret test results (Fig. 24.16). For example, a patient normally is able to generate more flexion torque in making the primary motion, compared with when this is due to coupled motion. If the patient generates only small flexion torque when asked to flex forward, but a higher flexion torque when asked to bend laterally to the right, it can be inferred that he or she is not giving full effort. Likewise, if a patient

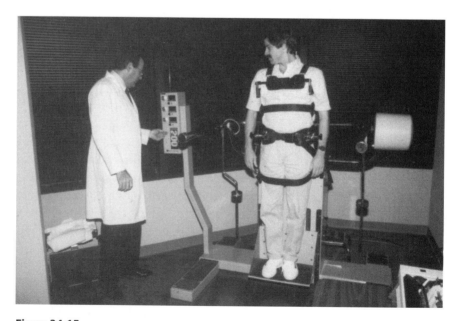

**Figure 24.15**
Patient in a dynamometer. This device measures velocity, range of motion, and torque in three axes simultaneously. This device can be used to identify patients who are not exerting maximal effort, but it is also useful for tracking a patient's progress during rehabilitation.

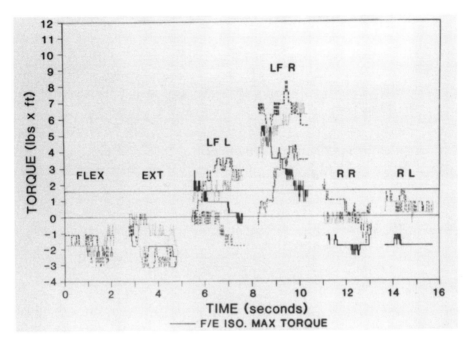

**Figure 24.16**

A torque-time reproduction from a patient who demonstrated poor effort. Essentially, no positive torques were generated in the flexion-extension torques (shown on left side of graph) because he complained of increasing pain with any attempt at forward flexion on the machine. The abbreviations LFL and LFR refer to lateral flexion to left and right, respectively; and the LFR torques are close to nine foot-pounds. Thus, the amount of torque exhibited with lateral flexion exceeded that demonstrated when the patient was asked to bend forward, suggesting maximal effort was not exerted in the latter case. By positive use of this information, the patient was encouraged to return to work within about 3 weeks. RR, rotation to right; RL, rotation to left.

demonstrates higher torques during dynamic motion than during isometric activity, a lack of full effort during the isometric testing can be presumed. Such comprehensive assessments can lower health care costs, assist with dispute resolution, and reduce the tendency to over-investigate and over-treat these patients.

## TREATMENT WITH SURGICAL INTERVENTION

Once the underlying mechanism of injury is recognized, a treatment plan can be developed and carried out with a high likelihood of success. For example, in a patient with a distraction injury, surgical management would include a construct that creates compression across the injured motion segments or FSU. Using a distraction construct across a spinal segment injured in a distraction mode would normally represent faulty decision analysis; however, there are exceptions. Harrington distraction rods may be used in the treatment of selected patients with "distraction" injuries because, if properly inserted, they act in four-point bending instead of pure distraction. Compression is then maintained by a loop of 18-gauge wire passed through the bases of the spinous processes above and below the distracted segment and tightened to prevent "over-distraction."

## TREATMENT WITHOUT SURGICAL INTERVENTION

Candidates for the nonoperative treatment of spinal deformity should have kyphotic deformity or lateral curvatures (scoliosis) of less than 15°. Anterior body height of the injured vertebra should be greater than 50% of the posterior height, and displacement should not exceed 5 mm. Finally, the patient should be neurologically intact.

## POTENTIAL DEFORMITIES

The potential for subsequent deformities also must be recognized and dealt with at the time of the initial intervention. For example, in the cervical spine, wide bilateral, multilevel laminectomies may result in "swan neck" deformities if fusion is not performed. This is difficult to treat once it develops (Fig. 24.17). Progressive deformity may be prevented by stabilizing the spine at the time of initial decompression.

## SPINAL INSTRUMENTATION

A few general remarks about spinal instrumentation may be of assistance to surgeons who are evaluating the multiple systems now available in the marketplace. A recommendation for any specific system is not intended.

Issues that must be considered before purchasing spinal instrumentation include effectiveness, ease of application (as judged from personal use and/or visits to surgeons who use the product), cost, reputation of the company that makes the product, and service by the company and the local company representative. A large choice of products is available, but there is no consensus regarding which one to choose.

Effectiveness may be difficult to assess because scientific publications that deal with specific products may be few. The company should identify what studies were done and by whom (e.g., an independent investigator or in-house testing). The stiffness of the construct compared with other products and with the intact human spine should be reviewed, and its performance under destructive testing and cyclic loading should be known. Design characteristics of hooks and screws should be evaluated with special attention to the perfor-mance of screws in cyclic loading. Long-term data (24 months or longer)

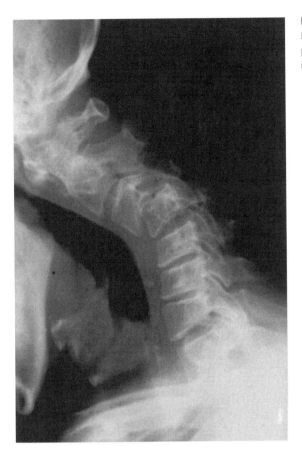

**Figure 24.17**
Lateral radiograph of a patient with significant kyphus after posterior injury to the cervical spine. This is often described as a swan neck deformity.

derived from clinical studies should be considered as well. Patient selection methodology, success of fusion, and complication rates contribute additional information. Knowledge of potential long-term problems, such as deformity at the motion segment above the rigid spinal construct, also must be considered.

Constructs that use pedicle screws have received some negative media coverage. Their insertion can lead to serious complications (e.g., neural injury), but provided the indications and technique for insertion are correct, the complication rates should be no higher than those of any other type of instrumentation.

Debate continues as to whether to rely primarily on anterior constructs, on posterior constructs, or on their combination for the greatest likelihood of fusion mass consolidation. A most important biomechanical factor for choosing an approach is knowing the source of instability: In a burst-fracture model of the lumbar spine using porcine spines, compressive axial stiffness and torsional rigidity were measured. Posterior instrumentation, posterior instrumentation with an anterior strut, and anterior instrumentation with an anterior strut were used from one level above to one level below the injury to restore the stiffness and rigidity of the spine. Anterior instrumentation with an anterior strut best restored the axial and bending stiffness of the spine. Posterior instrumentation with an anterior strut graft was less effective, and posterior instrumentation alone was worst. It seems that reconstruction of the anterior column with a rigid strut is an important consideration in anterior and/or posterior instrumentation. For the treatment of spinal deformities, the placement of instrumentation depends on the degree of kyphosis and/or scoliosis and on the number of FSUs to be stabilized by the instrumentation.

### Graft Placement

A graft should be positioned to prevent or reduce movement about every axis of motion. The instantaneous axis of rotation of an FSU is the axis about which the vertebra is rotating at an instant of time. Theoretically, placement of a fusion mass at the maximum distance from the instantaneous axis of rotation will be most effective to prevent movement around that axis. Therefore, because it is closer to the axes of motion, an anterior bone graft has less leverage than a posterior one with regard to its efficiency in stabilizing the spine, and should be less effective at preventing movement (flexion/extension, axial rotation, lateral flexion) than a posterior bone graft.

Lee and Langrana studied the clinical application of these principles in a biomechanical study of cadaver spines following spine fusion. They found that posterior fusion allowed anterior motion. Bilateral lateral fusion afforded good stabilization. Axial and bending stiffnesses were greatest for anterior fusions, intermediate for bilateral lateral fusions, and least for posterior fusions. Their results also suggest that the joints of adjacent unfused segments experience greater than normal stresses, especially facet joints. Posterior fusion generated the highest stresses on the adjacent unfused segments and bilateral lateral fusion affected adjacent unfused segments least.

To eradicate motion at the disc interspace, a graft construct that prevents motion between the fused vertebrae must be employed. An interbody fusion eliminates movement between vertebrae but also removes all or part of the intervertebral disc. When an anterior graft is used as an interbody spacer or to resist compressive forces, it should be placed more anteriorly.

A graft used to correct kyphotic deformity cannot effectively stabilize or stiffen the spine if it is too close to the apex of a deformation curve. This applies whether the graft is in tension or compression. If a strut graft is required for maintenance of stability against progressive kyphotic deformity, a long strut located far from the apex of the deformity curve offers greater leverage than a shorter, nearer strut (Fig. 24.18). A long strut, however, may fail because of buckling. The best solution may be to use two struts: one at a moderate distance (to share the load and prevent buckling) and one at a greater distance (for greater leverage).

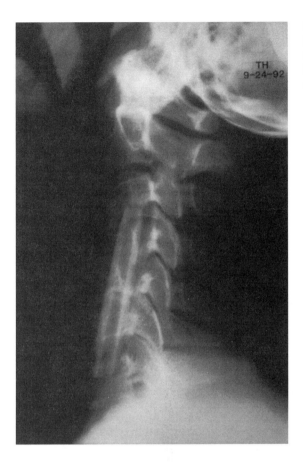

**Figure 24.18**
Lateral radiograph of a man who underwent an anterior fibular graft following decompression from C3 to C7. The graft is able to resist compression and is placed anteriorly within the vertebral column.

# References/Further Reading

Brown T, Hanson RJ, Yorra AJ. Some mechanical tests on the lumbosacral spine with particular reference to the intervertebral discs. J Bone Joint Surg [Am] 1957;39:1135–1164.

Denis F. The three column spine and its significance in the classification of acute thoracolumbar spinal injuries. Spine 1993;8:817–831.

Galante JO. Tensile properties of the human lumbar annulus fibrosus. Acta Orthop Scand 1967;Suppl 100:1–91.

Gertzbein SD. Spine update. Classification of thoracic and lumbar fractures. Spine 1994; 19:626–628.

Gurwitz GS, Dawson JM, McNamara MJ, et al. Biomechanical analysis of three surgical approaches for lumbar burst fractures using short-segment instrumentation. Spine 1993;18:977–982.

Keller TS, Szpalski M, Spengler DM, Hayez JP. Interpretation and parameterization of dynamic trunk isoinertial movements using an ensemble averaging technique. Clin Biomech 1993;8:220–222.

King AI, Prasad P, Ewing CL. Mechanics of spinal injury due to caudocephalad acceleration. Orthop Clin North Am 1975;6:19–31.

Lee CK, Langrana NA. Lumbosacral spinal fusion. A biomechanical study. Spine 1984; 9:574–581.

McElhaney JH. Dynamic response of bone and muscle tissue. J Appl Physiol 1966;21:1231–1236.

Panjabi MM, White AA. Biomechanics A to Z. In: White AA, Panjabi MM, eds. Clinical biomechanics of the spine. 2nd ed. Philadelphia: Lippincott, 1990:635–696.

Panjabi MM, White AA. Physical properties and functional biomechanics of the spine. In: White AA, Panjabi MM, eds. Clinical biomechanics of the spine. 2nd ed. Philadelphia: Lippincott, 1990:1–83.

Rockoff SD, Sweet E, Bleustein J. The relative contribution of trabecular and cortical bone to the strength of human lumbar vertebrae. Calcif Tissue Res 1969;3:163–175.

Urban J, Holm SH, Lipson SJ. Biochemistry. In: Weinstein JN, Wiesel SW, eds. The lumbar spine. Philadelphia: Saunders, 1990:231–265.

# A Rational Approach to Internal Fixation of the Cervical Spine

Paul R. Cooper

The use of internal fixation represents one of the most important advances in the management of disorders of the cervical spine during the past 50 years. Internal fixation effectively stabilizes the spine while bony healing takes place in patients with posttraumatic instability, after debridement and bone grafting in patients with osteomyelitis, and after decompression and bony fusion for degenerative disease.

Because the theoretical and practical considerations of the upper cervical spine (atlantooccipital and atlantoaxial articulation) are so different from those of the lower cervical spine (the third through the seventh cervical vertebrae), this chapter is divided in two distinct sections. Technical considerations are addressed, but the emphasis will be on decision making in the choice of fixation devices. Wiring techniques have been available for many decades and are well described in the literature and are not discussed. Although occipitocervical fixation devices are sometimes used for stabilizing the upper cervical spine, space constraints preclude their inclusion in this chapter.

## General Goals of Management

The goals of management consist of (1) prevention of injury to the spinal cord by immobilizing the unstable spine and reducing subluxations; (2) amelioration of the effects of acute spinal cord injury by administering high-dose corticosteroids, maintaining adequate systemic BP and oxygenation; (3) removal of bone or soft tissue compressing the spinal cord or nerve roots; (4) establishing a diagnosis in patients with instability resulting from nontraumatic conditions; (5) restoration of the mechanical stability of the spine; and (6) prevention of late spinal deformity.

## Imaging Assessment

### PLAIN FILMS

Initial assessment begins with plain films. It is essential that the patient's shoulders be pulled caudally to enable visualization of the C7 vertebral body in the lateral projection. If this is not possible, a swimmer's view should be obtained. Oblique or anteroposterior views may sometimes disclose rotatory subluxations or fractures that were not appreciated on the lateral views. An open-mouth view of the upper cervical spine should be obtained in all patients with suspected instability, especially when there is a history of trauma; this view is frequently the only means of diagnosing a fracture of the odontoid process of the dens.

In patients who present with instability in the absence of trauma, disc space collapse with adjacent vertebral body destruction and soft tissue swelling is suggestive of an infectious process. Although metastatic disease may involve adjacent vertebral bodies, the intervening disc space is preserved and involvement of the posterior elements is more likely to be present than is the case with infection. Routine computed tomography (CT) scanning usually helps to define the etiology of destructive processes.

### MAGNETIC RESONANCE IMAGING AND COMPUTED TOMOGRAPHY MYELOGRAPHY

Patients with evidence of a cervical spine fracture or dislocation or vertebral destruction from neoplasm or infection who are neurologically intact or have residual neurologic function below the level of the injury should have magnetic resonance imaging (MRI) performed to ascertain the presence of dural compression by a herniated disc, hematoma, bone, or soft tissue. If MRI is not available or is precluded by clinical circumstances, myelography with postmyelography CT scanning is essential.

In patients with complete neurologic deficit below the level of vertebral involvement, this author does not perform CT/myelography or MRI. It is unlikely that operative decompression of patients with complete deficit improves neurologic function and it is therefore unimportant to assess the presence of continued compression by bone or soft tissue with CT/myelography or MRI. This is a controversial issue and there are others who believe that operative decompression is indicated regardless of the patient's neurologic status.

### COMPUTED TOMOGRAPHY

The bony anatomy of a fracture or vertebral involvement from neoplasia or infection must be defined fully at some point by a CT scan with bone windows. This study is generally adequate to assess the extent and type of bony injury or destruction to determine the most appropriate type of stabilization procedure. Sagittal or coronal CT reconstruction of the axial CT will readily demonstrate almost all occult fractures of the dens and clarify the anatomy of other fractures.

### FLEXION AND EXTENSION FILMS

Flexion and extension films are obtained if there is doubt about the stability of the cervical spine after other imaging studies have been performed. Dynamic studies, however, are not appropriate in patients with subluxation visualized on plain x-rays, fractures or vertebral destruction demonstrated on plain x-rays or CT, which are likely to render the spine unstable, or in patients with spinal cord compression demonstrated on MRI or postmyelography CT scanning.

## Internal Fixation of the Lower Cervical Spine

The anatomic structure of the third through the seventh cervical vertebrae makes them ideally suited to the use of internal fixation devices. Anteriorly, screws may be used to anchor cervical plates to the vertebral bodies; posteriorly, well-defined lateral masses provide secure attachments for screws and plates that may be used to immobilize the facet joints.

Most often, instability of the lower cervical spine is caused by trauma. However, spinal reconstruction for iatrogenic instability produced as a result of extensive decompression for degenerative conditions frequently requires internal fixation to prevent deformity or postoperative instability. Similar considerations apply for reconstruction after decompression for osteomyelitis or neoplastic disease. Although less attention will be paid to the use of plating for these nontraumatic conditions, the principles relevant to the use of internal fixation for trauma also apply.

### TRAUMA

The choice of approach for stabilizing the cervical spine after trauma depends on (1) the mechanism and type of spinal injury; (2) the presence of residual spinal cord compression; (3) the type of neurologic deficit, if any; and (4) the skills and preferences of the surgeon. Frequently, either an anterior or posterior approach is satisfactory to achieve immediate stability and prevent long-term spinal deformity.

In patients with posttraumatic instability, the only *absolute* indication for anterior instrumentation is the presence of anterior compression of the spinal cord with preservation of neurologic function below the level of the injury. The anterior approach appears to be a more appropriate means to stabilize injuries involving the vertebral body. For example, vertebral body compression fractures with angulation are more logically treated with vertebrectomy, bone grafting to restore the height of the anterior column, and anterior cervical plating than with posterior instrumentation. Although posterior stabilization procedures may be successful for these injuries, instrumentation failure may occur as the spine assumes a kyphotic position due to lack of anterior support.

The only absolute indication for posterior stabilization is posterior spinal cord compression from bone or soft tissue. In general, it is most appropriate to use the posterior approach to treat instability resulting from injuries to the bony and ligamentous structures of the facet joints.

Occasionally, both anterior and posterior fixation may be required to ensure fixation when there is three-column spinal disruption. This situation should be suspected when there is (1) retrolisthesis and angulation of the superior vertebra on the next inferior one, suggesting disruption of the anterior and posterior longitudinal ligaments; (2) facet dislocation; and (3) distraction of the posterior interspinous ligaments with an anterior shear dislocation of one vertebra on another. Patients with a disc herniation requiring anterior cervical decompression and fusion, who also have ligamentous disruption, also need anterior and posterior instrumentation to minimize the chances of instrumentation failure.

## NONTRAUMATIC CONDITIONS

In patients with cervical instability from neoplastic or infectious processes, the extent and location of bony involvement determines the most appropriate means of internal fixation. For example, in patients with metastatic tumors with vertebral body collapse, kyphotic deformity, and anterior spinal cord compression by bone or soft tissue, anterior instrumentation with vertebrectomy and vertebral body replacement is indicated. When a posterior tumor produces dorsal compression of the spinal cord and destruction of the facet joints, posterior instrumentation utilizing lateral mass plating is indicated.

## POSTERIOR STABILIZATION PROCEDURES

### Lateral Mass Plates and Screws

The application of plates and screws similar to those originally described and popularized by Roy-Camille is technically simple, produces immediate stability of the cervical spine from C2 to C7 without the need for complex orthoses such as the halo vest, and is not dependent on the integrity of the laminae or spinous processes (Fig. 25.1).

Posterior plating is ideal for facet dislocations caused by fractures or ligamentous injuries. It is especially appropriate for patients with laminar and spinous process fractures when these structures cannot be wired. Although

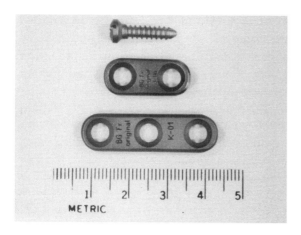

**Figure 25.1**
Original Roy-Camille plates with two and three holes, which are 13 mm apart (center hole to center hole), for stabilizing one and two motion segments, respectively. Plates are also available with four or five holes. Screws are 16 mm in length and are used to secure the plate to the lateral mass. The Haid plates are similar to the Roy-Camille plates, except that they are made with an inter-hole distance that is either 13 or 15.5 mm in length.

posterior plates also may be used to stabilize the cervical spine for injuries to the vertebral bodies, anterior plates provide superior fixation.

Posterior plates and screws should not be used in patients with soft bone because screw pull-out and loss of reduction is likely. Posterior plating also is contraindicated in patients with residual neurologic function and persistent anterior compression of the spinal cord by bone, disc, or soft tissue.

Posterior plates and screws are usually ineffective in the treatment of fixed or progressive kyphotic deformities; patients with such conditions are generally better managed by an anterior approach with reduction of the kyphus, bone grafting, and anterior plating. Such patients, however, also may need supplemental posterior plates and screws.

No plating system has been approved by the Food and Drug Administration for cervical spine fixation. All are safe and effective, however, and are widely used for posterior cervical fixation. Plating systems that are widely used for lateral mass fixation by spinal surgeons and that are commercially available in the United States are the Haid Universal Bone Plate (American Medical Electronics, Dallas, TX), titanium reconstruction plates (Synthes U.S.A., Paoli, PA), and the Axis fixation system (Danek, Memphis, TN).

The Haid Universal Bone Plate is made of titanium and contains two, three, four, or five holes to stabilize one, two, three, or four motion segments, respectively. The distance from the center of one hole to the center of the adjacent hole is 13 or 15.5 mm. The plates are held in place by titanium screws 3.5 mm in diameter. The initially marketed screws were 16 mm in length, but screws of multiple lengths are now available. The plates cannot be bent in the operating room but are curved slightly to restore the normal lordotic curve of the cervical spine.

Synthes U.S.A. markets titanium reconstruction plates 9 mm wide and 2.7 mm thick. Two varieties of plates are marketed, one with a distance of 8 mm from the center of adjacent holes and another with a 12-mm distance.

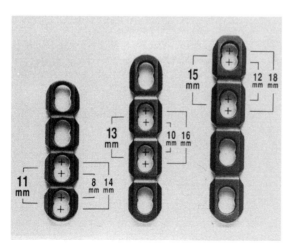

The plates are made in a variety of lengths and may be cut to the appropriate length, depending on clinical circumstances. They also may be bent to fit the bony anatomy of the lateral masses and restore the lordotic curvature of the cervical spine. The cortical or cancellous titanium screws are 3.5 mm in diameter and are available in a variety of lengths.

Sofamor Danek (Memphis, TN) has recently marketed the axis fixation system (Fig. 25.2). These contourable plates are made of a titanium alloy with 11-, 13-, and 15-mm inter-hole distances (center hole to center hole). Each hole allows for two positions of screw placement, allowing for considerable variability. The plates are marketed with two, four, six, and eight holes. Both cortical and cancellous screws 3.5 or 4 mm in diameter are available in a variety of lengths.

Patients are brought to the operating room in cervical traction. In patients with neurologic function below the level of their injury, electrodes are placed for evoked potential monitoring. After awake fiberoptic intubation, the patient is turned to the prone position in a three-pin head holder, and an x-ray is taken to confirm that alignment has been maintained.

A standard midline incision is made, and the muscles are dissected off the bone far laterally to expose the entire lateral mass of the vertebrae to be plated. If the spine is subluxed because of jumped facets, intraoperative reduction is achieved.

The center of the lateral mass to be plated is located midway between the superior and inferior facet joints and between the lateral edge of the lateral mass and the lateral extent of the lamina. A hole is drilled 10 mm in depth in the center of the lateral mass.

## Haid Plates

If Haid plates are to be used, a 2.7-mm drill bit is directed approximately 20° laterally and 10° to 20° rostrally. This latter trajectory places the screws well

**Figure 25.2**
Axis plate manufactured by Sofamor Danek. Plates are made with two, four, six, or eight holes. Each hole is notched, allowing the screw to fit in either the upper or lower part of the hole. This configuration allows for a variety of distances between screws, a useful feature when stabilizing more than one motion segment.

away from the lower facet joint and avoids including an additional motion segment in the construct (Fig. 25.3). The screws are placed in the lateral masses of the vertebrae above and below the motion segment to be stabilized. Thus, if the C4-5 motion segment is to be stabilized, the screws are inserted in the lateral masses of C4 and C5.

Although screws of multiple lengths are available, we use 16-mm-long screws in all patients. Two-hole plates with a 13-mm inter-hole distance are almost always appropriate for stabilizing one motion segment. In all cases, plate placement must be bilaterally symmetrical. Thus if a two-hole plate is used on one side, a two-hole plate must be used contralaterally and at the identical level.

If a three-hole plate is to be used to stabilize two motion segments, a plate with a 13-mm inter-hole distance is usually satisfactory. Occasionally in large patients, a three-hole plate with a 15.5-mm inter-hole distance is necessary to stabilize two motion segments. If four- or five-hole plates are needed to stabilize three or four motion segments, it is essential that ones with a 15.5-mm inter-hole distance be used. In utilizing plates with three holes, the most rostral and caudal holes are placed first and the screws are tightened. The middle hole is then drilled in the lateral mass, the exact point being determined by the position of the holes in the plates. These, however, should be as close as possible to the center of the lateral mass. The middle screw is then placed and tightened.

Posterior plates stabilize the spine so effectively that spontaneous fusion occurs at the facet joint or sites of bony injury. Therefore, in patients who have sustained trauma within several weeks of operation, bone grafts are not

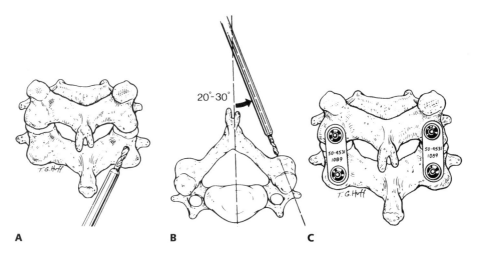

**Figure 25.3**
(A) Point of entry in the lateral mass for the drill hole. Note that the drill is angled 10° to 20° rostrally. (B) Axial view shows 20° lateral direction of the drill so that the screw avoids the vertebral artery. (C) Placement of the plates after the screws have been secured into the lateral mass.

used. Although we have had patients who have lost reduction, this has always occurred during the first weeks after operation as a result of faulty screw placement or placement of the plates at inappropriate levels.

## Synthes Reconstruction Plates

The technique for the insertion of the Synthes reconstruction plates is identical to the technique for the use of the Haid Universal Bone plates, with a few exceptions. The holes for the 3.5-mm diameter screws used with the Synthes reconstruction plate are drilled to a depth of 10 mm with a 2-mm bit, followed by a 3.5-mm cancellous tap. Cancellous screws 16 mm in length are used. For stabilization of one motion segment using two-hole plates, the plates with a distance of 12 mm from the center of adjacent holes are ideal. For stabilization of two or more motion segments, the use of alternate holes of the plate with 8-mm inter-hole spacing is ideal. The Synthes reconstruction plates are bent to fit securely against the lateral masses.

## Decision Making: Length and Location of Plates

Two-hole plates are ideal for patients with single-level subluxations or facet dislocations (Fig. 25.4). Anterior instrumentation and stabilization are usually

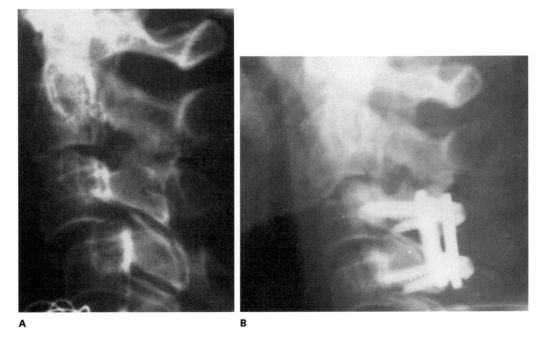

**A**                                    **B**

**Figure 25.4**
(A) Lateral cervical spine x-ray shows a C3-4 subluxation in a patient with a facet dislocation sustained as a result of trauma. (B) Postoperative x-ray shows excellent alignment at C3-4 with lateral mass plates in place.

preferred when there is injury to the vertebral body. If posterior plating is carried out in patients with vertebral body injury, three-hole plates should be used to stabilize both motion segments adjacent to the injured vertebral body, even though there may be a subluxation at only one level.

When there is a fracture of the lateral mass or pedicle, plating of that lateral mass will do nothing to restore stability, as it has lost its connection with the vertebral body. In this situation, an additional adjacent motion segment must be stabilized using a three-hole plate (Fig. 25.5). Thus, if there is a subluxation at C4-5, with an associated fracture of the lateral mass of C4, application of three-hole plates from C3-5 will be necessary. The use of three-hole plates also is appropriate when there is instability at two adjacent motion segments.

The need to stabilize three or four motion segments using four- or five-hole plates is infrequent but if necessary, these longer plates provide excellent posterior fixation. They may be particularly useful when applied prophylactically after laminectomy in patients who might be expected to develop a kyphotic deformity.

## Special Considerations for C2-3 Instability

If internal fixation is required for C2-3 instability, posterior plates and screws may be used. Because C2 has a large pedicle, this structure is used for engaging the screw rather than the lateral mass. The drill hole is made in C2 3–4 mm lateral to the medial border of the C2-3 facet joint and just rostral to the C2-3 articulation and directed superiorly at a very acute angle. The medial-lateral orientation is just lateral to the C2 pedicle to avoid the vertebral artery. Screws 20–24 mm in length are used to engage the pedicle of C2. Screws that are too long will pass through the C1-2 articulation and engage the C-1 lateral mass. At C3, 16-mm screws are placed in the lateral mass, as described previously.

The neck is immobilized using a Philadelphia collar. Lateral cervical spine films are taken once or twice during the first week after operation and monthly until the collar is removed. At the end of 3 months, the collar is removed and flexion and extension films of the cervical spine are taken to confirm stability. Lateral mass plating of the cervical spine is most appropriate for the management of fractures and subluxations from C3 to C7.

## ANTERIOR CERVICAL PLATES

### Indications

Anterior plating to stabilize the cervical spine is indicated when instability results from trauma to the vertebral body or adjacent discs, when there is destruction of the vertebral body by neoplastic disease or infection, or when extensive degenerative disease necessitates vertebrectomy and bone grafting. In

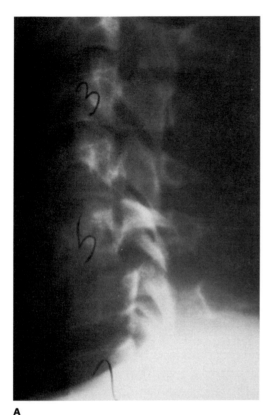

A

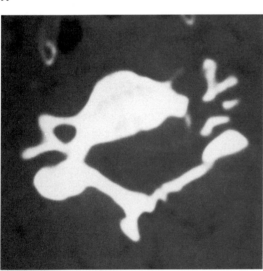

B

**Figure 25.5**
(A) Lateral cervical spine x-ray shows a C5-6 subluxation secondary to a facet dislocation. (B) Axial CT through the C5 vertebra shows a fracture of the C5 lateral mass. (C) Postoperative x-ray shows three-hole lateral mass plates that have been placed from C4 to C6. Inclusion of C4 was necessary because no screw could be placed in the fractured lateral mass of C5 on the left.

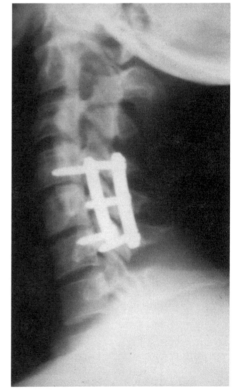

C

each of these instances, anterior plating prevents graft extrusion after removal of the disc or vertebral body.

## Technical Considerations

A standard approach with dissection of the soft tissues of the neck is used and has been well described in the literature. For patients with a posttraumatic cervical disc herniation and instability, a bone graft is placed after the disc is removed. If the vertebral bodies adjacent to the disc herniation are intact, minimal bone removal is necessary. When vertebrectomy is indicated, the adjacent discs are removed along with the cortical bone of the vertebral bodies above and below the resected vertebra.

Secure graft placement is enhanced by distraction of the vertebral bodies adjacent to the graft, using the distraction device designed by Caspar or by manual traction on the neck. A tricortical bone graft from the iliac crest is generally satisfactory for two-level vertebrectomies and is impacted into the area of the vertebrectomy, after which distraction is released, securing the graft in place. For three-level vertebrectomies, an iliac crest graft may be too short, so fibula (autograft or allograft) must be used. After the graft is impacted, distraction is released.

Two basic types of plating systems are available for stabilizing the cervical spine from an anterior approach: those with screws that engage the posterior vertebral cortex, such as the Caspar or Orozco plates, and the cervical spine locking plates marketed by Synthes, with screws that do not penetrate the posterior vertebral cortex.

The plates designed by Caspar, marketed by Aesculap (Tuttlingen, Germany), are made of titanium or stainless steel and are secured in place by 3.5-mm diameter cortical screws (Fig. 25.6). The plates are available in a number of lengths and are secured in place with the end holes over the mid-portion of the vertebral bodies to be plated.

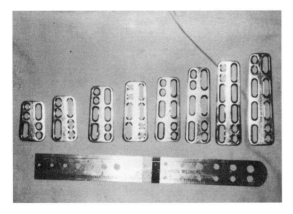

**Figure 25.6**
Caspar plates in a variety of lengths for stabilizing one or more motion segments of the cervical spine.

The plate is contoured and placed over the graft and adjacent vertebral bodies; its rostral-caudal position in relation to the vertebral bodies is confirmed utilizing intraoperative fluoroscopy.

Holes for the screws are drilled in the vertebral bodies at least 3–4 mm away from the rostral or caudal margins of the vertebral body, until the posterior cortex is reached. The first screw is tightened in place. The second screw hole is placed in similar fashion diagonally opposite the first, and a screw of appropriate length is tightened, followed by the remaining two screws. Ideally, the screw tips should just penetrate the posterior vertebral cortex (Fig. 25.7).

The cervical spine locking plates marketed by Synthes obviate some of the disadvantages of the Caspar plating system. In particular, fixation does not depend on engaging the posterior vertebral cortex, minimizing concern about

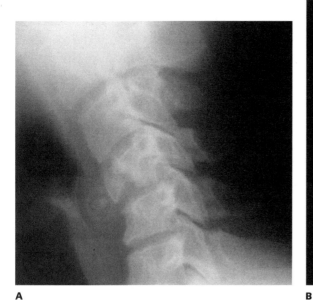
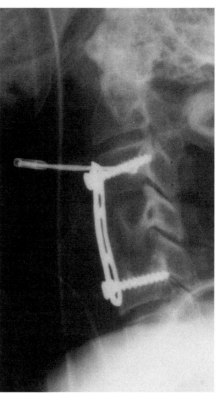

A                                                    B

**Figure 25.7**
(A) Lateral cervical spine x-ray shows kyphotic deformity of the cervical spine secondary to fracture of the C4 vertebral body. (B) Postoperative x-ray after a C4 vertebrectomy was performed. A bone graft has been inserted at the site of the vertebrectomy and a Caspar plate has been placed from C3 to C5. Note that the screws engage the posterior vertebral cortex.

penetration of the spinal canal. The screws and plates are made of titanium and permit imaging using MRI. The locking plates come in a variety of sizes. If placed correctly, the screws are locked to the plate. Although the screw-plate construct can back out of the bone, individual screws cannot disengage from the plate.

The soft tissue dissection in the neck is identical to that described for the Caspar plates. Fluoroscopy is used to confirm the relationship of the plate to the vertebral body. Meticulous placement of the plate in relation to the underlying vertebral bodies is essential, because the angle of the screw to the plate is fixed and cannot be altered to avoid a disc space or the edge of a remaining vertebral body adjacent to the bone graft.

For the screws to lock to the plate, the screws holes must be drilled into the vertebral bodies at a fixed angle in relation to the plate. A drill guide fits into the hole in the plate, ensuring that the holes are correctly angled (Fig. 25.8).

The 14-mm screws are 4.0 mm in diameter and fit precisely into the holes in the plate, to which they are locked by small screws that expand the head of the larger screws. Screws 4.5 mm in diameter are now available for longer plates, which are used for two- or three-level vertebrectomies.

## Internal Fixation of the Upper Cervical Spine

### CAUSES OF INSTABILITY

Atlantoaxial instability may result from trauma, degenerative disease, congenital or developmental disorders, neoplastic processes, and infection, and following removal of the dens to treat tumors or bony compression.

**Figure 25.8**
Caspar plate with drill guide in place. The drill guide engages the plate hole at a predetermined angle to ensure precise angulation of screw in relation to plate, a feature that is essential for locking of the screw to the plate.

## IMAGING EVALUATION

The evaluation of patients with actual or presumed instability of the C1-2 articulation should include studies to define the degree of subluxation, the presence of instability, the nature of bony injury or destruction, and the presence of spinal canal impingement or spinal cord compression by bone or soft tissue.

Evaluation begins with plain films, with open mouth views to visualize the odontoid and the C1-2 articulation. Computed tomography is useful to define bony destruction in patients with osteomyelitis or neoplastic disease and to define fractures of C1 and C2. Axial scans supplemented by sagittal or coronal reconstructions accurately define most bony abnormalities of this region. A subsequent MRI will define the relationship of the spinal cord to soft tissue within the spinal canal and has largely replaced CT/myelography for the study of lesions in this region (Fig. 25.9).

## OPERATIVE MANAGEMENT

### Pre-operative Management

An attempt should be made to reduce and stabilize all patients by utilizing cervical traction prior to operation. Stabilization with traction results in cessation of repetitive spinal cord trauma; simultaneous reduction relieves pressure against the spinal cord from bone and soft tissue.

### Choice of Operative Procedure

The specific internal fixation techniques used to stabilize the atlantoaxial articulation depend on (1) the etiology of the instability, (2) the anatomy of the bony and ligamentous destruction, (3) the reducibility of the subluxation, (4) the presence of a narrowed canal by bone or soft tissue, and (5) the skills and preference of the surgeon. Regardless of the specific technique employed, the success of the fusion depends on the use of bone grafts to achieve fusion. Internal fixation serves only to immobilize the spine until bony fusion occurs. If bony fusion does not take place, instrumentation failure and recurrence of instability are likely.

### Halifax Clamps

Halifax clamps consist of two hooks connected by a threaded screw (Fig. 25.10). The hooks are secured bilaterally around the arch of C1 and lamina of C2 or C3 to achieve stabilization of the atlantoaxial articulation.

Halifax clamps can be used only when the posterior elements of the levels to be stabilized are intact. They are easy to apply, are effective in maintaining alignment until fusion takes place, and, providing that there is adequate room within the spinal canal, carry little risk of neurologic injury.

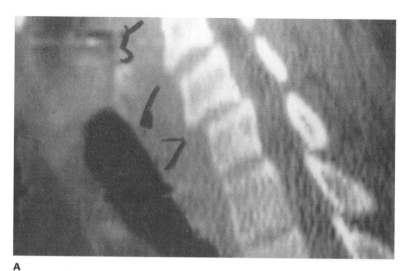

A

**Figure 25.9**
(A) Sagittal CT reconstruction of a patient with a C6-7 subluxation secondary to trauma. (B) Sagittal MRI shows C6 to C7 disc herniation and an abnormal signal within the adjacent spinal cord consistent with spinal cord injury. (C) Postoperative lateral x-ray after C6 to C7 discectomy, bone grafting, and plating with Synthes cervical spine locking plate.

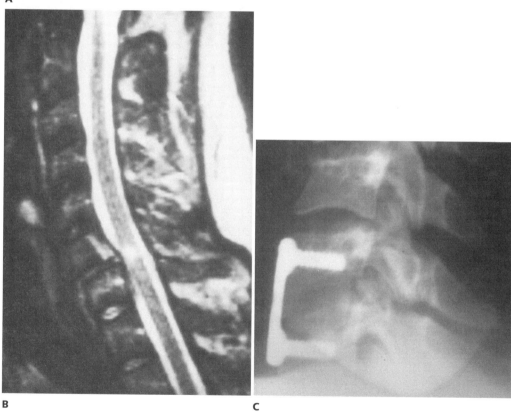

B                                    C

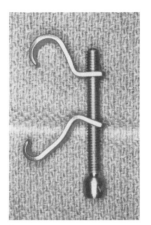

**Figure 25.10**

(A) Halifax clamp consists of two hooks on a threaded screw. The upper, more rounded hook engages the arch of C1, and the squarer inferior hook is used to engage the lamina of the lower cervical vertebra. (B) Lateral cervical spine x-ray in flexion of a patient with C1-2 instability secondary to rheumatoid disease. (C) Lateral x-ray taken after placement of Halifax clamps shows excellent alignment.

A

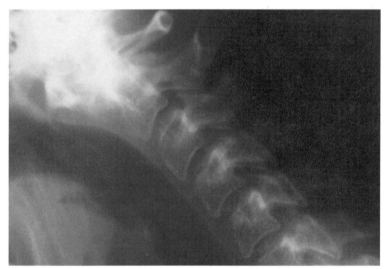

B

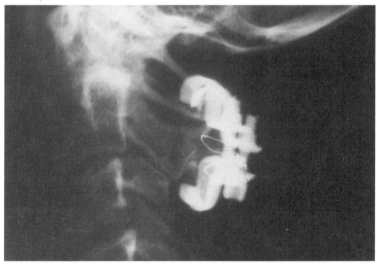

C

With the patient in the prone position and the head fixed to the table using a three-pin head holder, a lateral x-ray is taken to confirm the alignment. The author prefers to place the hooks bilaterally from C1 to C3 rather than C2, because hooks placed around C2 have a tendency to work their way free. The posterior elements of C1 to C3 are exposed. Soft tissue is stripped from the undersurface of the C1 arch and the lamina of C3. A rounded, threaded hook is passed around the arch of C1 and an unthreaded, more square hook is placed at the inferior aspect of the lamina of C3. The screw is tightened, bringing the two hooks and posterior elements of C1 and C3 closer to each another, reducing residual anterior subluxations of C1 on C2.

Bone grafts are placed between the posterior elements of the levels to be stabilized before the clamps and screw are tightened, or are wedged between the bone and clamps after the clamps are tightened. The identical procedure is carried out on the contralateral side. Postoperatively, patients are kept in a cervical collar for 3 months (see Fig. 25.10).

## Transarticular Screw Fixation

Stabilization of the atlantoaxial articulation may be accomplished by the bilateral placement of screws in the pedicle of C2, across the C1-2 articulation, and into the lateral mass of C1. Because this technique does not require passage of wires or other hardware into the spinal canal, it is particularly useful for patients with C1-2 instability with spinal canal compromise by pannus, fibrous tissue, or irreducible subluxations. It also is efficacious in the presence of fractures or neoplastic destruction of the posterior elements of C1 and C2, which preclude wiring or the use of Halifax clamps.

After the patient is positioned prone, the C-arm fluoroscopy is adjusted so that images of the C1-2 articulation may be obtained in the lateral projection. The position of the head is adjusted to achieve maximal reduction. A midline incision is made to expose the posterior elements of C1 and C2. The presence of the paraspinous muscles below C2 make it impossible to drill a hole in C2 and pass a screw at the very acute angle needed. This problem may be obviated by percutaneous passage of a drill guide, with a small incision bilaterally on either side of the midline at T1 through which a 2.7-mm drill bit of appropriate length is passed.

A drill hole is made at the inferior-medial aspect of the lateral mass of C2 2 mm above the facet joint and 2 mm medial to the center of the joint. In the medial-lateral plane the hole must be directed just lateral to the medial border of the pedicle (Fig. 25.11). If the hole is directed too medially, the spinal canal may be entered. If the hole is too lateral, the screw will not engage the lateral mass of C1, and penetration of the vertebral artery may occur. Patients must be studied preoperatively with CT scanning of the C1-2 region to assess variations in vertebral artery position that may preclude safe performance of this procedure.

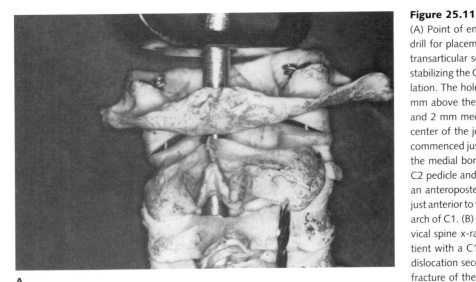

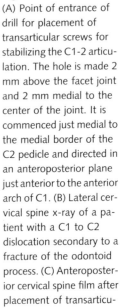

**A**

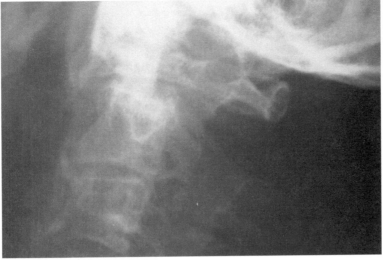

**B**

**Figure 25.11**
(A) Point of entrance of drill for placement of transarticular screws for stabilizing the C1-2 articulation. The hole is made 2 mm above the facet joint and 2 mm medial to the center of the joint. It is commenced just medial to the medial border of the C2 pedicle and directed in an anteroposterior plane just anterior to the anterior arch of C1. (B) Lateral cervical spine x-ray of a patient with a C1 to C2 dislocation secondary to a fracture of the odontoid process. (C) Anteroposterior cervical spine film after placement of transarticular screws. Note that screws engage the center of the lateral mass of C1 bilaterally. (D) Lateral cervical spine x-ray shows correct angulation of screws in the anteroposterior plane.

Lateral fluoroscopic guidance is essential to determine the optimal anteroposterior trajectory of the drill. The trajectory of the drill should be in a plane just posterior to the anterior arch of C1. Drilling the screw hole in the correct angle in the anteroposterior plane may be helped by pulling the spinous process of C2 posteriorly. To avoid penetrating the superior aspect of the lateral mass of C1, the hole should not be drilled rostral to the level of the anterior arch of C1. A second drill hole is made in identical fashion to the first hole. This author uses a 3.5-mm diameter self-tapping cortical screw. The screws range in length from 35–50 mm.

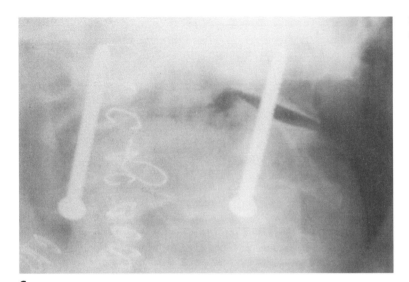

C

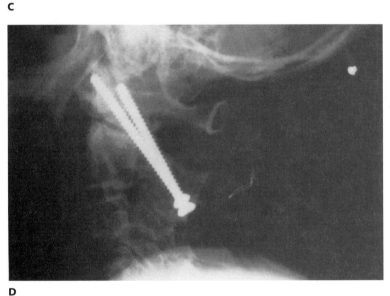

D

**Figure 25.11**
*Continued*

Transarticular screws must be supplemented by other posterior fixation techniques, such as wiring or the placement of Halifax clamps, to preclude screw breakage and loss of fixation. Corticocancellous bone grafts harvested from the iliac crest are placed between the posterior elements of C1 and C2 before the wires or clamps are tightened. Postoperatively, patients are managed in a Philadelphia collar for 10 to 12 weeks, after which time flexion-extension x-rays of the cervical spine in the lateral projection are obtained to confirm stability.

Transarticular screw fixation appears to be the ideal means of stabilizing the upper cervical spine when the spinal canal is compromised by rheumatoid pannus or irreducible subluxations or when the posterior elements of C1 and C2 are damaged or missing, precluding the use of wiring techniques. Unfortunately, because these situations preclude the use of wiring techniques, screw fixation cannot be supplemented by additional fixation. Thus, there is a considerable risk of screw breakage unless patients are provided with secure external fixation.

## Anterior Odontoid Screw Fixation

Anterior odontoid fixation obviates the limitation of rotation that is a result of all posterior fixation techniques at C1-2. By screwing the fractured odontoid to the body of C2, anterior odontoid fixation produces stability and restores normal anatomic relationships. Screw fixation of the odontoid process is indicated for type II odontoid fractures, which have a high rate of non-union when treated with external fixation devices.

The technique of anterior odontoid screw fixation is well described in a number of publications and is briefly outlined here (Fig. 25.12). The patient

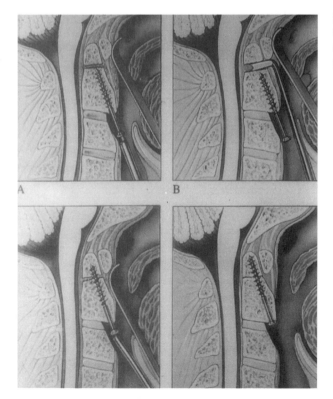

**Figure 25.12**

(A, B) The technique for anterior odontoid screw fixation for fractures of the odontoid process. (See text for explanation.)

is placed on the operating table supine, with the neck in extension for the usual anteriorly displaced odontoid fracture. Two fluoroscopic units are placed to obtain biplane fluoroscopic images of the C1-2 region. The patient's mouth is kept open with a mouth gag to optimize visualization of the odontoid in the anteroposterior projection. A standard right-sided transverse incision is made at the C5 level. A modified Caspar retractor is placed with a specially designed, angled retractor blade that attaches to the lateral retractor, enabling visualization of the C2-3 disc space. Using biplane fluoroscopic control and a series of instruments designed by Apfelbaum, a hole is drilled and tapped, starting at the anterior inferior body of C2 and directed superiorly into the odontoid. A partially threaded titanium screw pulls the odontoid back toward the body of C2. Stability is confirmed by flexing and extending the neck under fluoroscopy.

## Summary

The intelligent use of internal fixation in the patient with instability of the cervical spine depends on a knowledge of the specific anatomy and location of the injury. A clear idea of the etiology of the instability, a definition of the pathologic anatomy, and a knowledge of the presence and extent of neural compression are essential for optimizing a successful outcome.

## References/Further Reading

Aebi M, Mohler J, Zach GA, et al. Indications, surgical technique, and results of 100 surgically-treated fractures and fracture-dislocations of the cervical spine. Clin Orthop 1986; 203:244–257.

An HS, Gordin R, Renner K. Anatomic considerations for plate-screw fixation of the cervical spine. Spine 1991;16:S548–S551.

Anderson PA, Budurick TE, Easton KB, et al. Failure of halo-vest to prevent in vivo motion in patients with injured cervical spines. Spine 1991;16:S501–S505.

Apfelbaum RI. Anterior screw fixation of odontoid fractures. In: Camins MD, O'Leary PF, eds. Disorders of the cervical spine. Baltimore: Williams & Wilkins, 1992:603–608.

Benzel EC, Larson SJ. Functional recovery after decompressive spine operation for cervical spine fractures. Neurosurgery 1987;20:742–746.

Braakman R. Some neurological and neurosurgical aspects of injuries to the lower cervical spine. Acta Neurochir (Wien) 1970;22:245–260.

Bracken MB, Shepard MJ, Collins WF, et al. A randomized, controlled trial of methylprednisolone or naloxone in the treatment of acute spinal cord injury: results of the Second National Acute Spinal Cord Injury Study. N Engl J Med 1990;322:1405–1411.

Caspar W, Barbier DD, Klara PM. Anterior cervical fusion and Caspar plate stabilization for cervical trauma. Neurosurgery 1989;25:491–502.

Cooper PR, Cohen A, Rosiello, et al. Posterior stabilization of cervical spine fractures and subluxations using plates and screws. Neurosurgery 1988;23:300–306.

Ducker TB, Bellegarrigue R, Salcman M, et al. Timing of operative care in cervical spinal cord injury. Spine 1984;9:525–531.

Eismont FJ, Bohlman HH. Posterior methylmethacrylate fixation for cervical trauma. Spine 1981;6:347–353.

Fehlings M, Errico T, Cooper P, et al. Occipito-cervical fusion using a 5 mm malleable rod and segmental fixation. Neurosurgery 1993;32:198–208.

Fehlings M, Cooper PR, Errico T. Posterior plates in the management of cervical instability: long-term results in 44 patients. J Neurosurg 1994;81:341–350.

Garvey TA, Eismont FJ, Roberti LJ. Anterior decompression, structural bone grafting, and Caspar plate stabilization for unstable cervical spine fractures and/or dislocations. Spine 1992;17:S431–S435.

Geisler FH, Cheng C, Poka K, et al. Anterior screw fixation of posteriorly displaced type II odontoid fractures. Neurosurgery 1989;25:30–38.

Grob D, Sturzenegger K, Wursch R, et al. The retrodental pannus after posterior atlanto-axial fusion in rheumatoid arthritis. Presented at the Cervical Spine Research Society Annual Meeting, November 30–December 2, 1994.

Grob D, Jeanneret B, Aebi M, Markwalder T. Atlanto-axial fusion with transarticular screw fixation. J Bone Joint Surg [Br] 1991;73:972–976.

Hadley MN, Fitzpatrick BC, Sonntag VKH, et al. Facet fracture-dislocation injuries of the cervical spine. Neurosurgery 1992;661–666.

Harkey HL. Synthes cervical spine locking plate (Mosher plate). Neurosurgery 1993;32:682–683.

Heiden JS, Weiss MH, Rosenberg AW, et al. Management of cervical spinal cord trauma in Southern California. J Neurosurg 1975;43:732–736.

Jonsson H, Cesarini K, Pettren-Mallmin M, et al. Locking screw-plate fixation of cervical spine fractures with and without ancillary posterior plating. Arch Orthop Trauma Surg 1991;111:1–12.

Levi L, Wolf A, Rigamonti D, et al. Anterior decompression in cervical spine trauma: does the timing of surgery affect the outcome? Neurosurgery 1991;29:216–222.

Magerl F, Grob D, Seemann D. Stable dorsal fusion of the cervical spine (C2-Th 1) using hook plates. In: I. Kehr P, Weidner A, eds. Cervical spine. New York: Springer, 1987;217–221.

Maiman DJ, Pintar FA, Yoganandan N, et al. Pull-out strength of Caspar cervical screws. Neurosurgery 1992;31:1097–1101.

Maiman DJ, Barolat G, Larson SJ. Management of bilateral locked facets of the cervical spine. Neurosurgery 1986;18:542–547.

Marshall LF, Knowlton S, Garfin S, et al. Deterioration following spinal cord injury. A multicenter study. J Neurosurg 1987;66:400–404.

O'Brien PJ, Schweigel JF, Thompson WJ. Dislocations of the lower cervical spine. J Trauma 1982;22:710–714.

Orozco Declos R, Llovet Topes J. Osteosintesas en las fracturas de raquis cervical: nota de technica. Revista Orthop Traumatol 1970;14:285–288.

Randle MJ, Wolfe A, Levi L, et al. The use of anterior Caspar plate fixation in acute cervical spine injury. Surg Neurol 1991;36:181–189.

Ripa DR, Kowall MG, Meyer PR, et al. Series of ninety-two traumatic cervical spine injuries stabilized with anterior ASIF plate fusion technique. Spine 1991;16(Suppl 3):S46–S55.

Roy-Camille R, Saillant G, Laville C, et al. Treatment of lower cervical spine injuries—C3–C7. Spine 1992;17(Suppl 10):S442–S446.

Roy-Camille R, Gagna G, Lazennec JY. L'arthrodese occipito-cervicale. In: Roy-Camille R, ed. Rachis cervical superieur. Paris: Masson, 1986:49–51.

Roy-Camille R, Saillant G, Judet T, et al. Traumatismes recents des cinq dernières vertèbres cervicales chez l'adulte (avec et sans complication neurologique). Sem Hôp Paris 1983; 59:1479–1488.

Roy-Camille R, Saillant G, Berteaux D, et al. Early management of spinal injuries. In: McKibbon B, ed. Recent advances in orthopedics. New York: Churchill Livingston, 1979:57–87.

Stillerman CB, Wilson JA. Atlanto-axial stabilization with posterior transarticular screw fixation: technical description and report of 22 cases. Neurosurgery 1993;32:948–955.

Wagner FC, Chehrazi B. Early decompression and neurological outcome in acute cervical spinal cord injuries. J Neurosurg 1982;56:699–705.

Wilberger JE. Diagnosis and management of spinal cord trauma. J Neurotrauma 1991;8(Suppl 1):21–28.

# IX APPLICATIONS OF NEURO-IMAGING

# Stereotactic Principles and Their Clinical Application

**26**

Allan J. Hamilton

Bruce A. Lulu

*Stereotaxis* is a term that defines a broad field of medicine based on localization using cartesian coordinate systems. The physician visualizes a procedure in one coordinate system (usually derived from one of various imaging modalities, such as computed tomography [CT], magnetic resonance imaging [MRI], angiography, or positron emission tomography [PET]). The procedure is then translated into a second coordinate system in which it is executed. The two systems are linked by common hardware.

## Physical Principles and Applications

Visualizing the three-dimensional data set on which the physician plans the treatment or procedure requires that a computer create a "virtual patient," for which the physician plans the treatment or surgical intervention. A surgical trajectory, a treatment volume, and the anatomic localization of tumor can be defined. The computer then makes a detailed, point-to-point translation of the plan from the virtual patient to the physical patient. The physical patient coordinate system is established by using stereotactic hardware, usually a ring attached to the patient's head either by cranial fixation pins or, more recently and noninvasively, by a system of clamps and head molds; with the latter system, repeated fixation is possible. It is this ring that defines the origin and axes of the patient's system (Fig. 26.1).

Various devices can be attached to the head ring to allow, for example, imaging by CT (Fig. 26.2), MRI (Fig. 26.3), and angiography (Fig. 26.4). In this manner, any point that can be seen in the virtual patient (i.e., the diagnostic study) can be assigned a unique set of coordinates to that point which determines its position along the x, y, and z axes relative to the patient's head ring.

An invasive procedure can be performed by attaching a stereotactic surgical platform to the head ring. This platform allows the neurosurgeon to

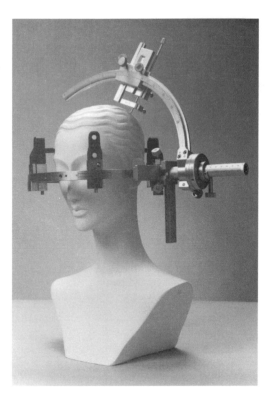

**Figure 26.1**
A stereotactic head ring affixed with four cranial fixation pins. To the left of the figure, a pin is seen in profile. The pins can rotate on their axes and move along the circumference of the ring for optimal positioning. A typical stereotactic platform is attached to the patient's left side of the ring, to permit a left parietal occipital entry site. (See text for further details.)

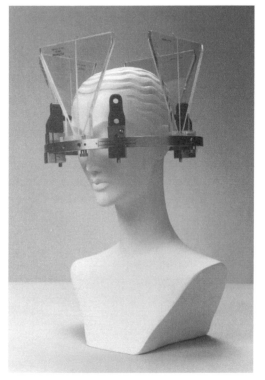

**Figure 26.2**
The same stereotactic head ring seen in Figure 26.1. The stereotactic platform has been removed and four stereotactic localizing plates have been attached to the head ring. These plates are designed for the purposes of imaging with CT.

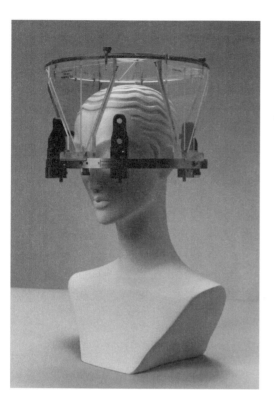

**Figure 26.3**
The same head ring shown in Figures 26.1 and 26.2, but now with the configuration used for visualizing stereotactic coordinates with MRI. Note the plate attached above the patient's head. This permits stereotactic coordinates to be derived from coronal and sagittal images just as they are traditionally obtained with MRI scan.

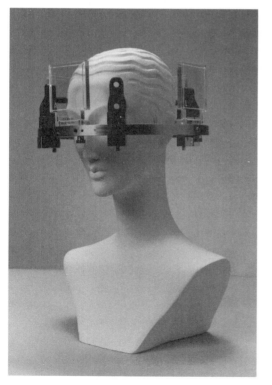

**Figure 26.4**
The same stereotactic head ring, to which have been affixed angiographic localizing plates such as would be used during contrast angiography for visualizing an AVM. The same plates also can be used intraoperatively for localization during a stereotactic procedure by obtaining intraoperative radiographs.

drill a small hole in the cranium and introduce a variety of probes, such as electrodes, radiofrequency lesioning probes, and biopsy probes (Fig. 26.5). For radiosurgery, the target can be placed at the center of many intersecting, highly collimated, megavoltage x-ray (or Cobalt 60 gamma ray) beams. These can be concentrated in doses high enough to ablate lesions in a specified region of the brain without the need for open surgery (Figs. 26.5, 26.6).

It is assumed, of course, that the data sets do not change between the time of diagnostic imaging and the time of treatment. This requires stereotactic physicists to carry out quality assurance of the highest order. For example, an MRI study that employs magnetically derived images could have geometric distortion from metal in the patient, leading to an error in translating data from the virtual to the physical patient; or if the stereotactic head ring were to move between the time of imaging and the time of treatment, the treatment would have a geometric error corresponding to the amount of head-ring shift. Either event could lead to a catastrophe.

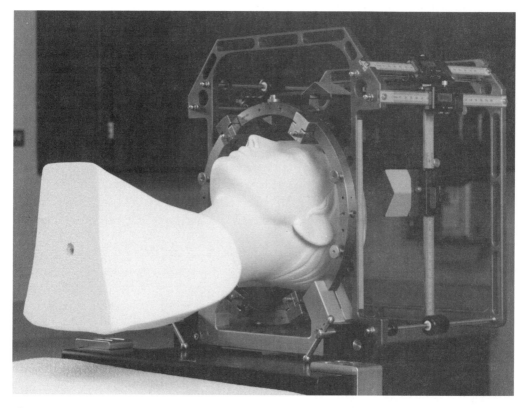

**Figure 26.5**
The same stereotactic head ring, now attached to a stereotactic head-holding stand and with the appropriate attachments for localization of an isocenter forced treatment with radiosurgery. Various vernier scales, seen above the patient, are used for localizing the isocenter via the use of an orthogonal laser sighting system.

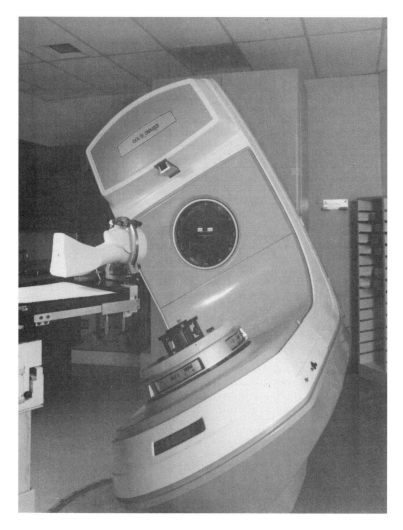

**Figure 26.6**
The patient in position, with the linear accelerator ready to rotate; it is at its lowest position and can rotate in an arc of nearly 360° around the patient. The small device at the bottom center of the photograph holds the collimators. These permit highly calibrated restriction of the x-ray beam so that it precisely fits the radiosurgery target.

Constructing a three-dimensional virtual patient from a three-dimensional data set such as a CT or an MRI data set is relatively straightforward, given the efficiency of current computer technology. Some brain lesions (e.g., arteriovenous malformations [AVMs]), however, are not visible on either CT or MRI, but can be seen only on contrast-enhanced radiographic studies. To identify the arterial nidus with certainty, the data set must be constructed with a contrasted arteriogram. One therefore must take two-dimensional orthogonal projections that are performed during the arteriography and reconstruct a three-dimensional volume for the nidus. In such situations, the physicians must use their best judgment. Usually, this involves outlining the nidus of the AVM from the arteriogram and then correcting the three-dimensional volume based on a CT or, preferably, an MRI scan.

## PATIENT COORDINATE SYSTEM

One can define the patient coordinate system in different ways, which vary according to the brand of stereotactic equipment used. For illustration, we define the x-y plane as the plane that contains the ring and bisects it in the superior-inferior direction. With the ring placed in the standard manner, the ring's positive x-axis is to the patient's right, the ring's positive y-axis is to the patient's anterior, and the ring's positive z-axis points to the patient's apex. The origin is in the center of the ring. If the ring is placed "backwards," positive x is toward the patient's left. Hence, the patient coordinate system is really the ring coordinate system.

Various commercial stereotactic systems allow different stereotactic volumes, some larger than others. It is important to ensure one has created a three-dimensional coordinate system that adequately straddles the region where the procedure is planned, and that encompasses all of the important structures that the neurosurgeon wishes to visualize.

## IMAGE COORDINATE SYSTEM

The image coordinate system is essentially the three-dimensional coordinate system of the virtual patient. To obtain a virtual patient of high quality, it is important that the physical patient not move or degrade the images during their acquisition. One needs to be aware of pin artifact and other metal artifacts, including dental orthoses, that can degrade the CT image and obscure the region of interest. The geometric accuracy of a CT image is typically on the order of 1 pixel in the transverse plane and half-a-slice thickness in the z (table travel) direction. Similarly, when MRI is used, geometric distortion of the image on the order of 5 mm can be due to poorly adjusted MRI magnets, ferromagnetic material in the patient, eddy currents in the stereotactic equipment, and possible chemical shift of localizer equipment rods.

The image coordinate origin is defined as the lower left-hand corner of the image. Positive x is to the viewer's right, positive y is up, and positive z is toward the viewer. Coordinate z is the same as table position, with perhaps a change in sign to give a right-handed coordinate system. Patient orientation on the image does not affect the image coordinate system.

## IMAGE TO STEREOTACTIC COORDINATE TRANSFORMATION

To relate what is seen on the CT or MRI to the stereotactic head ring, part of the stereotactic equipment must be visualized in the imaging study. Typically, an assortment of plastic rods, roughly parallel to the z-axis, is attached to the head ring before imaging. This ensures that rod images will show up on an arbitrary image. Even when the stereotactic frame is not aligned precisely with the CT coordinate axes before imaging, some trigonometry and simple algebra will allow the computing of stereotactic coordinates of at least three points. If the coordinates of three non-collinear points are known in both coordinate systems, the coordinate transformation between the two systems can be constructed explicitly by use of vector analysis.

## TARGET POSITION FROM RADIOGRAPHS

Determining the position of a test target, such as a ball bearing, can be performed accurately. This is achieved because the center of the ball bearing can be seen easily in both radiographs. Four fiducial marks (lead balls) are placed in the pattern of a square in a plastic plate, and four such plates are attached to the head ring (see Figs. 26.2, 26.4). In an anteroposterior (AP) or lateral radiograph, eight balls should be visible. Use of projective geometry permits the calculation of the stereotactic coordinates of a line passing through any point identified on the radiograph; film angle, gantry angle, film magnification, and angle between views are irrelevant. Given an AP and lateral view, with the same point identified in both views, two lines in space should intersect exactly, but because of measurement errors, inability to precisely identify the same physical point in both views, manufacturing tolerances, and so on, this does not happen. The point coordinates are then defined as the center of the line segment that connects the nonintersecting lines at their position of closest approach. For a test target, 0.5 mm or better is not uncommon. For an anatomic target such as an AVM nidus, selecting the same point in both views is extremely difficult if not impossible. Miss distances of several millimeters may be observed. Determining the surface of the nidus is even more problematic. Thus, radiographic coordinates must serve as a guide to the clinician, not be regarded as fact.

## STEREOTACTIC TRAJECTORIES

Once the stereotactic coordinates of a target point visualized on CT or MRI are calculated, an approach to that target point is desired. An approach can be determined by selecting an entry point, either on CT or directly on the patient's scalp. Modern stereotactic frames allow the translation of the frame in (x, y, z) to place the target point at the center of an arc system. This arc system can then pivot in azimuth and declination (latitude and longitude) to select an arbitrary approach direction to the target point. Thus, one can change the entry point by visual inspection, altering only the azimuth and declination rotation settings. Depth remains fixed, and no calculations are required.

## Clinical Applications

### STEREOTACTIC IMAGING DATA ACQUISITION

Some kind of anatomic atlas needs to be developed in reference to an orthogonal coordinate system. Until recently, this frame of reference was always one that employed rigid skeletal fixation. More recently, some nonskeletal fixation systems have been developed, as has frameless stereotaxy, which employs external fiducial markers. Nonetheless, some frame of reference *must* be established. Atlases can be created from almost any database but most commonly are derived from CT, MRI, or angiography. Once the images have been obtained, they are entered into a computer and important structures can be digitized and even their volumes rendered in three-dimensional displays for better manipulation by the surgeon and stereotactic team.

Once the stereotactic atlas has been created for the individual patient, the neurosurgeon decides exactly which procedure may be required. A variety of stereotactic procedures are now feasible; they include such diverse procedures as stereotactic biopsy, stereotactic (volumetric) resection of tumors, stereotactic radiosurgery, radiofrequency thermocoagulation, and stereotactic endoscopy. Most recently, stereotactic devices have even been developed for spinal stereotactic procedures. Each of these procedures is discussed briefly in the section that follows.

### STEREOTACTIC FRAME PLACEMENT

Stereotactic techniques can employ any standard computerized imaging database as long as both the target (e.g., tumor) and the stereotactic frame of reference can be visualized with that modality. For example, if one intends to image a tumor using a CT database, then one would need to ensure that both the target and the stereotactic frame were sufficiently dense to absorb x-rays. This can be achieved by using an aluminum or carbon fiber stereotactic frame and administering iodinated contrast to make the tumor appear as a distinct, enhanced mass. By contrast, metal (not carbon fiber) would be a poor choice of material for a magnetic stereotactic frame of reference, and a paramagnetic contrast agent would need to be employed for MRI. Currently, CT and MRI are the most commonly used modalities, but digital subtraction angiography and PET also are feasible.

Once the appropriate imaging database has been selected, one still can choose from skeletal fixation versus non-skeletal fixation. Non-skeletal fixation can be employed when errors of 2–3 mm would be acceptable. In most instances, because stereotactic procedures are ideally suited to either deep-seated lesions or lesions in eloquent areas, submillimeter accuracy is required, and one must resort to skeletal fixation.

There are a wide variety of frames currently on the market and approved for clinical use in the United States by the Food and Drug Administration.

All frames on the market permit the surgeon a limited variability in the size, position, and placement of the skeletal fixation pins. A minimum of three pins is required, but it is customary to place four pins for extra security, because any inadvertent movement of the stereotactic frame would invalidate all measurements.

As with any standard neurosurgical procedure, it is important that the surgeon give forethought to such matters as the orientation of the procedure and right- or left-hand dominance of the surgeon. It is usually far preferable, for example, to perform stereotactic procedures in the posterior fossa with the patient in the prone or lateral decubitus position rather than the standard supine position. Because most frames have only one portion of the ring that can be attached to the operating table, it is important to know the desired position of the patient with respect to the operating table prior to placing the frame. Most modern frames allow flexibility in positioning the frame and patient to permit a large number of surgical orientations (Fig. 26.7).

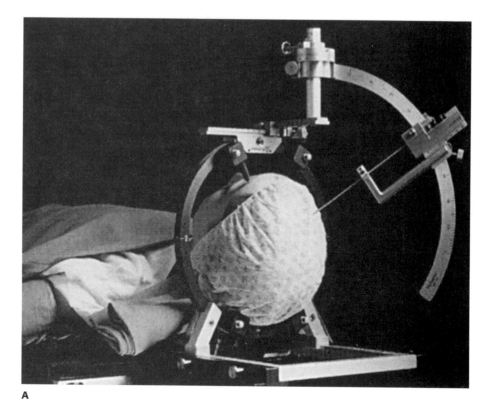

**A**

**Figure 26.7**
(A) Typical isocentric stereotactic system positioned for a biopsy in the right frontal region. The isocentric design permits a surgeon to swing the arm out of the way or move it into a better position for localization while still being centered on the target. (Photo courtesy of Fischer, Inc., Dallas, TX.)

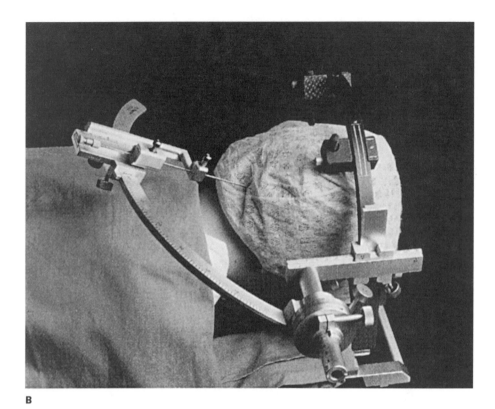

**B**

**Figure 26.7** *Continued*
(B) The same frame positioned for a posterior fossa biopsy. (Photo courtesy of Fischer,Inc., Dallas, TX.)

The authors always try to obtain the imaging database in a position as close as possible to that in which the patient will be positioned in the operating room. So, for example, if one will be positioning the patient in the prone position for a stereotactic procedure on the posterior fossa, it is advisable to position the patient prone for the preoperative MRI or CT. When this is not feasible, one should be aware that the brain and surrounding structures are bathed in CSF and there may be subtle changes in the actual location of intracranial structures when a patient moves from, for example, the supine to the prone position. In addition, the materials from which the frames and pins are constructed may "give" slightly when the full weight of the patient's head and neck is being supported in the operating room. The authors have found that this "sag" with the full weight of the patient's head can result in an inaccuracy of approximately 1 mm (unpublished data).

Finally, intracranial shifts in position and location can occur as the result of the stereotactic procedure. If a tumor is located adjacent to a fluid-filled structure such as a ventricle or a cyst, for example, the surgeon should attempt to develop a stereotactic trajectory that does not endanger that fluid-filled volume until all other stereotactic manipulations are completed. This obviates inadvertent decompression, which would cause a shift of the intracranial contents. A second danger is that of triggering the development of a new mass lesion. This can occur when significant intracranial bleeding is engendered by the stereotactic procedure. As the bleeding accumulates in the brain parenchyma, it forms a new mass lesion, which progressively shifts the original mass to new and unknown positions. It then becomes dangerous to proceed with further stereotaxis because the pre-operative images on which the procedure is based are no longer valid.

## Pediatric Patients

Special problems surround pediatric candidates for stereotaxis for a variety of reasons. First, their cranial circumferences are smaller and so special, longer pins are required. Second, few children under the age of 12 to 13 years of age can tolerate rigid immobilization, so endotracheal intubation and the administration of general anesthesia are required before placing the stereotactic head ring and proceeding with the stereotactic procedure. Specialized head rings have been developed that permit skeletal fixation but also allow the anesthesiologist ready access to the mouth and endotracheal tube (Fig. 26.8).

The pediatric patient must have a fully fused skull before one can consider placing a stereotactic head ring. The authors usually do not recommend skeletal fixation in patients under the age of 3 years. It is also important to carefully assess skull thickness pre-operatively. In chronic hydrocephalus, the skull is often thinned out by the chronic outward pressure of tense ventricles. Placing skeletal pins in such a very thin cranial bone involves running the risk of their penetrating both the bone and the underlying dura. If a stereotactic procedure is necessary in such a situation, then nonskeletal stereotactic frames, or even frameless systems, may be preferable, even though they are inherently less accurate than frames that employ cranial fixation.

## Stereotactic Imaging Database Manipulation

With modern, stereotactic, surgical-planning computers and software, the surgeon can manipulate one or more imaging databases to achieve the fullest potential for viewing the target and potential trajectories (Fig. 26.9), including all the structures that may have to be traversed to reach the target. Alternative targets and even trajectories that intercept multiple targets are feasible. The key to successful stereotaxis is navigation. The surgeon can and should plan the operation so completely on the computer that the actual surgical procedure is an anticlimax.

*Multimodality Databases*    Databases can be fused. For example, a surgeon can take a CT database and fuse it with an MRI database, and even with a third angiographic database and a fourth database derived from PET. One

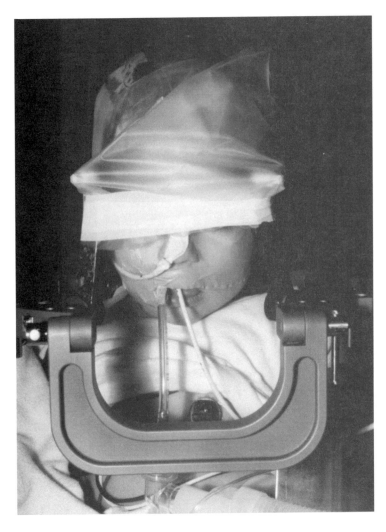

**Figure 26.8**
A small child undergoing stereotactic biopsy under general
anesthesia. The frame has a
swing-away portion that permits treatment of patients
under general anesthesia without any undue deviation or
warping of the frame.

might decide to biopsy or resect the most metabolically active site of a tumor as identified by PET, determine the major vascular structures that may impede the approach from angiography, and decide what anatomic structures are most at risk from MRI. As part of the pre-operative planning, new techniques in so-called functional MRI superimpose movement, speech, vision, seizure activity, and so on, onto the MRI to permit stereotactic mapping of eloquent areas of the brain.

## STEREOTACTIC BIOPSY

The primary goal of a stereotactic biopsy is to achieve maximum accuracy in histologic diagnosis with a minimum of parenchymal disruption. Stereotactic

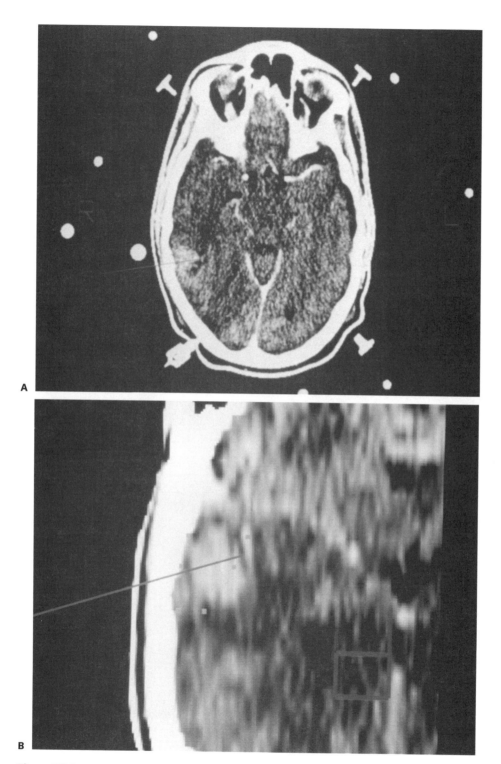

**Figure 26.9**

(A) Stereotactic CT in the axial orientation showing an enhancing lesion in the mid-right temporal region. The lesion was an anaplastic astrocytoma. (B) Data manipulations allow a surgeon to reconstruct additional views, such as this coronal view showing the lesion in its relationship to the tentorium in a potential surgical trajectory to the lesion.

473

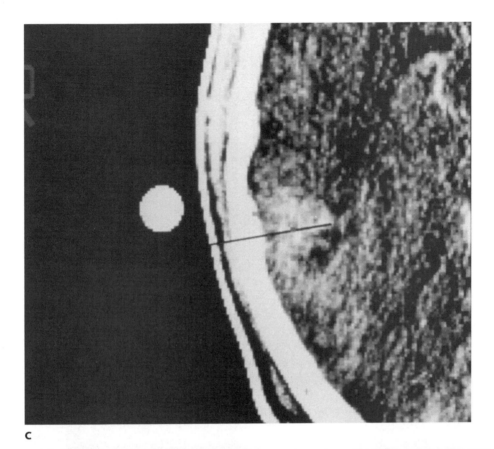

C

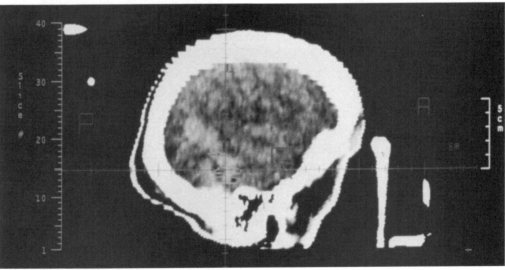

D

**Figure 26.9** *Continued*
(C) View, in the axial plane, of the surgical orientation seen in (B). (D) Computerized manipulation of the data depicting the lesion in its orientation along the planned surgical trajectory.

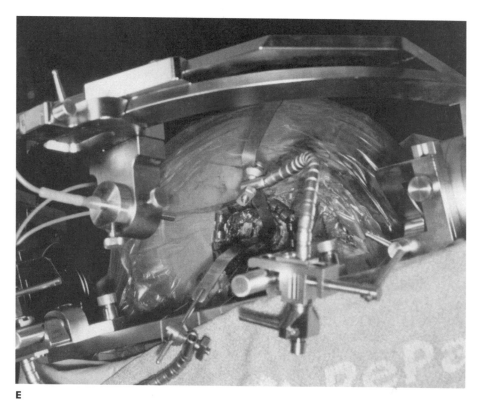

E

**Figure 26.9** *Continued*
(E) Intraoperative photograph showing the stereotactic craniotomy being performed. Note the small exposure and the stereotactic retractors in position. The fiberoptic cable seen on the left side of the photograph is attached to a laser that highlights the stereotactic trajectory.

biopsy is ideally suited for lesions that are deep-seated and in instances in which classic surgical exposure would entail excessive morbidity. It also is helpful for lesions in which surgical extirpation is not a high priority, for example, in a patient with numerous, diffuse intracranial lesions, for which surgical resection will not add to therapeutic outcome and the choice of radiation or chemotherapy awaits tissue diagnosis.

The target first must be visualized clearly, and the lesion or lesions that can be approached most safely and with the highest likelihood of diagnostic yield must be selected. Such factors as the presence of eloquent areas of the brain, major vascular structures, crossing of pial planes, and adjacent or associated fluid-filled structures also must factor into the selection.

Once the surgeon is confident of the stereotactic plan, the coordinates and images are then transferred to the operating room, either physically as hard copy and radiographic images or through fiber optic telephone lines to a second computer. It is advisable that a second set of confirmatory hand calculations always be carried out, in case of computer malfunction. The stereotactic software provides a number of coordinates that tell the surgeon exactly in what position the stereotactic surgical platform must be placed for it to be precisely and accurately aimed at the selected target. While the data are being transferred and the coordinates verified, the entry point for the stereotactic biopsy is selected in accordance with the pre-operative stereotactic plan.

Many, but not all, stereotactic frame systems are isocentric, that is, once the stereotactic platform is attached to the head ring and aimed a given target point, it can continue to be rotated around one or more axes while remaining fixed on the same target. Hence, minor adjustments can be made in the position of the entry point without any change in the target point.

Thereafter a burr hole is made, the dura opened, and the stereotactic biopsy probe introduced to the depth dictated by the computer coordinates. One tissue sample may suffice for diagnosis of an intracerebral abscess, but as many as 20 biopsies may be necessary for accurate histologic sampling of a pleomorphic tumor.

## Diagnostic Accuracy and Morbidity

In the last 70 consecutive cases at the authors' institution, there has been a 99% success rate in achieving a histologic diagnosis, with only a single case of brainstem biopsy being aborted because of trigeminal pain. There has been no permanent morbidity and no mortality. Similarly, low morbidity and mortality rates have been reported by others.

Stereotactic biopsy therefore has several advantages. It is minimally invasive, with very low morbidity and mortality as well as decreased costs in terms of, for example, operating time and length of hospital stay. Stereotactic techniques can be used to diagnose lesions in otherwise inaccessible locations, such as the deep central nuclei and the brainstem. Finally, in patients who will not require a craniotomy (e.g., patients with multiple metastases), it offers a rapid and safe method for obtaining a tissue diagnosis without subjecting the patient to major surgery or general anesthesia.

Stereotactic biopsy has some disadvantages as well: (1) Principal among these is sampling error. Stereotactic biopsy permits only very limited access to a portion of the lesion. Because the specimens obtained are quite small, the histologic data represent a sampling of only a minute portion of the whole tumor volume. As a result, histologic grading may underestimate the malignancy of a tumor if, for example, a more differentiated, benign area is sampled. Open craniotomy permits the surgeon to remove substantially larger volumes of tissue. (2) The most accurate stereotactic procedures require placement of a stereotactic frame with cranial fixation, which can be uncomfortable for the patient. The authors employ some short-acting intravenous sedation in addition to local anesthesia to make the patient as comfortable as possible during the placement of the stereotactic cranial fixation pins. (3) Because the operative site is not visualized directly, the surgeon cannot tell whether there is bleeding at the biopsy site(s). Hence, some neurosurgeons obtain a CT scan immediately after biopsy to ensure there is no active bleeding; most surgeons, however, advise against this because the incidence of hemorrhage is very small.

## STEREOTACTIC CRANIOTOMY

Using the same three-dimensional computerized imagery, a neurosurgeon can also perform a craniotomy. The highly accurate orthogonal stereotactic localization indicates exactly where the lesion is located, and by manipulating both CT and MRI data sets, the neurosurgeon can carefully plan the trajectory to the target, taking account of eloquent areas, blood vessels, pial planes, and locations of cysts and other fluid-filled spaces.

Furthermore, the three-dimensional data also can yield volumetric information (Fig. 26.10). The surgeon also can gain additional information about navigation. For example, to help guide the resection, the surgeon can visualize

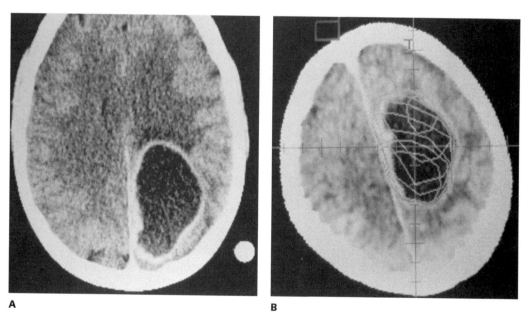

A                                                          B

**Figure 26.10**

(A) Enhanced CT scan of a 29-year-old patient with metastatic rhabdomyosarcoma. A large cystic lesion occupies the posterior parietal and occipital lobes. (B) Computerized single "wire diagram" depicting the volume of the cyst. The data can be manipulated to show any orientation, allowing the surgeon to orient the patient's positioning as well as the trajectory of approach.

C

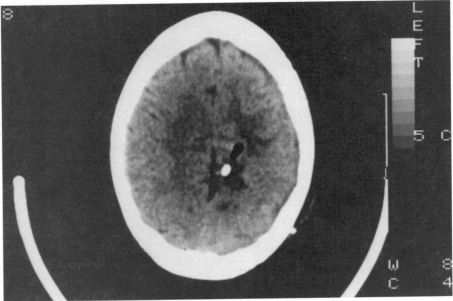

D

**Figure 26.10** *Continued*
(C) Intraoperative photograph showing the stereotactic craniotomy of the patient. Note the size of the opening relative to the size of the retractor blades. The cylindrical device at the superior portion of the photograph is a small laser that shows the surgeon the computerized trajectory. (D) Enhanced CT scan of the same patient, showing complete removal of the cyst and the tip of drainage catheter left for possible administration of intracystic chemotherapy.

the distance from the center of the trajectory of approach to the edges of the tumor margins. In addition, because the tumor's location, depth, and dimensions are so clearly established, the size of the craniotomy can be minimalized. Often, a simple trephine hole may permit adequate visualization for small tumors. It is the authors' preference to create beam's-eye views of the trajectory and then project the largest circumference of the tumor onto the entry site (see Fig. 26.9D). In this fashion, the surgeon essentially creates a template on the cranium prior to the actual craniotomy.

Because it is easy to become disoriented during neurosurgical dissection due to the almost uniform appearance of the white matter, sophisticated devices such as small laser beams can be used to guide the surgeon during the dissection. With stereotactic guidance, there is no such problem about disorientation because the stereotactic coordinates can be used for guidance, like a compass. Actual removal of a tumor is done by using the usual techniques, including bipolar cautery, laser coagulation, suction, and irrigation.

A stereotactic craniotomy permits a minimally invasive approach. Because the exposure is small, there is less bleeding and a shorter period of convalescence. In the authors' experience, most stereotactic craniotomy patients can be discharged from the hospital within 1 or 2 postoperative days; this is in contrast to the standard craniotomy patient who needs 7 to 10 days of hospitalization for recovery. One of the authors' adages is as follows: "When a stereotactic craniotomy is well planned, carrying out the procedure is nothing more than the physical realization of what has already been determined on the computer screen."

There are some disadvantages to stereotactic craniotomy. It requires deep retraction and only permits a very small field of view through the small cylindrical retractors that are used for the procedure. For this reason, the neurosurgeon needs to use a great deal of judgment in the pre-operative planning to ensure being able to reach the tumor and resect it with very limited exposure. Another disadvantage is the expense of acquiring a set of highly specialized surgical instruments if they are not available in a hospital where such procedures have not been performed, as well as a fairly large investment in computer equipment to permit detailed, three-dimensional database analyses.

## STEREOTACTIC WANDS

During the past decade, a variety of so-called stereotactic wands has become available. These allow the surgeon to re-register the position of certain anatomic landmarks or fiducial landmarkers once the patient's head has been fixed in the chosen head holder. The wand then correlates these with the three-dimensional database, permitting the surgeon to manipulate this in the operating room. For example, the surgeon can use the stereotactic wand to touch a wall of the tumor. By re-registering the wall of the tumor, the surgeon can compare it with the pre-operative MRI or CT database. The resection can be guided by the MRIs rather than by simply evaluating the tissue seen under the operating microscope. In addition, a surgeon can use the three-dimensional stereotactic wand to navigate. For example, a surgeon can touch a blood vessel with the tip of the wand, then refer back to a digital subtraction angiographic study and confirm its identity: In resection of an AVM, one can confidently identify the feeders that may need to be clipped.

Stereotactic wands also permit the surgeon to employ stereotaxis without skeletal fixation. Most recent stereotactic systems allow the surgeon to re-register anatomic landmarks such as the bridge of the nose, lateral or medial canthus, or external auditory meatus. When one or more of these points is re-registered and identified on the corresponding CT, MRI, or angiographic study, the computer can re-register the whole of the pre-stored imaging databases. Although these systems have a lower accuracy than skeletal fixation systems, their precision is nonetheless high enough to be useful during surgery. Furthermore, for lesions that are large enough to permit a 2- or 3-mm error, stereotactic wands offer an easy way in which to obtain stereotactic biopsies, without requiring that the patient be placed in a stereotactic head ring with cranial fixation pins and then undergo a CT scan with the frame in place. These methods are currently undergoing evaluation, but are likely to become standard fare in neurosurgery within the next decade.

## FUNCTIONAL STEREOTAXIS

In the field of functional neurosurgery, by employing stereotactic coordinates, the neurosurgeon can insert and direct a probe for the purposes of making a lesion in the chosen nucleus or tract. This becomes particularly important in such ablative procedures as thalamotomy and pallidotomy. Stereotactic techniques have been used for both radiofrequency and laser lesioning of such structures as the globus pallidus, nuclei of the thalamus, and cingular gyrus. Additional ablative procedures are being evaluated for the purposes of epilepsy surgery.

## BRACHYTHERAPY

Stereotactic coordinates can be used to guide the implantation of catheters that are loaded with radioactive seeds (Fig. 26.11A). This technique allows the tumor bed to receive a high dose of radiation (compared with the surrounding brain), the rate of fall-off being inversely proportional to the square of the distance from the implant seeds. The procedure can be used after standard external fractionated radiation therapy, radio-resistant tumors receiving an

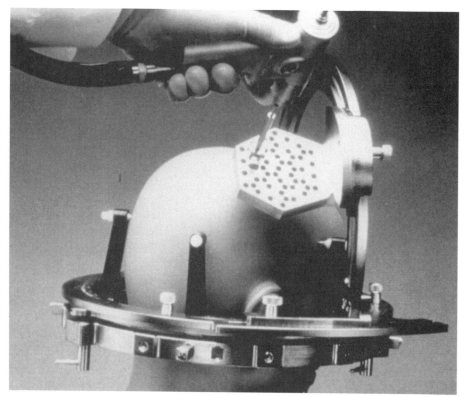

A

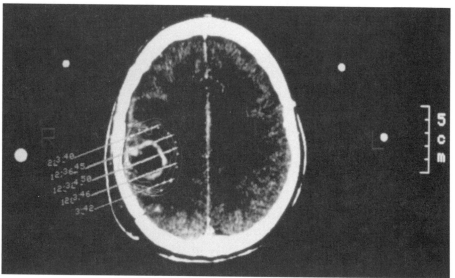

B

**Figure 26.11**

(A) Brown-Roberts-Wells mount with a modified template for passing brachytherapy catheters. (B) Tumor outlined, with the appropriate trajectories of the various catheters superimposed as well as the isodose contours for radiation treatment.

additional boost to produce necrosis in the tumor bed. Using the stereotactic three-dimensional information, the volume of the implant and the required number of catheters can be calculated pre-operatively and their position predetermined with a high degree of precision (see Fig. 26.11B).

## RADIOSURGERY

The essence of radiosurgery is the shooting of multiple radiation beams at a target from many different directions. Where the beams overlap, the dose is high, but outside that region it falls off rapidly. Because the radiation collimators are circular, the high-dose region is roughly spherical in shape. This works well for lesions that are roughly spherical, such as metastatic lesions. For irregularly shaped lesions, one must attempt coverage by using multiple foci of radiation.

The original technique of modern radiosurgery used the gamma knife, invented by Lars Leksell. This has 201 Cobalt-60 sources arranged in a hemispheric layout. A stereotactic head frame provides target translation to the center of focus of the Cobalt-60 sources. The half-life of Cobalt 60 is 5.26 years, so the sources must be replaced every 5 to 10 years at considerable cost, regardless of how many patients are treated. Available collimators are 4, 8, 14, and 18 mm in diameter, so the gamma knife cannot treat large intracranial lesions. However, the standard linear accelerator (LINAC) used in radiation oncology can be adapted for radiosurgery use at about one tenth the cost of a gamma knife. For LINAC-based radiosurgery, the LINAC table provides the translation of the target to the center of the LINAC radiation beam (isocenter). Gantry rotation adjusts the declination, and table rotation the azimuth. Linear accelerator collimators range from 4–80 mm in diameter. Irregularly shaped collimation of LINAC beams is accomplished easily and is being explored clinically, allowing more sparing of normal brain tissue than is possible with circular collimators alone.

Stereotactic techniques also can be employed to guide non-coplanar beams of radiation. These beams can be accurately localized with an error of 2 mm, and one is able to deposit very high levels of radiation in a tumor with a very rapid fall-off in adjacent normal tissues. An additional advantage of radiosurgery over brachytherapy is that it is essentially a noninvasive technique that does not require intracranial catheter placement.

The disadvantage of stereotactic radiosurgery has been that because one is employing cranial fixation, treatment must be completed at a single session. Recently, however, new frames employing such devices as dental molds or fiberglass mask systems permit accurate positioning of the patient's head in a frame. Thus, fractionated stereotactic radiotherapy is possible.

Stereotactic radiosurgery has proven to be extremely successful in the treatment of inoperable AVMs, with an 80% angiographic obliteration rate at 3 years (Fig. 26.12), as well as in the treatment of metastatic neoplasms and malignant gliomas. It has also proven effective in the treatment of acoustic neuromas and meningiomas. The exact role of stereotactic radiotherapy (as opposed to single-fraction stereotactic radiosurgery) is unknown, but is undergoing extensive evaluation. In addition, new modalities have been developed that permit stereotactic radiosurgery in locations outside of the cranium. At the authors' institution, research is being carried out on the use of stereotactic radiosurgery for the treatment of spinal and paraspinal neoplastic disease.

## Conclusion

There is no doubt in the authors' minds that stereotaxis is here to stay. It will become one of the major bulwarks of the neurosurgical specialty as increasingly

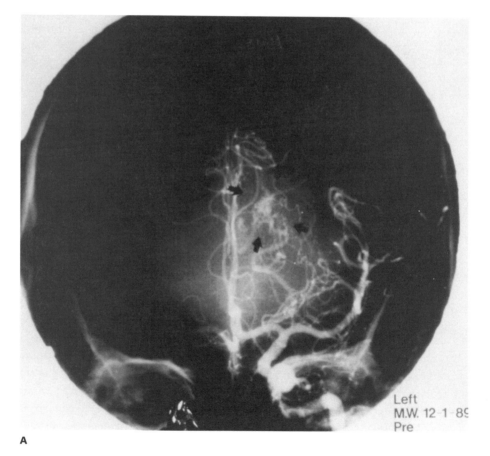

**A**

**Figure 26.12**

(A) Anteroposterior angiogram revealing an AVM *(arrows)* fed by the anterior and middle cerebral arteries on the left-hand side.

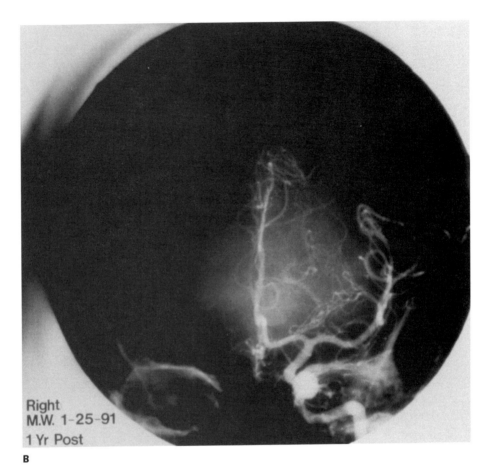

**B**

**Figure 26.12** *Continued*
(B) The same patient at 1 year postradiosurgery with complete angiographic obliteration of the AVM after a single isocentric surgical treatment.

sophisticated imagery drives the field toward less and less invasive procedures. (Stereotactic endoscopy is mentioned in chapter 28.)

With the increasing sophistication of computer hardware, the re-registration of fiducial markers has permitted frameless stereotaxy; the applications of this field are likely to expand. The use of stereotactic wands will allow surgeons to minimize the size of craniotomies as well as the extent of dissections. Furthermore, the use of stereotaxis has now been combined with neurosurgical microscopy to permit computerized navigation through the eyepiece of the microscope. This is very much akin to the "heads up" display that fighter pilots use in their aircraft. In a similar fashion, stereotactic data for trajectory, localization of the tumor, and tumor volumes, can all be projected simultaneously in one of the eyepieces of the microscope. This allows the surgeon to carry out a microsurgical dissection with the computerized imaging data constantly available. Virtual reality visual guidance systems are now under development, and will allow a neurosurgical procedure to be planned and, in effect, practiced, with a favorable effect on morbidity and mortality.

## Acknowledgments

Dr. Hamilton received a Clinical Investigators Career Development Award from the American Cancer Society (No. 93-21) and a stereotactic research grant from the Fischer-Leibinger Companies.

## References/Further Reading

Alexander E III, Coffey R, Loeffler JS. Radiosurgery for gliomas. In: Alexander E III, Loeffler JS, Lunsford CD, eds. Stereotactic radiosurgery. New York: McGraw-Hill, 1993:207–220.

Drake JM, Prudencio J, Holowka RT. A comparison of the PUMA robotic system and the ISG viewing wand for neurosurgery. In: Maciunas RD, ed. Interactive image-guided neurosurgery. Chicago: American Association of Neurological Surgeons, 1993:121–134.

Hamilton AJ, Lulu BA, Fosmire H, et al. Early clinical experience with LINAC-based spinal stereotactic radiosurgery. Neurosurgery 1995;36:311–319.

Hitchon PW, VanGilder JC, Wen BC, et al. Brachytherapy for malignant recurrent and untreated gliomas. Stereotactic and Functional Neurosurgery 1992;59(1–4):174–8.

Kelly PJ. Tumor stereotaxis. Philadelphia: Saunders, 1991:183–223, 268–295.

Kondziolka D, Lunsford LD, Linskey ME, Flickinger JC. Skull base radiosurgery. In: Alexander E III, Loeffler JS, Lunsford CD, eds. Stereotactic radiosurgery. New York: McGraw-Hill,1993:178–188.

Loeffler JS, Alexander A III. Radiosurgery for the treatment of intracranial metastases. In: Alexander E III, Loeffler JS, Lunsford CD, eds. Stereotactic radiosurgery. New York: McGraw-Hill, 1993:197–206.

Lunsford LD, Martinez AJ. Stereotactic exploration of the brain in the era of computed tomography. Surg Neurol 1984;22:222–230.

Manwaring K. Intraoperative microendoscopy. In: Maciunas RJ, ed. Interactive image-guided neurosurgery. Chicago: American Association of Neurological Surgeons 1993:217–232.

Maurer CR Jr, Fitzpatrick JM. A review of medical image registration. In: Maciunas RJ, ed. Interactive image-guided neurosurgery. Chicago: American Association of Neurological Surgeons, 1993:17–44.

# The Role of Ultrasonography in the Evaluation of Cerebrovascular Disease

Damirez Fossett

Laligam Sekhar

Cerebrovascular disease remains one of the leading causes of mortality and morbidity in the United States. Two-thirds of individuals sustaining strokes are under the age of 65, and almost half of these individuals have a stroke as the result of a surgically accessible lesion in the extracranial carotid artery. No therapeutic intervention other than rehabilitation can be offered once a completed cerebral infarct has occurred; however, medical and/or surgical therapy designed to minimize the likelihood of transient ischemic attacks (TIAs) and/or completed infarcts may be offered to individuals identified as being at risk of these, and several therapeutic options are now available to treat acute cerebrovascular occlusion prior to the onset of completed infarction. It is therefore of more than academic importance to be able to evaluate the cerebral circulation while it is still functionally intact.

Cerebral arteriography is still considered the definitive diagnostic tool in the evaluation of both extra- and intracranial carotid or vertebral occlusive disease; conventional angiography, however, is an invasive, relatively expensive study that is not without some associated morbidity and mortality, and although the risks of angiography may be outweighed by its benefits in the symptomatic patient, it is not clear that the patient with an asymptomatic bruit should be subjected to the potential risk of complications of conventional angiography. Angiography is also a limited study in that, despite biplanar visualization of both the carotid and vertebral arteries, it can define only the anatomy, not the physiology, of the circulation in question. In addition, shallow, ulcerated lesions which can be the source of cerebral emboli are occasionally not demonstrated by angiography. For these reasons, newer, noninvasive, less expensive tests have been developed to identify cerebrovascular occlusive lesions and aid in the selection of patients for conventional angiography and possible surgery.

Noninvasive tests can be grouped into two categories: indirect and direct. Indirect tests allow the physician to infer some degree of carotid stenosis by noting hemodynamic alterations as manifested by changes in BP or arterial pulsations within the ophthalmic artery or one of its branches; they include techniques such as ophthalmodynamometry, oculoplethysmography, and peri-orbital directional Doppler sonography. Such tests, however, in and of themselves, are of limited value in the evaluation of carotid occlusive disease for several reasons: (1) They are not positive unless the lesion is hemodynamically significant—the degree of stenosis must be at least 50%; (2) they cannot distinguish between extremely high-grade stenosis and complete occlusion; (3) their results may be invalidated by significant bilateral carotid occlusive disease, or they may be read as normal if there is adequate collateral blood supply to the ophthalmic artery; (4) they provide nonspecific information regarding the location of the lesion, which may lie anywhere between the aortic valve and the ophthalmic artery; (5) they do not distinguish between single and multiple lesions; and (6) they cannot define ulcerative plaques, which may be of extreme clinical significance.

Direct tests involve the use of high-frequency ultrasonic imaging to both visualize the extracranial carotid and vertebral arteries and evaluate alterations in blood flow within them. They include brightness modulation (B-mode) ultrasonography, M-mode ultrasonography, Doppler ultrasonography, color Doppler imaging, and duplex Doppler imaging. Direct, noninvasive studies are more sensitive than indirect studies to earlier and less severe lesions in the region of the bifurcation of the common carotid artery. These studies, however, provide little to no information about more distal lesions, such as those in the carotid siphon or in other areas within the intracranial circulation. Direct techniques also be may limited by anatomic constraints, being sometimes unable to image a carotid bifurcation that lies either deep or high in the neck, and by physiologic constraints, such as the inability of ultrasonic waves to penetrate calcified or atherosclerotic plaque.

Transcranial Doppler (TCD) techniques use low-frequency ultrasonic waves in a fashion similar to other ultrasonic techniques, to evaluate alterations in blood velocity within the intracranial circle of Willis. Occlusive lesions within these vessels, or spasm resulting from subarachnoid blood, may alter the normal sonographic representation of blood velocity and thereby suggest diminished or absent blood flow to areas of cerebral tissue. The identification of these types of intracranial flow-limiting lesions is important because they may give rise to signs or symptoms of ischemia. Transcranial Doppler imaging is an excellent noninvasive bedside technique to use in assessing and quantifying intracranial blood flow.

## General Ultrasonic Principles

Ultrasonic waves are sound waves at frequencies too high to be detected by the human ear (20,000 Hz or greater). Ultrasonic imaging techniques are based on differences between incident and reflected high-frequency sound waves as they interact with tissues of different densities. Differences in the acoustic impedance of various bodily tissues produce sound wave reflections that can be reconstructed to give anatomic detail about tissue structure, and in vascular imaging, can also be analyzed to supply data regarding the physiology of blood flow. High-frequency waves provide excellent resolution but poor tissue penetration. In contrast, low-frequency waves provide poorer resolution but allow the insonation of deeper vessels, such as those located within the cranium.

In practice, sound is emitted via an ultrasonic transducer. Contained within the transducer is a piezoelectric crystal connected to electronic circuitry. Short pulses of current emitted by these electrodes cause deformation of the crystal, producing high-frequency sound waves. The sound waves are transmitted through liquid or tissue as waves and are reflected back to the crystal by their interactions with tissues of various acoustic impedances. If the depth and direction of these tissue interfaces are known, a sonographic image can be formed and displayed on a video monitor. The data may be displayed in one of several different formats: amplitude modulation (A-mode), static B-mode, real-time B-mode, Doppler imaging, duplex Doppler imaging, or color-duplex Doppler imaging.

Amplitude modulation is the simplest form of ultrasonic imaging. Sonic echoes reflected from a tissue interface are displayed as spikes projecting from a baseline on the video monitor. The height of the spike is proportional to the intensity of the reflected beam. The degree of separation of the various spikes is proportional to the amount of time it takes for the sound to be emitted from, and returned to, the transducer. Because the velocity of sound in soft tissues is known (approximately 1540 m/sec), the depth of the insonated structure can be determined. Amplitude modulation is suitable for the imaging of stationary structures, for example, determining the presence or absence of a midline shift by imaging the falx cerebri, but it is not generally useful in the evaluation of cerebral vascular lesions. The development of computed tomography (CT) and magnetic resonance imaging (MRI) has made A-mode imaging nearly obsolete in neurosurgery.

Brightness modulation ultrasonography differs from A-mode in that each spike is squeezed into a small dot, which is visible on the video monitor. The brightness of each dot is proportional to the intensity of the reflected wave. Brightness modulation imaging provides excellent anatomic images of the carotid or vertebral arteries, though giving no flow-related information. The ultrasound waves are recorded as they are reflected back from tissue surfaces of different acoustic impedances, and produce a persistent anatomic image of the tissues insonated. These images can then be displayed on a video monitor. Brightness modulation scans can be performed in both the longitudinal and the transverse planes, and hence allow the observer to measure the degree of vascular stenosis with a resolution of between 0.5 and 1.0 mm. The introduction of real-time imaging has enhanced the informational content gleaned from B-mode imaging because it permits rapid processing of reflected ultrasound waves, making possible visualization of the physiologic motion of the imaged structure. A two-dimensional, real-time reconstruction of the anatomic structure can then be projected on the video monitor.

Doppler ultrasonic imaging is based on the principles of the Doppler Effect. First described by Doppler in 1842, this principle states that the frequency of sound emitted from a moving source changes in proportion to its velocity, the difference in frequency between the emitted ultrasonic signal and that reflected back being called the Doppler shift frequency. Doppler ultrasonography determines information regarding blood flow by detecting signals reflected from moving RBCs.

Doppler instrumentation allows either continuous waves or pulsed waves to be emitted. Continuous-wave Doppler registers flow characteristics of all vessels located within the depth of the field. It employs two piezoelectric elements, both contained within a single ultrasonic probe. The ultrasonic beam is transmitted from one crystal, and the beam is back-scattered to the second crystal. A pulsed-wave Doppler signal can be range-gated so that only a small sample of the field under the ultrasonic probe will be monitored (e.g., a single vessel at a certain depth). Intermittent short bursts of ultrasound energy are emitted from a single crystal mounted at the end of a transducer, and during the quiescent period, the same crystal that emitted the sound wave is used to receive back-scattered signals. The encounter with moving RBCs changes the frequency of the back-scattered wave, creating a Doppler shift.

Evaluation of the retrieved data for clinical purposes can be accomplished in several different ways. First, because the Doppler shift is generally in the audible range, adequate clinical interpretations can often be made simply by listening for abnormalities in flow signals. Next, moving the transducer back and forth over a column of moving blood in a vessel can create a pictorial image, not of the vessel wall itself, but of the column of flowing blood within the vessel—ultrasonic angiography. Finally, because flow in blood vessels is usually laminar rather than uniform (i.e., all RBCs are not moving at the same speed), a range of Doppler shift sound frequencies can be obtained in vessels with normal flow. Through spectral analysis, the signal received by the Doppler transducer can be separated into its component sound frequencies. With disturbed or turbulent flow, there is an increase in the range of Doppler shift sound frequencies, called *spectral broadening*. Concomitantly, there is an increase in the peak systolic velocity and in the end-diastolic velocity. These spectral wave forms can be displayed on the video monitor in real-time. Spectral broadening and increased peak systolic and end-diastolic velocities are all indicators of turbulent flow.

Color-flow Doppler imaging is used to demonstrate the direction of blood flow. Shades of color demonstrate flow velocity. By convention, flow away from the ultrasonic probe is blue, and flow toward the probe is red. Color imaging aids in detecting narrow channels of blood flow in areas of tight vascular stenosis and in pinpointing regions with very slow flow. Detection of tight stenosis as opposed to complete occlusion has great clinical significance; this crucial determination is more accurate utilizing color Doppler techniques.

Duplex sonography combines high-resolution, real-time B-mode imaging with Doppler techniques, overcoming many of the limitations of each technique. Brightness modulation defines the vessel anatomy and allows accurate placement of the pulsed Doppler into the vessel lumen to accurately measure changes in flow velocity. If color imaging is added, the direction of flow and the ability to visualize a tight stenosis is enhanced significantly.

## Ultrasonography in Extracranial Carotid Artery Disease

The pathophysiology of atherosclerosis, with special reference to its role in causing events within the cerebral circulation, is considered in detail in chapters 14 and 15, so only a brief outline of the factors relevant to the use of ultrasonography in cerebrovascular disease is attempted here.

### PLAQUE ANALYSIS

The basic structure of an uncomplicated atherosclerotic plaque consists of a fibrous cap and a lipid core, both of which are located within the intima of the artery. Fully developed plaques may undergo changes resulting in more complicated plaques. These changes include calcification, ulceration of the plaques' surface, or thrombus formation on fissures.

The ultrasonic technique best suited for the evaluation of atherosclerotic plaques is B-mode ultrasonography. It displays the best image of the luminal characteristics of an artery and of the constituents involved in the composition of a plaque. Though both longitudinal and cross-sectional views can be obtained using B-mode imaging, the degree of stenosis may be difficult to determine. Doppler spectral frequency analysis yields excellent quantitative information about the extent of stenosis.

With B-mode imaging, the different layers of the arterial wall have characteristic echogenic signals; the lumen, for example, is echo-free. The various components of the plaque also have characteristic echogenic signals. Simple plaques, smooth surface plaques of fibrous tissue origin, generally reflect homogeneous echo patterns. Plaques with high amounts of lipid may not be echogenic enough to be accurately detected. Calcified plaques are characterized by highly echo-reflective areas. Calcific plaques often have irregular surfaces and generally produce acoustic shadowing. Complex plaques or mixed plaques, generally regarded as ulcerative plaques, demonstrate heterogeneous echogenicity. The ability of real-time B-mode imaging to detect ulcerations accurately, however, has been disappointing. Ulcerations, when visible, generally appear as a cavitation or niche within the surface of a plaque, but technical and instrumentation problems with B-mode sonography occasionally lead to misleading plaque appearances. In this setting, color flow imaging may be of assistance, because flow reversal typically occurs in ulcer niches.

Intraplaque hemorrhage can be identified by the presence of an irregular anechoic area within the plaque. Plaque hemorrhage has been implicated in rapid progression from a lesser degree of stenosis to a higher degree of stenosis and even to total occlusion. It may also be a source of cerebral microemboli if the hemorrhage ruptures through the intima. Fresh thrombus is difficult to recognize with B-mode imaging because its echogenicity is quite similar to that of blood. If more than 24 hours old, it becomes organized and B-mode imaging may detect its presence. Old chronic atherosclerotic plaques, on the other hand, may become isoechoic to the surrounding soft tissues of the neck, rendering sonographic imaging of the carotid impossible.

## THE RELATIVE VALUE OF DIFFERENT NONINVASIVE IMAGING TECHNIQUES IN CEREBROVASCULAR DISEASE

Conventional angiography remains definitive for evaluation of the extracranial carotid arteries. Initial imaging can be performed by using ultrasonic techniques. Brightness modulation examination provides data regarding arterial and plaque anatomy, and Doppler imaging provides data regarding flow characteristics within the artery in question. If a stenotic lesion is found, Doppler techniques can quantify the degree of stenosis prior to angiography. If noninvasive testing does not disclose a lesion, conventional angiography should still be completed because it may uncover a lesion in the intracranial internal carotid artery (ICA) or embolic sources within the heart or aortic arch.

A number of recent reports have advocated the use of magnetic resonance angiography (MRA) in the noninvasive evaluation of cerebrovascular occlusive disease. Studies have produced conflicting findings regarding correlation of MRA with duplex ultrasonic imaging results. Documented limitations of MRA include its tendency to overestimate the degree of stenosis at the carotid bifurcation. Also, artifacts, including motion, swallowing, and flow anomalies from vessel loops, may lead to signal loss and the overestimation of stenosis; MRA is also costly. Benefits include the fact that MRA can be obtained easily at the end of a standard MRI. The diagnostic accuracy of MRA plus ultrasonic imaging may ultimately supplant invasive angiography in assessment of patients with cerebrovascular occlusive disease.

In patients with typical hemispheric symptoms, some have argued that there is no need for noninvasive cerebrovascular testing. For patients with typical TIAs, conventional angiography remains the standard test for complete evaluation prior to medical therapy or carotid endarterectomy. Ultrasonography may yet aid in focusing attention to a particular location, provide data about the histologic composition of the lesion, and establish a suitable baseline for follow-up after endarterectomy. Ultrasonic imaging is crucial for patients with serious allergy to contrast material or with renal failure.

For patients who have some residual neurologic deficit after an ischemic insult, B-mode scanning is quite useful. These patients may not be suitable for urgent carotid angiography because of the risk of worsening their neurologic deficit. The discovery of a pre-occlusive stenosis, intraplaque thrombus, or large ulcerating plaque on B-mode imaging in these patients may influence the timing of surgery.

The detection of a cervical bruit is not an uncommon clinical finding in an otherwise asymptomatic patient. It does not correlate necessarily with atherosclerotic carotid disease because many other clinical entities can create a carotid bruit (e.g., hyperthyroidism, anemia, fibromuscular dysplasia), while a near totally occluded carotid artery may produce no bruit at all. Duplex imaging is the ideal screening test for patients with a cervical bruit. Patients with low-grade stenosis can be followed with serial duplex imaging. Patients with high-grade stenoses should be considered for evaluation by angiography. Patients presenting with atypical, nonhemispheric cerebral symptoms such as transient dizziness, lightheadedness, blackouts, and bilateral visual symptoms are ideal candidates for noninvasive testing, because a variety of nonvascular causes may be responsible for these symptoms.

Intraoperative ultrasonography can be used to evaluate the results of carotid endarterectomy. Technical problems such as intimal flaps or stenosis at the arteriotomy site can be identified and rectified before the patient leaves the operating theater. As well, an immediate postendarterectomy baseline for the vessel can be established. Postoperatively, duplex scanning can then be utilized to identify early signs of narrowing at the site of repair, and to follow its progression, sometimes to recurrent stenosis or to the benign, mild postoperative stenosis caused by myointimal hyperplasia. Intraoperative ultrasonography also may be useful in evaluating saphenous vein bypass grafts in the neck. Stenosis at the anastomoses site, strictures along the course of the graft, and clot formation within the graft may be identified and rectified prior to wound closure.

## ACCURACY OF ULTRASONIC IMAGING

Most reports on the accuracy of noninvasive imaging use conventional angiography as the standard against which other studies should be measured. It must be remembered, however, that angiography can supply only anatomic, not physiologic, data and that angiography cannot precisely quantify the degree of stenosis. Therefore, these comparative studies will invariably lead to discrepancies.

The accuracy of any test is best expressed in terms of its sensitivity (defined as the probability that the test result will be positive if disease is present) and its specificity (the probability that the test will be negative if disease is not present). The false-positive rate is defined as the probability of a test being positive when the disease is absent, and the false-negative rate is defined as the probability that the test will be negative when the disease is present. The positive predictive value is the probability that the disease is present when the test result is positive, and the negative predictive value is the probability that the disease is not present when the test is negative. Importantly, the usefulness of any test may differ depending on the patient population to which it is applied. For example, if a test with a high sensitivity and a low specificity is performed on a population of patients in whom the incidence of severe carotid disease is high, the overall accuracy will appear to be very high, although the false-positive rate also may be high. Conversely, if the same test is performed on asymptomatic patients, in whom the incidence of disease may be low, the test will have an unacceptable number of false-positive results and unsatisfactory overall accuracy.

In summary, if the patient is symptomatic, the detection of even a minimal degree of carotid bifurcation disease may be important; therefore, a test with a very high sensitivity and a high negative predictive value should be used. Lesser degrees of specificity are acceptable under these circumstances, because it would be more dangerous to miss disease in such a patient than to subject an occasional patient to an unnecessary angiogram. By contrast, if the patient is asymptomatic, it is important to minimize unnecessary angiography, and a test with high specificity and a high positive predictive value, but perhaps a lesser degree of sensitivity, is appropriate.

Doppler spectral-frequency analysis, in experienced hands, may detect hemodynamically significant lesions of the ICA with a sensitivity ranging from 87% to 95%, and a mean specificity of 83%. In contrast, the sensitivity of B-mode imaging ranges from 57% to 82%, with a mean sensitivity of 75%. The mean real-time B-mode specificity is about 87%, though it ranges in the literature from 36% to 98%. For the diagnosis of stenosis, pulsed Doppler frequency–spectral analysis appears to yield better results than the B-mode scan. Brightness modulation scans do not clearly differentiate fresh thrombus from flowing blood and are unable to penetrate heavily calcified plaque, and they also have difficulty in identifying plaque ulceration. When used alone, they do not reliably identify complete occlusions, and their sensitivity and specificity in detecting greater than 50% stenoses is limited. Duplex imaging combines B-mode scanning with pulsed Doppler imaging; hence, it has a sensitivity and specificity greater than that of either technique alone. The reported sensitivity of duplex scanning has approached 99% in several recent studies, while the mean specificity has been reported to be 91% with a range from 85% to 97%. An important implication of these figures is that the accuracy of duplex scanning may exceed that obtained with angiography.

## Transcranial Doppler

Transcranial Doppler uses directional, pulsed, range-gated, low-frequency (2 MHz) ultrasonic waves to sample blood flow velocities in the large basal arteries forming the circle of Willis. Through the window formed by the thin plate of the squamous temporal bone, the distal ICA, the middle cerebral artery (MCA), the proximal anterior cerebral artery (ACA), and the posterior cerebral artery can all be insonated. Through the transorbital window, the carotid siphon and ophthalmic artery can be insonated, and via a transoccipital approach, the basilar and intracranial vertebral arteries can be insonated. Spectral wave forms are displayed on a video monitor, as are mean blood flow velocities. In 1975, Pourcelot defined the resistive index (RI) and showed it to be an excellent indicator of cerebral vascular resistance. Mathematically, it is obtained by dividing the difference between peak systolic and end-diastolic velocity by the peak systolic velocity. The RI compensates for variations in estimates of blood flow velocities due to differing degrees of angulation between the Doppler probe and the vessel being insonated; it has been shown to be a reproducible measure of cerebral vascular resistance and correlates better with blood flow than does mean blood velocity.

A pattern of increasing mean velocity, elevated resistive indices, and spectral broadening imply a physiologic change in blood flow which is likely proportional to changes in luminal size, and hence to cerebral vascular resistance, within the basal arteries forming the circle of Willis. Transcranial Doppler thus provides a reproducible, reliable, noninvasive method to quantitate changes in cerebral blood flow dynamics. The presence of intracranial vascular occlusive disease can be determined with a sensitivity of about 94% in the ICA and 100% in the MCA. The accuracy of TCD measurements diminishes in more distal vessels of the intracranial circulation. Data obtained from insonation of the ACA are also less reliable in establishing the existence of vascular occlusive disease because of collateral flow from the contralateral circulation through the anterior communicating artery.

Transcranial Doppler is of great value in symptomatic patients in whom other noninvasive extracranial vascular studies are normal. It offers a noninvasive way of evaluating the intracranial vasculature in search of flow-limiting lesions. Changes in mean velocity, in RI, and in the direction of blood flow (e.g., the presence of collateral flow patterns toward an involved artery) confirm a stenotic or occlusive lesion. Transcranial Doppler evaluation of patients who have normal, noninvasive, extracranial vascular studies may identify individuals with occlusive disease of the MCA or basilar artery. These occlusive lesions, if discovered in the acute phase, may be amenable to urokinase therapy, or in some cases of focal proximal MCA occlusion, to vascular bypass.

Patients sustaining aneurysmal subarachnoid hemorrhage have a propensity to develop vasospasm between 4 and 14 days posthemorrhage. They develop symptoms of ischemia and can go on to develop completed infarcts if the spasm goes unrecognized and untreated. Transcranial Doppler analysis of the large basal arteries forming the circle of Willis has become an increasingly popular way of quantifying blood flow in these patients. Repeated examinations can be performed at the bedside, and hence trends in the mean velocity and RI can be followed. Increasing mean velocities and, more reliably, increasing RIs correlate with luminal narrowing and suggest vasospasm. Unfortunately, spasm in the small, more distally located arteries may go undetected, but TCD is still clinically useful in following patients with subarachnoid hemorrhage.

Transcranial Doppler may be utilized for intraoperative evaluation during vascular neurosurgery. By utilizing TCD, one can assess flow velocities in the MCA and in other segments of the circle of Willis both pre- and postoperatively in patients undergoing extracranial or intracranial bypass surgery, or saphenous vein grafting to bypass occluded segments of the carotid artery. In a similar fashion, pre-, intra-, and postoperative TCD comparisons may be used to determine continued patency of the carotid after carotid endarterectomy.

Transcranial Doppler has been used both in the setting of trauma and in patients with hydrocephalus requiring ventriculoperitoneal shunts to demonstrate elevations in intracranial pressure. Elevations in the RI correlate well with increased intracranial pressure. This may be helpful in bedside, noninvasive monitoring of intracranial pressure in trauma patients. Investigation is ongoing to evaluate changes in the TCD spectral wave form, which might occur because of alterations in intracranial dynamics related to hyperventilation, and dehydration with osmotic diuretics. Further analysis of these effects must be undertaken to better interpret data obtained from TCD examinations in the setting of trauma.

The cessation of total brain function (brain death) generally correlates with cessation of blood flow to the brain. The cerebral perfusion pressure becomes zero when the mean arterial BP is the same as the intracranial pressure. A number of reports describing Doppler studies in brain-dead individuals have shown extremely high RIs, and also that a reversal of diastolic flow is characteristic of cerebral circulatory arrest. Caution, however, must be exercised in utilizing these TCD criteria to document brain death in infants, because studies have shown some inconsistencies. In the adult population, utilization of this bedside, noninvasive technique may help aid in earlier determination of brain death. Such determination may be important, especially if organ donation is a consideration.

## Ultrasonic Imaging of the Vertebral Artery

Ultrasonic imaging of the vertebral artery is more difficult than imaging of the carotid artery because of its topography and the tremendous amount of variation in its course. In addition, the artery is located relatively deep in the neck and is obscured, in some places, by overlying transverse processes and the base of the skull.

The diagnosis of vertebral artery stenosis or occlusion is difficult at best. Generally, the symptoms of vertebrobasilar insufficiency are vague, and it is not easy to correlate them with pathology found in the vertebral artery. Furthermore, there is little interest in the surgical correction of these lesions. For these reasons, in addition to variability in the course of the vertebral arteries and the inability to insonate these through the surrounding bony structures, ultrasonic imaging of the vertebral arteries has not been as extensively studied as that of the carotid arteries.

Despite the inherent problems in insonating vertebral arteries, duplex ultrasonography and color Doppler imaging of the vertebral arteries may aid in the diagnosis of the subclavian steal syndrome, in which stenosis of the proximal subclavian artery limits blood flow to the arm. The vertebral artery, in this situation, serves as collateral blood supply to the arm. The syndrome is generally clinically silent until the arm is exercised or used. The patient then develops symptoms of vertebrobasilar insufficiency as blood is shunted from the vertebrobasilar system into the arm. With ultrasonic techniques, the presence and direction of flow within the vertebral artery can be established. Reversal of flow (i.e., retrograde flow in the vertebral artery in a patient with clear-cut symptomatology) should suggest subclavian steal syndrome. Sometimes, this may be demonstrated only if the test is performed while the patient is exercising or utilizing the ipsilateral arm.

## Conclusion

Ultrasonography is the most useful noninvasive technique available to aid in the evaluation of both the intracranial and extracranial circulation in patients with cerebrovascular disease. Ultrasonic waves emitted from a transducer are reflected back to either the same transducer or a second transducer when they encounter moving objects, or when they interface with tissues of different densities. When the reflected waves are converted back into electronic signals, they can be processed to display anatomy, or as a Doppler shift, which allows determination of blood velocity.

In the assessment of lesions of the extracranial circulation, real-time B-mode imaging provides patho-anatomic information regarding the artery in question, and can define plaque morphology, whereas Doppler imaging determines changes in flow velocity, and utilizing spectral analysis can accurately delineate the severity of stenosis. Both modalities can be combined within a single duplex unit, which can display information in a color-coded format. Transcranial Doppler, in a similar fashion, allows the insonation of intracranial vessels in a noninvasive manner to quantitate changes in blood flow and to correlate these with intracranial vascular occlusive lesions. Despite recent advances in technology, which have brought MRA and three-dimensional CT angiography to our diagnostic arsenal, vascular ultrasonic imaging still provides a tremendous amount of useful data for the screening of patients considered at risk and in the evaluation of patients with symptomatic cerebrovascular occlusive disease.

## References/Further Reading

Aaslid R, Huber P, Nornes H. Evaluation of cerebrovascular spasm with transcranial Doppler ultrasound. J Neurosurg 1984;60:37–41.

Anderson CM, Saloner D, Lee RE et al. Assessment of carotid artery stenosis by MR angiography: comparison with x-ray angiography and color-coded Doppler ultrasound. AJNR 1992; 13:989–1003.

Barnes RW, Rittgers SE, Putney WW. Real-time Doppler spectrum analysis: predictive value in defining operable carotid artery disease. Arch Surg 1982;117:52–57.

Bluth EI, Mcvay LV, Merritt CRB, et al. The identification of ulcerative plaque with high resolution duplex carotid scanning. J Ultrasound Med 1988;73–76.

Chadduck WM, Crabtree HM, Blankenship JB, et al. Transcranial Doppler ultrasonography for the evaluation of shunt malfunction in pediatric patients. Childs Nerv Syst 1991;7: 27–30.

Chadduck WM, Seibert JJ, Adametz J, et al. Cranial Doppler ultrasonography correlates with criteria for ventriculoperitoneal shunting. Surg Neurol 1989;31:122–128.

Earnest F IV, Forbes G, Sandok BA, et al. Complications of cerebral angiography: prospective assessment of risk. AJNR 1983;4:1191–1197.

Harders A. Neurosurgical applications of transcranial Doppler sonography. New York: Springer, 1986.

Laster RL Jr, Acker JD, et al. Assessment of MR angiography versus arteriography for evaluation of cervical carotid bifurcation disease. AJNR 1993;14:681–688.

Lustgarten JH, Solomon RA, et al. Carotid endarterectomy after noninvasive evaluation by duplex ultrasonography and magnetic resonance angiography. Neurosurgery 1994; 34:612–619.

McGahan JP, Lindfors KK, Carroll BA. Diagnostic ultrasound in neurological surgery. In: Youmans JR, ed. Neurological surgery. vol 1. Philadelphia: Saunders, 1991:187–203.

O'Donnell TF, Erdoes L, Mackey WC, et al. Correlation of B-mode ultrasound imaging and arteriography with pathologic findings at carotid endarterectomy. Arch Surg 1985; 120:443–449.

Padayachee TS, Gosling RG, Bishop CC, et al. Monitoring middle cerebral artery blood velocities during carotid endarterectomy. Br J Surg 1986;73:98–100.

Sekhar LN, Wechler LA, Yonas H, et al. Value of transcranial Doppler evaluation in the diagnosis of cerebral vasospasm after subarachnoid hemorrhage. Neurosurgery 1988;22:813–821.

Sloan MA, Haley EC Jr, Kassell NF, et al. Sensitivity and specificity of transcranial Doppler ultrasonography in the diagnosis of vasospasm following subarachnoid hemorrhage. Neurology 1989;39:1514–1518.

Thiele BL, Strandness DE Jr. Duplex scanning and ultrasonic arteriography in the detection of carotid disease. In: Kempczinski RF, Yao JST, eds. Practical non-invasive vascular diagnosis. Chicago: Year Book Medical, 1987:339–361.

# Endoscopy of the Central Nervous System

Kim H. Manwaring

The first neurosurgical endoscopic procedure was performed in 1910 by a urologist, Lespinasse, of Chicago. He fulgurated the choroid plexus in two infants in an effort to diminish CSF production. Walter Dandy and others followed his lead, while W. J. Mixter performed a third ventriculostomy in 1923; that is, he created a bypass pathway from the third ventricle to the subarachnoid space.

In the 1970s, technologic improvements in the optical quality of endoscopes and cameras, and in the instrumentation for tissue dissection, ushered in an era of "video monitor–based" surgery. The previous surgical procedures of fulguration of the choroid plexus and third ventriculostomy were then, in a real sense, re-discovered and good results achieved. More recently, impressive outcomes from the endoscopic management of obstructive hydrocephalus due to colloid cyst have been described by several neurosurgeons, their results being comparable to those of craniotomy with microsurgery.

The coupling of a frame-based stereotactic system to a rigid lenscope has led to an extensive series of tumor biopsies, some partial removals, and a few complete resections. In 1991, a system was described in which a frameless stereotaxy technique was coupled with the neuroendoscope, allowing freehand guidance, to minimize the risks of disorientation and bleeding.

Extracranial applications of endoscopy in neurosurgery have included exploration of the extradural spinal canal and the intervertebral foramina, removal of intervertebral disc tissue, and management of carpal tunnel syndrome. The lysis of syringomyelia-related loculations with effective symptomatic relief, using a fine flexible neuroendoscope, was described by Hüwel and colleagues in 1992.

## Types of Endoscope

Four groups of neuroendoscopes have been developed for applications in the craniospinal access. Each has benefits and limitations in regard to imaging quality, ease of surgical handling, maneuverability in the surgical field, and specific design for a surgical procedure.

## LENSCOPES

The straight, rigid lenscope is presently the most extensively utilized instrument in neuroendoscopy. A typical construction consists of a stainless steel tube varying between 2 and 10 mm in diameter and containing glass rod lenses alternating with air pockets. This design provides a wide angle of vision, superior illumination, and a flat undistorted image.

The optical eyepiece is adaptable for direct viewing with the eye but a C-mount adapter is generally employed to connect also a small light-weight camera. A 0-degree lenscope allows direct forward viewing, while commonly available angled lens systems allow imaging at 30, 70, and 120 degrees off of the axis of the scope. Due to the high quality of the image that can be obtained, the diameter of the lenscope can be reduced to 1 mm. Advantages, compared with other types of endoscope, include image quality, degree of miniaturization, ruggedness, tolerance of handling, and sterilization. The high optical resolution and straight design of the lenscope greatly improve the neurosurgeon's ability to recognize anatomic structures and allow a rapid learning curve in mastery of orientation. By contrast, the light channel of a flexible fiberscope can rotate, so that the image as viewed on the monitor will rotate also, possibly disorientating the surgeon. Further, the rigid design of the lenscope is easily adapted to both frame and frameless stereotaxy systems (Fig. 28.1).

The disadvantages of the lenscope include the necessity either to continually reposition it within a retractor, to immobilize it within a stereotaxic frame, or to hold it in one's hands; satisfactory retractor systems to fixate the endoscope or allow it to "float" like a microscope are not yet available. Fortunately, due to the speed with which the target tissue can be accessed, many endoscope procedures are of brief duration. The necessity to hold the endoscope in the hands is not particularly troublesome if the procedure is limited to fewer than 30 minutes. If the lenscope is not handled carefully between surgical procedures, flexing of the rigid tube can easily dislodge the glass rod lenses, resulting in a need for costly repair.

### Illustrative Surgical Application

Endoscopic third ventriculostomy has been described by numerous authors as an effective alternative to shunting in the control of obstructive hydrocephalus in appropriately selected patients. When a magnetic resonance imaging (MRI) scan demonstrates obstruction at the level of the aqueduct or severe stenosis with enlarged third ventricle and small fourth ventricle, the floor of the third ventricle is often seen to herniate downward behind the posterior clinoid

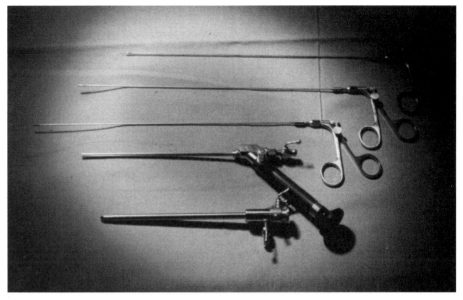

**Figure 28.1**
The rigid lenscope provides superior optical resolution. A further advantage is that rigid or straight instruments such as microscissors, biopsy forceps, and graspers can be controlled more precisely. Due to its weight, the lenscope must be held by a retractor system for prolonged surgical dissection procedures *and* frequently requires minor repositioning.

processess. A patient more than 2 years old who presents with symptomatic obstructive hydrocephalus and has not been shunted previously, or suffered meningitis, may be an excellent candidate for this surgical procedure. An endoscope is passed by a free-hand technique from a coronal burr hole 3 cm off of the midline. After a sample of CSF has been obtained for routine studies, the rigid lenscope with light cable and camera attached is passed through a peel-away sheath into the lateral ventricle. The landmarks around the foramen of Monro, including choroid plexus, thalamostriate and septal veins, fornical column, and margin of the frontal horn, are readily identified.

The lenscope is then passed through the foramen of Monro, and the thinned-out floor of the third ventricle is visualized directly ahead. The groove between the mammillary bodies lies posterior to the hypophyseal recess, and in obstructive hydrocephalus this region of the ventricular floor bulges down behind the posterior clinoid processess into the interpeduncular cistern. The ventricular floor is often sufficiently thin to be translucent. The floor of the ventricle is then perforated, using, for example, the ventriculoscope as a blunt dissector and creating a round opening through which CSF can escape readily into the subarachnoid space. Electrosurgical and laser techniques also have been described. On completion of fenestration, CSF can often be seen to flow to and fro across it. A postoperative MRI scan will usually show the newly created flow void, especially if enhanced by cine-MRI technique.

Normalization of intracranial hypertension and relief of symptoms and signs will follow in at least 60% of appropriately selected patients, and although ventricle size may not return to normal, the subarachnoid spaces and sulci widen. (Many neurosurgeons believe the improved cortical mantle appearance when such a patient is shunted, instead of treated by ventriculostomy, represents an artifact of excessively low intracranial pressure created by the siphoning effect of some shunts.)

## FIBERSCOPES

The fiberscope consists of a distal objective lens, a coherently organized bundle of light fibers, and an eyepiece incorporating a focusing lens. (The phrase *coherent organization* refers to the precise alignment relationship between bundles at the two ends of the scope. While light can be delivered to the surgical field through a noncoherent bundle, such as the fiberoptic light source used in surgical head lights, the coherent bundle reproduces the image through individual pixels of light.) The tightness of packing of a coherently organized bundle is referred to as the packing fraction. A typical 3-mm bundle many have up to 40,000 individual fibers, but because of the spaces between fibers, only approximately 60% of the cross-sectional area of the bundle transmits the image, the quality of which is, therefore, inferior to that of a lenscope. On the other hand, the fiber bundle confers flexibility: The steerable tip of a fiberscope commonly will pass through an arc of 180 degrees. The width of the viewing angle is limited in relation to the diameter of the endoscope to avoid excessive "fish eye" distortion at its tip. Fiberscopes for neurosurgery are available commonly in a range from 0.5–5.0 mm (Fig. 28.2).

By selecting appropriate viewing angles, the rigid lenscope can be made to provide visualization to one side, or even behind its tip. It is only the fiberscope, however, that allows an instrument to protrude through the working channel beyond the lens to give the neurosurgeon capability to operate "around the corner."

A further significant advantage of the fiberscope is the availability of fixed retractor systems that immobilize the body of the endoscope, together with a steering trigger and attached camera. This leaves the flexible light channel as the only moving part, and this can be easily manipulated by finger control only. The neurosurgeon can therefore undertake lengthy endoscopic dissection procedures without fatigue or the inconvenience of the multiple repositioning associated with use of a rigid scope.

The steerable fiberscopes in the 3- to 5-mm range provide excellent exploratory capability in the intraventricular and intracystic spaces. From a

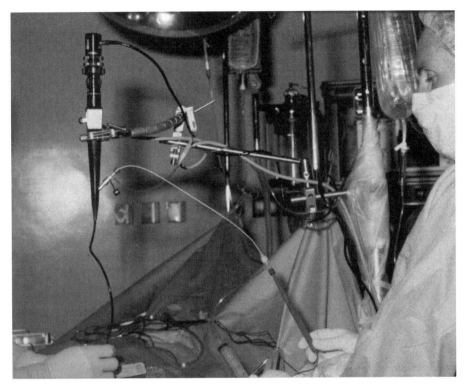

**Figure 28.2**
The steerable fiberscope can be mounted on a fixed retractor, leaving the neurosurgeon free to control the light viewing and instrumentation channel. Although its optical resolution is inferior to the rigid lenscope, the steerability of the tip through a 180-degree arc extends ventriculoscopic diagnostic capabilities and provides the opportunity to operate around corners and in recesses, for example, in the third ventricle.

coronal approach, it is practicable for the neurosurgeon to pass the fiberscope from the lateral ventricle through the foramen of Monro to the third ventricle, steer posteriorly 90 degrees, and directly view the opening of the aqueduct, posterior commissure, and suprapineal recess. A small flexible "daughter" fiberscope can then be passed through the instrument channel to visualize the length of the aqueduct and enter the fourth ventricle, or a balloon catheter can be used to dilate the aqueduct in cases of acquired aqueduct stenosis. Similarly, a 2-mm flexible fiberscope has been employed through a dorsal myelotomy in the spinal cord in patients with septated or noncommunicating hydromyelia to mechanically perforate membranes and then place a syrinx to peritoneal shunt.

Fiberscopes that are rigid like lenscopes also have been developed for neurosurgery. Although the image they display is not of the same high quality as that of lenscopes, rigid fiberscopes can be made very small, allowing them to pass through 2-mm shunt catheters. The rigid fiberscope also is fairly robust and will tolerate mild bending of the rigid tube. Steerable fiberscopes with instrument channels are generally more expensive than lenscopes and require greater expertise and care in interoperative cleaning and sterilization.

## Surgical Application

*Ventricular Biopsy and Septostomy*    The fiberscope is well suited to exploration of the ventricular system. Lesions that remain too small for identification by MRI, computed tomography (CT), or ultrasound imaging may be recognized by diagnostic ventriculoscopy. In contrast to the lenscope, the fiberscope permits access to the aqueduct without deforming the sensitive structures of the foramen of Monro. Neoplasms in the third ventricle or lateral ventricle recesses can be visualized easily and biopsied directly, and hemostasis can be achieved reliably. The transventricular approach to the aqueduct described has also been used recently in the management of hydrocephalus due to late presenting or acquired aqueduct stenosis by balloon dilation of the aqueduct, a so-called aqueductoplasty.

## OPTICALLY GUIDED INSTRUMENTS

Several conventional neurosurgical instruments (e.g., nerve hooks, microscissors, Penfield dissectors, and biopsy forceps) have been coupled with small fiberscopes to allow direct imaging of the tissue-instrument interface and allow the neurosurgeon increased visibility compared with "along the shaft sighting" through an operating microscope (Fig. 28.3).

## Surgical Application

*Optically Guided Instrumentation as an Adjunct to Open Craniotomy*    At open craniotomy with the operating microscope it is often difficult to visualize a recess adequately prior to dissection of brain tissue. Retraction of the brain is therefore necessary; this entails exerting deforming pressure on the brain, or the operating microscope must be positioned at an awkward angle. Alternatively, as in the field of skull-base surgery, only an extensive exposure permits straight-on surgical access. An optically guided nerve hook is an example of adjunctive visualization around a corner, which allows the neurosurgeon to watch the tissue part at the tip of a 90-degree curved dissector passed behind a blood vessel or in a recess of retracted brain tissue. A potential disadvantage is that minimal bleeding obscures the lens at the tip of the fiberscope, though it can sometimes be cleared by irrigation.

## STEREOENDOSCOPES

In an effort to conserve optical quality and minimize size, the majority of neuroendoscopes have been monoscopic in design. Depth of working field

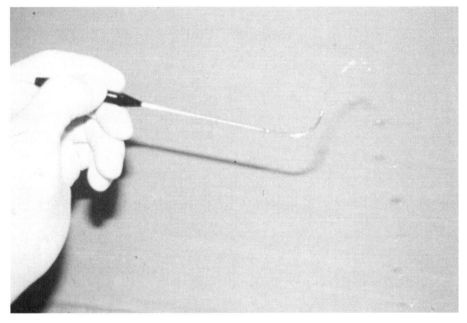

**Figure 28.3**
The dissecting endoscope is an example of an optically guided instrument that enables the surgeon to see into a recess or behind a vascular structure such as the neck of an aneurysm, thus obviating the need for more extensive craniotomy exposure or retraction of the brain.

must then be judged by monoscopic cues, that is, by noting how fast an object's size enlarges as it is approached or by making gentle contact with it. Monoscopic dissection, therefore, can be slow, risky, and tedious.

Over the past 5 years, a variety of stereoendoscopes have been introduced that provide true three-dimensional imaging through the rigid telescope by a variety of schemes. With the first, two separate parallel lenscopes can each be coupled with a camera and the images presented separately to the surgeon's eyes by either polarizing eyeglass technology or oscillating liquid crystal lenses in the eyeglasses. With the second, the alternating images from the two light channels are presented at a rate in excess of the ocular fusion frequency, giving the neurosurgeon the sense of smoothly reconstructed three-dimensional space. These endoscopes do not presently incorporate instrument channels.

A variation on the rod lens stereoendoscope is the stereoelectronic videoscope. The small charge-coupled devices (CCD) light-sensitive chips can be stacked vertically in the tip of a tube situated directly behind the distal objective lens. Because the CCD chip is brought directly into the surgical field without intervening rod lenses or fiber bundles, the illumination and sharpness of the image are exceptional. A three-dimensional effect can be achieved as described earlier. At present, due to size constraints in chip technology, electronic videoscopes are about 15 mm in diameter, though extremely light-weight and rugged.

## Surgical Application

*Intrathoracic Approach to the Spine*   A subaxillary intercostal approach to the spine after collapse of the lung allows visualization over a wide segment of the lateral vertebrocostal angle and sympathetic ganglion chain. Passage of dissecting scissors and of monopolar, bipolar, and laser energy sources through a separate intercostal approach allows direct interruption of the upper sympathetic chain; even vertebrectomy can be performed through a very limited exposure. Use of a stereoendoscope requires expensive supplemental equipment, including cameras, monitors, and eyewear.

*Endoscopic Instruments*   The development of new instruments of both rigid and flexible type to accommodate the working channel of the neuroendoscope represents an area of active research and design by many neurosurgeons and manufacturers as endoscopic techniques are adapted to the management of different disease entities. For details of these, the current specialized literature should be consulted.

*Frame and Frameless Stereotaxy Guidance Systems*   A significant limitation to endoscope-based approaches to the intracranial cavity is disorientation due to unusual anatomic appearance, deformation by the disease process, or obscuration by bleeding. Most tissue bleeding ceases without difficulty or consequence to the patient in the presence of continuous, warm, physiologic mock CSF or lactated Ringer's solution irrigation. Such bleeding, however, most commonly occurs when a blood vessel is divided during dissection but is not recognized due to limited visualization through the neuroendoscope. Further, the comparative simplicity of current endoscopic dissection procedures has been dictated by the lack of orienting anatomic information that is available at open craniotomy. Therefore, several investigators have combined the neuroendoscope with conventional fixed-frame stereotaxy systems and, more recently, with frameless stereotaxy systems.

When combined with the frame stereotaxy system, the endoscope has been used in an open-air-based system for removal of deep-seated tumors. In addition, removal of brainstem hematomas and angiomas has recently been described utilizing a frame-stereotaxy–guided cannula with endoscopic visualization of the passage of grasping forceps and laser probes.

The author has had experience with magnetic-field–guided frameles stereotaxy coupled with the neuroendoscope in a variety of ventricular and parenchymal conditions in which such guidance appeared critical to the intraoperative decision-making process and outcome. The monitor display presented to the neurosurgeon in this system (Research grant, Johnson &

Johnson Professional, Inc., Randolph, MA, to K. H. Manwaring and M. L. Manwaring) comprises the endoscope image, the CT or MRI image, and guidance bars. The trajectory between the burr hole and a target is indicated and a "glide path" shows the neurosurgeon how to pass from one to the other through the least injurious course. Such guidance makes it possible to approach posterior third ventricle colloid cysts through an interseptal, partially transcallosal approach with a 3-mm endoscope (Fig. 28.4), remove intracerebral hematomas more completely, and perform partial lobectomy (Fig. 28.5).

Intraoperative sonography through an open fontanelle or through a small craniotomy allows the benefit of real-time localization. This technique has facilitated dissection of intraventricular cysts and removal of intraventricular tumor. In addition, the simultaneous Doppler presentation often allows recognition and avoidance of arteries in the subjacent tissue.

## Set-up of the Neuroendoscopic Suite

Most neuroendoscopic procedures require small incisions and are relatively brief due to the rapid access to the site of dissection, biopsy, or excision. Several pieces of equipment, however, are necessary in addition to conventional instruments. Further, due to the potential for difficulty in controlling bleeding, their availability in the room of a craniotomy set-up is always desirable.

*Monitor*   High-resolution video monitors are available in most surgery departments due to the rapid acceptance of endoscopic procedures. The same small CCD cameras can be coupled to the C-mount of a rigid lenscope or fiberscope used in neurosurgery; gas sterilization is advisable for both camera and neuroendoscope.

Some neurosurgeons have preferred to look directly into the eyepiece for visualization. This risks contaminating the endoscope and, therefore, the surgical field, and it follows that other personnel in the surgical suite cannot observe the surgical procedure. When the monitor is used, the anesthesiologist, scrub nurse, and other assisting physicians can readily appreciate each step of surgical management and participate more meaningfully in the event of difficulties being encountered.

*Light Sources*   Three light sources are conventionally available for endoscopes. Xenon is the brightest light and also the most expensive. It is superb for interoperative documentation by photography, video cassette recorder, or thermal printers. Mercury arc vapor lights are less expensive and slightly less bright, but are generally adequate for most neuroendoscopic applications. Conventional halogen head-light sources are least expensive but yield inadequate illumination for many intraventricular applications, and particularly in the presence of blood clot.

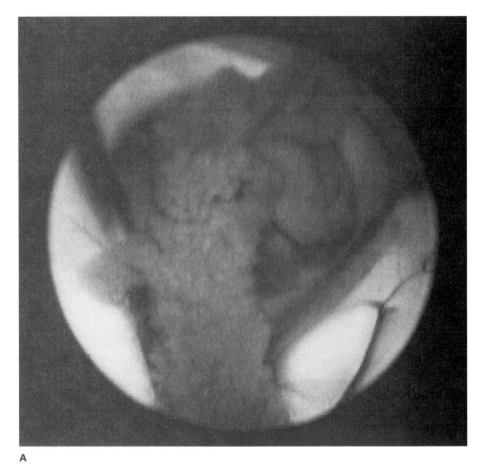

**A**

**Figure 28.4**

(A) The endoscopic appearance of the colloid cyst causing obstruction at the foramen of Monro shows that choroid plexus often overlies the dome. From an endoscopic approach, the steps of removal include coagulation of the choroid plexus overlying the tumor cyst capsule, coagulation of small capsular vessels, incision of the capsule, aspiration evacuation of the cyst, devascularization shrinkage of the remaining capsule membrane or its removal by biopsy forceps, and postremoval inspection to assure CSF pathway patency to the aqueduct.

*Irrigation* Access to the intracranial cavity is normally achieved using a peel-away sheath as a cannula for the neuroendoscope. The sheath serves four purposes of safety: (1) Because it is normally selected to be larger in diameter than the neuroendoscope, irrigation fluid from within the endoscope can readily pass out of the ventricular system or cyst cavity through the residual diameter of the sheath, thus avoiding intracranial hypertension. (2) The sheath provides a conduit through which the neurosurgeon can return directly to the tissue which he or she is dissecting, so that repeated passages of the endoscope or of instruments within it impose no new injury on the brain. (3) When a

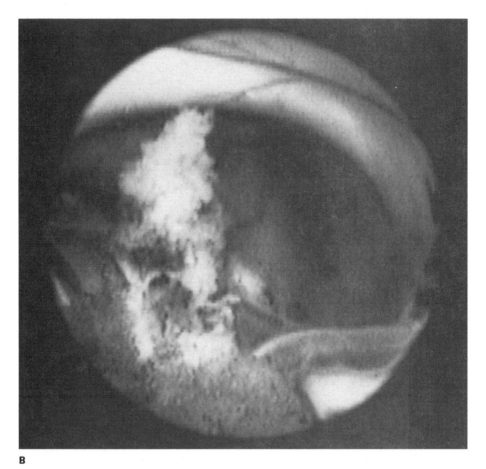

B

**Figure 28.4** *Continued*
(B) The postoperative appearance of the foramen of Monro demonstrates the cyst to be entirely removed. The overlying choroid plexus has been coagulated and need not be further disturbed.

fiberscope is used, its tip often will be directed "around the corner." The possibility of a significant complication due to the neurosurgeon's neglecting to straighten the tip of the endoscope, thus lacerating the brain as the fiberscope is removed, is avoided by the presence of the peel-away sheath, which halts the withdrawal until the tip has been steered correctly. For example, for procedures within the third ventricle, the peel-away sheath is normally carried below the foramen of Monro. This protects the thalamostriate vein and the fornix from injury. (4) If it is necessary or desirable to conclude the operation by inserting a shunt, the peel-away sheath allows precise positioning of the catheter at the selected site under direct endoscopic visualization. This enhances significantly optimal positioning and, therefore, longevity of ventriculoperitoneal shunt function.

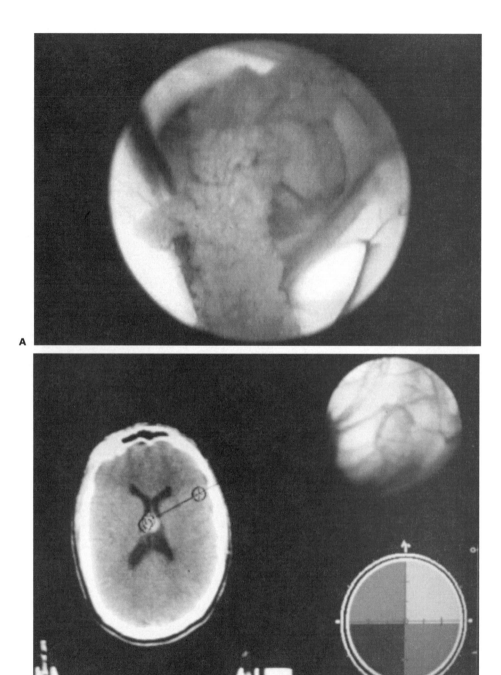

**Figure 28.5** (A) Obstructive colloid cysts may be anteriorly positioned at the foramen of Monro or more posteriorly in the roof of the third ventricle. The most posteriorly placed colloid cysts are not easily reached through the foramen of Monro, which often lies in a deep recess due to mass effect above it. (B) Combined magnetic field frameless stereotaxy and endoscope imaging allows targeting an approach through the paramedian ventricular cavity and partially across the corpus callosum to the interseptal space. By hydrojet dissection, the walls of the septum are separated over the dome of the colloid cyst, which can then be excised directly without passing through the foramen of Monro or transgressing the column of the fornix. The trajectory on CT scan demonstrates the approach into the interseptal space. The endoscopic image shows the cyst membrane after aspiration of its contents but before devascularization of the membrane. The guidance bars in the lower right corner guide the endoscope along a glide path trajectory from the coronal burr hole to the tumor.

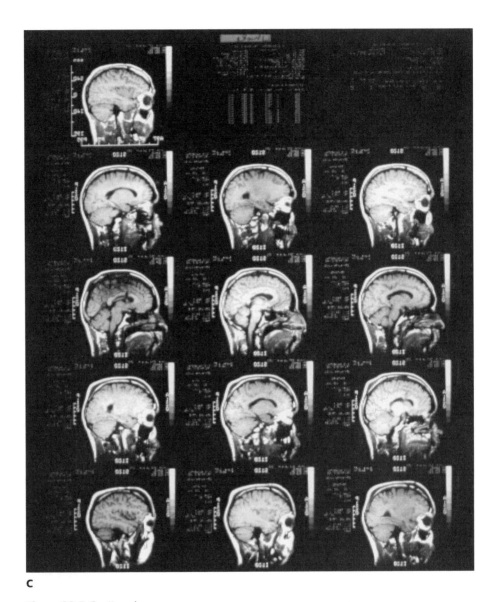

**C**

**Figure 28.5** *Continued*
(C) The postoperative sagittal MRI scan demonstrates the practicability of this approach to the colloid cyst by using a combined endoscopic and frameless stereotaxy technique.

In the author's experience, hydrojet dissection is an effective technique to safely skeletonize vessels of cyst and ventricle walls as well as to dissect tumor capsules. The technique employs rapid irrigation through a port at the endoscope lens tip as the endoscope is brought to the interface of vessel and tissue. Irrigation with warm, lactated Ringer's solution dissects white matter away without tearing blood vessels (which have a higher tensile strength). Subsequently, wide fenestration can be achieved using contact electrosurgical or laser probes. While the safety of this simple technique rests on avoidance of intracranial hypertension by continuous use of an appropriately selected peel-away sheath cannula, it is safer to use an intracranial pressure monitor in the surgical field or in the subdural space for lengthy or complicated procedures (Fig. 28.6).

Because monitor equipment, frame or frameless stereotaxy equipment, and energy sources such as laser and electrosurgery crowd the neuroendoscopic suite, it is vital to provide equipment orientation so that all key surgical personnel can maintain visualization of the monitor. With simultaneous presentation of frameless stereotaxy and guidance information on the monitor, we have found it helpful to bring a liquid crystal display of high-resolution type directly in front of the surgeon's eyes as a "head's up" monitor.

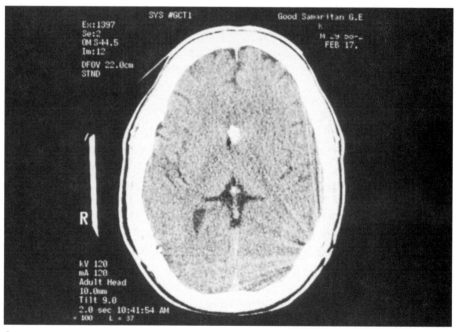

A

**Figure 28.6**

(A) Patients presenting after infancy with obstructive hydrocephalus due to primary aqueductal stenosis or acquired aqueductal stenosis may be managed effectively by endoscopic third ventriculostomy. Patients who were previously shunted and became symptomatic for shunt failure or inadequacy of shunt function may subsequently be made shunt-free by third ventriculostomy. This 29-year-old man presented with slit ventricles and recurrent headaches (slit ventricle syndrome) 6 years postplacement of a shunt with intervening revisions for aqueductal stenosis.

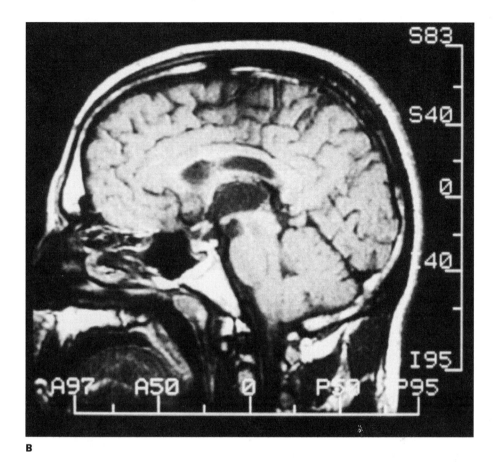

**B**

**Figure 28.6** *Continued*
(B) Removal of the ventriculoperitoneal shunt and placement of external ventricular drains allowed sufficient gradual ventricular enlargement to provide access for safe endoscopic third ventriculostomy. The patient is now, 2 years after third ventriculostomy, free of a shunt and of any symptoms of intracranial hypertension.

## Surgical Procedures and Results

A significant variety of neuroendoscopic operative procedures have been described in the literature (Table 28.1). Several neurosurgical centers have reported on series of endoscopic procedures that exceed 250 cases. Lewis and Crone noted, for example, a reduced incidence of shunt revision to 7.5% over 2 years in 80 shunt procedures that were placed using endoscopic positioning. In a series with patients with loculated hydrocephalus, the shunt revision rate decreased from 3.04 per year before endoscopy to 0.25 per year after endoscopy, with a mean follow-up of 26 months. In several series, there have been no reports of recurrences of colloid cysts following endoscopic removal in their experience or the author's. Finally, Lewis and Crone cited a complication

**Table 28.1**   Neurosurgical Procedures Performed Endoscopically

Diagnostic
  Ventriculoscopy
  Cisternoscopy
  Biopsy in the intraventricular intercystic or cisternal spaces
  Adjunct endoscopic inspection at microsurgical craniotomy
Therapeutic
  Ventriculoperitoneal shunt
  Excision of encysted shunt catheters
  Third ventriculostomy
  Septostomy of septum pellucidum
  Fenestration or membranectomy of loculating ventricular membranes
  Aqueductoplasty
  Foramenoplasty
  Evacuation of chronic subdural hematomas
  Evacuation of intraventricular hematomas
  Evacuation of parenchymal intracerebral hematomas
  Excision of intraventricular tumors, particularly colloid cyst
  Parenchymal endoscopic biopsy
  Parenchymal partial lobectomy
  Fenestration of loculated hydromyelia
  Excision of disc nucleus pulposus
  Release of the carpal tunnel

incidence of 12 (2%) of 550 patients over 6 years. In Kelly's reported series of stereotactic endoscopic ventriculostomy of 16 patients, there was only one failure. Oka et al. reported a similar success rate in obstructive hydrocephalus patients in which third ventriculostomy and aqueductoplasty were performed.

## Conclusion

Neuroendoscopy is a rediscovered surgical technique due to improved instruments, imaging technology, and preoperative imaging capabilities. Its application has been limited largely to the clear fluid spaces of the brain, principally ventricles, cisterns, and pathologic cysts. In these surgical applications, the results appear clearly promising, given the avoidance of implanted shunts, simplification of shunts, or avoidance of more invasive craniotomies. More extensive dissection procedures have been few but appear clearly feasible when coupled with guidance technologies such as frame and frameless stereotaxy systems. This newer technology may significantly reduce the limitations of the neuroendoscope, specifically disorientation and the resultant difficulty with hemostasis. Nonetheless, there is still room for advances in instrumentation.

## References/Further Reading

Crone KR. Endoscopic technique for removal of adherent ventricular catheters. In: Manwaring KH, Crone KR, eds. Neuroendoscopy, vol I. New York: Mary Ann Liebert, 1992:41–46.

Hirsch JF. Percutaneous ventriculocisternostomies in noncommunicating hydrocephalus. Monogr Neural Sci 1982;8:170–178.

Hüwel N, Perneczky A, Urban V, Fries G. Neuroendoscopic technique for the operative treatment of septated syringomyelia. Acta Neurochir (Wien), Supplementum. 1992;54:59–62.

Jones RFT, Stening WA, Brydon M. Endoscopic third ventriculostomy. Neurosurgery 1990; 26:86–92.

Kelly PJ. Stereotactic third ventriculostomy in patients with non-tumoral adolescent/adult onset aqueductal stenosis and symptomatic hydrocephalus. J Neurosurg 1991;75:865–873.

Lewis AI, Crone KR. Advances in neurosurgery. Contemp Neurosurg 1994;16:19.

Lewis AI, Crone KR, Taha J, et al. Surgical resection of third ventricle colloid cysts. J Neurosurg 1994;81:174–178.

Manwaring KH. Intraoperative microendoscopy. In: Maciunas RJ, ed. Interactive image-guided neurosurgery. Park Ridge, IL: American Association of Neurological Surgeons, 1993: 217–232.

Manwaring KH. Endoscope-guided placement of the ventriculoperitoneal shunt. In: Manwaring KH, Crone KR, eds. Neuroendoscopy, vol. I. New York: Mary Ann Liebert, 1992:29–40.

Manwaring KH. Endoscopic ventricular fenestration. In: Manwaring KH, Crone KR, eds. Neuroendoscopy, vol. I. New York: Mary Ann Liebert, 1992:79–89.

Manwaring KH, Manwaring ML, Moss SD. Magnetic field guided endoscopic dissection through a burr hole may avoid more invasive craniotomies—A preliminary report. Vienna: Springer, Minimally Invasive Neurosurgery, vol II. 61:18–19, 1994.

Oka K, Yamamoto M, Ikeda K, Tomonaga M. Flexible endoneurosurgical therapy for aqueductal stenosis. Neurosurgery 1993;33:236–243.

Otsuki T, Yoshimoto T. Endoscopic resection of a subthalamic cavernous angioma: technical case report. Neurosurgery 1994;35:751–754.

Otsuki T, Yoshimoto T. A new approach for the endoscopic stereotactic brain surgery using high-power laser. Optical Fibers Med VI SPIE 1991;14/20:220–224.

Otsuki T, Yoshimoto T, Jokura H, Katakura R. Stereotaxic laser surgery for deep-seated brain tumors by open-system endoscopy. Stereotact Funct Neurosurg 1990;54-55:404–408.

Sainte-Rose C. Third ventriculostomy. In: Manwaring KH, Crone KR, eds. Neuroendoscopy, vol.I. New York: Mary Ann Liebert, 1992:47–62.

Zamorano L, Chavantes C, Dujovny M, Ausman J. Stereotactic endoscopic interventions in cystic and intraventricular brain lesions. Acta Neurochir (Wien) 1992;54:59–76.

Zamorano L, Chavantes C, Dujovny M, Malik G. Image guided stereotactic resection of intracranial lesions: endoscopic and laser technique. In: Dyck P, Bouzaglou A, eds. Neurosurgery: state of the art reviews. Philadelphia: Hanley and Belfuss, 1989:105–118.

# Index

HLA-DR2, in multiple sclerosis, 36
HMSN (hereditary motor and sensory neuropathy).
      *See* Charcot-Marie-Tooth neuropathy
      (CMT); Déjérine-Sottas syndrome
HNPP (hereditary neuropathy with liability to
      pressure palsies), genetic relationship to
      CMT1A, 55–56
Hormones, deficiencies, neural transplantation for,
      213
Hormones, manipulation, for cancer-related pain,
      118
HTLV-1
   infections, 68. *See also* Adult T-cell leukemia
      (ATL); Tropical spastic paraparesis (TSP)
      adult T-cell leukemia and, 77
      cellular involvement in, 78
      of central nervous system, 85–86
      neurologic complications, risk for, 78
      tropical spastic paraparesis and, 77–78
   long-terminal repeat, 73–74
   neurologic dysfunction and, 76–79
   structure of, 70–71
   transmission of, 77
Human immune globulin (HIG), 150, 162
Human immunodeficiency virus (HIV). *See* HIV-1
Human leukocyte antigens (HLAs), processing and
      presentation, 33, 36
Human T-cell lymphotropic virus type III. *See* HIV-1
Human T-cell lymphotropic viruses, 68. *See also*
      HTLV-1
Hunt-Hess classification of subarachnoid
      hemorrhage, 271
Huntington's disease
   animal models of, 198, 211
   basal ganglia feedback loop dysfunction in,
      195–196
   brain damage in, 186
   CAG repeats in, 197
   CAG trinucleotide repeat in, 174
   caudate nucleus in, 186
   expanded trinucleotide repeats in, 60
   eye movement abnormalities in, 186
   fetal tissue transplants for, 211
   genetics of, 59, 167
   irritability/lability in, 186
   pathogenesis of
      glutamine repeats and, 197
      mechanisms in, 197–199
Hydrocephalus
   after subarachnoid hemorrhage, 275, 276
   after syringomyelia repair, 356

from aqueductal stenosis, endoscopic third
      ventriculostomy for, 512–513
   communicating or extraventricular obstructive,
      329–330
   experimental, 328
   normal-pressure, 332–333, 334
   ventricular shunting techniques for, 349, 351
   ventriculo-extrathecal shunting for, 347
6-hydroxydopamine (6-OHDA), 206
Hypercholesterolemia, atherosclerosis and, 239
Hypercoagulable states
   primary, 248
   secondary, 248
Hyperglycemia, pre-ischemic, 15, 17
Hypernatremia, after subarachnoid hemorrhage,
      278
Hypertension
   after subarachnoid hemorrhage, 280–281
   atherosclerosis and, 239
   atherothrombotic stroke in, 252
   neurogenic, 126
   posttraumatic intracranial, cerebral hemodynamics
      and, 286
   stroke risk and, 258
Hypertensive hypervolemic hemodilution, for
      vasospasm, 274
Hyperventilation
   for intracranial pressure reduction, 276–277
   optimized *vs.* conventional, 293–294
Hypoglycemia
   loss of mitochondrial function and, 4–5
   pathophysiologic mechanisms of, 19–20
Hyponatremia
   after subarachnoid hemorrhage, 278
   hypovolemic, 273
Hypophysectomy, for cancer-related pain, 118
Hypothermia, 295–296
Hypoxia, loss of mitochondrial function in, 4–5

ICA (internal carotid artery), artery-to-artery brain
      embolization from, 255
ICAM-1, 42, 43
ICP. *See* Intracranial pressure (ICP)
IGF-1 (insulin-like growth factor 1), 241–242
IGFs (insulin-like growth factors), 241–242, 368
Image coordinate system, for stereotaxis, 466
Immune suppression, for Lambert-Eaton myasthenic
      syndrome, 163–164
Immune system
   components of, 32
      antigen-specific, 32–33, 36, 38